Special Publication No. 7
of the Society for Geology
Applied to Mineral Deposits

Lanthanides, Tantalum and Niobium

Mineralogy, Geochemistry, Characteristics of Primary Ore Deposits, Prospecting, Processing and Applications

Proceedings of a workshop in Berlin, November 1986

Edited by

Peter Möller, Petr Černý and Francis Saupé

With 137 Figures

Springer-Verlag Berlin Heidelberg GmbH

Prof. Dr. PETER MÖLLER
Hahn-Meitner-Institut Berlin GmbH
Bereich Kernchemie und Reaktor
Postfach 390128, D-1000 Berlin 30 FRG

Prof. Dr. PETR ČERNÝ
Department of Earth Science
University of Manitoba, Winnipeg
Manitoba R3T 2N2 Canada

Prof. Dr. FRANCIS SAUPÉ
Centre de Recherches Pétrographiques et Géochimiques
B.P. 20, F-54501 Vandoevre les Nancy Cedex

ISBN 978-3-642-87264-8 **ISBN 978-3-642-87262-4 (eBook)**
DOI 10.1007/978-3-642-87262-4

Library of Congress Cataloging-in-Publication Data. Lanthanides, tantalum, and niobium: mineralogy, geochemistry, characterisitcs of primary ore deposits, prospecting, processing, and applications/edited by Peter Möller, Petr Černý, and Francis Saupé.

(Special publication no. 7 of the Society for Geology Applied to Mineral Deposits) Proceedings of a workshop held in Berlin, FRG from Nov. 26–28, 1986. Includes index.

ISBN 978-3-642-87264-8

1. Rare earth metals-Congresses. 2. Tantalum-Congresses. 3. Niobium-Congresses. I. Möller, P. (Peter) II. Černý, P. III. Saupé, Francis. IV. Series: Special publication ... of the Society for Geology Applied to Mineral Deposits; no. 7. TN490.A2L36 1989 669'.29–dc19 88-31553

Originally published by Springer-Verlag Berlin Heidelberg New York 1989
Softcover reprint of the hardcover 1st edition 1989

Typesetting: International Typesetters, Inc., Makati, Philippines; Printing: Mercedes-Druck GmbH, Berlin; Binding: Lüderitz & Bauer GmbH, Berlin;
2132/3020-543210 – Printed on acid-free paper

Preface

At a time of surplus of nearly all metals and the inevitable closing of many mining operations, it is a matter of the industry's survival to seek new commodities that are in high demand. The lanthanides, better known as the rare-earth elements (REE), tantalum (Ta) and niobium (Nb) are among the promising alternatives. Bibliographies of the metallogeny and related aspects of these elements are surprisingly short, despite their rapidly increasing importance in modern technology. A need to review and condense the available expertise has been emerging in the recent past.

Two principal aims of the Society for Geology Applied to Mineral Deposits (SGA) are to disseminate the scientific knowledge of mineral resources, and to cultivate personal contacts between economic geologists. Workshops are instrumental in reaching these goals; thus the suggestion of Prof. Pierre Evrard, President of SGA from 1984–1985, to devote the second workshop of the Society to REE, Ta and Nb was unanimously supported by the SGA Council. From the onset, it was decided to limit the topics to those that are of interest to industry, and to include the wide range of fields indicated in the subtitle. Consequently, the REE as tracers of geological phenomena were not included. This aspect has received much attention by geochemists, and the corresponding literature is far more abundant than that covering the metallogeny of REE, Ta and Nb. Thanks to this decision, not only is a handy volume offered to the mining and processing community but also duplication of more fundamental books (such as "Rare earth elements geochemistry"; Henderson, P., editor) is avoided. The series "Special Publications of the Society for Geology Applied to Mineral Deposits" ensures a wide circulation of this compendium.

The rare-earth elements include the following series of 15 metals, with the atomic numbers (Z) 57 through 71:

Lanthanum	(La, Z = 57)	Terbium	(Tb, Z = 65)
Cerium	(Ce, Z = 58)	Dysprosium	(Dy, Z = 66)
Praseodymium	(Pr, Z = 59)	Holmium	(Ho, Z = 67)
Neodymium	(Nd, Z = 60)	Erbium	(Er, Z = 68)
Prometheum	(Pm, Z = 61)	Thulium	(Tm, Z = 69)

Samarium	(Sm, Z = 62)	Ytterbium	(Yb, Z = 70)
Europium	(Eu, Z = 63)	Lutetium	(Lu, Z = 71)
Gadolinium	(Gd, Z = 64)		

Yttrium (Y, Z = 39), although not a lanthanide, is conveniently included with the REE because of its chemical similarity. It is actually this close similarity in chemical behaviour among the REE plus Y, and between Nb and Ta, that strongly influences their migration through the geochemical cycle. This similarity has also complicated the classic wet analysis of these elements, which was plagued with difficulties since the first attempts to isolate them.

The first REE mineral, "cerite", was discovered by Cronstedt in 1751. Half a century later, two chemists, the German Klaproth and the Swedish Berzelius, isolated an unknown oxide from this mineral, called at the time "ceria". In 1794, a second Swedish chemist, Gadolin, isolated another oxide, "yttria", from minerals found in the pegmatite at Ytterby, Sweden. Both supposed oxides were later found to be mixtures of several different true oxides. It still took over a century to separate lutetium, the last of the 15 individual REE.

Niobium (Nb, Z = 41) and tantalum (Ta, Z = 73) are transition elements and extremely close chemical analogues, as can be deduced from their position to the right of the REE in the periodic table and as a consequence of the "lanthanide contraction". Because of their similarity, development of a separation process also took much time. "Tantalum" was discovered in 1802 by the Swedish chemist Ekeberg in the course of analyzing pegmatite minerals from Ytterby, Sweden and Kimito, Finland. The new "element" was named after the Greek mythological hero Tantalus because the separation was a "tantalizing" procedure. What Ekeberg actually discovered was a mixture of Nb and Ta. One year earlier, the British chemist Hatchett isolated an element in minerals from New London, Connecticut and called it "columbium". However, this also turned out to be a mixture of "columbium" with tantalum. Their separation was achieved for the first time by the German chemist Rose in 1844. He retained the name tantalum for one element and called the other after Niobe, the daughter of Tantalus. In 1866, niobium and columbium were found to be identical and the International Chemical Union adopted the first of the two synonyms in 1950. The second is still occasionally used in North America.

Thus, the entire suite of elements dealt with in this volume was first discovered in minerals from the same locality, the famous pegmatite at Ytterby, Sweden. This fact stresses the geochemical kinship between all these elements. Indeed, as is well known by prospectors, they often occur together in some pegmatites, alkaline granites and carbonatites with their associated alkaline rocks.

Other pegmatites and alkaline rocks are barren and their REE, Ta and Nb contents may be below the Clarke values. This could result from fractionation, and it was one of the aims of the workshop to come to a satisfactory explanation of this difference. Second-order differences are also known: REE and Nb are more common in carbonatites and associated alkaline rocks, whereas Ta is mostly recovered from pegmatites and greisens associated with peraluminous granitoids. Tantalum and niobium are chemically as similar as are the REE among themselves. Then, why and under which conditions do they fractionate? Understanding the fractionation conditions will help in evaluating the ore potential of the above mentioned rock types.

Only small tonnages of rare earths, tantalum and niobium are produced. Their tremendous importance is derived from their specific high technology applications. The extent of industrial uses of the REE is enormous; it ranges from glass industry through metallurgy, catalytic effects in oil refineries, luminescence activation and elektroceramic compounds to, i. a. the recent explosively growing field of high temperature superconductors. Tantalum and niobium are refractory metals. Small-size, large-surface capacitors based on tantalum have the most reliable performance in electronic systems, and tantalum-bearing alloys are especially valued for their thermal and chemical stability in, e. g. jet engines, gas turbines and prosthetic devices. Niobium also has applications in superconducting alloys, special steels, jet engine parts, components of nuclear reactors, steel-cutting carbide tools and other products.

The extensive applications of the metals considered here have stimulated the inclusion of their coverage into this volume, along with ore benefication and metallurgy. Juxtaposition of these topics with sensu lato geological aspects of REE, Nb and Ta should be of mutual benefit to the geological, ore-processing, metal-refining and metal-consuming industries. Unfortunate as it is, our experience indicates that communication among these seemingly divergent enterprisers is rather limited.

Some 60 scientists gathered in Berlin (FRG), from November 26 through 28, 1986, to present and discuss their experience and knowledge in mineralogy, primary ore deposits, prospecting, processing and applications of REE, tantalum and niobium. The various contributions, converted into manuscripts, were reviewed by at least two different colleagues. Longer comments, if submitted in writing, were appended to the corresponding articles; shorter and oral comments were included in the articles with references to their authors. The editors wish to express their gratitude to the authors of both the contributions and comments, as well as to the reviewers who accepted the unrewarding but necessary task. Last but not least, thanks are due to the publisher who produced this volume in a short time with the famous Springer quality.

The workshop session was greatly facilitated by material assistance and space liberally provided by the Board of the Hahn-Meitner-Institute in Berlin (FRG). The Commission of European Communities and the International Union of Geological Sciences generously subsidized the workshop. The editors and the Council of the Society owe a full measure of thanks to these institutions.

January 1989 P. MÖLLER, P. ČERNÝ, F. SAUPÉ

Contents

Contributors

You will find the addresses at the beginning of the respective contribution

Part I Mineralogy

Mineralogy of the Rare-Earth Elements

F. P. CESBRON

1 Introduction

The history of rare earths began in 1787 when Lt C.A. Arrhenius observed an unknown mineral in a quarry located at Ytterby, near Stockholm, and worked for feldspars used in porcelain manufacture. Johann Gadolin, former student of Torbern Bergman at the Uppsala University, analyzed this sample in 1794, finding a "new earth". In 1797, Anders Gustaf Ekeberg again examined it and gave the name of "yttria" to the new element (in fact, a mixture of several rare earths) discovered by Gadolin. It was only in 1800 that M.H. Klaproth gave this first rare-earth mineral its present name, gadolinite, called earlier "ytterbite".

In 1802 Ekeberg, both a poet and assistant professor of chemistry at Uppsala, also discovered tantalum in two Ytterby minerals that he named "yttrotantal" (now yttrotantalite) and "tantalit". Thus, from the very beginning, yttrium and tantalum were closely associated.

Since the gadolinite discovery, more than 160 mineral species with REE constituting a major component have been described but only three or four of them are currently recovered economically. Of course the list is not complete: new species are discovered each year and their descriptions submitted for approval to the Commission on New Minerals and Mineral Names of the International Mineralogical Association. One of the latest, minasgeraisite, is a silicate with REE found for the first time in association with bismuth, but only several hundred milligrams of it are available.

An updated list of these minerals is given in the Appendix, but it must be kept in mind that many more can accept minor or trace amounts of REE in their structure, without being considered REE minerals. On the other hand, some minerals like apatite, fluorite, titanite, uraninite, vesuvianite, zircon etc, which theoretically are not REE minerals, can sometimes contain important concentrations. For instance, an exceptional vesuvianite, occurring on fractures in altered greenstones from San Benito County, California, is reported with 20.57 wt.% RE_2O_3, being only slightly enriched in light REE (Crook and Oswald 1979).

This list only mentions the name and formula of each mineral; the interested reader will find the crystal symmetry and one of the latest available references for instance in Fleischer's Glossary of Mineral Species (1987).

Laboratoire de Minéralogie-Cristallographie associé au CNRS, Université P. et M. Curie, Tour 16, 4 Place Jussieu, 75252 Paris Cedex 05, France

Lanthanides, Tantalum and Niobium
P. Möller, P. Černý and F. Saupé (Eds.)

Besides the lanthanides, IUPAC (International Union of Pure and Applied Chemistry) also includes yttrium and scandium as rare-earth elements (which, as is well known, are not rare at all; the average abundance of cerium in the crust is higher than those of copper or lead!). Only the mineralogy of Y and lanthanides will be considered here.

2 Nomenclature

For quite a long time it was believed that Ce and Y, with the highest abundances among REE, were the predominant elements in REE minerals. Improved analytical techniques, particularly optical spectrography, X-ray fluorescence, electron microprobe and neutron activation, allowed the analysis of very numerous minerals and demonstrated the existence of several analogues with other rare earths of the lanthanide series predominant. These are usually La, Nd and sometimes Gd, but with the recent discovery of hingganite $(Yb,Y)BeSiO_4(OH)$ and keivyite $Yb_2Si_2O_7$, we may also suppose the existence of Sm-, Pr- and Dy-rich phases, yet to be discovered, as their respective average abundances, ranging from 7.3 to 5.2 ppm (Turekian and Wedepohl 1964), are higher than that of ytterbium (3 ppm).

In order to avoid the creation of a multitude of new names, Levinson proposed in 1966 a system of nomenclature since adopted by the International Mineralogical Association (IMA). A group name, bastnaesite, for instance, is proposed for a mineral in which the REE distribution is not determined. When this is known, the species name is obtained by addition of the chemical symbol of the predominant element to the group name: bastnaesite-(La), bastnaesite-(Y) etc. Table 1 gives these group names with the different predominating elements.

Table 1. Group names of rare-earth minerals

Halides		*Phosphates*	
Fluocerite	Ce,La	Florencite	Ce,La,Nd
		Monazite	Ce,La,Nd
Carbonates		Rhabdophane	Ce,La,Nd
Bastnaesite	Ce,La,Y		
Hydroxyl-bastn.	Ce,Nd	*Arsenates*	
Lanthanite	Ce,La,Nd	Retzian	Ce,La,Nd
Synchisite	Ce,Nd,Y		
		Silicates	
Oxides		Allanite	Ce,La,Y
Aeschynite	Ce,Nd,Y	Britholite	Ce,Y
Davidite	Ce,La,Y	Gadolinite	Ce,Y
Fergusonite-ß	Ce,Nd,Y	Nordite	Ce,La
		Törnebohmite	Ce,La
Sulphates		Tritomite	Ce,Y
Chukhrovite	Ce,Y	Tundrite	Ce,Nd
Vanadates			
Wakefieldite	Ce,Y		

3 Crystal Chemistry

3.1 Effective Ionic Radii

As lanthanides present very similar chemical properties, they can very easily substitute for one another and, for a long time, their separation was difficult. This of course originates in their external electronic shells (5s, 5p, 6s) which are identical. On the other hand, they present unfilled 4f-orbitals and, starting from zero electron (lanthanum), this 4f sub-shell is progressively filled up to 14 electrons (lutetium) as the atomic number increases (Moeller 1961). This filling is accompanied by a regular decrease of the ionic radius, known as the "lanthanide contraction".

According to Shannon (1976), the effective ionic radii, for a regular oxidation state of 3 and a coordination number (CN) of 8 with respect to oxygen, are:

Light REE or "Cerium group"			Heavy REE or "Yttrium group"		
	Z	r^{3+}		Z	r^{3+}
Lanthanum	57	1.160 Å	[Yttrium	39	1.019 Å]
Cerium	58	1.143	Gadolinium	64	1.053
Praseodymium	59	1.126	Terbium	65	1.040
Neodymium	60	1.109	Dysprosium	66	1.027
Promethium*	61	1.093	Holmium	67	1.015
Samarium	62	1.079	Erbium	68	1.004
Europium	63	1.066	Thulium	69	0.994
			Ytterbium	70	0.985
			Lutetium	71	0.977

*artificial

3.2 Oxidation State and Bonding

The regular oxidation state is 3^+ in nearly all the natural mineral species; divalent ions are easily oxidized to a tri-positive state and only Eu^{2+}, Sm^{2+} and Yb^{2+} are sufficiently stable to exist in aqueous solutions. In minerals, Eu^{2+} is known to be easily accepted in various mineral structures and, for instance, can substitute for Ca^{2+} in scheelites formed under reducing conditions or in plagioclases. Though not yet reported, it could be present in some apatites since the synthetic compound $Eu_5(PO_4)_3OH$ is known. In fluorites, Sm^{2+} is present as a result of natural irradiation and imparts a light green colour to the crystals, while Eu^{2+} is responsible for the strong blue-violet fluorescence of the Cumberland fluorites (Bill et al. 1967; Bill and Calas 1978).

As for the tetravalent state, although some artificial heteropolyanions are able to stabilize Pr^{4+} and Tb^{4+} (Sinha 1982), it seems that the only tetrapositive REE occurring in nature is Ce^{4+} ($r^{4+} = 0.97$ Å for CN = 8), associated, for instance, with Th^{4+} ($r^{4+} = 1.05$ Å) in cerianite (Ce, Th)O_2 and with U^{4+} ($r^{4+} = 1.00$ Å) in uraninite. According to Fleischer and Altschuler (1969), oxidation in sea water of Ce^{3+} into Ce^{4+} and its incorporation in the Mn-oxides/hydroxides of nodules would account for the marked impoverishment of Ce in sedimentary apatites of marine origin.

All REE are strongly lithophile in character; their predominantly ionic bonding in minerals is mainly with fluorine and oxygen, rarely with chlorine.

3.3 Coordination Numbers

In mineral structures, coordination numbers (CN) from 6 to 12 have been observed but, as many of these polyhedra of oxygens around the REE are usually strongly distorted, it is difficult to describe them easily. Theoretically, accommodation of REE in a site should be governed by their size; so, heavy REE should be mainly 6-fold coordinated and light REE 10- to 12-fold coordinated, with elements of intermediate size (or mixed heavy and light REE) being accommodated in polyhedra with seven to nine oxygens. In fact, if it is beyond doubt that the size of the available site can sometimes play an important role, it is also evident that the composition of the forming solutions and, as a general rule, the genetic conditions, are of importance on the site occupation.

CN=6. This coordination is mainly found in the structure of artificial oxides of heavy REE like Dy_2O_3 which are unstable under normal conditions and rapidly alter into hydrous carbonates. Still, this coordination is observed in the structure of minerals with close- or quasi-close-packed assemblages of oxygens; they mostly are Ti-oxides. Thus in davidite $(La,Ce)(Y,U,Fe^{2+})(Ti,Fe^{3+})_{20}(O,OH)_{38}$, with a structure deriving from that of crichtonite, some of the lanthanides and yttrium reside in quite regular octahedral sites (M_1) with an average distance RE-O of 2.243 Å (Gatehouse et al. 1979); some Ce^{4+} is apparently associated with U^{4+} at the same site. An identical coordination occurs in loveringite $(Ca,Ce)(Ti,Fe^{2+},Cr,Mg)_{21}O_{38}$ which belongs to the same structural family.

CN=7. REE artificial oxides with intermediate to large ionic radii present this kind of coordination. In Sm_2O_3, besides an Sm ion in an octahedral site, there are two other ions with a seven-fold coordination and the corresponding polyhedra (Fig. 1b) are trigonal prisms with a seventh oxygen coordinating through one prism face (Cromer 1957).

This coordination is also frequent around REE substituting for Ca^{2+} in accessory rock-forming minerals such as apatite (one of the two available sites, the other one is 9-coordinated) or titanite and its yttrian varieties ("keilhauite" with up to 12 wt.% RE_2O_3).

CN=8. The coordination polyhedron is perfectly regular and cubic in yttrian and cerian fluorites where these rare earths can reach more than 10 wt.% as YF_3 or CeF_3, as also in cerianite (fluorite structure) and REE-rich pyrochlores (derived fluorite structure). Conversely, the polyhedron is a square-based antiprism (Fig. 1c) in nordite, a complex silicate where only traces of Ca substitute for REE, also in cappelenite $(Ba,Ca,Na)(Y,La,Ce)_6B_6Si_3(O,OH)_{27}$, hellandite $(Ca,Y)_6(Al,Fe^{3+})Si_4B_4O_{20}(OH)_4$ and tadzhikite $Ca_3(Ce,Y)_2(Ti,Al,Fe^{3+})Si_4B_4O_{22}$.

Another classical polyhedron for this coordination is the triangular dodecahedron, composed of 12 triangular faces; it is found in fergusonite $YNbO_4$ and formanite $YTaO_4$, xenotime YPO_4, REE-rich zircons etc (Fig. 1d).

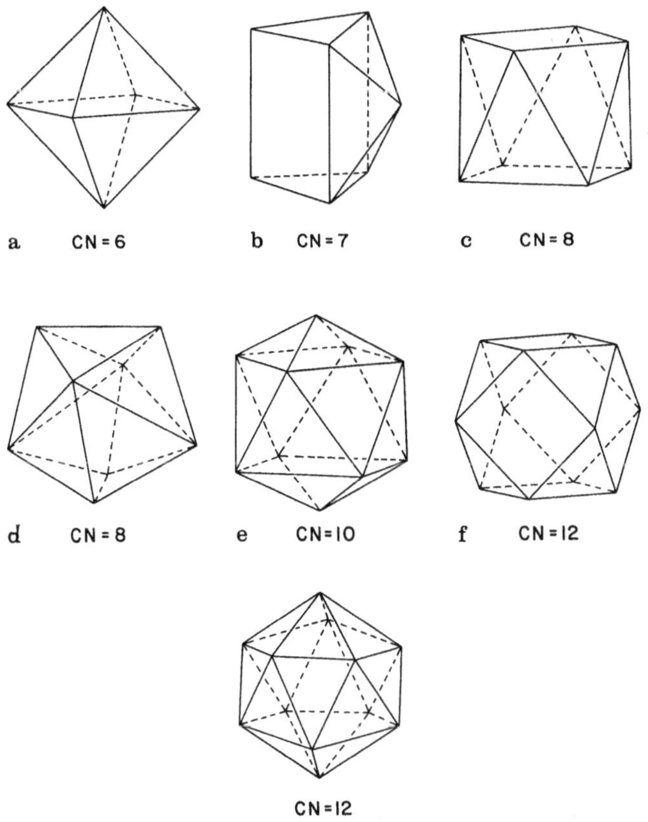

Fig. 1. Idealized coordination polyhedra around REE. **a** Octahedron; **b** monocapped-trigonal prism; **c** square-based antiprism; **d** triangular dodecahedron; **e** bicapped-square antiprism; **f** cuboctahedron; **g** icosahedron

The eight-fold coordination is probably the most common but, as the complexity of polyhedra increases with the number of vertices, their description usually is difficult. For instance, 257 polyhedra with eight vertices can be mathematically described (Britton and Dunitz 1973). Various eight-fold coordinated polyhedra are also present in churchite $YPO_4.2H_2O$ (isostructural with gypsum), gadolinite $Y_2FeBe_2Si_2O_{10}$, euxenite $(Y,Ca,Ce,U,Th)(Nb,Ta,Ti)_2O_6$ and polycrase $(Y,Ca,Ce,U,Th)(Ti,Nb,Ta)_2O_6$, kainosite $Ca_2(Y,Ce)_2Si_4O_{12}(CO_3).H_2O$, vyuntspakhkite $(Y,Yb)_4(Al,Si)Si_5O_{18}(OH)_5$ etc.

CN = 9. According to Sinha (1982), distorted tricapped trigonal prisms and monocapped square antiprisms are found in the structures of artificial hydrated sulphates of neodymium or samarium. In the first of these, three additional RE-O bonds "pierce" the middle of the prism faces; in the second, one square base of the antiprism is topped with a ninth oxygen: then a four-facet pyramid replaces the original face.

In natural species, nine-fold coordination is observed in apatite, with its second set of Ca-sites accommodating larger REE as the Ca-O distances range from about 2.41 to 2.81 Å (Sudarsanan and Young 1969). In cerite $(Ce,Ca)_9(Mg,Fe)Si_7(O,OH,F)_{28}$, the different polyhedra are all derived from a triangular dodecahedron (eight vertices and twelve triangular facets) with a ninth oxygen, belonging to an RE-OH bond, linked through a rhombus-shaped window formed by two adjacent and nearly coplanar triangles (Moore and Shen 1983). Monazite $(Ce,La)PO_4$, cheralite $(Ca,Ce,Th)(P,Si)O_4$, tundrite $Na_3(Ce,La)_4(Ti,Nb)_2(SiO_4)_2(CO_3)_3O_4(OH).2H_2O$ and gagarinite $NaCaY(F,Cl)_6$ also present a nine-fold coordination around the REE.

CN = 10. Many different polyhedra are possible, including bicapped square prisms and antiprisms (Fig. 1e), pentagonal prisms and antiprisms, tetradecahedra and various irregular convex polyhedra.

Ten-fold coordination is known in carbonates like ancylite $SrCe(CO_3)_2(OH).H_2O$, burbankite $(Na,Ca)_3(Sr,Ba,Ce)_3(CO_3)_5$ and lanthanite $(La,Ce)_2(CO_3)_3.8H_2O$, and in silicates like chevkinite $(Ca,Ce,Th)_4(Fe^{2+},Mg)_2(Ti, Fe^{3+})_3Si_4O_{22}$. Lanthanite has a structure with two very similar but quite distorted bicapped square antiprisms, with RE-O average distances respectively of 2.584 and 2.582 Å (Dal Negro et al. 1977), allowing accommodation of large REE: lanthanite-(La), -(Ce) and -(Nd).

CN = 11. Known in artificial $La(NO_3)_3.6H_2O$, this coordination is rare and, with the exception of bastnaesite $(Ce,La)(CO_3)F$, hydroxyl-bastnaesite and, possibly, allanite, does not seem to have been found in REE minerals. It is interesting to note that this is a quite large coordination for medium-sized rare earths like yttrium in bastnaesite-(Y).

In allanite, according to Dollase (1971), an eleventh oxygen at 3.13 Å "should perhaps be included" to the ten oxygen atoms surrounding the A_2 site in the epidote structure, mostly occupied by lanthanides replacing calcium. These are never found in the A_1 site, 9-coordinated.

CN = 12. This coordination is restricted to large REE. One example is a slightly distorted icosahedron (Fig. 1g), in artificial $(NH_4)_2La(NO_3)_5 \cdot 3\text{-}4H_2O$. Obviously, the ionic radius of the rare earth will be close to the oxygen one (1.38 Å): 1.36 Å for La^{3+} and 1.34 Å for Ce^{3+}.

In loparite $(Ce,Na,Ca)_2(Ti,Nb)_2O_6$, which belongs to the perovskite group, cerium is surrounded by 12 oxygens constituting a regular cuboctahedro (Fig. 1f). This polyhedron, more distorted, is also present in davidite where the oxygens form a close-packed assemblage: the site M_0, accommodating not only the large Ce and La ions but also some calcium, strontium lead and thorium (Gatehouse et al. 1979), just replaces an oxygen in their assemblage (Y, smaller REE and probably Ce^{4+}) are in the M_1 octahedral sites).

3.4 Isomorphic Substitutions

This wide range of possible coordinations not only allows substitutions of REE between sites, but also with various other elements. As these can be univalent to quadrivalent, an appropriate electrostatic charge balance will have to be achieved. By far the most common replacement is of REE by calcium, followed in an approximative decreasing order by thorium, uranium, sodium, iron, strontium, lead

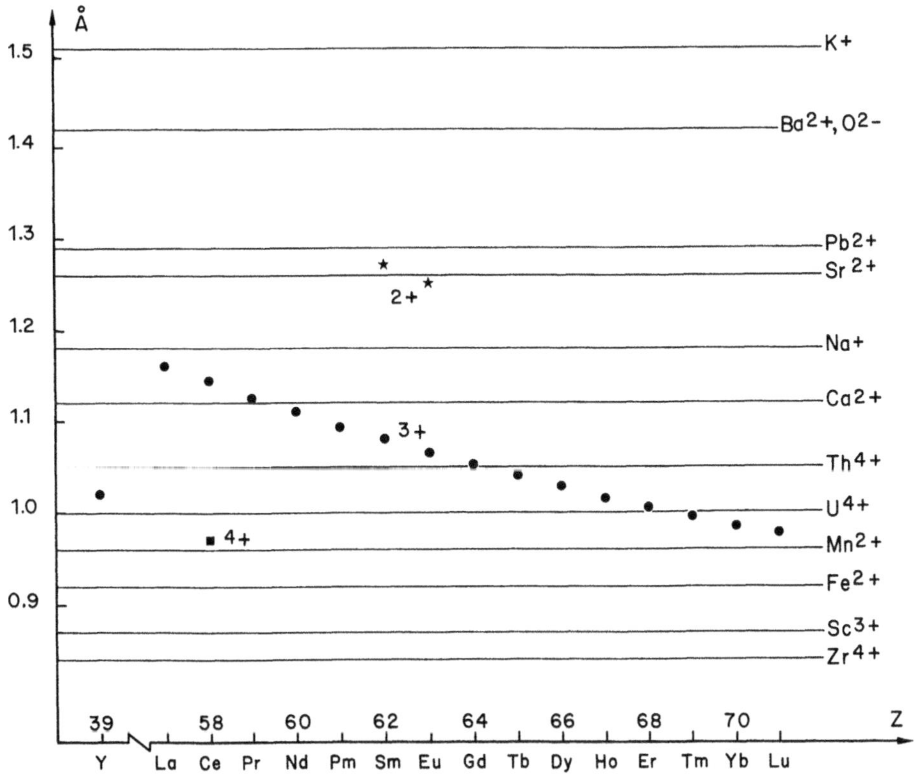

Fig. 2. Effective ionic radii for REE and their most common substituting ions (CN = 8)

and barium, and then zirconium, potassium, manganese, scandium etc. Figure 2 shows the atomic radii of the REE as compared with those of these elements, all of them for CN = 8.

Calcium. With a radius of 1.12 Å, Ca^{2+} would ideally replace Pr^{3+} and Nd^{3+}. Light REE effectively substitute for calcium in the structure of the epidote-allanite series, though Y also is present; the charge balance is obtained by $REE^{3+} + Fe^{2+} \rightarrow Ca^{2+} + Al^{3+}$.

In some carbonates, oxides and silicates, the charge balance can be achieved by the introduction of a monovalent element such as Na^+ with its ionic radius slightly higher than that of La^{3+}. In carbocernaite $(Ca,Na)(Sr,Ce,Ba)(CO_3)_2$ and in burbankite $(Na,Ca)_3(Sr,Ba,Ce)_3(CO_3)_5$, the exchange is of the type $Ce^{3+} + Na^+ \rightarrow 2(Ca,Sr,Ba)^{2+}$; it is of the same kind in loparite $(Ce,Na,Ca)_2$-$(Ti,Nb)_2O_6$, yttropyrochlore $(Y,Na,Ca,U)_{1\text{-}2}(Nb,Ta,Ti)_2(O,OH)_7$, mosandrite $(Na,Ce,Ca)_3Ti(SiO_4)_2F$ etc. In fluorite, the coupled substitution $REE^{3+} + Na^+ \rightarrow 2Ca^{2+}$ was recently confirmed by Möller and Dulski (1985-6).

In monazite, REE are often largely replaced by actinides and one of the two possible replacements is $2REE^{3+} \rightarrow Ca^{2+} + (Th,U)^{4+}$. Figure 3 shows how Pr and particularly Nd are preferentially substituted by Ca in monazites from Alps fissures and pegmatites (data from Mannucci et al. 1986).

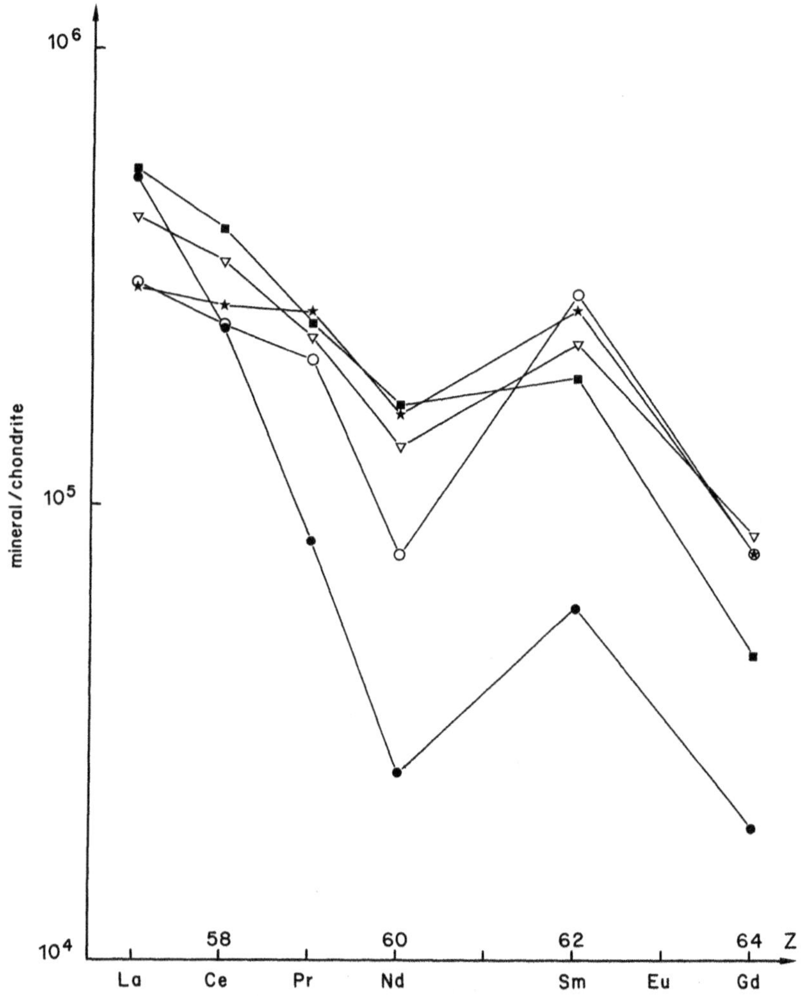

Fig. 3. Chondrite-normalized REE patterns of pegmatitic and fissure monazites (Alps). Data from Mannucci et al. (1986).

■ Fissure monazite:	$(Th,U)O_2$	0.9%,	CaO	0.1%,	SiO_2	0.1%
★ Pegmatite monazite:	$(Th,U)O_2$	19.2%,	CaO	0.6%,	SiO_2	2.8%
▽ Pegmatite monazite:	$(Th,U)O_2$	7.0%,	CaO	1.4%,	SiO_2	0.2%
○ Pegmatite monazite:	$(Th,U)O_2$	21.1%,	CaO	3.3%,	SiO_2	0.8%
● Pegmatite monazite:	$(Th,U)O_2$	22.5%,	CaO	4.7%,	SiO_2	0.1%

Pegmatite monazites are richer in Si, Ca and particularly Th and U

In some exceptional titanites, where Exley (1980) found up to 87% Ca replaced by Ce, La and Nd, as also in the yttrian varieties ("keilhauite"), the balance is obtained by $2REE^{3+} + Fe^{2+} \rightarrow 2Ca^{2+} + Ti^{4+}$, while in the euxenite-polycrase series one of the two possible substitutions is $REE^{3+} + Ti^{4+} \rightarrow Ca^{2+} + (Nb,Ta)^{5+}$.

Substitutions involving two sites of very different size are also possible as, for instance, in apatite with the balance $Ca^{2+} + P^{5+} \rightarrow REE^{3+} + Si^{4+}$. The same exchange occurs also in britholite where REE and Si predominate over Ca and P. In the metamict silicate melanocerite $(Ce,Ca)_5(Si,B)_3O_{12}(OH,F).nH_2O$, Vlasov (1966) suggests the replacement $Ce^{3+} + B^{3+} \rightarrow Ca^{2+} + Si^{4+}$.

Fluorite offers various kinds of substitutions involving cations and anions as in the exchange $Y^{3+} + O^{2-} \rightarrow Ca^{2+} + F^-$ (responsible for the pink or red colour of Alpine fluorites), or even interstitial anions with $REE^{3+} + F^- \rightarrow Ca^{2+}$.

Actinides. U and Th are mainly found in appreciable amounts in REE oxides and phosphates, particularly when they occur in pegmatites. In addition to the exchange $2REE^{3+} \rightarrow Ca^{2+} + (Th,U)^{4+}$, there is also a possibility of a balance involving small ions as in $REE^{3+} + P^{5+} \rightarrow (Th,U)^{4+} + Si^{4+}$. Mannucci et al. (1986) observed both in monazites from Alps pegmatites, enriched in Si, Ca, Th and U as compared with crystals from Alpine veins. The same kind of substitution, involving Zr^{4+} and Hf^{4+}, in addition to actinides, occurs frequently in minerals like zircon and xenotime.

REE can also be present in actinide minerals like uraninite UO_2 and its varieties "cleveite" and "nivenite" (one uraninite from Arendal, Norway, contains 9.40 wt.% ΣY_2O_3 and 2.25 wt.% ΣCe_2O_3), thorianite ThO_2, thorite $ThSiO_4$ and its monoclinic dimorph huttonite. In the monazite-huttonite series, close to the $ThSiO_4$-$REPO_4$ boundary, the balance also requires the entrance of Ca^{2+}, F^- and $(OH)^-$ (Kucha 1980).

According to their size, actinides should preferentially replace the heavy REE but Murata et al. (1953) found no correlation between them. On the contrary, Gramaccioli and Segalstad (1978) observed a depletion of Nd and Sm in a monazite from an Alpine pegmatite, containing 16 wt.% UO_2 and 11 wt.% ThO_2.

Zirconium. Although zircon is not, generally speaking, an REE mineral, many crystals can contain up to several percent of rare-earth oxides with the yttrium group predominating over the cerium group, especially in the more or less metamict and hydrated varieties such as "cyrtolite" and malacon". Charge balance is maintained with the various substitutions $Y^{3+} + P^{5+} \rightarrow Zr^{4+} + Si^{4+}$ (limited solid solution with xenotime), $2Y^{3+} \rightarrow Zr^{4+} + Ca^{2+}$ and probably also $Y^{3+} + OH^- \rightarrow Zr^{4+} + O^{2-}$ in the metamict varieties. In tranquillityite $Fe_8(Zr,Y)_2Ti_3Si_3O_{24}$, this substitution is also limited with only 2.76 wt.% Y_2O_3.

Manganese. Garnets are known to concentrate heavy REE but their content is usually low (some hundreds of ppm or less); still, amounts of about 2.01 wt.% Y_2O_3, with traces of Yb, Er, Dy and Ho, have been found in spessartines associated with pegmatites (Elk Mountain, New Mexico) and Jaffe (1951) proposed the replacement $Y^{3+} + Al^{3+} \rightarrow Mn^{2+} + Si^{4+}$.

Scandium. Scandium was discovered in 1879 by L.F. Nilson during the separation of the different rare earths of the erbium-ytterbium group obtained from euxenite and gadolinite. With the exception of thortveitite $(Sc,Y)_2Si_2O_7$ and, to a much less extent, of euxenite and gadolinite, the exchange $Y^{3+} \rightarrow Sc^{3+}$ is limited and amounts of Sc_2O_3 rarely exceed 0.2 wt.%. Nevertheless, Phan (1967) mentions a xenotime with 1.5 wt.% scandia.

These substitutions do not usually change the symmetry of the crystal, and accommodation of large REE is eventually accomplished by distortion of the crystallographic site, especially when the coordination number is high. If the site is quite small, however, as in xenotime (CN = 8), substitution of many LREE for

HREE will be difficult and then the stable structure will be that of monazite: for this reason there is no xenotime-(Ce). Rarely, as sometimes happens in apatite, exsolutions of new phases such as xenotime and monazite can appear (Amli 1975) with depletion of REE in the host crystal.

A mineral presently under investigation, however, burbankite $A_3B_3(CO_3)_5$ with A = Na, Ca and B = Sr, Ca, Ba, Ce, La, shows a progressive change from hexagonal (optically uniaxial) to monoclinic symmetry (strongly biaxial) as the REE content increases. As there are ten negative charges, electrostatic neutrality can be obtained only by completely filling the A site with Na; if more REE are available, then some divalent ions in the B site will have to be replaced by more Na. Exactly when the crystal becomes monoclinic is hard to tell. Optics are too sensitive to slight deformations and, for some composition domains, refinement of the crystal structure can be achieved in both systems with much the same results.

4 Metamictization

Several REE minerals can contain various amounts of uranium and thorium. The natural radioactive decay of these actinides will induce damage in the structure (displaced and interstitial atoms, vacancies) leading, if exposure to radiation is sufficiently long, to an "amorphization" eventually associated with new microcrystalline phases. This phenomenon is called "metamictization" and depends, for a given radiation flux, on the bonding strength; weak ionic structures will be more easily damaged, especially if they contain one or several ions readily susceptible to change of oxidation state. Other factors can also play a role, such as the different ionic radii of REE (Holmes 1931), the coordination numbers (Machatschki 1941) etc.

Progressive destruction of the structure will be accompanied by an increase in the cell parameters and a decrease in density, refractive indices and birefringence, hardness, resistance to chemical attack etc. (Mitchell 1973). The mineral will frequently absorb water or fix hydroxyl groups.

The original structure can frequently be restored by more or less prolonged heating at a selected temperature but, sometimes, the mineral recrystallizes as a mixture of several crystalline phases requiring less energy to form than the original one: Lima de Faria (1964) found more than six phases developing on heating of allanite but, after 1 h at 1000°C in air, the main phase was cubic CeO_2.

REE minerals known to occur in the metamict state are quite numerous; they are mainly oxides or silicates, and are given in Table 2 (names of very metamict species are underscored).

Some actinide-rich REE minerals are usually unaffected, although exceptional occurrences of metamict crystals have been reported: these are cerianite $(Ce,Th)O_2$ with ThO_2 up to about 5 wt.%, xenotime YPO_4 which can contain up to 5 wt.% UO_2 and 3,3 wt.% ThO_2 (Heinrich 1958) and especially monazite $(Ce,La)PO_4$. This mineral is usually quite rich in thorium since ThO_2 contents of 10 wt.% are common and values as high as 28.20 wt.% (Wadia and Fernando 1945) and even 31.50 wt.% (Bowie and Horne 1953) have been reported (transition toward cheralite); con-

Table 2. Metamict rare-earth minerals

Oxides		Silicates	
Aeschynite	Loparite	Allanite	Mosandrite
Brannerite	Polycrase	Cappelenite	Perrierite
Davidite	Polymignite	Chevkinite	Rowlandite
Euxenite	Samarskite	Eudialyte	Steenstrupine
Fergusonite	Tanteuxenite	Gadolinite	Thalenite
Fersmite	Yttrocrasite	Hellandite	Tritomite
Formanite	Yttrotantalite	Ilimaussite	Tritomite-(Y)
Kobeite	Zirkelite	Melanocerite	Yttrialite
Phospho-silicates			
Britholite			

centrations of uranium can reach 15 wt.% UO_2 (Gramaccioli and Segalstad 1978). Strong ionic bondings in the structure of these minerals are very likely responsible for this resistance to metamictization.

An interesting case is that of an apatite from the Adirondack Mountains, New York; containing 36.7 wt.% RE_2O_3 and only 0.93 wt.% ThO_2, it is reported by Lindberg and Ingram (1964) as partially metamict, and the sharpness of the X-ray lines in the powder diagram could be increased by heating. Another one is the metamict core of a vesuvianite crystal, previously mistaken for allanite, containing 6.06 wt.% RE_2O_3, 2.69 wt.% ThO_2 and 0.84 wt.% UO_2 (Himmelberg and Miller 1980).

5 Alteration

5.1 Hydrothermal Alteration

Rare-earth minerals can undergo transformations under hydrothermal conditions. For instance, monazite in hydrothermal-greisen deposits from Soviet Central Asia is often replaced by fluocerite and yttrian parisite (Vlasov 1966). Maruejol and Cuney (1985) mention the destabilization of monazite and the new crystallization of parisite, yttrian parisite and fluorite, and betafite, during the magmatic evolution and the hydrothermal alteration of the Xihuashan tungsten-bearing granites (Jiangxi, China). In the Vallorcine granite (Alps) and the associated metamorphic rocks, monazite grains present a rim (corona texture) with development of allanite and apatite; the reaction would have been contemporaneous with a retrograde metamorphism (Cuney 1986; pers. commun.). This mineral is known for its ability to move in hydrothermal systems. The best examples are the monazites from Alpine veins and those, associated with cassiterite, quartz and sometimes tourmaline, found by Bandy (1944) in every vein system of the Llallagua tin porphyry (Bolivia). Some monazite was also observed in the gold-bearing quartz veins cutting tourmalinized sandstones at Loulo (Mali), probably originating in the recrystallization of some detrital monazite (Cesbron et al. 1984).

Fluocerite, present in deposits genetically related to syenites and alkali granites like the Bastnäs skarns (Sweden) and the Mountain Pass carbonatites (California), is frequently altered into parisite and bastnaesite (Vlasov 1966). Furthermore, bastnaesite often replaces other cerium minerals, such as allanite, under epithermal conditions (Sverdrup et al. 1959).

Other carbonates, like synchisite-(Y) or ancylite, can be derived from pre-existing REE minerals; the first one during pneumatolytic-hydrothermal alteration of monazite and kainosite (Vlasov 1966), the second by action of late hydrothermal solutions on burbankite in veins of ankerite carbonatites. Parisite and synchisite were also observed by Hogarth et al. (1985) as the result of hydrothermal action of CO_2-rich waters on an early generation of Ce-rich apatites (1.22–1.72 wt.% Ce_2O_3) in carbonatites cutting Precambrian fenitized biotite gneisses and quartzites at Gatineau (Quebec); during the process, a new generation of REE-free apatite was also formed.

Finally, several REE-rich varieties of zircon were recorded in the past under the names "alvite", "hagatalite", "naegite", "nogizawalite", "oyamalite" and "yamaguchilite". Frequently highly metamict, they were mainly observed in granites and granitic pegmatites from Japan (see Nambu 1970, for general bibliography and data) and their content of RE_2O_3 and P_2O_5 can reach up to about 26 and 9.8 wt.% respectively ("nogizawalite"). Robinson (1979) demonstrated, using electron-microprobe analyses, Guinier photographs and X-ray precession and Weissenberg diagrams, that these "varieties" were, in fact, intergrowths of zircon and xenotime, sometimes in perfect epitaxial orientation and most probably formed by late hydrothermal solutions enriched in Y, P and lanthanides, pervading former metamict zircons. In "alvites" from Norway, the intergrowths are relatively coarse, with xenotime crystals commonly exceeding 25–100 μm. These domains are smaller and fewer in number in "hagatalite" and "yamaguchilite", while no discrete phase boundaries are discernible in "oyamalite" and "naegite".

5.2 Supergene Alteration

Exposed to supergene conditions, many REE mineral species are susceptible to more or less important alterations. Obviously, metamict minerals will be more sensitive and they frequently present an oxidation crust enriched in H_2O, iron and titanium oxides (rutile, leucoxene, etc.), and depleted in silicon, uranium and rare earths. REE liberated during weathering can precipitate as secondary minerals. Most common are carbonates such as ancylite and bastnaesite (also found as primary minerals), lanthanite, tengerite $CaY_3(CO_3)_4(OH)_3{\cdot}4H_2O$ etc, oxides [cerianite CeO_2 or cerotungstite $CeW_2O_6(OH)_3$], phosphates [churchite $YPO_4{\cdot}2H_2O$, rhabdophane $(Ce,La)PO_4{\cdot}H_2O$] and, of course, some rare sulphates, vanadates and arsenates mainly found in the oxide zone of various metallic deposits. Some uranyl-carbonates and uranyl-silico-carbonates of Y, Dy and Gd have recently been found in the lower oxide zone of Shinkolobwe, Zaire (Deliens and Piret 1982).

Some of these secondary minerals, like bastnaesite $(Ce,La)(CO_3)F$, can themselves undergo further alteration. Late hydrothermal or supergene, this fluocarbonate is frequently seen as an alteration product of silicates like allanite, britholite, cerite, chevkinite, gadolinite, thalenite, yttrialite or of other primary carbonates such as parisite. By hydration and loss of fluorine, bastnaesite alters to lanthanite and rhabdophane; by loss of the REE with the exception of Ce^{3+}, oxidized to Ce^{4+}, it leads to cerianite $(Ce,Th)O_2$ (Vlasov 1966).

When REE do not form secondary minerals in the proximity of their source, they can be adsorbed on the surface of clay minerals formed during weathering; several hundred ppm of REE can be trapped in this manner. Exceptionally they can travel quite far before being deposited as carbonates. For instance, the source of the REE in the lanthanite-(Nd) deposited in the sedimentary basin of Curitiba, Brazil (Cesbron et al. 1979) is still problematical considering the present geological knowledge of the area.

Figure 4 shows the chondrite-normalized REE partial patterns of lanthanite-(Ce) from the Britannia Mine, Wales, UK (Bevins et al. 1985) and of lanthanite-(Nd) from Curitiba. The first one displays no anomaly and probably represents the REE composition of the allanite identified in the nearby acid pyroclastics, from which they may have been leached. The second shows a very pronounced cerium depletion and this could be the consequence of the oxidation of Ce^{3+} to Ce^{4+} and its incorporation in another phase; transportation and precipitation were accomplished under oxidizing conditions, after the sedimentation of the host rock (carbonate-silty clays). It should be observed that this lanthanite-(Nd) is one of the minerals richest in europium (1.64 wt.% Eu_2O_3); previously a rhabdophane in limonitic beds from Salisbury, Connecticut, USA (Hildebrand et al. 1957), was reported with 1.08 wt.% Eu_2O_3.

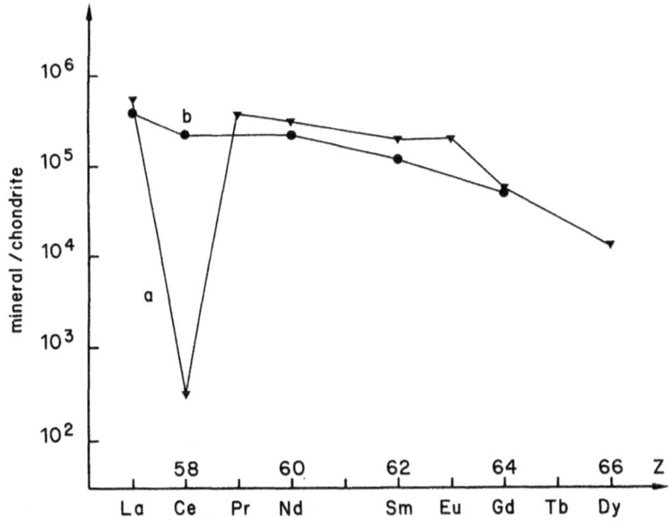

Fig. 4. Chondrite-normalized REE patterns of lanthanide-(Nd) from Curitiba, Brazil (**a**) and lanthanide-(Ce) from Britannia Mine, UK (**b**)

6 REE Minerals and Related Species

Although crystal chemistry controls in part the composition of the REE in minerals, paragenetic conditions are quite as important. Fleischer (1965) and Fleischer and Altschuler (1969) remarked that, generally, silica-rich igneous rocks concentrate yttrium and heavy lanthanides, while those with low silica and high carbonate contents are enriched in light REE (cerium group).

The numerous REE silicates such as cerite, chevkinite, eudialyte, ilimaussite, karnasurtite, melanocerite, mosandrite, rinkolite, steenstrupine etc., related to alkali granites and syenites, nepheline-syenites etc. and their pegmatites, are cerium- and lanthanum-rich. There are some exceptions, like cappelenite $(Ba,Ca,Na)(Y,La,Ce)_6B_6Si_3(O,OH)_{27}$, found in nepheline-syenite pegmatites from Norway, with 52.62 wt.% Y_2O_3 and only 4.20 wt.% $(La,Ce)_2O_3$, or tritomite-(Y) $(Y,Ca,La,Fe)_5(Si,B,Al)_3(O,OH,F)_{13}$ observed both in granite- and nepheline-syenite pegmatites.

Figure 5 shows the chondrite-normalized REE patterns of bastnaesites from various occurrences (average of several analyses), established from Fleischer's data (1978). The enrichment in Y and heavy lanthanides is apparent for bastnaesite and bastnaesite-(Y) from granite pegmatites and granites, while hydrothermal and alkaline rock bastnaesites are depleted in these and enriched in light REE. Additional REE distribution patterns are given by P. Möller (this Vol.).

Clark (1984) defined three groups of minerals according to their REE content: (1) minerals with very low amounts of REE, (2) minerals containing minor but not essential contents of REE and (3) minerals with major concentrations of REE (see updated list in Appendix). In this group, Clark also included some minerals belonging to the two previous ones but presenting unusually high REE contents.

It would be irrelevant to present all rare-earth minerals here; the interested reader can refer to Clark (1984) and Vlasov (1966) even if, as in the case of the latter, the nomenclature has since undergone some modifications with several species no longer considered valid. On the other hand, it is interesting, for each class of minerals, to say a word about the usual occurrence of its members, as also about minerals with unusually high REE concentrations.

6.1 Halides

They are usually found in granitic pegmatites (fluocerite from Österby, Sweden), albitized alkali granites and albitites emanating from alkali syenites (gagarinite from subalkali-quartz syenites). Other occurrences are in hydrothermal deposits related to alaskites (fluocerite from Central Asia), alkali granites and syenites (fluocerite from Pikes Peak, Colorado, and Mountain Pass, California etc.).

Among halides, fluorite is not in itself an REE mineral but trace or minor amounts of them are frequently present and can impart different colorations to the crystals. Thus yttrium is responsible for the pink colour through the (YO_2) centre, consisting of an O_2^{3-} molecular ion stabilized by an adjacent Y^{3+} ion (Bill 1969). Blue and yellowish-green fluorites owe their colour, respectively, to yttrium-associated F centres and co-existing yttrium- and cerium-associated F centres; Sm^{2+} gives a light

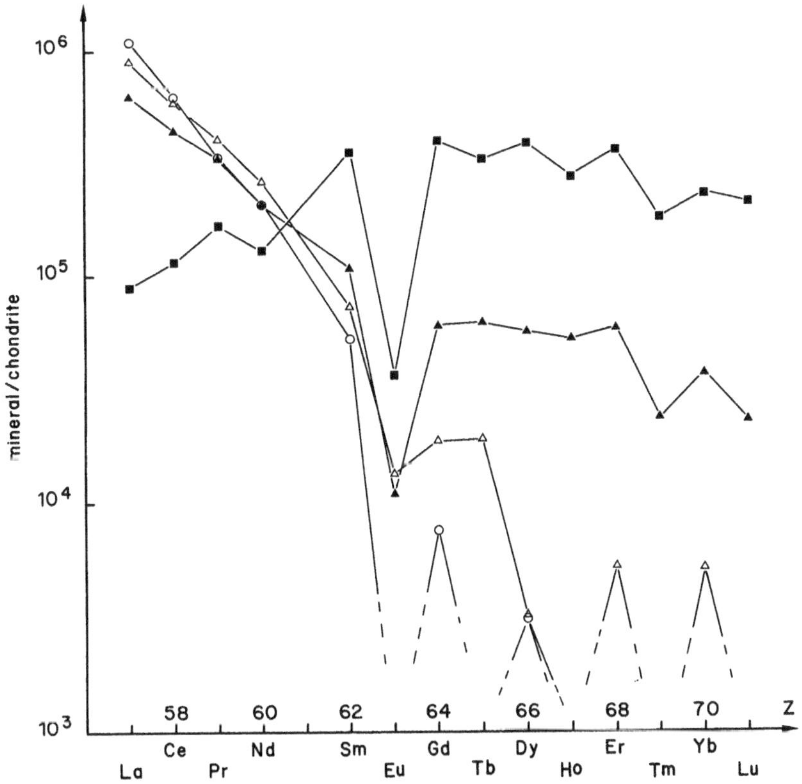

Fig. 5. Chondrite-normalized REE patterns of bastnaesites from various environments (average of several analyses). Data from Fleischer (1978).
■ Bastnaesite-(Y): granite pegmatites and granites (Y = 35.3%, Ln = 64.7%)
▲ Bastnaesite-(La,Ce): granite pegmatites and granites (Y = 23%, Ln = 77%)
△ Bastnaesite-(Ce,La): hydrothermal systems (Y = 5.1%, Ln = 94.9%)
○ Bastnaesite-(Ce,La): alkaline rocks (Y = 2.1%, Ln = 97.9%)

green colour and Eu^{2+} a blue-violet fluorescence (Bill and Calas 1978). Exceptionally, fluorite can contain important amounts of REE. Sverdrup (1968), in Norwegian pegmatites genetically related to granites, found an yttrian fluorite with 18.08 wt.% YF_3 (of a total of 4.65 wt.% light REF_3 and 23.0 wt.% heavy REF_3) and a cerian fluorite (formerly yttrocerite) with 16.84 wt.% CeF_3 (of a total of 34.33 wt.% light REF_3 and 13.62 wt.% heavy REF_3). As usual, heavy REE are found in fluorites occurring in granite pegmatites and light REE in those from alkaline rocks.

6.2 Carbonates

More than 40 REE minerals with $(CO_3)^{2-}$ anionic complexes, often associated with F^- ions (bastnaesite, synchisite, parisite etc.) and occasionally with $(SiO_4)^{4-}$ (caysichite, iimoriite, kainosite, tundrite), are reported; uranyl-carbonates, uranyl-silico-carbonates and boro-carbonates are rare.

Many of them are of primary origin and usually related to nepheline-syenites and their pegmatites, alkali granites and pegmatites, various carbonatites etc. Pegmatitic and hydrothermal depositions are common processes and some of these carbonates can be found in varying environments such as riebeckite-granites (parisite from Montorfano, Italy), bituminous limestones (parisite from Muzo emerald deposit, Columbia) or talc deposits (parisite from Luzenac, France). Supergene formation also is common; for instance, the only boro-carbonate, moydite (RE,Ca) $[B(OH)_4]$ (CO_3), derived directly from an altered pegmatitic hellandite (an REE-rich boro-silicate) and kimuraite $CaY_2(CO_3)_4.6H_2O$, recently discovered in the fissures of an alkali olivine basalt (Nagashima et al. 1986), is a hydration product of lokkaite $(Y,Ca)_2(CO_3)_3 \cdot 2H_2O$.

The most widespread carbonate, calcite, does not contain significant amounts of rare earths. Headden (1906) reported 0.03 wt.% RE_2O_3 in a calcite from Joplin, Missouri; more recently, Eby (1975) analyzed a maximum of 3300 ppm REE, with a marked enrichment for the cerium group, in calcites from the Oka carbonatite complex.

6.3 Oxides

In 39 of their oxides/hydroxides, from a total of 42, REE are associated with Ti and/or Ta-Nb; uranium also can be present, leading to metamictization. These minerals usually occur as accessory phases in granites, nepheline syenites and carbonatites but can be more abundant in pegmatites. Several are resistant enough to be found later as detrital material in placer deposits (aeschynite, euxenite-polycrase, fergusonite, samarskite, etc.).

REE usually are present in uraninite and thorianite (see Sect. 3.4: isomorphic substitutions – actinides), cerium being the most common element: cerianite CeO_2 has the same crystal structure as UO_2 and ThO_2 and Ce can substitute for Th in thorianite up to at least 8 wt.% $(Ce,La)_2O_3$. Conversely, REE are not very abundant in pitchblendes; still heavy lanthanides have been economically recovered from concentrates produced from spent uranium processing solutions (Stamper and Chin 1976).

Mostly Ce (up to 11.3 wt.% Ce_2O_3), but also Y, La, etc. can substitute for Ca in perovskite $CaTiO_3$, according to the exchange $Ce^{3+} + Na^+ \rightarrow 2Ca^{2+}$, but perfect isomorphism with loparite $(Ce,Na,Ca)_2(Ti,Nb)_2O_6$ does not occur (Povarennykh 1972). Latrappite (Ca,Na)(Nb,Ti,Fe)O_3 and lueshite $NaNbO_3$, both members of the perovskite group and found in carbonatites, can also accept 2–3 wt.% RE_2O_3.

6.4 Phosphates

Phosphates constitute a less important class than carbonates but they include two major REE ores, xenotime and especially monazite. They certainly are the most widespread of all rare-earth minerals, occurring in granites and granitoides, alkaline rocks, carbonatites, metamorphic and sedimentary rocks (placer deposits) and hydrothermal systems (monazite is present in porphyry copper deposits, molybdenum and tin porphyries, Alpine veins etc.).

There are also several silico-phosphates such as britholite (isostructural with apatite), karnasurtite, laplandite, phosinaite, saryarkite, steenstrupine and cheralite, this later forming a series with monazite; a limited series also exists between monazite and huttonite $ThSiO_4$.

Phosphates usually are hypogene with the exception of churchite $YPO_4.2H_2O$, brockite $(Ca,Th,Ce)PO_4.H_2O$ and rhabdophane $(Ce,La)PO_4.H_2O$. Like the carbonates kimuraite and lanthanite-(Nd), rhabdophane-(Nd) can be more or less depleted in Ce if oxidation to Ce^{4+} occurs before the mineral crystallization. The chondrite-normalized REE pattern of rhabdophane-(Nd), from Salisbury limonite ores (Hildebrand et al. 1957), displays, in addition, an important negative anomaly in Pr; with the exception of a possible analytical error, a Pr^{4+} loss could account for this anomaly as Pr^{4+} compounds are known, at least artificially.

Rare earths, partially replacing Ca, are common in apatites $(Ca_5(PO_4)_3(F,Cl,OH)$; the most common substitution is the exchange $Ca^{2+} + P^{5+} \rightarrow REE^{3+} + Si^{4+}$ and it is unquestionable that apatites from alkaline rocks and carbonatites, usually presenting high amounts of REE, are enriched in Si. Other substitutions such as $2Ca^{2+} \rightarrow REE^{3+} + Na^{+}$ and possibly $Ca^{2+} + OH^{-} \rightarrow REE^{3+} + O^{2-}$ play a minor role.

Apatites frequently contain a few hundred to several thousand ppm REE. For instance, apatites in "Kiruna-type iron ores" from Lapland have a rare-earth metal content ranging from 1545 to 8855 ppm (Parak 1973). Of course, much higher concentrations have been observed; porphyry-copper apatites frequently contain up to 15 000 ppm and those from Oka carbonatite up to 40 000 ppm (Eby 1975). Fleischer and Altschuler (1969) report as much as 12 wt.% RE_2O_3 in apatites from alkaline igneous rocks, with much higher yttrium concentrations than in monazite from similar rocks. Lindberg and Ingram (1964) analyzed a sample from Adirondack Mountain, close to the apatite-britholite boundary, with 36.7 wt.% RE_2O_3 (including 12.9 wt.% Ce_2O_3) and 12.9 wt.% SiO_2, giving a ratio $P/Si = 3.19/2.81$. In apatite from the Mineville magnetite ore, McKeown and Klemic (1956) found up to 32.4 wt.% RE_2O_3 with extremely fine-grained and intergrown monazite and bastnaesite inclusions.

Furthermore, an interesting characteristic of apatite is its tendency to contain inclusions (or exsolutions?) of rare-earth minerals such as xenotime and monazite (Pigorini and Veniale 1968, Åmli 1975), monazite and bastnaesite (McKeown and Klemic 1956), monazite and allanite (Parak 1973) etc. These inclusions occur in very REE-rich apatites (several wt.% RE_2O_3) as well as in crystals with only several thousand ppm REE, as in Kiruna-type iron ores, but their presence is insufficient to account for the total REE content. Åmli (1975) observed an REE depletion around these inclusions; similarly, in apatite crystals from the Guadalupe porphyry-copper deposit (Sonora, Mexico), the REE content drops sharply when monazite "inclusions" begin to appear.

6.5 Silicates

Besides silico-carbonates and silico-phosphates, rare-earth silicates also include several boro-silicates such as cappelenite, hellandite, melanocerite, okanoganite, stillwellite, tadzhikite, tritomite-(Ce) and tritomite-(Y); all of these occur in pegmatites of nepheline-syenites, granites and alkali granites, with the exception of stillwellite, found in metasomatically altered carbonate rocks.

Many calcium-bearing, rock-forming silicates can accept rare earths in their structure but concentrations do not usually exceed several hundred ppm. For instance, in garnets, melilites, biotites and pyroxenes from the Oka carbonatite complex, the maximum found by Eby (1975) was 3600, 2300, 650 and 780 ppm respectively. Yet, common silicates such as epidote, titanite, vesuvianite and zircon can have larger amounts exceeding 1 wt.% RE_2O_3. Exceptionally, rarer species can also be enriched in REE: "schizolite" from a Greenland nepheline-syenite, in fact a manganoan pectolite $Na(Ca,Mn)_2Si_3O_8(OH)$, was reported by Winther (1901) with 2.40 wt.% Y_2O_3; serandite $Na(Mn,Ca)_2Si_3O_8(OH)$, which forms an isostructural series with pectolite, can also contain significant amounts of REE (Semenov et al. 1976).

In epidote, only the 10-coordinated A_2 site can accommodate REE and Rudashevskii (1969) reported up to 6.9 wt.% RE_2O_3 in a partially metamictized crystal from a skarn. Similar concentrations were also found in a clinozoisite (Maaskant et al. 1980).

According to Henderson (1980), when titanite and apatite co-exist, light REE are preferentially accommodated by the latter: the analyzed titanite had only 2.04 wt.% RE_2O_3, but much higher values were reported in the varieties "keilhauite" and "yttrotitanite" [up to 12 wt.% $(Y,Ce)_2O_3$]. Titanites developing in endoskarns are sometimes enriched in rare earths and display a distinctive pink pleochroism.

The square-based antiprisms of oxygens surrounding calcium ions in vesuvianite can accommodate not only REE but also Th and U, sometimes inducing partial metamictization. As previously mentioned, Himmelberg and Miller (1980) found up to 6.06 wt.% RE_2O_3 and Crook and Oswald (1979) a surprising value of 20.57 wt.%.

Owing to the fact that zircon is isostructural with xenotime, rare earths of the yttrium group (HREE) will preferentially enter in its structure according to the substitution $Zr^{4+} + Si^{4+} \rightarrow Y^{3+} + P^{5+}$; Vlasov (1966), however, mentions some metamict zircons ("cyrtolite") from North Karelia with a predominance of light REE. The oxide total content can reach up to several percent and Görz and White reported a maximum of 6.10 wt.% Y_2O_3. Robinson (1979) proved that richer zircons are, in fact, intergrowths of zircon and xenotime.

7 Conclusions

Many more new species will undoubtedly be discovered in the future and mineralogy of the rare-earth elements will continue its present fast evolution owing to improved analytical techniques. Still, real progress will only be achieved when it is possible to routinely analyze, with fairly good accuracy, all the elements of the series. At present, this is not the case and the exact composition of many of the 160 REE minerals is still unknown or at least imprecise. It will also be necessary to greatly improve our knowledge of the geometry and occupancy of the sites accommodating rare earths in the crystal structure and, if possible, to relate them to the genetic conditions of crystallization. A lot of work is still ahead...

References

Åmli R (1975) Mineralogy and rare earth geochemistry of apatite and xenotime from the Gloserheia granite pegmatite, Froland, Southern Norway. Am Mineral 60:607–620
Bakakin VV, Belov NV, Borisov SV, Solovyeva LP (1970) The crystal structure of nordite and its relationship to melilite and datolite-gadolinite. Am Mineral 55:1167–1181
Bandy MC (1944) Mineralogy of Llallagua. Patino Mines Bolivia. La Paz, Bolivia. Tucson Gem and Mineral Soc (1976)
Bevins RE, Rowbotham G, Stephens FS, Turgoose S, Williams PA (1985) Lanthanite-(Ce), $(Ce,La,Nd)_2(CO_3)_3.8H_2O$, a new mineral from Wales, U.K. Am Mineral 70:411–413
Bill H (1969) Investigation of colour centres in alkaline earth fluorides. Helv Phys Acta 42:771–797
Bill H, Calas G (1978) Color centers, associated rare-earth ions and the origin of coloration in natural fluorites. Phys Chem Mineral 3:117–131
Bill H, Sierro J, Lacroix R (1967) Origin of coloration in some fluorites. Am Mineral 52:1003–1008
Bowie SHU, Horne JET (1953) Cheralite, a new mineral of the monazite group. Mineral Mag 30:93–99
Britton D, Dunitz JD (1973) A complete catalogue of polyhedra with eight or fewer vertices. Acta Crystallogr A29:362–371
Cesbron F, Sichere MC, Vachey H, Cassedanne JP, Cassedane JO (1979) La lanthanite à europium de Curitiba, Paraná, Brésil. Bull Mineral 102:342–347
Cesbron F, Kosakevitch A, Picot P (1984) Résultat d'étude d'échantillons du gisement de Loulo (Mali). Rapp BRGM 84 SGN 211 MGA, Orléans
Clark AM (1984) Mineralogy of the rare earth elements. In: Henderson P (ed) Rare earth element geochemistry. Elsevier, Amsterdam Oxford New York, pp 33–61
Cromer DI (1957) The crystal structure of monoclinic Sm_2O_3. J Phys Chem 61:753
Crook WW, Oswald SG (1979) New data on cerian vesuvianite from San Benito County, California. Am Mineral 64:367–368
Dal Negro A, Rossi G, Tazzoli V (1977) The crystal structure of lanthanite. Am Mineral 62:142–146
Deliens M, Piret P (1982) Bijvoetite et lepersonnite, carbonates hydratés d'uranyle et de terres rares de Shinkolobwe, Zaïre. Can Mineral 20:231–238
Dollase WA (1971) Refinement of the crystal structures of epidote, allanite and hancockite. Am Mineral 56:447–464
Eby GN (1975) Abundance and distribution of the rare-earth elements and yttrium in the rocks and minerals of the Oka carbonatite complex, Quebec. Geochim Cosmochim Acta 39:597–620
Exley RA (1980) Microprobe studies of REE-rich accessory minerals: implications for Skye granite petrogenesis and REE mobility in hydrothermal systems. Earth Planet Sci Lett 48:97–110
Fleischer M (1965) Some aspects of the geochemistry of yttrium and the lanthanides. Geochim Cosmochim Acta 29:755–772
Fleischer M (1978) Relative proportions of the lanthanides in minerals of the bastnaesite group. Can Mineral 16:361–363
Fleischer M (1987) Glossary of mineral species, 5th edn. Min Rec Tuscon, 227 pp
Fleischer M, Altschuler ZS (1969) The relationship of the rare-earth composition of minerals to geological environment. Geochim Cosmochim Acta 33:725–732
Gatehouse BM, Grey IE, Kelly PR (1979) The crystal structure of davidite. Am Mineral 64:1010–1017
Görz H, White EW (1970) Minor and trace elements in HF-soluble zircons. Contrib Mineral Petrol 29:180–182
Gramaccioli CM, Segalstad TV (1978) A uranium- and thorium-rich monazite from a south-alpine pegmatite at Piona, Italy. Am Mineral 63:757–761
Headden (1906) Am J Sci 21:301
Heinrich EW (1958) Mineralogy and geology of radioactive raw materials. McGraw-Hill, New York
Henderson P (1980) Rare earth element partition between sphene, apatite and other coexisting minerals of the Kangerdlugssuaq intrusion, E. Greeland. Contrib Mineral Petrol 72:81–85
Hildebrand FA, Carron MK, Rose HJ (1957) Re-examination of rhabdophane (scovillite) from Salisbury. Connecticut. Geol Soc Am Bull 68:1744–1745
Himmelberg GR, Miller TP (1980) Uranium- and thorium-rich vesuvianite from the Seward Peninsula, Alaska. Am Mineral 65:1020–1025
Hogarth DD, Hartree R, Loop J, Solberg TN (1985) Rare-earth element minerals in four carbonatites near Gatineau, Quebec. Am Mineral 70:1135–1142
Holmes A (1931) Radioactive minerals. Nat Res Council Bull 80:229–272

Jaffe HW (1951) The role of yttrium and other minor elements in the garnet group. Am Mineral 36:133–155
Kucha H (1980) Continuity in the monazite-huttonite series. Mineral Mag 43:1031–1034
Levinson AA (1966) A system of nomenclature for rare-earth minerals. Am Mineral 51:152–158
Lima de Faria J (1964) Identification of metamict minerals by X-ray powder photographs. Junta Invest Ultramar, Estud Ensaios Doc 112, Lisbon
Lindberg ML, Ingram B (1964) Rare-earth silicatian apatite from the Adirondack Mountains, New York. US Geol Surv Prof Pap 501-B:64–65
Maaskant P, Coolen JJMMM, Burke EAJ (1980) Hibonite and coexisting zoisite and clinozoisite in a calc-silicate granulite from southern Tanzania. Min Mag 43:995–1003
Machatschki F (1941) Zur Frage der Stabilität des Zirkongitters. Zentr Mineral A:38–40
Maksimovic' Z (1985) Hydroxyl-bastnaesite-(Nd), a new mineral from Montenegro, Yugoslavia. Min Mag 49:717–720
Mannucci G, Diella V, Gramaccioli CM, Pilati T (1986) A comparative study of some pegmatitic and fissure monazite from the Alps. Can Mineral 24:469–474
Maruejol P, Cuney M (1985) Xihuashan tungsten-bearing granites (Jiangxi, China): mineralogical controls on REE, Y, Th, U mobility during magmatic evolution and hydrothermal alteration. Fortschr Mineral 63:149
McKeown FA, Klemic H (1956) Rare-earth-bearing apatite at Mineville, Essex County, New York. US Geol Surv Bull 1046-B:9–23
Mitchell RS (1973) Metamict minerals: a review. Mineral Rec 4:177–182 (pt 1) 214–223 (pt 2)
Moeller T (1961) The chemistry of rare earths. In: Spedding FH, Daane AH (ed) The rare earths. John Wiley & Sons, New York London, p 33
Möller P, Dulski P (1985–6) Substitution of Ca^{2+} by trivalent rare earths in natural fluorite. Lanthanide Actinide Res 1:225–232
Moore PB, Shen J (1983) Cerite, $RE_9(Fe^{3+},Mg)(SiO_4)_6 \cdot (SiO_3OH)(OH)_3$: its crystal structure and relation to whitlockite. Am Mineral 68:996–1003
Murata KJ, Rose HJ, Carbon MK (1953) Systematic variation of rare earths in monazite. Geochim Cosmochim Acta 4:292–300
Nagashima K, Miyawaki R, Takase J, Nakai I, Sakurai K, Matsubara S, Kato A, Iwano S (1986) Kimuraite, $CaY_2(CO_3)_4 \cdot 6H_2O$, a new mineral from fissures in an alkali basalt from Saga Prefecture, Japan, and new data on lokkaite. Am Mineral 71:1028–1033
Nambu M (ed) (1970) Introduction to Japanese minerals. IMA-IAGOD Meet 70, Geol Surv Jpn
Parak T (1973) Rare earths in the apatite iron ores of Lapland together with some data about the Sr, Th and U content of these ores. Econ Geol 68:210–221
Phan KD (1967) Le scandium dans les minéraux et les roches encaissantes de certaines pegmatites malgaches. Bull Bur Rech Geol Min Paris 3:78–97
Pigorini B, Veniale F (1968) L'apatite accessoria nella diverse facies litologiche delle formazioni granitoidi della Val Sersera (Vercelli). Rendiconti Soc It Mineral Petrol 24:1–32
Povarennykh AS (1972) Crystal chemical classification of minerals. Plenum, New York London
Robinson GW (1979) The occurrence of rare-earth elements in zircon. PhD, Queen's Univ, Kingston, Ont
Rudashevskii NS (1969) Epidote-allanite from southern Siberia metasomatites. Zap Vses Min Obshch 98:739–745
Semenov EI, Maksimyuk IE, Arkangelskaya VN (1976) On the minerals of the pectolite-serandite group. Zap Vses Min Obshch 104:154–163
Shannon RD (1976) Revised effective ionic radii and systematic studies of interatomic distances in halides and chalcogenides. Acta Crystallogr A32:751–767
Sinha PS (1982) Systematics and the properties of the lanthanides. Reidel, Dordrecht Lancaster, p 127
Stamper JW, Chin E (1976) Rare earth elements. In: Mineral facts and problems. US Dep Inter, Bur Mines Bull 650, p 682
Sudarsanan K, Young RA (1969) Significant precision in crystal structure details: Holly Springs hydroxyapatite. Acta Crystallogr B25:1534–1543
Sverdrup TL (1968) Yttrofluorite-yttrocerite-cerfluorite in Norwegian pegmatites. Norsk Geol Tidsskr 48:245–252
Sverdrup TL, Bryn KØ, Saebø P (1959) Contributions to the mineralogy of Norway. N° 2. Bastnaesite, a new mineral for Norway. Norsk Geol Tidsskr 39:237–247
Turekian KK, Wedepohl KH (1964) Distribution of elements in some major units of the earth's crust. Geol Soc Am Bull 72:186

Vlasov KA (ed) (1966) Geochemistry and mineralogy of rare elements and genetic types of their deposits, vol 2: Mineralogy of rare elements. Isr Progr Sci Transl, Jerusalem, 945 pp
Wadia DN, Fernando LJD (1945) Ilmenite, monazite and zircon, gems and semi-precious stones of Ceylon. Ceylon Rec Dep Mineral Prof Pap 2
Winther C (1901) Schizolite, a new mineral. Meddel Grønl 24:196

Appendix: Rare-Earth Minerals

HALIDES

Fluocerite	$(Ce,La)F_3$
Fluocerite-(La)	$(La,Ce)F_3$
Gagarinite	$NaCaY(F,Cl)_6$
Tveitite	$Ca_{1-x}Y_xF_{2+x}$ x=0.3

CARBONATES

Ancylite	$SrCe(CO_3)_2(OH).H_2O$
Bastnaesite	$(Ce,La)(CO_3)F$
Bastnaesite-(La)	$(La,Ce)(CO_3)F$
Bastnaesite-(Y)	$(Y,Ce)(CO_3)F$
Burbankite	$(Na,Ca)_3(Sr,Ba,Ce)_3(CO_3)_5$
Calcioancylite	$(Ca,Sr)Ce(CO_3)_2(OH).H_2O$
Calkinsite	$(Ce,La)_2(CO_3)_3.4H_2O$
Carbocernaite	$(Ca,Na)(Sr,Ce,Ba)(CO_3)_2$
Cebaite	$Ba_3Ce_2(CO_3)_5F_2$
Cordylite	$Ba(Ce,La)_2(CO_3)_3F_2$
Daqingshanite	$(Sr,Ca,Ba)_3(Ce,La)(PO_4)(CO_3)_{3-x}(OH,F)_x$
Donnayite	$Sr_3NaCaY(CO_3)_6.3H_2O$
Ewaldite	$Ba(Ca,Y,Na,K)(CO_3)_2$
Gysinite	$Pb(Nd,La)(CO_3)_2(OH).H_2O$
Huanghoite	$BaCe(CO_3)_2F$
Hydroxylbastnaesite	$(Ce,La)(CO_3)(OH,F)$
Hydroxylbastnaesite-(Nd)	$(Nd,Ce)(CO_3)(OH,F)$
Khanneshite	$(Na,Ca)_3(Ba,Sr,Ce,Ca)_3(CO_3)_5$
Kimuraite	$CaY_2(CO_3)_4.6H_2O$
Lanthanite	$(La,Ce)_2(CO_3)_3.8H_2O$
Lanthanite-(Ce)	$(Ce,La,Nd)_2(CO_3)_3.8H_2O$
Lanthanite-(Nd)	$(Nd,La)_2(CO_3)_3.8H_2O$
Lokkaite	$(Y,Ca)_2(CO_3)_3.2H_2O$
Mckelveyite	$Ba_3Na(Ca,U)Y(CO_3)_6.3H_2O$
Parisite	$(Ce,La)_2Ca(CO_3)_3F_2$
Röntgenite	$Ca_2(Ce,La)_3(CO_3)_5F_3$
Sahamalite	$(Mg,Fe^{2+})Ce_2(CO_3)_4$
Schuilingite	$PbCu(Nd,Gd,Sm,Y)(CO_3)_3(OH).1,5H_2O$
Synchysite	$(Ce,La)Ca(CO_3)_2F$
Synchysite-(Nd)	$(Nd,La)Ca(CO_3)_2F$
Synchysite-(Y)	$(Y,Ce)Ca(CO_3)_2F$
Tengerite	$CaY_3(CO_3)_4(OH)_3.3H_2O$
Thorbastnaesite	$Th(Ca,Ce)(CO_3)_2F_2.3H_2O$
Zhonghuacerite	$Ba_2Ce(CO_3)_3F$

OXIDES

Aeschynite	$(Ce,Ca,Fe,Th)(Ti,Nb)_2(O,OH)_6$
Aeschynite-(Nd)	$(Nd,Ce,Ca,Th)(Ti,Nb)_2(O,OH)_6$
Aeschynite-(Y)	$(Y,Ca,Fe,Th)(Ti,Nb)_2(O,OH)_6$
Brannerite	$(U,Ca,Y,Ce)(Ti,Fe)_2O_6$

Calciobetafite	$(Ca,RE,Th,U)_2(Nb,Ta,Ti)_2O_7$
Cerianite	$(Ce^{4+},Th)O_2$
Ceriopyrochlore	$(Ce,Ca,Y)_2(Nb,Ta)_2O_6(OH,F)$
Cerotungstite	$CeW_2O_6(OH)_3$
Chrichtonite	$(Sr,La,Ce,Y)(Ti,Fe^{3+},Mn)_{21}O_{38}$
Davidite	$(La,Ce)(Y,U,Fe^{2+})(Ti,Fe^{3+})_{20}(O,OH)_{38}$
Davidite-(Ce)	$(Ce,La)(Y,U,Fe^{2+})(Ti,Fe^{3+})_{20}(O,OH)_{38}$
Euxenite	$(Y,Ca,Ce,U,Th)(Nb,Ta,Ti)_2O_6$
Fergusonite	$YNbO_4$
Fergusonite beta	$YNbO_4$
Fergusonite beta-(Nd)	$(Nd,Ce)NbO_4$
Fergusonite beta-(Ce)	$(Ce,La,Nd)NbO_4$
Fersmite	$(Ca,Ce,Na)(Nb,Ta,Ti)_2(O,OH,F)_6$
Formanite	$YTaO_4$
Hibonite	$(Ca,Ce)(Al,Ti,Mg)_{12}O_{19}$
Kobeite	$(Y,U)(Ti,Nb)_2(O,OH)_6$
Loparite	$(Ce,Na,Ca)_2(Ti,Nb)_2O_6$
Loranskite	$(Y,Ce,Ca)ZrTaO_6$
Loveringite	$(Ca,Ce)(Ti,Fe^{3+},Cr,Mg)_{21}O_{38}$
Murataite	$(Na,Y)_4(Zn,Fe)_3(Ti,Nb)_6O_{18}(F,OH)_4$
Niobo-aeschynite	$(Ce,Ca,Th)(Nb,Ti)_2(O,OH)_6$
Plumbopyrochlore	$(Pb,Y,U,Ca)_{2-x}Nb_2O_6(OH)$
Polycrase	$(Y,Ca,Ce,U,Th)(Ti,Nb,Ta)_2O_6$
Polymignite	$(Ca,Fe,Y,Th)(Nb,Ti,Ta,Zr)O_4$
Samarskite	$(Y,Ce,U,Ca,Pb)(Nb,Ta,Ti,Sn)_2O_6$
Scheteligite	$(Ca,Y,Sb,Mn)_2(Ti,Ta,Nb,W)_2O_6(O,OH)$
Tantalaeschynite-(Y)	$(Y,Ce,Ca)(Ta,Ti,Nb)_2O_6$
Tanteuxenite	$(Y,Ce,Ca)(Ta,Nb,Ti)_2(O,OH)_6$
Uranmicrolite	$(U,Ca,Ce)_2(Ta,Nb)_2O_6(OH,F)$
Uranpyrochlore	$(U,Ca,Ce)_2(Nb,Ta)_2O_6(OH,F)$
Vigezzite	$(Ca,Ce)(Nb,Ta,Ti)_2O_6$
Yttrobetafite	$(Y,U,Ce)_2(Ti,Nb,Ta)_2O_6(OH)$
Yttrocolumbite	$(Y,U,Fe^{2+})(Nb,Ta)O_4$
Yttrocrasite	$(Y,Th,Ca,U)(Ti,Fe^{3+})_2(O,OH)_6$
Yttropyrochlore	$(Y,Na,Ca,U)_{1-2}(Nb,Ta,Ti)_2(O,OH)_7$
Yttrotantalite	$(Y,U,Fe^{2+})(Ta,Nb)_2O_4$
Yttrotungstite	$YW_2O_6(OH)_3$
Zirkelite	$(Ca,Th,Ce)Zr(Ti,Nb)_2O_7$

BORATES

Braitschite	$(Ca,Na)_7(Ce,La)_2B_{22}O_{43}.7H_2O$
Moydite	$(Y,Gd,Dy,Sm,Nd,Ca)[B(OH)_4](CO_3)$

ARSENATES-SULPHATES-VANADATES

Agardite	$(Y,Ca)Cu_6(AsO_4)_3(OH)_6.3H_2O$
Chernovite	$YAsO_4$
Chukhrovite	$Ca_3(Y,Ce)Al_2(SO_4)F_{13}.10H_2O$
Chukhrovite-(Ce)	$Ca_3(Ce,Y)Al_2(SO_4)F_{13}.10H_2O$
Goudeyite	$(Al,Y)Cu_6(AsO_4)_3(OH)_6.3H_2O$
Kemmlitzite	$(Sr,Ce)Al_3(AsO_4)(SO_4)(OH)_6$
Retzian	$Mn_2Ce(AsO_4)(OH)_4$
Retzian-(La)	$Mn_2La(AsO_4)(OH)_4$
Retzian-(Nd)	$Mn_2Nd(AsO_4)(OH)_4$
Wakefieldite	YVO_4
Wakefieldite-(Ce)	$(Ce,Pb^{2+},Pb^{4+})VO_4$

PHOSPHATES

Belovite	$(Sr,Ce,Na,Ca)_5(PO_4)_3(OH)$
Brockite	$(Ca,Th,Ce)(PO_4).H_2O$
Cheralite	$(Ca,Ce,Th)(P,Si)O_4$
Churchite	$YPO_4.2H_2O$
Florencite	$CeAl_3(PO_4)_2(OH)_6$
Florencite-(La)	$(La,Ce)Al_3(PO_4)_2(OH)_6$
Florencite-(Nd)	$(Nd,Ce)Al_3(PO_4)_2(OH)_6$
Lermontovite	$(U,Ca,Ce)_3(PO_4)_4.6H_2O$
Monazite	$(Ce,La)PO_4$
Monazite-(La)	$(La,Ce,Nd)PO_4$
Monazite-(Nd)	$(Nd,Ce,La)PO_4$
Ningyoite	$(U,Ca,Ce)_2(PO_4)_2.1\text{-}2H_2O$
Petersite	$(Ca,Fe,Y,Ce)_2Cu_{12}(PO_4)_6(OH)_{12}.6H_2O$
Phosinaite	$Na_3(Ca,Ce)PSiO_7$
Rhabdophane	$(Ce,La)PO_4.H_2O$
Rhabdophane-(La)	$(La,Ce)PO_4.H_2O$
Rhabdophane-(Nd)	$(Nd,La)PO_4.H_2O$
Vitusite	$Na_3(Ce,La,Nd)(PO_4)_2$
Xenotime	YPO_4

URANYL-CARBONATES and URANYL-SILICATES

Bijvoetite	$(Y,Dy,Gd)_2O_3.4UO_3.4CO_2.14H_2O$
Kamotoite-(Y)	$(Y,Nd,Gd,Sm,Dy)_2O_3.4UO_3.3CO_2.14.5H_2O$
Lepersonnite	$CaO.(Gd,Dy,Y)_2O_3.24UO_3.8CO_2.4SiO_2.60H_2O$

SILICATES

Allanite	$(Ce,Ca,Y)_2(Al,Fe^{3+})_3(SiO_4)(Si_2O_7)O(OH)$
Allanite-(La)	$(La,Ca,Y)_2(Al,Fe^{3+})_3(SiO_4)(Si_2O_7)O(OH)$
Agrellite	$Na(Ca,RE)_2Si_4O_{10}F$
Ashcroftine	$KNaCaY_2Si_6O_{12}(OH)_{10}.4H_2O$
Britholite	$(Ce,Ca)_5(SiO_4,PO_4)_3(OH,F)$
Britholite-(Y)	$(Y,Ca)_5(SiO_4,PO_4)_3(OH,F)$
Cappelenite	$Ba(Y,Ce)_6Si_3B_6O_{24}F_2$
Caysichite	$(Y,Ca)_4Si_4O_{10}(CO_3)_3.4H_2O$
Cerite	$(Ce,Ca)_9(Mg,Fe)Si_7(O,OH,F)_{28}$
Chevkinite	$(Ca,Ce,Th)_4(Fe^{2+},Mg)_2(Ti,Fe^{3+})_3Si_4O_{22}$
Eudialyte	$Na_4(Ca,Ce,Fe,Mn)_2ZrSi_6O_{17}(OH,Cl)_2$
Gadolinite	$Y_2Fe^{2+}Be_2Si_2O_{10}$
Gadolinite-(Ce)	$(Ce,La,Nd,Y)_2Fe^{2+}Be_2Si_2O_{10}$
Hellandite	$(Ca,Y)_6(Al,Fe^{3+})Si_4B_4O_{20}(OH)_4$
Hingganite-(Yb)	$(Yb,Y)BeSiO_4(OH)$
Iimoriite	$Y_2(SiO_4)(CO_3)$
Ilimaussite	$Ba_2Na_4CeFe^{3+}Nb_2Si_8O_{28}.5H_2O$
Ilmajokite	$(Na,Ce,Ba)_2TiSi_3O_5(OH)_{10}.nH_2O$
Iraqite	$(K,La,Ce,Th)(Ca,Na,La)_2Si_8O_{20}$
Joaquinite	$Ba_2NaCe_2Fe^{2+}(Ti,Nb)_2Si_8O_{26}(OH,F).H_2O$
Kainosite	$Ca_2(Y,Ce)_2Si_4O_{12}(CO_3).H_2O$
Karnasurtite	$(Ce,La,Th)(Ti,Nb)(Al,Fe^{3+})(Si,P)_2O_7(OH)_4.3H_2O$
Keivyite	$Yb_2Si_2O_7$
Laplandite	$Na_4CeTiPSi_7O_{22}.5H_2O$
Melanocerite	$(Ce,La,Y)_4CaBSi_2O_{12}(F,OH).nH_2O$
Minasgeraisite	$(Y,Yb,Ca,Bi)_2(Ca,Mn)Be_2Si_2O_{10}$
Miserite	$K(Ca,Ce)_4Si_5O_{13}(OH)_3$
Monteregianite	$(Na,K)_6(Y,Ca)_2Si_{16}O_{38}.10H_2O$

Mosandrite	$(Na,Ca,Ce)_3Ti(SiO_4)_2F$
Nordite	$(La,Ce)(Sr,Ca)Na_2(Na,Mn)(Zn,Mg)Si_6O_{17}$
Nordite-(Ce)	$(Ce,La)(Sr,Ca)Na_2(Na,Mn)(Zn,Mg)Si_6O_{17}$
Okanoganite	$(Na,Ca)_3(Y,Ce)_{12}Si_6B_2O_{27}F_{14}$
Perrierite	$(Ca,Ce,Th)_4(Mg,Fe^{2+})_2(Ti,Fe^{3+})_3Si_4O_{22}$
Rowlandite	$Y_4Fe^{2+}Si_4O_{14}F_2$
Saryarkite	$Ca(Y,Th)Al_5(SiO_4)_2(PO_4,SO_4)_2(OH)_7.6H_2O$
Sazhinite	$Na_2CeSi_6O_{14}(OH).5H_2O$
Semenovite	$(Ca,Ce,La,Na)_{10-12}(Fe^{2+},Mn)(Si,Be)_{20}(O,OH,F)_{48}$
Steenstrupine	$(Ce,La,Na,Mn)_6(Si,P)_6O_{18}(OH)$
Stillwellite	$(Ce,La,Ca)BSiO_5$
Strontio-chevkinite	$(Sr,La,Ce,Ca)_4(Fe^{2+},Fe^{3+})(Ti,Zr)_2Ti_2Si_4O_{22}$
Tadzhikite	$Ca_3(Ce,Y)_2(Ti,Al,Fe^{3+})B_4Si_4O_{22}$
Thalenite	$Y_3Si_3O_{10}(OH)$
Thortveitite	$(Sc,Y)_2Si_2O_7$
Tombarthite	$Y_4(Si,H_4)_4O_{12-x}(OH)_{4+2x}$
Törnebohmite	$(Ce,La)_3Si_2O_8(OH)$
Törnebohmite-(La)	$(La,Ce)_3Si_2O_8(OH)$
Tranquillityite	$Fe_8^{2+}(Zr,Y)_2Ti_3Si_3O_{24}$
Tritomite	$(Ce,La,Y,Th)_5(Si,B)_3(O,OH,F)_{13}$
Tritomite-(Y)	$(Y,Ca,La,Fe^{2+})_5(Si,B,Al)_3(O,OH,F)_{13}$
Tundrite	$Na_3(Ce,La)_4(Ti,Nb)_2(SiO_4)_2(CO_3)_3O_4(OH).2H_2O$
Tundrite-(Nd)	$Na_3(Nd,La)_4(Ti,Nb)_2(SiO_4)_2(CO_3)_3O_4(OH).2H_2O$
Vyuntspakhite	$(Y,Yb)_4(Al,Si)_3Si_5O_{18}(OH)_5$
Yftisite	$(Y,Dy,Er)_4(Ti,Sn)O(SiO_4)_2(F,OH)_6$
Yttrialite	$(Y,Ce,Th)_2Si_2O_7$

Mineralogy of Niobium and Tantalum: Crystal Chemical Relationships, Paragenetic Aspects and Their Economic Implications

P. ČERNÝ[1] and T.S. ERCIT[2]

Abstract

A systematic review of the oxide minerals of niobium and tantalum is presented, including varietal names, crystal chemistry, structural features, relationships to other phases and paragenesis of each mineral group, series or species. A separate section deals with oxide minerals that occur exclusively as secondary, low-temperature phases of largely metasomatic origin. Oxide minerals of Sn, Ti and W that carry substantial Nb and Ta are discussed as well. A new crystal chemical classification of the oxide minerals of Nb and Ta is presented.

A brief review is presented of the crystal chemistry of the silicates of Nb and Ta, and of the silicates of Ti and Zr that exhibit substitution of these elements by Nb and Ta. Paragenetic affiliation of these silicates is reviewed. Miscellaneous minerals of Nb and Ta (borates, phosphates, borosilicates) are also mentioned.

Most of the oxide minerals of Nb and Ta typically occur in granitic rocks of orogenic affiliation, particularly in rare-element granitic pegmatites. However, some of these are also found in, and a few are characteristic of, anorogenic alkaline igneous sequences from carbonatites and gabbros to nepheline syenites and alkaline granites. In contrast, all silicates of Nb ($\pm$Ta)- and (Nb,Ta)-bearing titanosilicates and zirconosilicates are related to the anorogenic parageneses.

At present, the oxide minerals are the only important sources of Nb and Ta. Niobosilicates, and mainly some of the titanosilicates and zirconosilicates with significant Nb substitution, represent potential ore minerals for the future. However, no alternate ore minerals seem to be available for extraction of tantalum.

1 Introduction

Thorough understanding of ore mineralogy is a conditio sine qua non in today's exploration, mining and processing of diverse ore types. Maximal utilization of even "simple" ores requires reliable identification of different mineral species, deciphering of their textural relationships and understanding the compositional var-

[1]Department of Geological Sciences, University of Manitoba, Winnipeg, Manitoba, Canada, R3T 2N2
[2]Mineral Sciences Division, National Museum of Natural Sciences, Ottawa, Ontario, Canada, KlA OM8

Lanthanides, Tantalum and Niobium
P. Möller, P. Černý and F. Saupé (Eds.)

iations in the major, minor and trace components of the individual minerals (e.g. Cabri 1987).

Such an understanding is all the more important in the case of an extremely diversified spectrum of minerals with complex compositional variability, which characterizes the ore mineralogy of niobium and tantalum. The ever increasing number of Nb, Ta-bearing species, recent recognition of extensive zoning and replacement phenomena, broad ranges in the Nb/Ta ratio, variable contents of "waste" and additional economically interesting components – all of these aspects affect the effectiveness of ore-mineral concentration and metallurgy, and ask for thorough mineralogical study at the level of individual deposits. Furthermore, understanding of the ore mineralogy of niobium and tantalum is frequently hampered by a lack of basic knowledge of some minerals, particularly those of complex chemistry and predominantly metamict structural state. The need for extensive "pure" research, which finds instant "practical" applications, is obvious.

It is not only the widespread and economically important mineral species that must be taken into account in studies of diverse types of mineralization. Minor mineral components, particularly late replacement products of the dominant primary phases may significantly influence the behaviour of the latter in milling, gravity separation or flotation. They may shift the Nb/Ta ratio of the ore to a significant degree, or introduce undesirable components such as Ti, Sn or Sb. Also, mineral species that are extremely rare to absent in most deposits of a given type (e.g. simpsonite or cesstibtantite in granitic pegmatites) may attain economic significance as major ore minerals in a few specific localities. Thus, painstaking attention must be paid to even the least conspicuous and volumetrically minor mineral components.

Most minerals of niobium and tantalum, and all economically important species, belong to the class of oxide minerals. These are widespread in all genetic classes of deposits, and the individual species or assemblages are commonly typical of specific types of mineralization. In contrast, silicates of niobium and tantalum occur exclusively in peralkaline igneous and metasomatic suites of anorogenic descent, and so far they are only rarely used as sources of these metals. Besides the oxides and silicates, only a single representative species (or series) is known in each of the borate and phosphate classes, and both of them are extremely rare.

In addition to the species in which they constitute essential and stoichiometric components, niobium and tantalum also substitute into a multitude of other minerals. However, the Nb, Ta contents are generally low, and so far only rutile and cassiterite have attained economic importance.

In view of the above relationships, the present chapter is aimed at a general review of the Nb,Ta-bearing minerals, subdivided into oxide, silicate and miscellaneous phases. A brief review of the structural classification of the oxide minerals precedes the oxide section. The oxide and silicate sections are complemented by reviews of the major minerals with appreciable Nb and Ta substitution. The final section is devoted to a brief paragenetic review.

2 Classification of Oxide Minerals of Niobium and Tantalum

Hawthorne (1983, 1985, 1986) has proposed that "structures may be ordered or classified according to the polymerization of those coordination polyhedra with higher bond-valences". Nb,Ta-oxide minerals are particularly suited to this approach for several reasons. The principal anionic groups, namely $(Nb,Ta)O_6$ octahedra and $(Nb,Ta)O_7$ pentagonal bipyramids, are typified by high bond valences (Pauling bond strengths of 5/6 and 5/7 v.u. respectively), yet not so high as to prevent polymerization of these groups (i.e. not greater than 1 v.u.). In all, except for a very few structure types, the bond strengths of Nb,Ta-O polyhedra are considerably greater than those of adjacent polyhedra; consequently, the classification need only involve Ta,Nb polyhedra. However, Ta^{5+} and Nb^{5+} show limited to substantial isomorphism with a restricted number of other highly charged, small cations as Ti^{4+}, Sn^{4+} and W^{6+}; that is, when these elements occur in Nb,Ta-oxide minerals, they tend to play the same crystallochemical rôle as Nb and Ta. For ease of discussion, we designate Ta, Nb and their isomorphic relatives as M-cations, and the anions they are bonded to as O-anions. The classification presented here is based on the polymerization of M-cation polyhedra in Ta,Nb-oxide minerals. In deriving the classification, only oxideminerals species with Ta and Nb as the dominant M-cations are considered; hence, several Nb,Ta-bearing (but not Nb, Ta-dominant) Ti,Zr-oxide minerals are not included.

The classification is summarized in Table 1. The minerals are divided into classes on the basis of the types of M-cation polyhedron clusters found in Nb,Ta-oxide minerals. Classes are ranked according to decreasing Lewis basicity of the characteristic cluster, and range from M_2O_5 at the most basic end to M_3O_{13} at the least basic end. At the present time, seven such classes are known; a somewhat arbitrary eighth class has been added to include inadequately characterized species.

Each class is divided into subclasses on the basis of the degrees of polymerization of the cluster type defining the class. Subclasses are ranked in order of increasing degrees of polymerization; i.e. finite clusters → infinite chains → infinite sheets → frameworks.

The third and last level of subdivision is by means of structure type. To date, 21 different structure types are known for Nb,Ta-oxide minerals.

In addition to the classification, Table 1 gives a description of the M-O structure module for each structure type, a list of the species which belong to each structure type and their formulae.

For further details on the classification, the reader is referred to Ercit (1989).

3 Major Oxide Minerals of Niobium and Tantalum

In contrast to the structural hierarchy shown in the preceding section, the brief descriptions of individual groups, series or species of Nb,Ta-bearing oxide minerals follow a much more "relaxed" sequence. Crystallochemical relationships are not strictly adhered to. Rather, the sequence progresses roughly from the most common

Table 1. Classification of Nb,Ta-oxide minerals

Class (Cluster type)	Subclass (Polymerization)	Structure Type	Description (of M-$\emptyset$ structure module)	Species	Formula
1. $[M_3\emptyset_{13}]$	i) Finite cluster	Simpsonite	Edge-sharing octahedral trimers	Simpsonite	$Al_4Ta_3O_{13}(OH)$
2. $[M\emptyset_4]$	i) Chain	ß-Fergusonite	Zig-zag chains of edge-sharing octahedra	ß-Fergusonite	$(Y,RE)NbO_4$
				ß-Fergusonite-Ce	$(Ce,Y)NbO_4$
				ß-Fergusonite-Nd	$(Nd,Y)NbO_4$
		Samarskite	Zig-zag chains of edge-sharing octahedra (distorted wolframite structure)	Samarskite	$(Y,Fe,U,RE)(Nb,Ta,Ti)O_4$
	ii) Sheet	Stibiotantalite	Perforated sheets of corner-sharing octahedra	Stibiocolumbite	$SbNbO_4$
				Stibiotantalite	$SbTaO_4$
				Bismutotantalite	$BiTaO_4$
		Alumotantite	Corrugated, perforated sheets of corner-sharing octahedra	Alumotantite	$AlTaO_4$
	iii) Undesignated	Fergusonite	Exact structure unknown	Fergusonite	$(Y,RE)NbO_4$
		UTa_2O_8	Exact structure unknown	Liandratite	$U^{6+}\square(Nb,Ta)_2O_8$
				Petscheckite	$U^{4+}Fe^{2+}(Nb,Ta)_2O_8$
3. $[M\emptyset_4]$ and $[M\emptyset_3]$	i) Chain plus Sheet	Qitianlingite	Isolated zig-zag chains of edge-sharing octahedra plus corrugated sheets made of similar chains linked by corners	Qitianlingite	$(Fe,Mn)_2(Nb,Ta)_2WO_{10}$
4. $[M\emptyset_3]$	i) Sheet	Columbite	Corrugated sheets made of zig-zag chains of edge-sharing octahedra linked by corners	Manganocolumbite	$MnNb_2O_6$
				Ferrocolumbite	$FeNb_2O_6$
				Manganotantalite	$MnTa_2O_6$
				Ferrotantalite	$(Fe,Mn)(Ta,Nb)_2O_6$
				Magnocolumbite	$MgNb_2O_6$
				Fersmite	$CaNb_2O_6$
				(Pseudo-ixiolite)	$(Ta,Nb,Mn,Fe)_3O_6$

	Euxenite	Distorted columbite structure	Euxenite	$(Y,RE)(Nb,Ti,Ta)_2O_6$
			Tanteuxenite	$(Y,RE)(Ta,Nb,Ti)_2O_6$
			Polycrase	$(Y,RE)(Ti,Nb,Ta)_2O_6$
	Thoreaulite	Perforated, double-octahedron thick sheets of edge- and corner-linked octahedra	Thoreaulite	$Sn^{2+}Ta_2O_6$
			Foordite	$Sn^{2+}Nb_2O_6$
ii) Framework	Tapiolite	Framework of edge- and corner-linked octahedra	Ferrotapiolite	$FeTa_2O_6$
			Manganotapiolite	$(Mn,Fe)Ta_2O_6$
			Staringite	$Sn_{3-3x}Fe_xTa_{2x}O_6$
			(Disordered ferrotapiolite)	$(Ta,Fe)_3O_6$
	Aeschynite	Framework of edge- and corner-linked octahedra	Aeschynite	$(Ce,Ca)(Ti,Nb)_2O_6$
			Aeschynite-Y	$(Y,Ca)(Ti,Nb)_2O_6$
			Aeschynite-Nd	$(Nd,Ce,Ca)(Ti,Nb)_2O_6$
			Nioboaeschynite	$(Ce,Ca)(Nb,Ti)_2O_6$
			Tantalaeschynite-Y	$(Y,Ce,Ca)(Ta,Ti,Nb)_2O_6$
			Rynersonite	$CaTa_2O_6$
			Vigezzite	$(Ca,Ce)(Nb,Ta,Ti)_2O_6$
	Changbaiite	Framework of edge- and corner-linked octahedra	Changbaiite	$PbNb_2O_6$
	Perovskite	Framework of corner-linked octahedra	Lueshite	$NaNbO_3$
			Natroniobite	$NaNbO_3$
			Latrappite	$(Ca,Na)(Nb,Ti)O_3$
	Rankamaite	Framework of corner-linked octahedra (tungsten-bronze derivative)	Rankamaite	$(Na,K)_{3-x}Al(Ta,Nb)_{1}$[illegible] $(O,OH)_{30}$
			Sosedkoite	$(K,Na)_{3-x}Al(Ta,Nb)_{1}$[illegible] $(O,OH)_{30}$

Table 1. *Cont.*

Class (Cluster type)	Subclass (Polymerization)	Structure Type	Description (of M-∅ structure module)	Species	Formula
		Normal Pyrochlore	Framework of corner-linked octahedra		$A_{2-m}B_2O_6\ (F,OH,O)_{1-n}\cdot pH_2O$
				a) *Pyrochlore subgroup*	$(B = Nb + Ta > 2Ti,\ Nb > Ta)$
				Pyrochlore	A = (Na,Ca)
				Kalipyrochlore	A > 20% K
				Bariopyrochlore	A > 20% Ba
				Plumbopyrochlore	A > 20% Pb
				Yttropyrochlore	A > 20% Ln with Y dominant
				Ceriopyrochlore	A > 20% Ln with Ce dominant
				Uranpyrochlore	A > 20% Ln with U dominant
				b) *Microlite subgroup*	$(B = Nb + Ta > 2Ti,\ Ta \geq Nb)$
				Microlite	A = (Na,Ca)
				Stannomicrolite	A > 20% Sn
				Bariomicrolite	A > 20% Ba
				Plumbomicrolite	A > 20% Pb
				Stibiomicrolite	A > 20% Sb
				Bismutomicrolite	A > 20% Bi
				Uranmicrolite	A > 20% U
				c) *Betafite subgroup*	$(B = 2Ti \geq Nb + Ta)$
				Calciobetafite	A = Ca
				Plumbobetafite	A > 20% Pb
				Stibiobetafite	A > 20% Sb
				Yttrobetafite	A > 20% Ln with Y dominant
				Betafite	A > 20% U
		Mixed inverse-normal pyrochlore	Framework of corner-linked octahedra	Cesstibtantite	$(Sb,Na)_{2-m}Ta_2(O,OH)_6(OH,Cs)_{1-n}$
		Parabario-microlite	Framework of corner-linked octahedra	Parabariomicrolite	$BaTa_4O_{10}(OH)_2\cdot 2H_2O$

5. $[M_4\phi_{11}]$	i) Framework	Calciotantite	Sheets of edge-sharing pentagonal bipyramids, cross-linked via corners by octahedra to form a framework	Calciotantite	$CaTa_4O_{11}$
		Natrotantite	Sheets of edge-sharing pentagonal bipyramids, cross-linked via corners by octahedra to form a framework	Natrotantite	$Na_{2-x}Ta_4O_{11-x}(OH)_x$
	ii) Undesignated	Irtyshite	Exact structure unknown (natrotantite polytypoid)	Irtyshite	$Na_2Ta_4O_{11}$
		Franconite	Unknown	Franconite Hochelagaite	$Na_2Nb_4O_{11}\cdot 9H_2O$ $CaNb_4O_{11}\cdot nH_2O$ ($n=8$)
6. $[M_3\phi_8]$	i) Framework	Wodginite	Zig-zag chains of edge-sharing octahedra cross-linked via corners by octahedra to form a framework	Wodginite Ferrowodginite Titanowodginite Tantalowodginite (ixiolite)	$MnSnTa_2O_8$ $FeSnTa_2O_8$ $MnTiTa_2O_8$ $(Mn_{0.5}\square_{0.5})TaTa_2O_8$ $(Ta,Nb,Mn,Fe,Sn,Ti)_4O_8$
		Lithiotantite	Zig-zag chains of edge-sharing octahedra, cross-linked via corners by octahedra to form a framework	Lithiotantite	$LiTa_3O_8$
7. $[M_2\phi_5]$	i) Framework	Tantite	Unknown	Tantite	Ta_2O_5
8. Inadequately characterized species				Gerasimovskite Ishikawaite Natrobistantite Scheteligite Yttrocolumbite Yttrotantalite	$(Mn,Ca)_{2-x}(Nb,Ti,Si)_5O_{12}\cdot nH_2O$ $(Fe,U,Y)(Nb,Ta)O_4$ $(Na,Bi)_{2-m}Ta_2O_6(OH,O)_{1-n}$ $(Ca,Fe,Mn,Sb,Y,Bi)_2$- $(Ti,Ta,Nb,W)_2(O,OH)_7$ $(Y,U,Fe)(Nb,Ta)O_4$ $(Y,U,Fe)(Ta,Nb)O_4$

to the not-so-widespread species, and from the simple and largely perfectly crystalline compounds with Na, Ca, Fe, Mn, Mg, Al, Ti and Sn to the complex and largely metamict REE, Y, U-bearing phases.

3.1 Columbite-Tantalite Group

The orthorhombic minerals of this group have a general formula AM_2O_6, where A = Fe, Mn, Mg and M = Nb, Ta. Minor quantities of Fe^{3+}, Sc, Ti, Sn and W are commonly present. End-member compositions include

ferrocolumbite	$FeNb_2O_6$,
manganocolumbite	$MnNb_2O_6$,
manganotantalite	$MnTa_2O_6$, and
magnocolumbite	$MgNb_2O_6$.

Compositions with a dominant ferrontantalite component $(Fe>Mn)(Ta>Nb)_2O_6$ are also known but the end member $FeTa_2O_6$ belongs to the tetragonal tapiolite series.

The compositional range of the (Fe,Mn)-based members of the group is shown in Fig. 1, modified from Černý and Ercit (1985). It covers most of the columbite quadrilateral and is separated by a miscibility gap from the tapiolite field.

Columbite-tantalite is known in a disordered form $(A,M_2)_{\Sigma 4}O_8$ (pseudo-oxiolite of Nickel et al. 1963b), and in intermediate structural states grading into essentially ordered phases $A_4M_8O_{24}$, with AMM AMM... cation ordering stacked in the X direction and consequent tripling of *a* (Sturdivant 1930). Disordered members can be converted into the ordered structure on extended heating at 1000°C.

The *a*/*c* ratio of columbite-tantalite was found a suitable order-disorder indicator by Komkov (1970): the unit cell expands along *c* and contracts along *a* with increasing disorder. Černý and Turnock (1971) introduced the *a* (3 × *a* for disordered phases) vs *c* diagram as a convenient graphic representation of the degree of order. This plot also indicates an approximate Fe/Mn ratio but is relatively insensitive to changes in Nb/Ta. In its present modification the *a*-*c* diagram is based on unit cell dimensions derived in part from Turnock's (1966) synthetic phases, in part from natural disordered minerals close to end-member compositions (cf. Wise et al. 1985; Černý and Ercit 1985; Ercit 1986 for further details). The data for disordered structural end members are relatively uncertain because of lack of well-documented synthetic phases. Based on the available data, the percentage of cation order may be calculated using Ercit's (1986) equation:

$$\% \text{ order} = 1727 - 941.6\,y, \text{ where } y = \mathbf{c} - 0.2329\,\mathbf{a};$$

precision is estimated at about ±5% for minerals poor in contaminating elements.

Figure 2 shows that disordered columbite-tantalites are more common than their ordered counterparts, and minerals of intermediate degree of order predominate. With the exception of manganotantalite (e.g. von Knorring et al. 1966), highly ordered natural phases are virtually non-existent. It must be noted, however, that the data shown in Fig. 2 are derived from columbite-tantalites of pegmatitic origin, and the structural state of columbite-tantalite from other parageneses is essentially unknown.

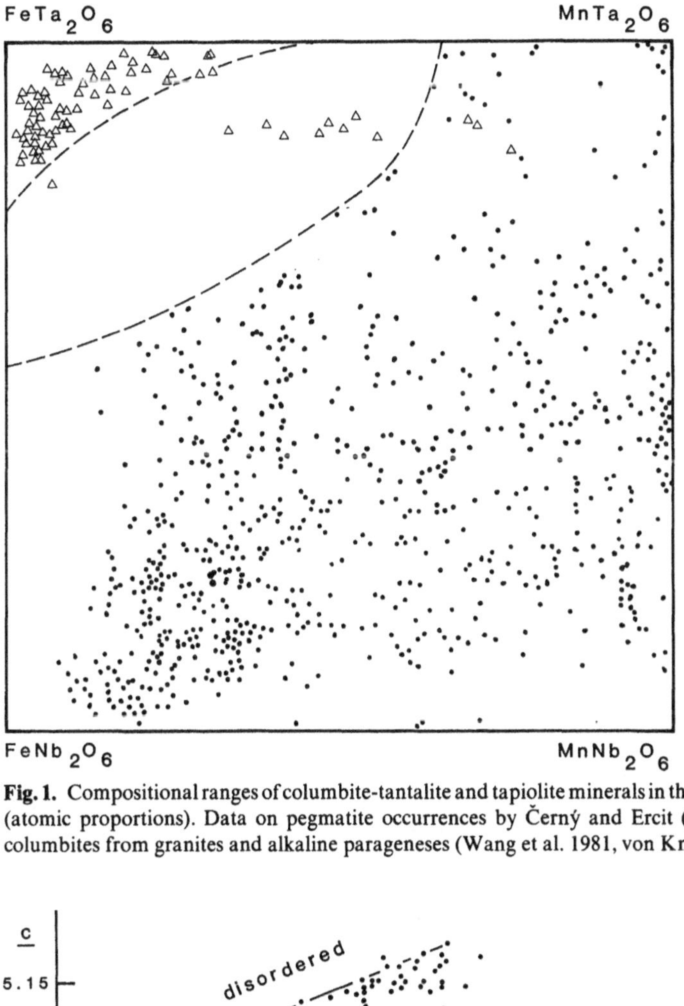

Fig. 1. Compositional ranges of columbite-tantalite and tapiolite minerals in the columbite quadrilateral (atomic proportions). Data on pegmatite occurrences by Černý and Ercit (1985) complemented by columbites from granites and alkaline parageneses (Wang et al. 1981, von Knorring and Fadipe 1981)

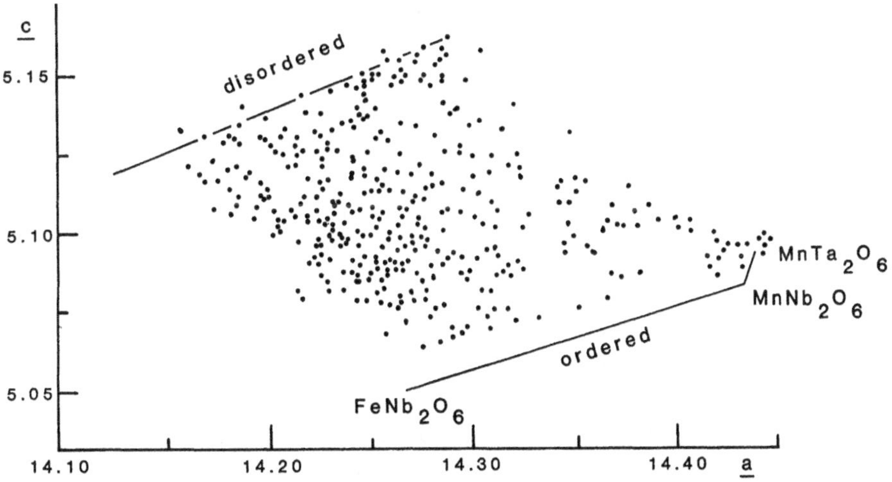

Fig. 2. The a–c diagram (in Å) of the columbite-tantalite minerals from granitic pegmatites, with a tripled for disordered phases. Based on Fig. 1A of Černý and Ercit (1985), adjusted for locality bias and complemented by unpublished data of the authors

Metamictization of columbite-tantalite was documented from many localities (Heinrich 1962; Ewing 1976a). The reported cases come from granitic pegmatites, but they may be expected to occur in other parageneses as well, in the presence of U or Th.

The composition of columbite-tantalite from granitic pegmatites covers most of the range shown in Fig. 1 (Černý and Ercit 1985) but it is much more restricted to the Nb- and Fe-rich parts of the quadrilateral for occurrences in alkaline igneous suites (Williams et al. 1956, von Knorring 1960).

Columbite-tantalite is decidedly the most widespread of all Nb,Ta-bearing minerals, found in substantial quantities in most types of deposits. It commonly occurs as a primary phase that may be replaced by late minerals (such as microlite or fersmite in granitic pegmatites; Černý and Turnock 1971), or it may originate as a replacement product of earlier minerals (e.g. pyrochlore in carbonatites; van d. Veen 1963). Zoning and replacement or veining by columbite-tantalite of variable composition seems to be widespread in pegmatites (Lahti 1987).

3.2 Tapiolite Series

Until recently, tetragonal tapiolite was known to have a rather uniform composition close to $FeTa_2O_6$, with negligible to subordinate substitutions by Mn and Nb. However, Lahti et al. (1983) described the first occurrence of a Mn-dominant phase, and the tapiolite minerals attained the status of a series with the end members ferrotapiolite $FeTa_2O_6$ and manganotapiolite $MnTa_2O_6$.

Despite the above development, the tapiolite series still remains compositionally constrained. Manganotapiolite and Mn-rich compositions violating the empirical two-phase region in the columbite quadrilateral have been found so far at only a single locality (Fig. 1). Clark and Fejer (1978), Lahti et al. (1983) and Černý et al. (1984) confirmed the previously recognized Mn/(Mn + Fe) and Ta/(Ta + Nb) ranges, as well as the very restricted incorporation of Sn and Ti. However, Khvostova and Arkhangelskaya (1970) and Wise (1987) have shown that Fe^{3+} is also an important component in some tapiolites.

The structural state of the tapiolite minerals varies from the cation-disordered monorutile structure $(Fe_1Ta_2)\Sigma_2O_4$ to the ordered trirutile structure $Fe_2Ta_4O_{12}$, with AMM AMM... stacking of cation positions in the Z direction and consequent tripling of the *c* dimension. Normalized to equal cell type, the *c* parameter increases with disorder much more than *a*. However, this change closely approximates the effect of Mn substitution (Fig. 3). Consequently, the degree of order in tapiolite minerals can be better estimated from the intensities of the trirutile-type diffractions. So far, metamictization of tapiolite minerals has not been reported.

The persistent differences in chemical composition, symmetry and structure between tapiolite and columbite-tantalites have been known for a long time. However, it was only recently that Weitzel (1976) provided a plausible explanation. He has suggested that the polarizing effect of Ta^{5+} on the surrounding oxygens, weaker in comparison with that of Nb^{5+}, favours the rutile structure for $FeTa_2O_6$. Nevertheless, $MnTa_2O_6$ crystallizes in the orthorhombic columbite structure as the ionic radius of Mn^{2+} is too large for the rutile structure. Thus, it is not surprising that

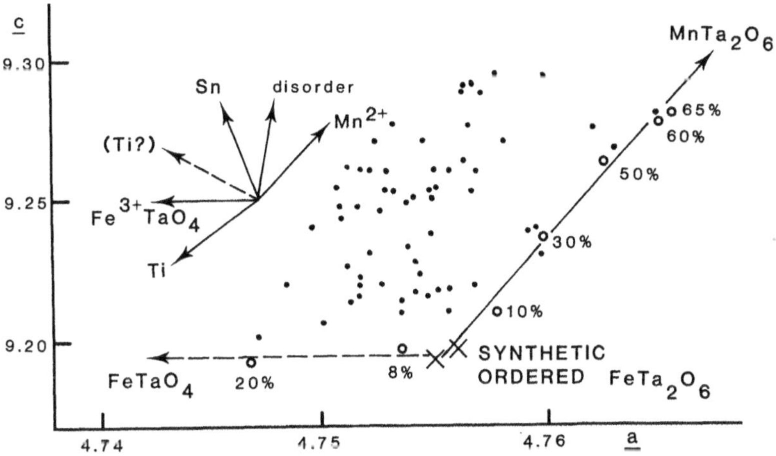

Fig. 3. Response of unit cell dimensions of tapiolite minerals to disordering and compositional changes. Synthetic ordered phases, $FeTaO_4$ and Mn substitutions of Turnock (1966) as refined by Wise (1987). Plots of natural tapiolite samples from the Yellowknife pegmatite field (Wise 1987)

recent re-examination of the alleged Nb-dominant members of the tapiolite structure type, mossite (Brøgger 1897) and manganomossite (Bowley 1923), has led to their discreditation (Ewing 1976a; Dunn et al. 1979).

Occurrences of tapiolite minerals are restricted to granitic pegmatites and some Ta-mineralized granites, commonly characterized by low *a* (F_2) and consequent poor fractionation of Fe and Mn. However, Mn-enriched ferrotapiolite also occurs in late assemblages of Ta minerals in complex and highly fractionated pegmatites; ferrotapiolite is generated during a late reversal in Fe-Mn fractionation (Ercit and Černý 1982; Ercit 1986). Tapiolite is known to replace simpsonite (Ercit 1986), and is in turn replaced by microlite (von Knorring 1974).

3.3 Ixiolite Group

As defined by Nickel et al. (1963b), ixiolite is a complex oxide with the formula $(Ta, Nb, Sn, Fe, Mn, Ti)_4O_8$ and an orthorhombic until cell corresponding to that of disordered columbite-tantalite. In contrast to the latter, heating promotes ordering into the wodginite structure rather than the ordered columbite structure (cf. Černý and Ercit 1985 for specific details).

The present-day ixiolite mineral family represents quite a spectrum of compositional varieties. Besides the type chemistry rich in Sn and Ta, ixiolites with substantial Sc^{3+}, Fe^{3+}, Ti^{4+}, W^{6+}, U and Nb $\gtrless$ Ta have been described during the last 2 decades:

(stannian) ixiolite	$(Ta,Nb,Sn,Fe \gtrless Mn,Ti)_4O_8$,
titanian ixiolite	$(Ta,Ti,Nb,Fe \gtrless Mn,Sn)_4O_8$,

scandian ixiolite $(Nb \gtrsim Ta,Fe^{3+},Fe^{2+},Sc,Ti,Zn,Sn,Mn,REE)_4O_8$, and
"wolframoixiolite" $(Nb,W,Fe,Mn,Ta,Zr,U,Ca,Mg,Ti)_4O_8$.

Titanian ixiolite $(Ta,Ti,Nb,Mn,Fe,Sn)_4O_8$, which forms a wodginite phase on heating, is known to contain up to 15.8% TiO_2 (unpublished data of the authors). The titanian "columbite" of Sveshnikova et al. (1965) and the "Ti-columbite" of von Knorring and Fadipe (1981; Table 1, anal. 9) most probably also belong to titanian ixiolite. This variety typically forms exsolutions in niobian rutile, as discussed later in Section 5.1.

Scandian ixiolite with over 10% Sc_2O_3 and $Fe_2O_3 > FeO$, $(Nb \gtrsim Ta, Fe^{3+}, Fe^{2+}, Sc, Ti, Zr, Sn, Mn, REE)_4O_8$, was described by von Knorring et al. (1969b) and Borisenko et al. (1969). Heating of this mineral produces a phase with the wolframite-type $Fe^{3+}NbO_4$ structure.

"Wolframoixiolite", $(Nb,W,Fe,Mn,Ta,Zr,U,Ca,Mg,Ti)_4O_8$ reported by Ginsburg et al. (1969) forms an ordered columbite-type phase and a UTa_2O_8-type phase on heating. Other W-rich phases such as the wolframian (tungstenian) columbite and ixiolite described by Amichba and Dubakina (1976a,b), Beddoe-Stephens and Fortey (1981) and Konovalenko et al. (1982) have not been structurally characterized, or they represent mixtures of orthorhombic phases with wolframite (Fleischer 1977, 1984).

The extensive flexibility of the ixiolite structure accommodates cations with broad ranges of ionic radius and valence. This has led, and undoubtedly will lead in the future, to attempts at establishing new mineral species based on compositional variation alone (cf. the recently discredited ashanite; Zhang et al. 1980, Foord 1982b, Foord pers. commun. 1983, 1987). It is becoming increasingly urgent to develop guidelines for crystallochemical classification and nomenclature in the columbite-type family of minerals, to prevent cluttering the terminology with superfluous and confusing names.

The ixiolite group especially requires rationalization because most of its members are being proven to be disordered analogues of ordered phases with well-defined stoichiometry and structure, and with lower than orthorhombic symmetry. Figure 4 indicates that most stannian and titanian ixiolites overlap the Mn/(Mn+Fe) and Ta/(Ta+Nb) range of wodginite. Figure 5 shows a similar compositional overlap in an (Sn,Ti)–(Nb,Ta)–(Fe,Mn) diagram.

When treated as orthorhombic, ixiolites plot along the disordered edge of the *a-c* diagram, with a typical scatter outside the columbite structural quadrilateral (Fig. 6). However, detailed examination of X-ray powder diffractograms and precession photographs of most stannian (and probably also titanian) ixiolites reveals them as monoclinic, pseudo-orthorhombic phases (Wise and Černý 1986, Wise 1987) analogous to minerals described as highly disordered wodginite (Ercit 1986). This similarity is strengthened by their behaviour upon heating (Fig. 7).

The scandian and wolframian (tungstenian) ixiolites may represent, in formal principle, disordered analogues of monoclinic wolframite-type phases. However, the non-stoichiometric ratios and appreciable number of cations found in natural minerals will probably generate multiphase heating products, instead of a single ordered compound, as already observed by Ginsburg et al. (1969).

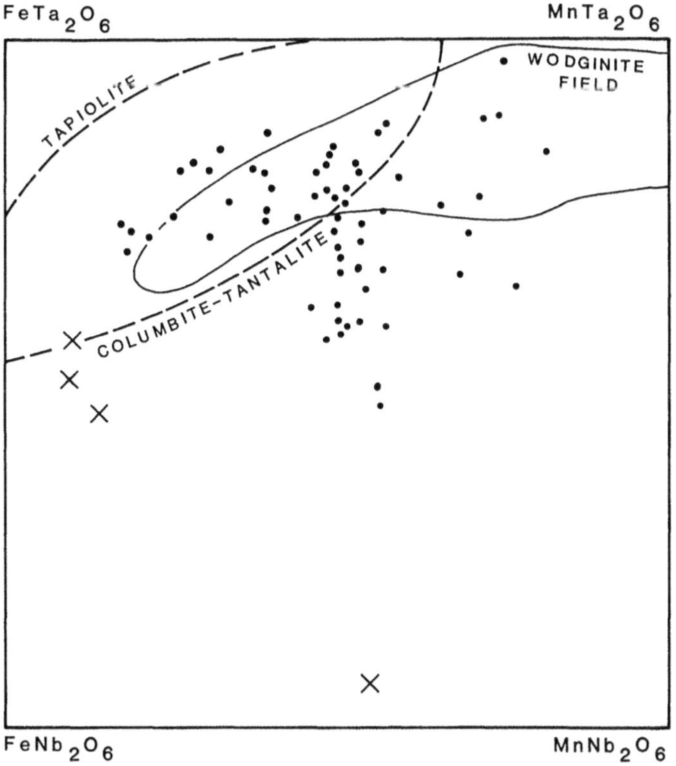

Fig. 4. Ixiolite compositions in the columbite quadrilateral (atomic proportions). *Dots* mark compositions of stannian and titanian ixiolite, *x* indicates scandian ixiolite composition. Modified from Fig. 1B of Černý and Ercit (1985). Compositional gap separating columbite-tantalite and tapiolite from Fig. 1, wodginite field from Fig. 9

To date, metamict ixiolite has not been reported. However, it is conceivable that it may occur, given the inherent instability of the disordered structure and diversity of cation population (including U and Th in some cases).

From the paragenetic viewpoint, ixiolites are found mainly in granitic pegmatites. Some varieties, particularly those rich in W, are reported from greisens and wolframite-bearing vein deposits (Amichba and Dubakina 1976a,b; Beddoe-Stephens and Fortey 1981). Ixiolite commonly occurs in intimate microscopic intergrowths with cassiterite, niobian rutile or microlite, and it is replaced by microlite, wodginite and fersmite (Sahama 1978; Paul et al. 1981; Khvostova et al. 1983; Ferreira 1984; Černý et al. 1986).

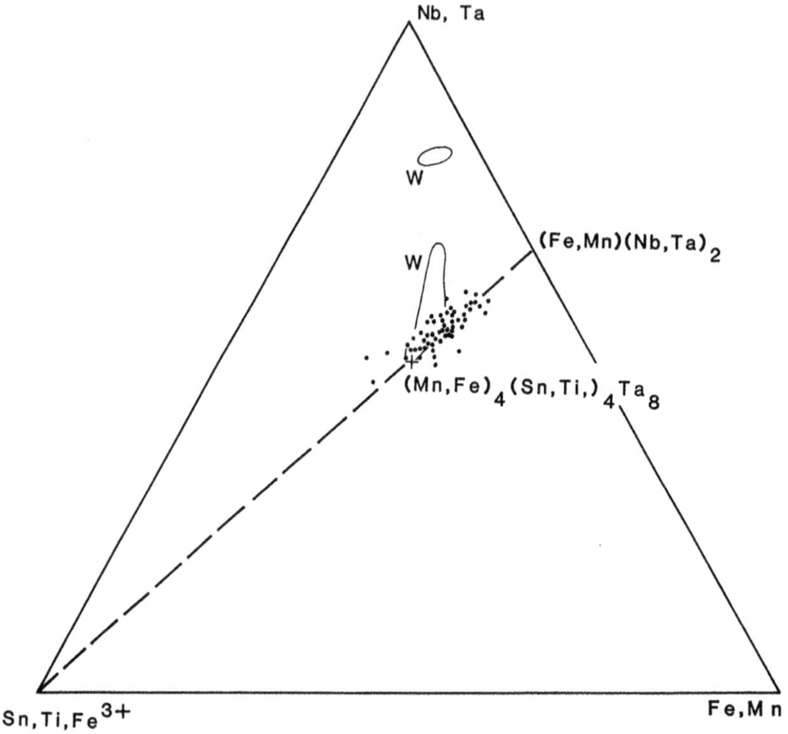

Fig. 5. Composition of stannian and titanian ixiolite calculated assuming full cation occupancy in the (Nb,Ta)–(Sn,Ti,Fe^{3+})–(Fe,Mn) diagram (atomic ratios). Adjusted from Fig. 2 of Černý and Ercit (1985) and complemented by unpublished data of P.Č. and M.A. Wise. Compositional field of wodginite from Fig. 10

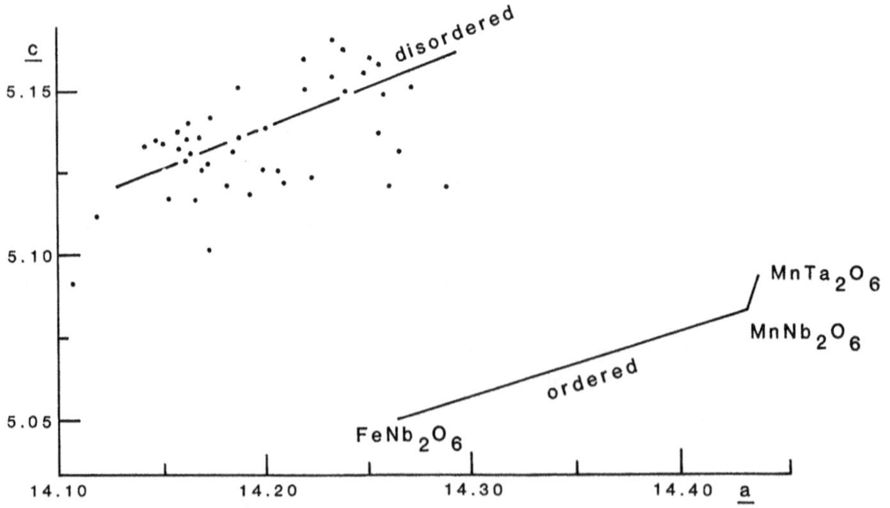

Fig. 6. Stannian and titanian ixiolite treated as orthorhombic phases in the *a*-*c* diagram, with *a* tripled to match that of ordered columbite-tantalite. Based on Fig. 1B of Černý and Ercit (1985), from unpublished data of P.Č. and M.A. Wise

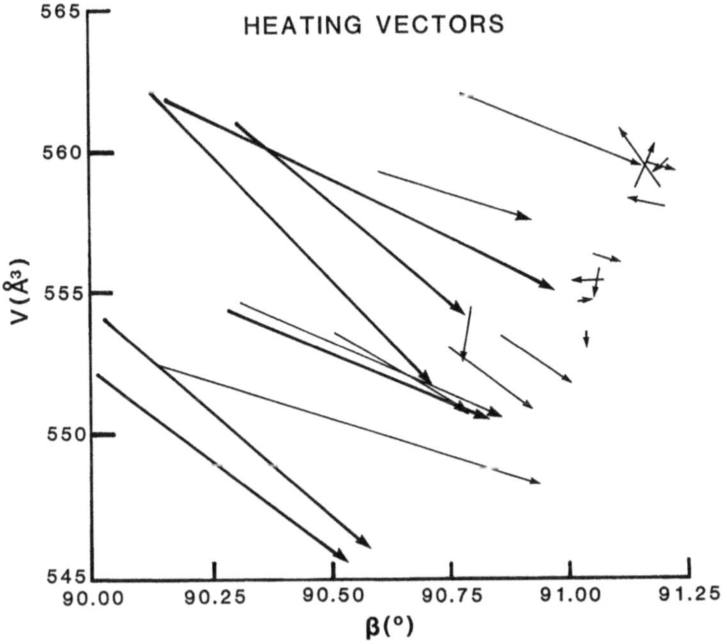

Fig. 7. Monoclinic stannian and titanian ixiolite in the V-β diagram (*heavy arrows*: natural state at *tail end*, heated at *arrowhead*). *Light arrows* represent wodginite data from Fig. 11. Adjusted from Ercit (1986) and Wise (1987)

3.4 Wodginite Group

Monoclinic wodginite is closely related to the structure of columbite-tantalite and ixiolite, despite its separation in the structural classification of Table 1 (cf. Ercit et al. 1984, Ercit 1986). Wodginite was originally described with an ixiolitelike formula based on 32 oxygens per unit cell (Nickel et al. 1963a). Progressive improvement of the structure refinement has led to the present ordered formula with three cation sites, $A_4C_4M_8O_{32}$, where $A = Mn,Fe^{2+}$, $C = Sn,Ti,Ta,Fe^{3+}$ and $M = Ta > Nb$ (Ferguson et al. 1976; Ercit 1986).

Most wodginites are close to the composition of the Mn,Sn,Ta-rich type specimens of Nickel et al. (1963a). However, extensive substitutions were recognized recently that led to the definition of new species (Ercit et al. 1984):

wodginite	$Mn_4Sn_4Ta_8O_{32}$,
ferrowodginite	$Fe_4Sn_4Ta_8O_{32}$,
titanowodginite	$Mn_4Ti_4Ta_8O_{32}$, and
tantalowodginite	$Mn_2\square_2Ta_4Ta_8O_{32}$.

Figure 8 shows the cation populations of the A- and C-sites, illustrating the general variability and the compositions of the natural representatives of the end-member species.

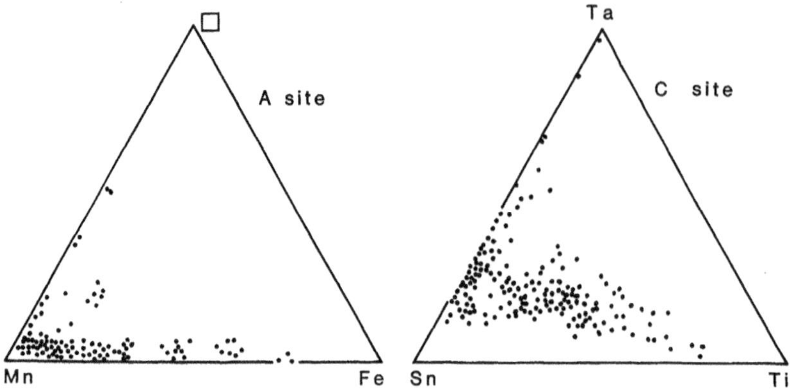

Fig. 8. A- and C-site populations (atomic ratios) in the structure of wodginite. Modified from Fig. 3 of Černý and Ercit (1985); cf. Ercit (1986)

The existence of a natural analogue to Turnock's (1966) synthetic Fe^{3+}-based phase is suggested by a mineral described by Kornetova et al. (1978), which may be interpreted as close to $(Fe,Mn)_4(Fe_2^{3+}Ta_2)_{\Sigma 4}(Ta,Nb)_8O_{32}$. However, this mineral occurs in an intimate intergrowth with an Sn-rich phase, and the X-ray diffraction data suggest the presence of a columbite-type phase. Thus, the data are not conclusive.

In the compositional columbite quadrilateral of Fig. 9, wodginite data closely overlap the main field of ixiolite, and both species violate the tantalite-tapiolite gap of Fig. 1. In Fig. 10, the overlap of wodginite and ixiolite is again extensive but stoichiometry requirements of the ordered wodginite structure constrain its compositional field in terms of maximal contents of the C-cations. In contrast, ixiolite possessing apparently a full cation occupancy does not extend toward the A-deficient compositions of tantalowodginite.

Partial disorder in wodginite was suggested by, e.g. Khvostova et al. (1966), because of variable β, occasionally approaching 90°. Complementary effects of compositional factors must also be considered when assessing order-disorder by means of unit cell parameters, such as substitution of Mn by Fe and of Sn by Ti (von Knorring et al. 1969a; Graham and Thornber 1974; Ercit et al. 1984; Ercit 1986). Nevertheless, the V vs β diagram of Fig. 11 serves well to generally separate fully ordered and incompletely disordered wodginites. The diagram also shows that the behaviour of disordered wodginite on heating is analogous to that of stannian and titanian ixiolite, emphasizing the relationship between the minerals. Wodginite has not been so far observed in any stage of metamict breakdown.

Fig. 9. Wodginite compositions in the columbite quadrilateral (atomic proportions). Data of Ercit (1986). Compositional gap separating columbite-tantalite and tapiolite from Fig. 1, ixiolite field from Fig. 4

Fig. 10. Composition of wodginite group minerals calculated with the assumption of full cation order in the (Nb,Ta)–(Sn,Ti,Fe^{3+})–(Fe,Mn) diagram (atomic ratios; cf. Ercit 1986 for data and method of calculation). Ixiolite field from Fig. 5

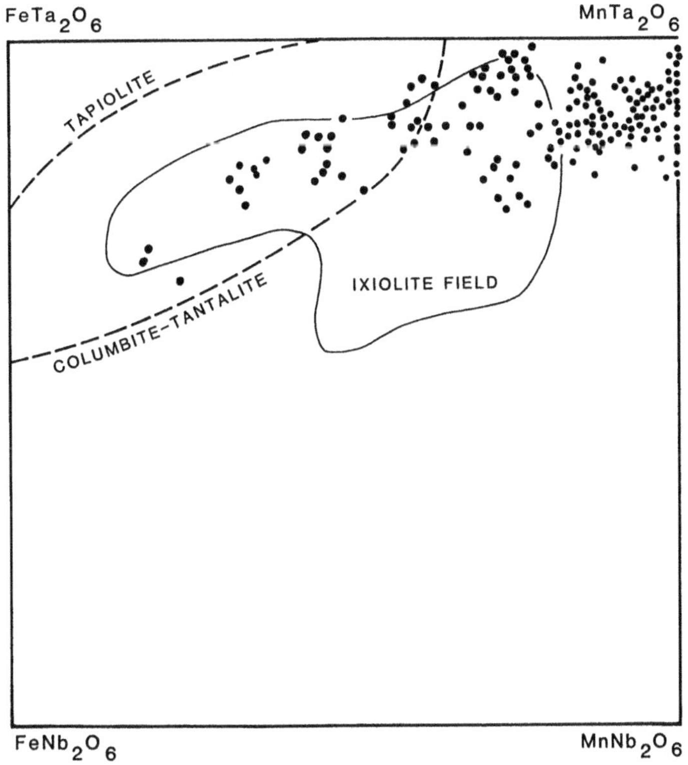
FeTa2O6
MnTa2O6
TAPIOLITE
IXIOLITE FIELD
COLUMBITE-TANTALITE
FeNb2O6
MnNb2O6

Fig. 9

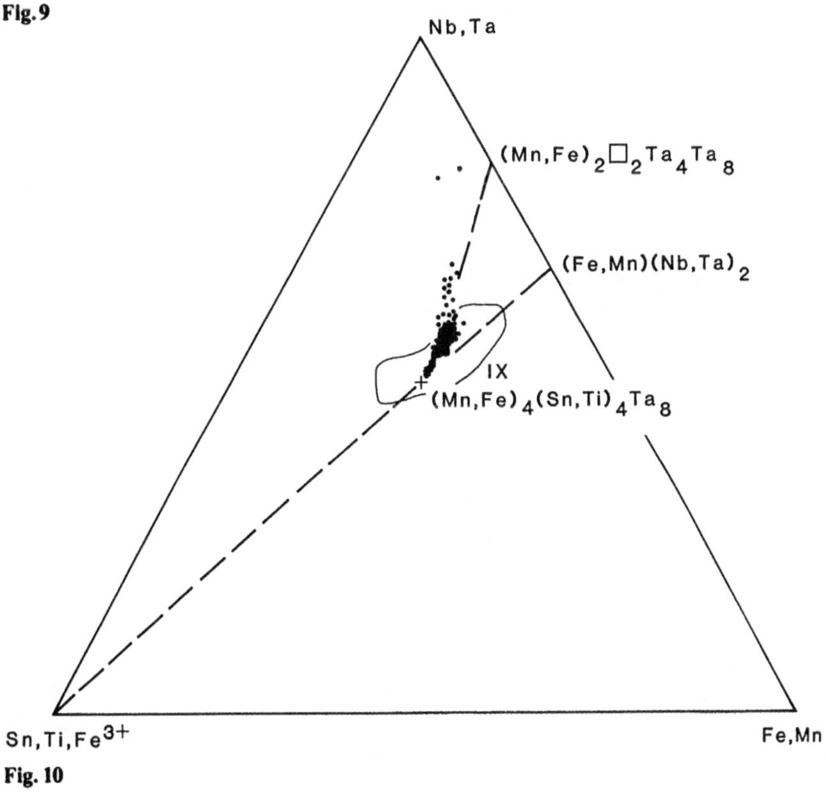
Nb,Ta
(Mn,Fe)2□2Ta4Ta8
(Fe,Mn)(Nb,Ta)2
IX
(Mn,Fe)4(Sn,Ti)4Ta8
Sn,Ti,Fe3+
Fe,Mn

Fig. 10

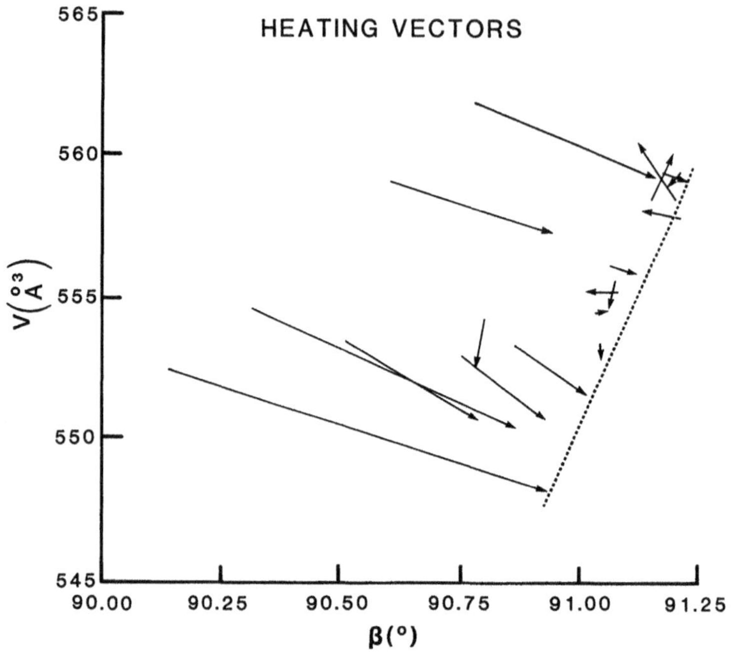

Fig. 11. Wodginite minerals in the V-β diagram. *Arrows* indicate the shift of cell dimensions from natural state (*tail ends*) to increased cation order after heating (*arrowheads*). The *dotted line* represents the loci of fully ordered wodginites of any composition (after Ercit 1986)

So far, wodginite occurrences are restricted to granitic pegmatites. The Fe- and Ti-enriched types tend to occur in geochemically primitive pegmatites but they are also found in the less fractionated assemblages of otherwise highly evolved deposits (Ercit 1986). Wodginite is commonly a primary mineral of Ta, and rarely undergoes replacement by later Ta-bearing phases such as manganotantalite and microlite Černý and Ercit 1985).

3.5 Qitianlingite

This mineral is known at present only from its type locality at Mt. Qitianling in the Hunan Province of China. It is an orthorhombic phase, structurally related to the columbite-ixiolite-wodginite family. It has an ordered structure with three distinct cation sites, corresponding to the main metals in its idealized formula of $Fe_2Nb_2WO_{10}$ (Wang et al. 1985). Significant Mn and Ta substitute for Fe and Nb respectively; minor Ti and Sn also are present.

Qitianlingite was found associated with quartz, K-feldspar, zinnwaldite, muscovite, cassiterite, wolframite and wolframoixiolite, probably in a greisen-type or vein-type tungsten deposit (Wang et al. unpublished).

3.6 Stibiotantalite Group

Orthorhombic minerals of this group have a general formula of AMO_4, where A = Sb,Bi and M = Nb,Ta. So far, only three species close to end-member composition have been found:

stibiocolumbite $SbNbO_4$,
stibiotantalite $SbTaO_4$, and
bismutotantalite $BiTaO_4$.

"Bismutocolumbite" chemistry has not been even remotely approached by any natural mineral, but it can be expected to crystallize under appropriate conditions of Bi and Nb activities (Foord 1982b). Figure 12 summarizes the compositions of all members of this group analyzed to date, as reviewed by Correia-Neves et al. (1974), with additional data from Voloshin et al. (1977), Foord (1981, 1982a) and Groat et al. (1987).

The vast majority of minerals belonging to this group is perfectly crystalline, but bismutotantalite from Uganda has been reported which may be partly metamict (Foord 1982b).

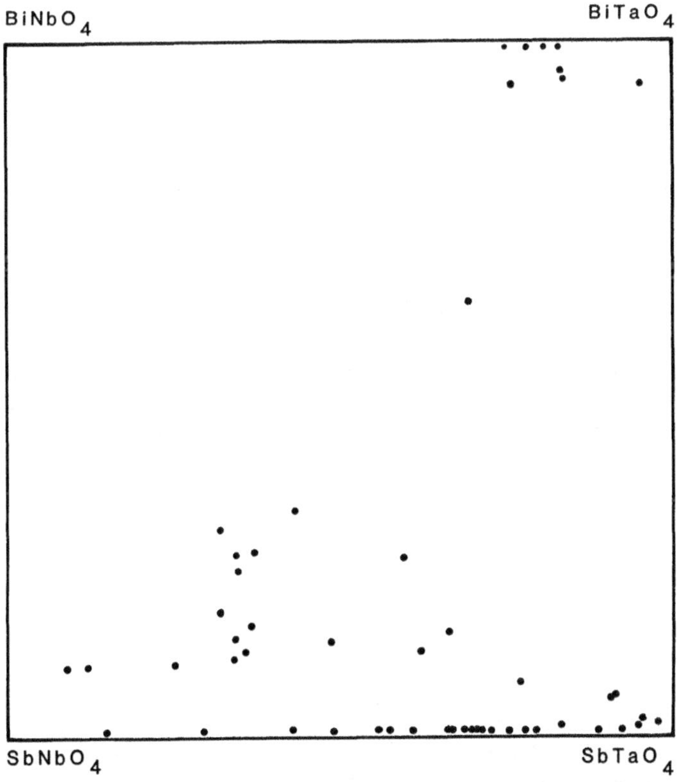

Fig. 12. Minerals of the stibiotantalite group in the stibiotantalite quadrilateral. Data collected by Correia-Neves et al. (1974) and Foord (1982a) complemented by those of Voloshin et al. (1983a), von Knorring and Fadipe (1981), Stern et al. (1986) and Groat et al. (1987)

Minerals of the stibiotantalite group are formed only in highly fractionated rare-element pegmatites, particularly in lepidolite-rich complex types. They are generated largely in late stages of pegmatite consolidation, occasionally replacing tantalite, simpsonite, holtite or microlite. In turn, they are commonly subject to replacement by other phases; increased activities of strong bases and sensitivity of Sb^{3+} and Bi^{3+} to redox conditions probably are the main reasons. Microlite, stibiomicrolite, rynersonite, cesstibtantite, sosedkoite and bismutomicrolite are the most common replacement products (cf. Černý and Ercit 1985; Groat et al. 1987).

3.7 Foordite-Thoreaulite Series

The tantalum-rich thoreaulite is historically the first mineral of this series discovered by Buttgenbach (1933), whereas foordite was defined only recently (Černy et al. 1988a). The two monoclinic end members have the compositions.

foordite $SnNb_2O_6$, and
thoreaulite $SnTa_2O_6$.

Natural minerals have been so far documented with 30 to 91 mol% of the thoreaulite component. Besides Sn, subordinate quantities of Pb are reported (up to 12.5 at% of the A-site population), and minor Sb is detectable at most localities. Foordite-thoreaulite minerals are one of the very few natural oxidic phases with divalent Sn, although some authors claim a quadrivalent state for up to 25% of the total Sn content (Nekrasov et al. 1982). However, structural and crystal chemical considerations rule out this possibility (Černý et al. 1988a, Ercit and Černý 1988).

Structurally related to the stibiotantalite group, the foordite-thoreaulite minerals have not been found in the metamict state. Smooth variations of unit cell dimensions with composition also suggest absence of disorder between the A- and M-site populations (Ercit and Černý 1988).

The foordite-thoreaulite minerals owe their origin to unusual conditions: oxygen fugacity must be low and activities of alkali elements and transition metals must also be suppressed, to prevent formation of Sn^{4+}-bearing phases on the one hand, and of microlite, columbite-tantalite, wodginite, etc. on the other. Thus, it is not surprising that members of the discussed series are known from only nine localities (although they are quantitatively important in the Manono pegmatites of Zaire; Thoreau 1949). Also, hydrothermal breakdown of foordite-thoreaulite under oxidizing conditions is widespread, with formation of cassiterite, "staringite", columbite-tantalite, ixiolite-wodginite, rankamaite and lithiotantite (Černý and Ercit 1985; Černý et al. 1988a).

3.8 Changbaiite

This mineral has been so far known only from its type locality at Tonghua, southeastern Kirin, China. Changbaiite is trigonal, close to the ideal composition of $PbNb_2O_6$, with only minor contents of Fe, Ti and Ta. It was found with quartz and K-feldspar in kaolinite veinlets, and in kaolinite-filled cavities, in a mineralized potassic granite (Chao 1979).

3.9 Pyrochlore Group

Minerals of this group are probably the second most abundant phase type concentrating Nb and Ta in all categories of their deposits, second only to columbite-tantalites. Nevertheless, it is only since the 1960s that the extensive compositional variability of this group has been recognized, specifically in its microlite subgroup. New Minerals are still being described, despite the restrictions imposed by the recommendations of the I.M.A. Commission on New Minerals and Mineral Names (Hogarth 1977).

The general formula of the group is $A_{2-m}M_2O_6(O,OH,F)_{1-n} \cdot pH_2O$, with the current limits of m = 0–2, n = 0–1 and p = 0–? (Foord 1982b; Groult et al. 1982). The subdivision of this group into the pyrochlore, microlite and betafite subgroups, and the nomenclature that follows the definition of species are self-evident from Table 2. Hogarth (1977) relates the numerous names applied to the natural pyrochlore-group minerals to the new nomenclature.

The cubic structure of the pyrochlore-group minerals is extremely accommodating to cations of diverse ionic radii and valences. Consequently, the number of subordinate and minor elements entering their composition is considerable. Besides the major cations shown in Table 2, minor quantities of Cs,Mg,Sr,Fe, Mn,Cu,Al,As,Mo,Th,Zr and W are known to be present (e.g. Mihálik 1967;

Table 2. Classification of pyrochlore group minerals (modified) from Hogarth 1977)

Subgroup:		Pyrochlore $Nb+Ta > 2Ti$ $Nb > Ta$	Microlite $Nb+Ta > 2Ti$ $Ta \geq Nb$	Betafite $2Ti \geq Nb+Ta$
No A-cation other than Na or Ca > 20% total A-cations		Pyrochlore	Microlite	Calciobetafite (Ca > Na)
Any A-cation, other than Na or Ca, > 20% of total A-cations	K	Kalipyrochlore		
	Sn		Stannomicrolite	
Name is for the dominant non-(Na,Ca) cation.	Ba	Bariopyrochlore	Bariomicrolite[b]	
	Pb	Plumbopyrochlore	Plumbomicrolite	Plumbobetafite
	Sb		Stibiomicrolite	Stibiobetafite
	Bi		Bismutomicrolite[c]	
	Y[a]	Yttropyrochlore		Yttrobetafite
	Ce[a]	Ceriopyrochlore		
	U	Uranpyrochlore	Uranmicrolite	Betafite

[a] Ce = La to Eu, Y = Y + Gd to Lu; the prefixes "cerio" or "yttro" are applied when Y + Ce > 20% total A-cations, not only when Y > 20% or Ce > 20%.
[b] Status uncertain; cf. Sect. 3.10 Parabariomicrolite.
[c] Probably includes natrobistantite of Voloshin et al. (1983c).

Mitchell and Zulkiewicz 1970; Eid and von Knorring 1976; Lumpkin et al. 1986; Ercit 1986).

Metamictization of the pyrochlore-group minerals is very common; subordinate to substantial U contents are widespread in pyrochlores, betafites and microlites of diverse origin.

Paragenetically, pyrochlore is typical of carbonatites and other deposits associated with anorogenic magmas. Betafite is also known from this environment, and from relatively primitive granitic pegmatites. Microlite varieties are widespread in specific types of metasomatized apogranites, and mainly in moderately to highly fractionated rare-element pegmatites of orogenic affiliation. Pyrochlores and betafites are largely primary minerals in alkaline rocks, replaced by columbite and other decomposition products (van d. Veen 1963). In peraluminous granites and rare-element pegmatites, microlite may occur as either primary mineral or as a replacement product of columbite-tantalite, tapiolite, ixiolite, simpsonite and other minerals. Internal constitution of crystals of all members of the pyrochlore group is commonly heterogeneous: complex zoning, or veining with diffuse boundaries, are rather common features (van d. Veen 1963, Knudsen, pers. commun. 1986). The most dramatic compositional complexity is encountered in pegmatitic microlite (Voloshin and Pakhomovskyi 1982; Lumpkin et al. 1986; Ercit 1986).

3.10 Parabariomicrolite

This mineral species has so far been identified from only a single occurrence at Alto do Giz, Brazil, where it forms as a replacement product of microlite, associated with simpsonite (Ercit et al. 1986a,b). It probably also occurs at Kivu, Zaire. As a rare secondary mineral, it should be possibly discussed in Section 4 but is mentioned here because of its structural relationship to the pyrochlore group.

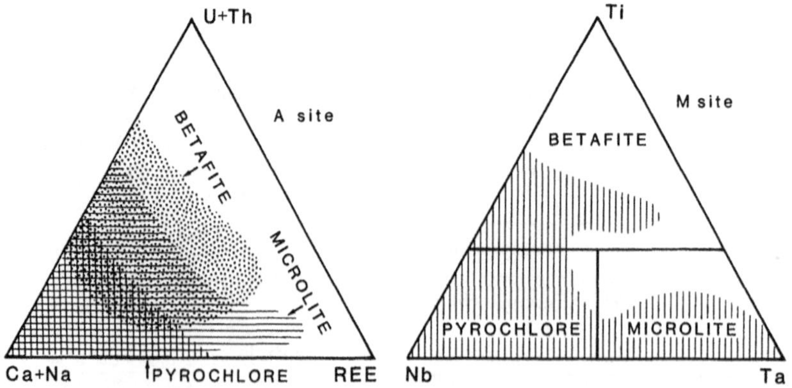

Fig. 13. A- and M-site populations (atomic ratios) in the minerals of the pyrochlore group. Modified from Gorzhevskaya et al. (1974), and complemented from Mihálik (1967), Mitchell and Zulkiewicz (1970), Petruk and Owens (1975), Eid and von Knorring (1976), Hogarth (1977), Lindquist and Rehtijärvi (1979), Černý et al. (1979), Voloshin and Pakhomovskyi (1982), Lumpkin et al. (1986), Ercit (1986), Wise (1987), Knudsen this Vol.

The chemical composition of parabariomicrolite corresponds to the formula $BaTa_4O_{10}(OH)_2.2H_2O$. The rhombohedral mineral has a structure derived by slight distortion of the pyrochlore structure along the [111] body diagonal, due to ordering of occupied and vacant A-cation sites in alternate polyhedral layers parallel to (0001).

In view of the above deviations of parabariomicrolite from the crystal chemistry and structure of true members of the pyrochlore group, the type bariomicrolite (originally rijkeboerite; van d. Veen 1963) requires re-examination.

3.11 Cesstibtantite

This mineral was first described from Kola Peninsula by Voloshin et al. (1981b) as a derivative of the microlite subgroup of the pyrochlore group, with an ideal formula $(Cs, Na)SbTa_4O_{12}$. The 1:1 ratio of (Cs, Na) to Sb in the A site was supposed to be the crystal chemical earmark of the new species. However, Ercit and Černý (1982) and Nickel and Robinson (1985) recognized cesstibtantite at additional localities where it displayed gross deviations from the above ratio. Furthermore, slight anisotropism was observed in one occurrence, and cesstibtantites from two localities displayed weak diffractions violating extinction conditions in the pyrochlore structure (Ercit et al. 1985b; Ercit 1986).

Refinement of cesstibtantite structure reported in the above publications led to the recognition of this mineral as the first occurrence of (partial) inverse pyrochlore structure in natural phases, a structure that has been long known in synthetic products: instead of the classic $AM_2O_6\emptyset$ formula type of pyrochlore (with large cations in A, small cations in M, and O,OH, F in $\emptyset$), inverse pyrochlores correspond to $M_2\emptyset_6A'$ (where M cations play the same structural role as M cations in the normal pyrochlore structure, but both O and $\emptyset$ anions are in the O sites of the normal pyrochlore structure, and the large monovalent A′ cations such as Cs are located in the $\emptyset$ sites of the normal pyrochlore structure).

Cesstibtantite is a late Ta-bearing mineral found only in highly fractionated, pollucite-bearing rare-element pegmatites. It has been so far found only at two localities (the Australian occurrence of Nickel and Robinson is secondary, due to weathering), but increased Cs_2O contents of microlite at other localities (Eid and von Knorring 1976) suggest that it could be discovered in additional pegmatite deposits.

3.12 Natrobistantite

Voloshin et al. (1983c) described this mineral from granitic pegmatites of Kjoktogoi, Sin-Czyan, People's Republic of China. In a manner similar to the treatment of cesstibtantite, by Voloshin et al. (1981b), they assigned a formula with (Na,Cs)/Bi of 1.0 to natrobistantite: $(Na,Cs)Bi(Ta,Nb,Sb)_4O_{12}$. In the absence of both crystal structure refinemet and additional finds of this mineral, it is questionable whether it should be considered an independent species. It is very close in composition to some Cs-bearing bismutomicrolites of Eid and von Knorring (1976). In view of the compositional diversity of cesstibtantite, and of the low Cs content of natrobistantite, this latter mineral may prove to be bismutomicrolite.

Natrobistantite occurs associated with bismuthinite, antimonian bismutotantalite, manganocolumbite and microlite. It replaces bismutotantalite, and is in turn replaced by heterogeneous microlite.

3.13 Scheteligite

This mineral was described by Bjørlykke (1937) as a new species of apparently orthorhombic (morphological) symmetry, with a chemical composition of $(Ca,Y,Sb,Mn)_2(Ti,Ta,Nb,W)_2O_6(O,OH)$. The chemistry suggests a relationship to the pyrochlore group, possibly close to stibiobetafite (Černý et al. 1979). However, re-investigation of the type material is required to clarify the status of this phase.

3.14 Simpsonite

This hexagonal Ta-oxide mineral has been recently reviewed by Ercit (1986). It has the chemical composition corresponding to $Al_4Ta_3O_{13}(OH)$, first recognized by von Knorring and Hornung (1963). It is modified by only a negligible substitution of Ta by Nb, and by the presence of minor Sn and traces of Ti, probably disordered over the Al and Ta sites of the structure. Substitution of F for OH was not encountered, and seems extremely unlikely on structural grounds.

Simpsonite has been found at only nine localities; however, it has had a status of an ore mineral of Ta in at least three occurrences where it occurs in substantial quantities. In any case, occurrences of simpsonite are restricted to highly fractionated complex granitic pegmatites; with one exception, all host pegmatites are of the low-pressure petalite subtype. The associations of Ta-oxide minerals preceding, or coexisting with simpsonite are very similar in all occurrences, including cassiterite, several species of microlite and wodginite minerals, and manganocolumbite-manganotantalite. Also, close similarities are shown by the late mineral associations that replace simpsonite (Ercit and Černý 1982; Ercit 1986).

3.15 Perovskite Group

In general, the crystal chemistry of this group is extremely diversified, particularly if synthetic members are taken into account. However, even the naturally occurring minerals are numerous, due to the flexibility of the structure which accommodates a variety of cations while undergoing changes in symmetry (Roth 1957). The ($Nb \gg Ta$) members of this group, derived from the ideal composition of orthorhombic perovskite $CaTiO_3$, are

loparite	$(Na,Ca,REE)(Ti,Nb)O_3$,
latrappite	$(Ca,Na)(Nb,Ti)O_3$, and
lueshite	$NaNbO_3$,

all of them also orthorhombic (Nickel 1964). A rare monoclinic form of $NaNbO_3$ bears the name of natroniobite (Bulakh et al. 1960).

The chemical composition of the perovskite-type minerals is highly variable. Historically it has generated a complex nomenclature which is in part still used, and which must be understood in order to deal with early literature. Figure 13, modified from Gorzhevskaya et al. (1974), illustrates the relationships among the recognized species and obsolete varieties.

The REE contents of all natural members of this group are conspicuously Ce,La-dominant (Fig. 14). Loparite and niobian loparite are particularly enriched in REE oxides (14–46 wt.%), whereas perovskite proper and its close derivatives rarely contain over 10 wt.% of them.

Despite its minor U and Th content, perovskite is usually well crystalline. However, loparite and niobian loparite are occasionally metamict.

Minerals of the perovskite group are typical constituents of alkalic parageneses. They occur as primary accessory minerals in ultrabasic to gabbroic peralkaline rocks, in carbonatites, and in agpaitic nepheline syenites. They may form at the expense of mosandrite or early pyrochlore, and are locally extensively replaced by pyrochlore, albite or natrolite. "Metaloparite" is an alkali-leached, A-deficient loparite with extensive substitution of (OH) for O.

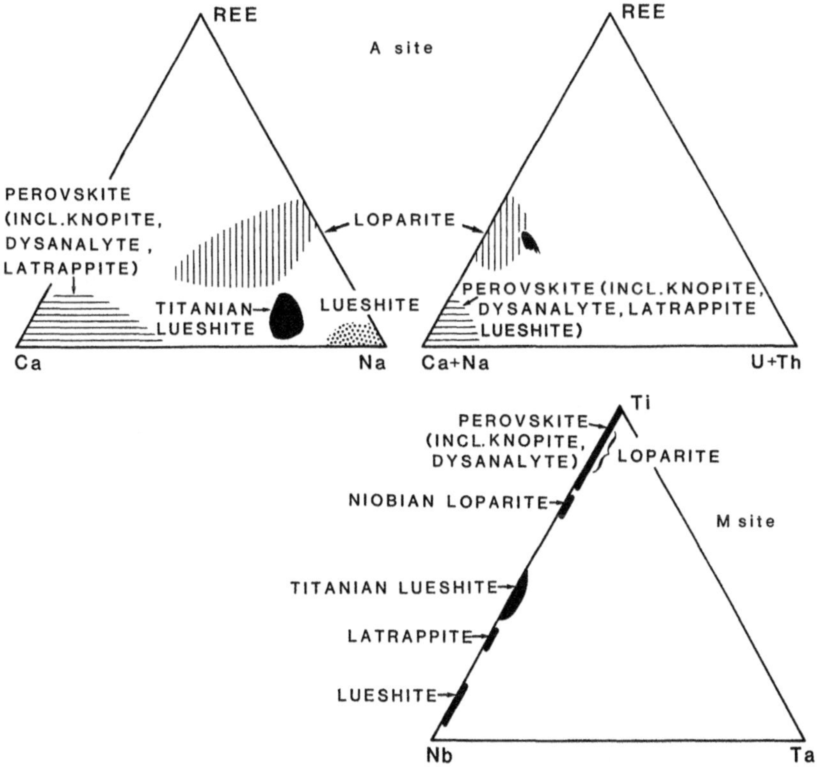

Fig. 14. A- and M-site populations (atomic ratios) in the perovskite-type minerals (modified from Gorzhevskaya et al. 1974)

3.16 Fergusonite Group

Minerals of the fergusonite group are largely tetragonal, with the scheelite-type structure and simple general formula AMO_4 but variable chemical composition. Recognized crystal chemical members are

β-fergusonite	$(Y,REE>Ca,U,Th)(Nb,Ta)O_4$,
β-fergusonite (Ce)	$(Y,REE>Ca,U,Th)(Nb,Ta)O_4$,
β-fergusonite (Nd)	$(Y,REE>Ca,U,Th)(Nb,Ta)O_4$,
α-fergusonite	(tetragonal, high-T polymorph), and
formanite	$(Y,REE>Ca,U,Th)(Ta,Nb)O_4$.

Occurrences of natural α-fergusonite are extremely rare, and virtually all synthesis studies have focussed on the β-fergusonite (monoclinic distortion of the scheelite structure). Consequently little is known about the α-fergusonite. Risörite is a Ti-enriched variety of fergusonite, not particularly different from the Ti-rich members of the fergusonite population shown in Fig. 15.

Yttrocolumbite (yttroniobite) and yttrotantalite have been quoted in the literature as orthorhombic analogues of the tetragonal fergusonite phases. However, re-investigation of some of the early described yttrocolumbites-yttrotantalites showed them to belong to the fergusonite group (Gorzhevskaya et al. 1974).

Figure 15 illustrates the general variations in the A- and M-site populations but cannot show the variability in the REE composition, which is indeed extremely diversified. The fergusonite structure can accommodate the whole range of REE cations, and in this respect the composition seems to be a function of the geochemistry of the parent environment, rather than of crystal chemical constraints. Besides the classic Y + HREE composition, fergusonites are known with Ce and Nd maxima, and Dy is commonly enriched as well (Gorzhevskaya et al. 1974).

A variety of names applied to the fergusonite minerals in the past – tirite, braggite, kochelite, sipylite, rutherfordite, arrhenite – arose from the above mentioned compositional variability, and from changes induced by metamictization and alteration. Metamict decay is widespread because of the ubiquitous subordinate contents of U and Th.

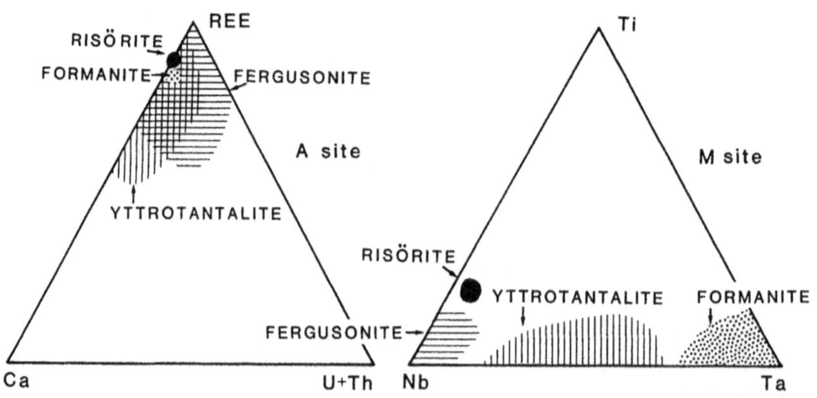

Fig. 15. A- and M-site populations (atomic ratios) in the minerals of the fergusonite group (modified from Gorzhevskaya et al. 1974)

Fergusonite minerals are found in dolomitic carbonatites, anorogenic granites and metasomatic deposits of peralkaline chemistry, and in anorogenic-related pegmatites. They are also found in rare-element mineralized orogenic, peraluminous granites and in the rare-earth type of pegmatites of the same affiliation. However, it is virtually absent from geochemically more evolved rare-element pegmatites of the orogenic kindred.

According to Gorzhevskaya et al. (1974), fergusonite from granites and granitic pegmatites that were not subject to albitization belongs to the Y+HREE type, whereas its counterpart from alkalic granites and pegmatites, and from albitized granitic pegmatites, is enriched in LREE.

Alteration of fergusonite is widespread, in part because of its metamict state and hydration, but it is poorly understood. Some old varietal names such as rutherfordite, arrhenite and kochelite were applied to altered fergusonite minerals.

3.17 Petscheckite and Liandratite

These two minerals were described as new species by Mücke and Strunz (1978) from granitic pegmatites of the Berere district, Madagascar. They are metamict, containing substantial quantities of uranium oxides:

petscheckite	$U^{4+}Fe(Nb,Ta)_2O_8$, and
liandratite	$U^{6+}\square(Nb,Ta)_2O_8$.

For petscheckite, two additional varieties are distinguished: oxypetscheckite with $(Fe^{3+}{}_{0.67}\square_{0.33})$ and hydroxypetscheckite with $(Fe^{3+}{}_{0.33}\square_{0.67})$ and $O_7(O,OH)$. In both petscheckite and liandratite, heating restores a crystalline state resembling that of synthetic $U^{6+}Ta_2O_8$.

So far, these two minerals are known only from their type locality, associated with columbite. It is conceivable that their future finds will be restricted to U-rich pegmatite parageneses as well.

3.18 Euxenite Group

Euxenite has a general formula of AM_2O_6 and an ordered columbite-type structure. In contrast to the columbite-tantalite group sensu stricto, the A cations are much larger and coordinated with eight oxygens.

Compositionally, the following members are considered well-established:

euxenite	$(Y,REE>U,Th)(Nb,Ti,Ta)_2O_6$,
tanteuxenite	$(Y,REE>U,Th)(Ta,Nb,Ti)_2O_6$, and
polycrase	$(Y,REE>U,Th)(Ti,Nb,Ta)_2O_6$.

Ampangabéite and loranskite are identical with euxenite. Yttrocrasite (and possibly kobeite) is apparently identical with polycrase, and the two names are used indiscriminately in the literature. Delorenzite and eschwegeite are synonymous with tanteuxenite. The principal modes of chemical substitution are $(Ca+Nb,Ta) \rightleftarrows (Y,REE)+Ti$ and $Ca+(U,Th) \rightleftarrows 2(Y,REE)$, the first one introducing substantial quantities of a fersmitelike component (Fig. 16).

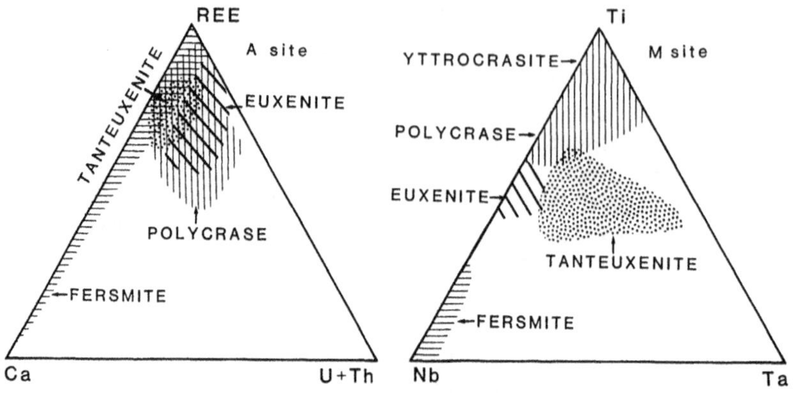

Fig. 16. A- and M-site populations (atomic ratios) in minerals of the euxenite group, and in fersmite. Data of Gorzhevskaya et al. (1974), complemented by those of Foord and Mrose (1978), Černý et al. (1981a) and Ewing (1974)

Crystal chemical relationships in the euxenite group sensu stricto and the identification of its members are complicated by the metamict state of most its members, and by different degrees of hypogene alteration and weathering effects (van Wambeke 1970; Ewing 1975). Deficiencies in the A site are common, coupled with the introduction and removal of diverse cation species. Consequently, heating experiments often fail to restore the original structure of the euxenite-type minerals because of these secondary deviations from the original stoichiometry.

Establishing the identity of euxenite minerals is also complicated by the close relationship to the structure and composition of the aeschynite group, which has the same general formula. The aeschynite structure is stable at low temperatures for all REE(Nb,Ta)TiO_6 compounds, and can be converted into the high-temperature euxenite structure only for HREE-based compositions (Komkov 1973). Similar relationships are observed in heating experiments on natural minerals, except that LREE-based compounds do convert to the euxenite structure but at temperatures higher than those required to convert the HREE-rich minerals (Ewing 1975; Ewing and Ehlmann 1975). The aeschynite minerals are not restricted to the "type" (Ca,LREE)-compositions, as will be shown below, and may approximate the Y,HREE-dominant compositions of natural euxenite rather closely. Thus, it is not surprising that Ewing (1976b) and Ewing et al. (1977) resorted to canonical discriminant analysis as a statistical means to separate populations of the two mineral groups.

Euxenite minerals are found in metasomatic assemblages associated with alkaline granitoid intrusions, and in rare-element granites of the orogenic association. They typically occur in Y,REE-rich pegmatites of both the anorogenic and orogenic kindred. Replacement of the euxenite minerals is widespread, with fersmite and pyrochlore phases as secondary products (van Wambeke 1970; Ewing 1974, 1975).

3.19 Fersmite

Fersmite is commonly quoted as a member of the euxenite family of minerals (e.g. Gorzhevskaya et al. 1974). This affiliation is justified on compositional grounds. The ideal chemical composition of fersmite is $Ca(Nb,Ta)_2O_6$, but it is commonly modified by the same substitutions as those quoted above for euxenite (Fig. 16). An extremely Ti,REE-enriched composition was described by Ewing (1974).

Structurally, however, fersmite possesses a true columbite structure with both A and M cations in octahedral coordination, in contrast to the eight-fold coordination of A cations in euxenite. To date, structural relationships along the fersmite-euxenite join have not been studied.

Fersmite has not been found in the metamict state, and it exhibits ordered structure in all examined occurrences. Apparently the ionic radius of Ca^{2+}, while small enough for octahedrally coordinated A sites, is too large for random interchange with the M cations in a disordered structure.

Primary fersmite is known from carbonatites and plumasitic nepheline syenites of the anorogenic suites, and from a variety of pegmatite types of the orogenic rare-element class (allanite, gadolinite and lepidolite subtypes). Fersmite is also found as a secondary phase, replacing euxenite, columbite, stibiotantalite and pyrochlore (Ewing 1974; Foord and Mrose 1978; Černý et al. 1981a; Foord 1982b).

3.20 Aeschynite Group

Aeschynite and related minerals have an ideal formula of $AMM'O_6$, with A = Y,Ce; M = Nb,Ta; and M′ = Ti. However, substitution of Ca + (Nb,Ta) for (Y,REE) + Ti is common, and it leads to the compositions of vigezzite and rynersonite with a columbitelike stoichiometry and cation population close to AB_2O_6:

aeschynite	$(Ce,Ca)(Ti,Nb)_2O_6$,
aeschynite-Y	$(Y,Ca)(Ti,Nb)_2O_6$,
aeschynite-(Nd)	$(Nd,Ce,Ca)(Ti,Nb)_2O_6$,
nioboaeschynite	$(Ce,Ca)(Nb,Ti)_2O_6$,
tantalaeschynite	$(Ca,REE)(Ta,Ti,Nb)_2O_6$,
vigezzite	$(Ca,Ce)(Nb,Ta,Ti)_2O_6$, and
rynersonite	$CaTa_2O_6$.

Lyndochite, sinicite, priorite and blomstrandine are synonymous with aeschynite, priorite being identical with aeschynite-(Y). Polymignite may also belong to the aeschynite minerals. Laboratory synthesis confirmed continuous substitution of Ce and Y (Komkov and Belopolskyi 1966). Figure 17 shows the broad extent of Nb-Ti substitution, and the singularity of the vigezzite and tantal-aeschynite compositions. In contrast, rynersonite occurrences are somewhat more widespread; they show an extensive range of Ta/Nb > 1 ratios but very calcic population of the A site.

The REE content of the aeschynite minerals is commonly dominated by the cerium group, but the proportion of HREE increases in the Y-rich (priorite) minerals. It is these Y-rich phases that also commonly show enrichment in Dy (Gorzhevskaya et al. 1974).

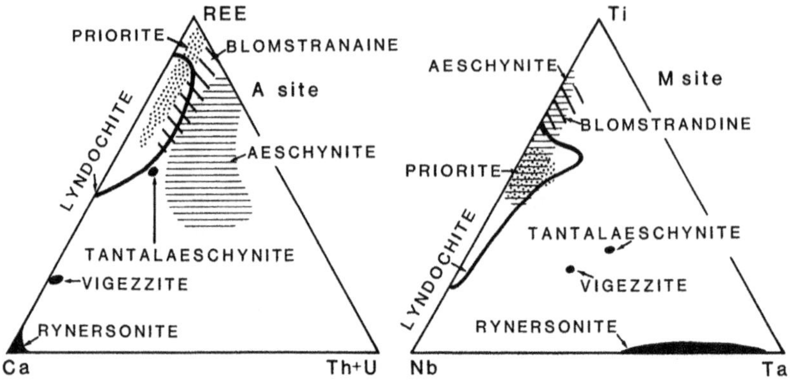

Fig. 17. A- and M-site populations (atomic ratios) in minerals of the aeschynite group. Data of Gorzhevskaya et al. (1974) complemented by those of Adusumili et al. (1974), Foord and Mrose (1978) and Graeser et al. (1979)

The metamict state is common in all types of aeschynite but vigezzite and rynersonite are very poor in U and Th, and consequently crystalline. The problems encountered in the identification of aeschynite were commented on in the preceding section on euxenite.

Aeschynite is a typical constituent of rare-element mineralization in plumasitic nepheline syenites, and in rare-element granites of the anorogenic alkaline affiliation. It is also encountered in REE-rich types of anorogenic and orogen-related granitic pegmatites. So far, tantal-aeschynite and vigezzite are known only from their type localities of relatively primitive geochemical types (Adusumili et al. 1974; Graeser et al. 1979). In contrast, rynersonite is formed in highly fractionated pegmatites, as reflected by its low Nb/Ta ratio.

Surprisingly, neither replacement of aeschynite by other phases nor metasomatism of Nb, Ta minerals by aeschynite have yet been reported. However, rynersonite is known to replace stibiotantalite and bismutotantalite, and is, in turn, replaced by fersmite and microlite (Foord and Mrose 1978; von Knorring and Sahama 1979).

3.21 Samarskite

In contrast to the euxenite and aeschynite minerals, samarskite is relatively easy to identify because of its characteristic Fe content. The total of $FeO + Fe_2O_3$ ranges from about 3.2 to as much as 23 wt. % (Kobayashi et al. 1970). However, this broad variation in Fe also indicates that the chemistry of samarskite is complex. Concentrations of U(4.0–29.6 wt. % U-oxides), Ti (to 5.5wt. % TiO_2) and Ca(0.2–8 wt. % CaO) confirm that the crystal structure of samarskite is extensively accommodating, and that derivation of a general formula could be a problem. This is compounded by the metamict state of virtually all natural samarskites. Metamict decay is usually so advanced that heating experiments (in air) generate mixtures of three to four phases, none of which is representative of the original structure.

The present understanding of samarskite crystal chemistry is based on the work of Komkov (1965) and Sugitani et al. (1985) who were fortunate to study relatively "fresh" specimens under carefully controlled heating conditions and, in the case of the Japanese authors, resorted to synthesis as well. Samarskite appears to have a general formula of AMO_4, and it may occur in a disordered ixiolite-type structure or in a low-temperature, ordered, topologically monoclinic but metrically orthogonal structure of the wolframite type. The order-disorder relationship is probably a function of compositional non-stoichiometry as well as P-T conditions; for example, Sugitani et al. (1985) synthetized samarskitelike compositions approaching $YFe_2Nb_5O_{16}$.

Figure 18 shows some of the most typical features of samarskite chemistry. Calciosamarskite compositions seem to be separated from samarskite proper, but the latter possibly grades into U-rich phases designated ishikawaite (although definite proof of ishikawaite identity with samarskite seems to be lacking). The Ti content is generally low but Ta may be considerably enriched.

Besides ishikawaite and calciosamarskite, other varieties include adelpholite, Ti-enriched khlopinite, vietinghofite or ferrosamarskite, hydrosamarskite, and possibly plumboniobite with 8 wt. % PbO.

The REE contents of samarskite are much less enriched in HREE than could be expected in a phase with substantial Y. In contrast, the intermediate REE (Dy,Gd and in part Sm) commonly show a prominent maximum (Gorzhevskaya et al. 1974).

Metasomatic mineralizations associated with anorogenic peralkaline granites may carry subordinate samarskite, but the main occurrences of this mineral are restricted to pegmatites. Most of the pegmatitic samarskite is found in anorogenic suites, but also in those associated with orogen-related granites that are characterized by elevated Y and REE contents. Replacement phenomena are not observed, but epitaxial intergrowths of samarskite with columbite are widespread, attesting to their structural similarity.

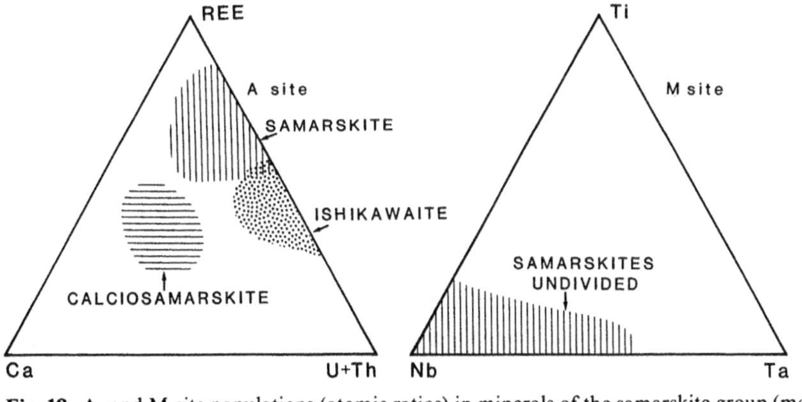

Fig. 18. A- and M-site populations (atomic ratios) in minerals of the samarskite group (modified from Gorzhevskaya et al. 1974)

4 Secondary Oxide Minerals of Niobium and Tantalum

The minerals described in this section are so far only minor to rare components of economic Nb,Ta-rich mineralizations. They are predominantly of pegmatitic origin, and they are formed in late stages, commonly by replacement of early Nb,Ta-bearing precursors. Nevertheless, these secondary phases have a place in the present review: if abundant, they may significantly modify the Nb,Ta contents of the primary assemblages, and they also may affect the behaviour of primary minerals in ore processing.

4.1 Rankamaite-Sosedkoite

Rankamaite and sosedkoite are extremely finely acicular to fibrous minerals. As they are not amenable to single-crystal X-ray diffraction techniques, their structure has not been determined, but von Knorring et al. (1969c) recognized the close similarity of the rankamaite powder pattern to those of "tetragonal tungsten bronze"-derived structures (e.g. K_xWO_3, $Na_2Ta_8O_{21}$). The compositions and stoichiometries that can be accommodated by this general structure type are quite flexible. Thus, it is not surprising that the complex arrays of major and minor components entering the chemistry of rankamaite and sosedkoite have led to several formulas (von Knorring et al. 1969c; Voloshin et al. 1982a; Ercit 1986), and the content of some elements is questionable because of impurity of the analyzed concentrates. The present formulas given by Ercit (1989) for the two orthorhombic minerals are

rankamaite $(K,Na)_{3-x}Al(Ta,Nb)_{10}(O,OH)_{30}$, and
sosedkoite $(Na,K)_{3-x}Al(Ta,Nb)_{10}(O,OH)_{30}$.

Minor Ca, Pb, Fe and Sb were detected.

Occurrences of both minerals are restricted to highly fractionated, simpsonite-bearing granitic pegmatites. Rankamaite is found as a felty replacement of simpsonite, and of cassiterite generated by oxidation of thoreaulite. Sosedkoite is known as a late post-simpsonite mineral along fractures and intergranular contacts in microlite, cesstibtantite and stibiotantalite.

4.2 Natrotantite – Irtyshite

Natrotantite was originally assigned the formula of $NaTa_3O_8$ and monoclinic symmetry (Voloshin et al. 1981a). However, an excellent match was found later between natural natrotantites and the synthetic hexagonal compound $Na_2Ta_4O_{11}$ (Ercit et al. 1985a). Deviations from this ideal composition could be due to the substitution of $Na^+ + O^{2-}$ by $\square + (OH)^-$. Irtyshite corresponds to the ideal anhydrous composition of $Na_2Ta_4O_{11}$ (Voloshin et al. 1985b) and it matches natrotantite in all respects except the space group, suggesting a polytypoid relationship between the two minerals (Ercit 1989).

Natrotantite is known only as a member of simpsonite paragenesis is highly fractionated granitic pegmatites (Ercit 1986). It occurs as a replacement product of simpsonite, alone or intergrown with microlite, and associated with alumotantite. In contrast, irtyshite was found associated with lithiotantite, manganocolumbite and cassiterite in veinlets crosscutting thoreaulite, or along contacts of lithiotantite and ixiolite (Voloshin et al. 1985b).

4.3 Alumotantite

The mineral alumotantite corresponds to the orthorhombic low-temperature polymorph of $AlTaO_4$ as opposed to the high-temperature synthetic phase with the tetragonal rutile-type structure. Alumotantite was first noticed as an unknown mineral by MacGregor (1946) at Bikita, Zimbabewe but described as a new species by Voloshin et al. (1981a) from the Kola Peninsula. Ercit et al. (1989) improved the crystallographic setting, identified the mineral in samples from Bikita and established its presence in the Alto do Giz pegmatite in Brazil.

So far, alumotantite shows only limited substitution of Sn which probably enters in equal amounts both Al and Ta sites. Nb contents are negligible, probably constrained for geochemical rather than crystallochemical reasons.

Alumotantite occurrences are restricted to highly fractionated rare-element pegmatites of orogenic affiliation. The mineral is found in them as a replacement product of simpsonite, associated with microlite, parabariomicrolite, natrotantite, manganotantalite, cesstibtantite and other products of simpsonite breakdown.

4.4 Calciotantite

Calciotantite $CaTa_4O_{11}$ corresponds to a hexagonal synthetic phase of the same composition, structurally related to natrotantite (Voloshin et al. 1982b). It occurs in a granitic pegmatite in the Kola Peninsula, as euhedral microscopic inclusions in microlite, along with apatite and scheelite. "Ungursaite" (Voloshin et al. 1985a) represents an Nb-rich variety of calciotantite, with minor substitution of $Na^+ + (OH)^-$ for $Ca^{2+} + O^{2-}$.

4.5 Lithiotantite

The monoclinic mineral lithiotantite has the composition $Li(Ta,Nb)_3O_8$, and it corresponds to the low-temperature structure of synthetic $LiTa_3O_8$ (Voloshin et al. 1983a). It occurs in highly fractionated granitic pegmatites of eastern Kazakhstan, replacing thoreaulite along with irtyshite, manganocolumbite, cassiterite, rankamaite and natrotantite.

4.6 Tantite

Triclinic Ta_2O_5 was described as a mineral from the granitic pegmatites of the Kola Peninsula (Voloshin et al. 1983b). It forms microscopic grains of interstitial to granular microlite, associated with stibiotantalite, holtite and "ungursaite".

4.7 Kimrobinsonite

White, cryptocrystalline aggregates of kimrobinsonite were described in an intimate intergrowth with black cesstibtantite from western Australia (Nickel and Robinson 1985). Both minerals formed by weathering of a Ta,Sb-bearing precursor, possibly stibiotantalite. Kimrobinsonite is cubic, with a general formula of $(Ta,Nb)(OH)_{5-2x}(O,CO_3)_x$, with x~1.2. Minor Na, Fe and Sb were detected as well.

4.8 Franconite-Hochelagaite

Franconite (Jambor et al. 1984) and hochelagaite (Jambor et al. 1986) are monoclinic minerals with similar general formulas and unit cell dimensions:

franconite $Na_2Nb_4O_{11}\cdot 9H_2O$, and
hochelagaite $CaNb_4O_{11}\cdot 8H_2O$.

Both phases show limited substitution of $2Na^+$ for Ca^{2+}, but they probably are not isostructural; a miscibility gap and a two-phase region are indicated for intermediate compositions (Jambor et al. 1986).

Microscopic globules of extremely fine-grained aggregates of both minerals occur in open vugs on weloganite, quartz and calcite, in dawsonite-bearing sills penetrating limestone in Montreal, Quebec. Both minerals are of low-temperature hydrothermal origin.

4.9 Gerasimovskite

Gerasimovskite is an X-ray amorphous mineral, possibly a hydroxide $Nb(OH)_5$, or a hydrated oxide $Nb_2O_5\cdot \sim 5H_2O$. The end-member composition is not known in nature; intermediate members of a series gerasimovskite-belyankinite $Ti(OH)_4$ or $TiO_2\cdot \sim 4H_2O$ have been analyzed, containing subordinate but persistent Ca and occasionally Mn (manganobelyankinite). Consequently, different methods may be used for calculating formulas which so far have no crystal chemical significance. Fleischer (1987) prefers the following formulations:

gerasimovskite $(Mn,Ca)(Nb,Ti)_5O_{12}\cdot 9H_2O$,
belyankinite $Ca_{1-2}(Ti,Zr,Nb)_5O_{12}\cdot 9H_2O$, and
manganbelyankinite $(Mn,Ca)(Ti,Nb)_5O_{12}\cdot 9H_2O$.

All the above minerals are found in alkaline pegmatites in the Kola Peninsula. Gerasimovskite is believed to have formed by hydrothermal alteration of minerals of the murmanite-lomonosovite series (Semenov 1957).

5 Substitution of Niobium and Tantalum in Oxide Minerals and Related Phases

Several oxide minerals incorporate minor to substantial quantities of Nb and/or Ta into their structures that substitute mainly for Sn^{4-}, Ti^{4-} or W^{6-}. These substitutions are largely coupled with the charge-balancing entry of Fe, Mn or Sc. Four minerals are discussed below that are of economic importance in diverse types of deposits.

5.1 Cassiterite and "Staringite"

Cassiterite incorporates a tapiolite component mainly by substituting $(Fe,Mn)^{2+} + 2(Ta,Nb)^{5+}$ for $3Sn^{4+}$. Figure 19 shows a random collection of data on pegmatite cassiterites from the literature (Clark et al. 1976; Dunn et al. 1978; Černý et al. 1981b, 1985, 1986, 1988b) and from unpublished files of the authors. The data indicate that the substitution closely conforms to the tapiolite stoichiometry, suggesting that the role of Fe^{3+} is probably negligible. It is also evident that pegmatitic cassiterites are Ta-dominant. Among the divalent cations, Fe predominates but in extremely fractionated pegmatites Mn may become the principal transition metal (unpublished data of the authors). Recent results of Ercit (1986) and Wise (1987) confirm the above generalizations.

Non-pegmatitic cassiterites from greisen deposits and Sn,W(Mo) veins in granitic intrusions show a greater scatter of (Fe,Mn)/(Nb,Ta) ratios, suggesting an important role of Fe^{3+} (Izoret et al. 1985) and indicating a broader range of Nb/Ta. However, Fe is strongly dominant over Mn in these environments. The entry of Fe,Nb and Ta into non-pegmatitic cassiterite is much debated in the literature; opinions range from assigning most of these elements to microinclusions of columbite-tantalite and tapiolite (Maksimyuk and Lebedeva 1968) to incorporating Fe-hydroxystannates and Fe-stannates (Grubb and Hannaford 1966; Voronina et al. 1979). Clark et al. (1976) and Cohen et al. (1985) established the presence of interstitial Sn^{2+} in natural and synthetic cassiterite, compensated by Al^{3+} and Fe^{4+} entering the regular Sn^{4+} sites. The stannate compounds are held responsible for generating magnetite and hematite respectively, on heating; this leads to significant

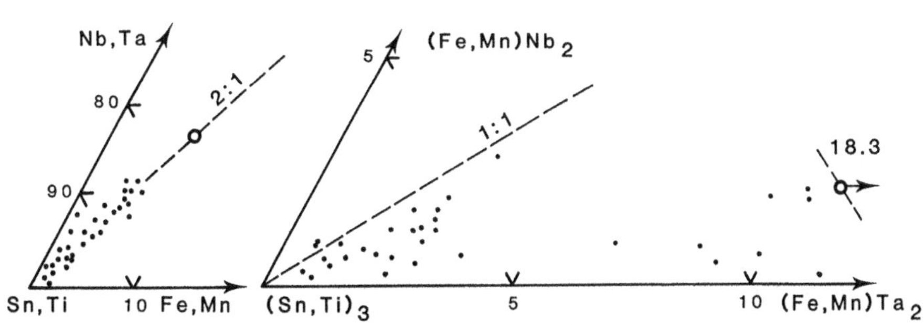

Fig. 19. Compositional relationships (atomic ratios) in pegmatitic cassiterite (after Fig. 4D,E of Černý and Ercit 1985). *Open circle* denotes the type staringite from Burke et al. (1969). In the (Nb,Ta)–(Sn,Ti)–(Fe,Mn) diagram, the cassiterite data are close to the "tapiolite" substitution trend. Tantalum is incorporated in preference to Nb

differences in magnetic susceptibility of roasted cassiterite, affecting its behaviour in ore beneficiation. Similar differences in magnetic properties are caused by primary inclusions and exsolutions as well (Ramdohr 1961). Summaries of the mineralogy and chemistry of cassiterite are provided by Rudakova (1972) and Taylor (1979).

"Staringite" is a name applied to Fe,Nb,Ta-rich cassiterites, containing up to about 28 mol% ferrotapiolite in their composition, and exhibiting a trirutile structure (Burke et al. 1969). Khvostova et al. (1983) designated cassiterites with intermediate Fe,Nb,Ta contents with the ill-defined term "ainalite" (Amark 1941). Recent results of Zayachkivsky (1985) and Kissin (pers. commun. 1986) suggest that modulation of monorutile and trirutile structures may be the best approximation at present; detailed study of "staringite" and "ainalite" is required.

5.2 Rutile

In Nb, Ta-bearing geochemical environments, rutile readily accepts Nb and Ta into its structure, mainly in tapiolitelike stoichiometry with (Fe$>$Mn) as shown above for cassiterite. However, deviations from this relationship are common, showing a slight to considerable excess of (Fe$>$Mn) over the 1:2 ratio to (Nb $\gtreqless$ Ta). This ratio varies from the ideal value of 0.5 to as high as 2.10. This may be caused by the incorporation of either the rutile-type compound $Fe^{3+}(Nb,Ta)O_4$, or Fe alone. Cation totals normalized to four oxygens usually exceed the ideal value of 2.00, when all Fe is calculated as Fe^{2+}. The range of totals is 1.97 to 2.16. Oxygen deficiency, incorporation of hydroxyl or substitutions of the $Ti^{4+} \rightleftarrows 2Fe^{2+}$ or $3Ti^{4+} \rightleftarrows 4Fe^{3+}$ type may be responsible. Compositions containing substantial Fe^{3+} (to the exclusion of Fe^{2+}) and incorporation of Fe^{3+} into interstitial positions are well documented (Lugovskoi and Stolyarova 1969; Ruck et al. 1986). Another compensating factor may be the presence of Ti^{3+} or Ti^{2+}, induced by the entrance of Fe^{3+} or Fe^{2+}, as proposed for Sn in cassiterite.

Figure 20A shows the bulk compositions of niobian and tantalian rutiles based on the data collected by Černý et al. (1981b) and complemented by those of Lebedeva (1968), Sahama (1978), Černý et al. (1986), and on unpublished data of the authors. Most phases with an appreciable content of substituting elements and with Nb predominant over Ta (ilmenorutile) are heterogeneous, containing exsolved titanian ixiolite or, rarely, ilmenite (Fig. 20b,c). In contrast, Ta-dominated minerals (strüverite) are homogeneous. Exceptions to this behaviour are very rare. The Nb/Ta and Mn/Fe ratios of the exsolved ixiolite are invariably higher than those of the parent (as well as exsolved) Fe-dominant rutile phase (Fig. 21a,b). If W and Sc are present in the rutile precursor, they also become enriched in the exsolved ixiolite (Sahama 1978; unpublished data of P.Č. show up to 23.3 wt% WO_3 and 9.3% Sc_2O_3).

The structural state of all specimens examined to date is disordered, with "monorutile" unit cell dimensions. The few titanian ixiolites studied have shown disordered ixiolite structure in the natural state, and wodginite diffraction patterns after heating (unpubl. data of P.Č.).

Further data and references on crystal chemistry of niobian and tantalian rutile can be found in Černý et al. (1981b) and Černý and Ercit (1985).

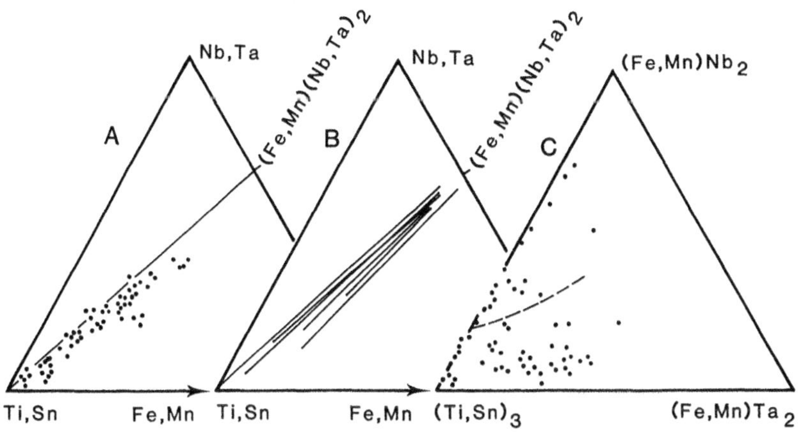

Fig. 20. Compositional relationships (atomic ratios) in niobian and tantalian rutile (after Fig. 4A,B,C of Černý and Ercit 1985). **A** A deviation from the rutile to columbite-tantalite trend in the direction of higher (Fe,Mn), suggestive of the presence of Fe^{3+}; **B** examples of exsolved rutile-ixiolite pairs. **C** Minerals plotting *below the dashed line* are homogeneous, those *above the dashed line* are largely exsolved

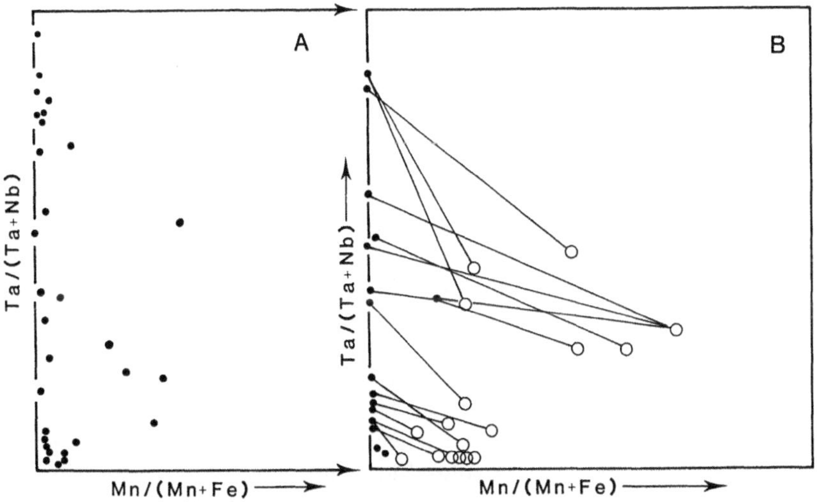

Fig. 21. Niobian and tantalian rutile composition in the columbite quadrilateral (after Černý et al. 1981b). **A** Presentation of most of the data; **B** selected examples of co-existing niobian-tantalian rutiles and exsolved ixiolites (*open circles*). Note the preference of the ixiolite phase for Nb and Mn

Niobian rutile is relatively widespread in terms of its diverse parageneses: it is found in mineralized miaskites and metasomatic deposits associated with plumasitic nepheline syenites; in anorogenic alkaline rare-element granites and associated metasomatic and vein deposits; in anorogenic rare-element pegmatites; in peraluminous, orogenic to post-orogenic rare-element granites; and in granitic rare-element pegmatites of orogenic affiliation. In contrast, tantalian rutile is found only in the last two types of deposits, in keeping with the overall Nb>Ta signature of anorogenic igneous sequences.

5.3 Ilmenite

In Nb,Ta-bearing environments, ilmenite $FeTiO_3$ shows subordinate substitution of Mn for Fe and only a limited extent of solid solution with Fe_2O_3. The entry of Nb, Ta is facilitated by substitutions that affect only the Ti sites: $(Fe,Mn)^{2+} + 2(Nb,Ta)^{5+}$ for $3Ti^{4+}$ and $Fe^{3+} + (Nb,Ta)^{5+}$ for $2Ti^{4+}$. The maximum known concentration of Nb_2O_5 is 3.80 wt%, and that of Ta_2O_5 is 0.08 wt%, in accessory ilmenite of biotite leucogranites (Kuzmenko and Eskova 1968).

Nb, Ta-bearing ilmenite occurs mainly in leucogranites and granitic pegmatites of calc-alkalic, peraluminous and alkaline sequences. However, its low concentration in these rocks, and its generally low contents of Nb and Ta make it a poor candidate for economic exploitation.

5.4 Wolframite Series

Members of the ferberite $FeWO_4$-hübnerite $MnWO_4$ series are well-known carriers of Nb and Ta. The pentavalent cations are incorporated into the structure by substituting $(Fe,Mn)^{2+} + W^{6+}$ by $(Fe,Sc)^{3+} + (Nb,Ta)^{5+}$; Sc is known to positively correlate with the (Nb + Ta) content, although it is subordinate to Fe^{3+}.

On the average, the highest Nb, Ta contents are found in wolframite of large-scale greisens related to Ta-bearing granite intrusions (to 5 wt% Nb_2O_5 and 0.8 wt% Ta_2O_5). In wolframites from quartz veins and surrounding greisen haloes associated with Sn,W(Mo) deposits in granites, the Nb and Ta contents are perceptibly lower; they decrease further to insignificant levels in wolframite from sulphide-enriched subvolcanic deposits (to 0.03 wt% Nb_2O_5, 0.002 wt% Ta_2O_5) (Kuzmenko and Eskova 1968).

All the above concentrations of Nb and Ta in wolframite are controlled by the geochemistry of their parent environment, rather than by crystal chemical factors. Occasional finds of Nb,Ta-rich wolframite in granitic pegmatites indicate that incorporation of these elements can reach 31.6 wt% Nb_2O_5 and 4.2 wt% Ta_2O_5 (Saari et al. 1968; and unpubl. data of the authors). As mentioned in the discussion of wolframian (tungstenian) ixiolite, crystal chemical relationships have not been sufficiently explored at intermediate columbite-wolframite bulk compositions. Consequently, the limits to incorporation of Nb and Ta into a true wolframite structure are so far unknown.

5.5 Calzirtite

This tetragonal mineral is essentially an oxide of Ca, Zr and Ti, with a formula of $Ca(Ca,Zr)_2Zr_4(Ti,Fe)_2O_{16}$. Minor Nb enters its composition (0.10 wt% Nb_2O_5), but the crystal chemical role of Nb is difficult to assess. It may substitute for either Ti or Zr, both of which are present in substantial amounts; the mineral seems to be amenable to several substitution mechanisms but the negligible quantity of Nb does not allow the identification of the proper substitution with any degree of certainty.

Calzirtite is known from metasomatic rocks consisting mainly of calcite, forsterite and magnetite, associated with an ultrabasic alkalic massif in eastern Siberia (Zdorik et al. 1961).

5.6 Zirkelite

This mineral has a general formula close to the stoichiometry of the pyrochlore group, $(Ca,Th,Ce)Zr(Ti,Nb)_2O_7$. However, its monoclinic symmetry excludes it from the cubic family of pyrochlore-microlite-betafite minerals. Zirconolite and niobzirconolite are identical with zirkelite (Hogarth 1977).

Niobium substitutes for Ti in zirkelite; the range of Nb_2O_5 contents is from 2.86 to 24.84 wt%, along with 0.09–2.00 wt% Ta_2O_5.

Zirkelite is known to occur in alkaline pyroxenites (jacupirangites) as a primary accessory phase, or in close association with metasomatic amphibole, phlogopite or carbonates (Pudovkina et al. 1974). Niobium-rich varieties ("niobozirconolite") are found in carbonatites (Vlasov 1964, 1966).

5.7 Murataite

Although no Nb-free analogue of murataite is known, the stoichiometry of this cubic mineral suggests that it is a derivative of a purely Ti-based compound with up to 10.0 wt% Nb_2O_5: $(Na,Y,Er)_4(Zn,Fe)_3(Ti_{5.13}Nb_{0.87})_{\Sigma 6.00}O_{18}$. Murataite is a rare accessory mineral in pegmatites hosted by a biotite-granite facies of the Pikes Peak batholith, near St. Peters Dome, Colorado. It is associated with riebeckite, astrophyllite, senaite and quartz. It is commonly replaced by zincian senaite (Foord et al. 1984; senaite is the mineral "Y" of Adams et al. 1974).

5.8 Crichtonite Group

Minerals of this group have a general formula $AM_{21}O_{38}$, with A – Ca, LREE, Pb, Sr, Na, K, Ba and M – Ti, Fe, Mn, Zn, Zr, Sn, HREE, U, Th, Cr, V and others (Gatehouse et al. 1979). Individual species are defined by the dominant cation in the A site:

crichtonite	Sr,
senaite	Pb,
davidite	LREE,
loveringite	Ca, and
landauite	Na.

Electron microprobe analysis shows Nb_2O_5 from 0.2 to 3.98 wt% in senaite from Colorado and Brazil (Foord et al. 1984). It is conceivable that minor quantities of Nb, dominant over Ta, may also be found in other members of this group, when associated with alkaline igneous complexes (e.g., in davidite; Gatehouse et al. 1979).

6 Silicates of Niobium and Tantalum

Only a few silicates contain Nb($\pm$subordinate Ta) as a specific cation, or a dominant one which is in part substituted by Ti. All these silicates shown in Table 3 are basically niobosilicates of Na,K,Ca and/or Ba, with negligible (if any) Al content and a low degree of tetrahedral polymerization.

From the paragenetic viewpoint, all these silicates occur in anorogenic alkaline igneous suites, typically in their pegmatite differentiates. Some of them belong to early primary phases (e.g. niocalite in carbonatites, nenadkevichite with K-feldspar, aegirine in pegmatites). Others such as bornemanite, vuonnemite and ilimaussite are late phases, formed at the hydrothermal stage. The general affiliation of all these silicates with anorogenic alkaline magmas explains their general Nb $\gg$ Ta signature.

None of the minerals listed in Table 3 have attained the status of a significant ore mineral of Nb. However, some of the other silicates described below have a potential economic importance.

7 Substitution of Niobium and Tantalum in Silicates

A great number of silicates accommodate Nb($\pm$Ta) as a substituent for Ti or Zr. With the exception of zircon, all of these silicates conform to the same crystal chemical type as those discussed in the preceding section. They belong to titanosilicates and zirconosilicates of Na, K, Ca and Ba, rarely with substantial Sr or REE; some contain (Fe,Mg,Mn)-populated sites. Aluminum is generally low or absent.

Incorporation of $(Nb,Ta)^{5+}$ is achieved by coupled substitutions balancing the differences in charge relative to $(Ti,Zr)^{4+}$:

$(Ti,Zr)^{4+} + Ca^{2+} \leftarrow (Nb,Ta)^{5+} + (Na,K)^{+}$ (nenadkevichite, lovozerite, fersmanite, låvenite)

$(Ti,Zr)^{4+} + Si^{4+} \leftarrow (Nb,Ta)^{5+} + Al^{3+}$ (titanite, kimzeyite)

$2(Ti,Zr)^{4+} \leftarrow (Nb,Ta)^{5+} + Fe^{3+}$ (shcherbakovite, baotite, chevkinite, titanite)

$2(Ti,Zr)^{4+} \leftarrow (Nb,Ta)^{5+} + (Y,REE)^{3+}$ (zircon, eudialyte, chevkinite)

$3(Ti,Zr)^{4+} \leftarrow (Nb,Ta)^{5+} + (Th,U)^{4+} + (Y,REE)^{3+}$ (zircon, chevkinite)

$Ti^{4+} \leftarrow (Nb,Ta)^{5+} + Cl^{-}$ (baotite)

$Zr^{4+} + F^{-} \leftarrow (Nb,Ta)^{5+} + O^{2-}$ (woehlerite, hiortdahlite, lavenite)

Tables 4 and 5 summarize the silicates exhibiting substitution of Nb($\pm$Ta) for Ti and Zr respectively. The listing is not complete but it covers all the important species with significant Nb($\pm$Ta) contents, and it also demonstrates the full range of different mineral types and their occurrences. Within each of the Tables 3, 4 and 5, a subdivision of the silicates is indicated according to tetrahedral polymerization, which is predominantly very low. Crystal chemical relationships, however, occasionally link minerals tabulated separately. Thus, lomonosovite belongs to the same interlayered titanosilicate-sodium phosphate type as bornemanite and vuonnemite; astrophyllite-kupletskite forms a limited solid solution series with niobophyllite and zircophyllite; murmanite is transitional to epistolite, and labuntsovite to nenadkevichite.

Table 3. Silicates with stoichiometric or dominant Nb component

Mineral	Formula	Nb_2O_5,wt.%	Ta_2O_5,wt.%	Nb_2O_5/Ta_2O_5	Paragenesis[a]
Nesosilicates					
Epistolite	$(Na,Ca)_2(Nb,Ti)_2Si_2O_9 \cdot nH_2O$	31.40	to 0.90	≥34.9	lp;r.s,ss
Bornemanite	$BaNa_4Ti_2NbSi_4O_{17}(F,OH) \cdot Na_3PO_4$	8.8–9.2	0.66	≥13.4	lp,h;ns
Vuonnemite	$Na_5Ti^{3+}Nb_2(Si_2O_8F)_2 \cdot 2Na_3PO_4$	23.16	?	—	h.r;ns
Sorosilicates					
Niocalite	$Ca_{14}Nb_2(Si_2O_7)_4O_6F_2$	16.56	?	—	em.c
Nenadkevichite	$(Na,Ca,K)(Nb,Ti)Si_2O_6(O,OH) \cdot 2H_2O$	24–25[b]		—	m,ep;s
Ilímaussite	$Ba_2Na_4CeFeNb_2Si_8O_{28} \cdot 5H_2O$	13.20	?	—	h,v;ss
Phyllosilicates					
Niobophyllite	$(K,Na)_3(Fe,Mn)_6(Nb,Ti)_2Si_8(O,OH,F)_{31}$	14.8–15.3	?	—	h;r.f
Unclassified					
Komarovite	$(Ca,Mn,H)_2Nb_2Si_2O_{10}(OH,F)_2 \cdot H_2O$	47.0	?	—	h;s
Na-komarovite	$(Na,Ca,H)_2Nb_2Si_2O_{10}(OH,F)_2 \cdot H_2O$	?	?	—	h;s

[a] e,Early; l,late; m,magmatic; p,pegmatitic; h,hydrothermal; r,metasomatic, replacement; v,vein; c,carbonatite; g,alkaline granite; s,alkaline syenite; ns,nepheline syenite; ss,sodalite syenite; f,fenite.
[b] Data between the Nb_2O_5 and Ta_2O_5 columns refer to bulk determinations of the total of both oxides.

Table 4. Silicates with Nb(Ta) substituting for Ti

Mineral	Formula	Nb_2O_5,wt.%	Ta_2O_5,wt.%	Nb_2O_5/Ta_2O_5	Paragenesis[a]
Nesosilicates					
Titanite	$(Ca,Na)(Ti,Nb,Al)SiO_5$	to 6.5	to 16.0	4.4–0.18	m,p
Lomonosovite	$Na_2(Ti,Nb)_2Si_2O_9 \cdot Na_3PO_4$	3.00–6.34[b]		—	m(ep);ns
Murmanite	$(Ca,Na)_2(Ti,Nb)_2Si_2O_9 \cdot nH_2O$	6–10	0.50–0.56	17.9–11.7	lm,lp,r;ns
Tundrite	$Na_3(Ce,La)_4(Ti,Nb)_2(SiO_4)_2(CO_3)_3O_4(OH) \cdot 2H_2O$	4.26	—	—	lp
Tundrite-(Nd)	$Na_3(Nd,La)_4(Ti,Nb)_2(SiO_4)_2(CO_3)_3O_4(OH) \cdot 2H_2O$	4.90–6.09	—	—	lp
Sorosilicates					
Mosandrite	$(Na,Ca,Ce)_3(Ti,Nb)(SiO_4)_2F$	1.62–5.5		—	e-l,m,p;ns
Batisite	$(Ba,K,Na)_3(Ti,Nb)_2Si_4O_{14}$	0.36		—	ep;ns
Shcherbakovite	$(K,Na,Ba)_3(Ti,Nb)_2Si_4O_{14}$	to 0.44		—	lh;s?,ns?
Bafertisite	$BAA(Fe,Mn)_2(Ti,Nb)Si_2O_7(O,OH)_2$	0.32–0.84		—	lh;v
Leucosphenite	$BaNa_4(Ti,Nb)_2B_2Si_{10}O_{30}$	0.1–0.77		—	m,p;s
Lamprophyllite	$Na_2(Sr,Ba)_2(Ti,Nb)_3(SiO_4)_4(OH,F)_2$	2.30		—	ep;s,ns
Chevkinite	$(Ca,Ce,Th)_4(Fe,Mg)_2(Ti,Nb,Fe)_3Si_4O_{22}$	to 7.40	0.002	—	m,p;g,s,ns
Vinogradovite	$(Na,Ca,K)_4(Ti,Nb)_4AlSi_6O_{23}(O,OH)_7.H_2O$	3.52		—	lp,r;ns
Labuntsovite	$(K,Ba,Na)(Ti,Nb)(Si,Al)_2(O,OH)_7 \cdot H_2O$	1.45–2.33		—	h,r;ns,s
Janhaugite	$(Na,Ca)_3(Mn,Fe)_3(Ti,Zr,Nb)_2(Si_2O_7)_2O_2(F,OH)_2$	5.00	0.26	19.2	e-l,m-p;g
Cyclosilicates					
Baotite	$Ba_4(Ti,Nb)_8Si_4O_{28}Cl$	10.8–11.5	—	—	hv;g,s
Inosilicates					
Ramsayite	$Na_2(Ti,Nb)_2Si_2O_9$	2.14–3.89	< 0.1	—	(m),ep;ns
Phyllosilicates					
Astrophyllite	$(K,Na)_3(Fe,Mn)_7(Ti,Nb)_2Si_8O_{24}(O,OH)_7$	0.07–2.15	0.008	—	(m)ep;s,ns
Kupletskite	$(K,Na)_3(Mn,Fe)_7(Ti,Zr,Nb)_2Si_8O_{24}(O,OH)_7$	0.66–2.48		—	p;ns

Unclassified (and in part questionable species)					
"Yuksporite"	$(Na,K)_4(Ca,Sr,Ba)_4(Ti,Al,Fe,Nb)_3Si_8O_{16}(F,Cl)_2 \cdot 4H_2O$	2.30[b]		–	lp
"Karnasurtite"	$(Ce,La,Th)(Ti,Nb)(Al,Fe)(Si,P)_2O_7(OH)_4 \cdot 3H_2O$	2.20–8.20		–	p;ns,ss
Fersmanite	$(Ca,Na)_4(Ti,Nb)_2Si_2O_{11}(F,OH)_2$	18.0–22.0	0.36	42	ep;ns
Laplandite	$Na_4Ce(Ti,Nb)PSi_7O_{22} \cdot 5H_2O$	1.88	–	–	lp;s
Koashvite	$Na_8Ca_2(Ti,Fe,Nb)Si_8O_{24}(\cdot H_2O?)$	0.40	–	–	lp,r:s
Penkvilksite	$(Na,Ca)_4(Ti,Zr,Nb)_2Si_8O_{22} \cdot 5H_2O$	0.93–1.08	to 0.06	≥18	lp,h;s
"Pao-t'ou-k'uang"	$Ba(Ti,Nb,Fe)_2SiO_7$	11.5	–	–	p,h,v;s,g
Kazakovite	$Na_6H_2(Ti,Nb)Si_6O_{18}$	1.40	–	–	m(p);ss

[a] e,Early; l,late; m,magmatic; p,pegmatitic; h,hydrothermal; r,metasomatic, replacement; v,vein; c,carbonatite; g,alkaline granite; s,alkaline syenite; ns,nepheline syenite; ss,sodalite syenite; f,fenite.

[b] Data between the Nb_2O_5 and Ta_2O_5 columns refer to bulk determinations of the total of both oxides.

Table 5. Silicates with Nb(Ta) substituting for Zr

Mineral	Formula	Nb_2O_5,wt.%		Ta_2O_5,wt.%	Nb_2O_5/Ta_2O_5	Paragenesis[a]
Nesosilicates						
Zircon	$ZrSiO_4$	to 2.7		to 0.61	7–0.5	m,p
"Gelzircon"	$ZrSiO_4 \cdot nH_2O$	0.17		0.03	5.5	lp,h,r;s
"Zirfesite"	$(ZrO_2,Fe_2O_3)SiO_2 \cdot nH_2O$		to 2.40[b]		—	lh;ns
Kimzeyite	$Ca_3(Zr,Ti,Nb)_2(Si,Al)_3O_{12}$	1.0		Not detected	—	m;c
Sorosilicates						
Hiortdahlite	$(Ca,Na)_3(Zr,Nb)Si_2O_7(O,OH,F)_2$		1.89		—	m,p,v;ns
Låvenite	$(Na,Ca)_3(Zr,Nb)Si_2O_7(O,OH,F)_2$		2.97–5.20		—	m,ep;s,ss,ns
Woehlerite	$NaCa_2(Zr,Nb)Si_2O_8(O,OH,F)$	11.6–14.5		0.1	≈ 50	m,p,e-l,r;ns
Rosenbuschite	$(Ca,Na)_3(Zr,Ti,Nb)Si_2O_8F$		0.57		—	m,ep;ns
Seidozerite	$(Na,Ca)_2(Zr,Ti,Mn,Nb)_2Si_2O_7(O,F)_2$		0.60		—	ep;ns,ss
Cyclosilicates						
Catapleiite	$Na_2(Zr,Nb)Si_3O_9 \cdot 2H_2O$	0.07–3.31		0.08	2.5	l,r,m-p;ns
Eudialyte	$Na_4(Ca,Ce,Fe)_2(Zr,Nb)Si_6O_{17}(OH,Cl)_2$	0.16–4.37		to 0.75	6	lm,ep;ns
Darapiosite	$KNa_2Li(Mn,Zn)_2(Zr,Nb)Si_{12}O_{30}$	0.90		—	—	p;s
Lovozerite	$Na_2Ca(Zr,Ti,Nb)Si_6(O,OH)_{18}$	0.06–0.08		0.047	1.7–1.3	e-lm,p;ns
Inosilicates						
Elpidite	$Na_2(Zr,Nb)Si_6O_{15} \cdot 3H_2O$	0.60–1.43		0.07	8.6	lm,p;ns
Gaidonnayite	$Na_2(Zr,Nb)Si_3O_9 \cdot 2H_2O$	3.00		—	—	lp,r;ns
Vlasovite	$Na_2(Zr,Nb)Si_4O_{11}$		0.33		—	lp,r;s,ns
Phyllosilicates						
Zircophyllite	$(K,Na,Ca)_3(Mn,Fe)_7(Zr,Nb)_2Si_8O_{27}(OH,F)_4$	2.30		—	—	lp;s
Lemoynite	$(Na,K)_6Ca(Zr,Nb)_6Si_{10}O_{26} \cdot 5\text{–}6H_2O$	2.218		—	—	lh,p;ns

[a] e,Early; l,late; m,magmatic; p,pegmatitic; h,hydrothermal; r,metasomatic, replacement; v,vein; c,carbonatite; g,alkaline granite; s,alkaline syenite; ns,nepheline syenite; ss,sodalite syenite; f,fenite.
[b] Data between the Nb_2O_5 and Ta_2O_5 columns refer to bulk determinations of the total of both oxides.

In general, substitution of Nb for Ti is more extensive than for Zr. As illustrated by the last two examples quoted above, more or less complete solid solution series are found between some titanosilicates and niobosilicates. In contrast, substitution of Nb for Zr rarely exceeds 10 at % of the latter, because of the significant difference in ionic radii (Kuzmenko and Eskova 1968).

Of the minerals listed in Tables 4 and 5, zircon and titanite are the only ones found in granitoid rocks and pegmatites of the calc-alkalic to peraluminous sequences. This accounts for their low Nb_2O_5/Ta_2O_5 ratios that occasionally decrease to < 1.0 (e.g. in titanite: Clark 1974; Paul et al. 1981; Groat et al. 1985). They also occur, of course, in igneous suites of the alkaline kindred that are parental to all the other silicates carrying Nb($\pm$Ta). In keeping with the geochemical features of these rocks, the Nb_2O_5/Ta_2O_5 ratios of all these silicates are rather high (although precise data are commonly not available, the minor Ta content having not been analyzed for; Tables 4,5).

Paragenetic affiliation of the Nb($\pm$Ta)-bearing silicates is diversified, and it is controlled mainly by the abundance and accumulation of Nb in different members of comagmatic igneous suites. High levels of Nb lead to significant contents of this element in accessory minerals of plutonic intrusions, particularly in agpaitic nepheline syenites of Nb-enriched igneous complexes (e.g. the Lovozero massif in the Kola Peninsula). However, most of the discussed minerals are formed in the early to late stages of pegmatite consolidation, including low-temperature hydrothermal phases; alteration and replacement is common. Occurrences in hydrothermal vein deposits associated with minerals of Y,REE,Ba,Sn and F are rare.

According to Ovchinnikov and Solodov (1980), only a few of the Nb($\pm$Ta)-carrier silicates have been considered, experimented with and extracted as ore minerals of Nb: eudialyte, lovozerite, murmanite and mosandrite.

8 Miscellaneous Minerals of Niobium and Tantalum

Several other minerals of Nb and Ta should be mentioned in conclusion, to round off the systematic review with the exotic species that also complement the geochemical features of these two metals.

8.1 Holtite

Holtite was first found in alluvial pebbles, associated with stibiotantalite and tantalite (Pryce 1971). A second occurrence was discovered in a pollucite-bearing complex granitic pegmatite of the spodumene subtype, in association with cleavelandite, lepidolite, zircon, rubellite and amblygonite (Voloshin et al. 1977). This orthorombic mineral is essentially a borosilicate with a complex formula:

$(Ta,Nb,Sb,Al,Fe,Mn,Mg,Ca)Al_{6-n}\,[\{(Al,As)_kSi_{1-k}\}\,O_4]_3[B_{1-m}O_3](O,OH)_{2to3}$, with $m+n < 1$, and $k \cong (m+n)/3$ (Voloshin et al. 1977). The structure of holtite has not been determined so far.

8.2 Behierite

The only borate of Ta, behierite $TaBO_4$, is a tetragonal mineral with a zircon-type structure (Mrose and Rose 1961). At present, it is known only from its type locality in Madagascar, from pegmatites that typically exhibit extreme activity of B during their late consolidation (reflected in the formation of hambergite, rhodizite and manandonite; Moore 1982).

8.3 Zimbabweite

Zimbabweite is an orthorhombic mineral, similar to stibiotantalite in appearance but unlike any other oxide species of (Nb,Ta) in crystal chemistry. Its composition can be expressed as

$(Na,K)_2PbAs_4(Ta,Nb,Ti)_4O_{18}$,

with minor Ba, Bi, U and Sn contents (Foord et al. 1986). In the absence of crystal structure refinement, the classification of this mineral is uncertain. It was found as loose crystals in a kaolinized pegmatite, which otherwise yielded aquamarine, topaz and verdelite.

8.4 Olmsteadite-Johnwalkite

The only phosphates of Nb and Ta known to date come from the granitic pegmatites of the southern Black Hills of South Dakota. They belong to a substitutional series of

olmsteadite $K_2(Fe^{3+},Fe^{2+},Mn)_4(Nb,Ta)_2(PO_4)_4O_4(H_2O,OH)_4$, and
johnwalkite $K_2(Mn,Fe^{3+},Fe^{2+})_4(Nb,Ta)_2(PO_4)_4O_4(H_2O,OH)_4$

(Moore et al. 1976; Dunn et al. 1986). Both minerals are late secondary phases, associated with goethite, amorphous Mn oxides and numerous other secondary phosphates.

9 Paragenetic Relationships

The following listing provides a simplified review of Nb,Ta-bearing major, (subordinate) and [minor] minerals in the main types of niobium and tantalum deposits. *Typical* minerals are *italicized*, disregarding their abundance.

1. Ultrabasic to gabbroic peralkaline association
   - A) Igneous rocks: perovskite
   - B) Metasomatic albitites: uranpyrochlore, (pyrochlore, betafite, columbite)
   - C) Metasomatic phoscorites, camaphorites: uranpyrochlore, pyrochlore (*zirkelite*, niobian ilmenite)
2. Carbonatite association
   - A) Calcite (+dolomite, apatite) carbonatites: *pyrochlore*, *latrappite* (uranpyrochlore, *niocalite*)

  B) Calcite + dolomite (+apatite) carbonatites: *fergusonite-(Ce), uranpyrochlore*, (columbite), [fersmite, zirkelite]
3. Agpaitic nepheline syenite association
  A) Urtites: *loparite, metaloparite*
  B) Luyavrites: *eudialyte, lovozerite, murmanite*, loparite, *mosandrite*
4. Plumasitic nepheline syenite association
  A) Miaskites: *pyrochlore*, fersmite, aeschynite, niobian rutile
  B) Metasomatites: *pyrochlore, fersmite, (aeschynite*, titanian columbite, niobian rutile)
5. Sub- to peralkaline anorogenic granite association
  A) Rare-element granites: columbite, pyrochlore, fergusonite, aeschynite-(Y), (titanian ixiolite, niobian rutile, cassiterite)
  B) Metasomatites: *ceriopyrochlore, yttropyrochlore, plumbopyrochlore, columbite, polycrase*, fergusonite, niobian rutile, (samarskite, euxenite)
  C) Pegmatites: *fergusonite*, columbite, samarskite, euxenite, polycrase, (niobian rutile, betafite) [*ilmenite*, microlite]
6. Calc-alkaline to peraluminous, orogenic to post-orogenic granite association
  A) Rare-element granites: columbite-*tantalite, microlite, tantalian rutile*, tantalian cassiterite, (niobian rutile, euxenite, fergusonite), [samarskite, loparite]
  B) Rare-element pegmatites: *columbite-tantalite* group, *wodginite* group, *microlite* subgroup, (*stibiotantalite* group, *tapiolite* group, *tantalian* and niobian *rutile, tantalian cassiterite*), [fergusonite, euxenite, titanian euxenite, aeschynite-(Y),samarskite, *staringite, foordite-thoreaulite, simpsonite, cesstibtantite, secondary Ta-oxide minerals*]

As mentioned above, this review is grossly simplified and reduced to the principal paragenetic features of generalized types of deposits. More detailed information can be found in Vlasov et al. (1964, 1966), Kuzmenko and Eskova (1968), Gorzhevskaya et al. (1974), Foord (1982b) and Černý and Ercit (1985).

10 Concluding Remarks

All currently important ore minerals of niobium and tantalum belong to the class of oxides. Their assemblages are considerably different in diverse types of deposits. The main distinctions in mineralogy are seen between the deposits of alkaline anorogenic affiliation, ranging from carbonatites and gabbros to alkaline granites and their pegmatites (typically associated with Ti,Zr,Y,REE), those related to calc-alkaline to peraluminous late-tectonic to post-orogenic granites (associated with Li,Sn,W), and those found in orogenic rare-element granitic pegmatites (accompanied by Li,Rb,Cs,Be,Sn,B). The first of the above deposit types produce virtually exclusively Nb, the second type has a potential for both Nb and Ta, and the last one is significant only in terms of Ta.

Niobosilicates, and mainly some locally abundant titanosilicates and zirconosilicates with appreciable Nb substitution, represent potential ore minerals for the future. Large-volume, low-concentration reserves of these minerals may render them economically significant, despite the difficulties encountered in their processing. However, no alternate ore minerals seem to be available for the future extraction of tantalum, unless it proves to be sufficiently enriched in some of the oxide minerals of Nb associated with alkaline gabbros and carbonatites.

Acknowledgements. The authors are indebted to Dr. L. Wambeke, C. Knudsen and other participants of the REE,Nb,Ta workshop for their discussion input into the present chapter, and to Dr. E.E. Foord for a critical review of the manuscript. Original research of the authors, A.J. Anderson, L.A. Groat and Dr. M.A. Wise incorporated into this presentation was performed under the programs funded by the National Science and Engineering Research Council of Canada Operating Grant A7948, Canada Department of Energy, Mines and Resources Research Agreements 1983–1986, Canada-Manitoba Mineral Development Agreements, Department of Indian and Northern Affairs (Yellowknife) Research Contracts, and Research Grants from the Tantalum Mining Corporation of Canada, Ltd., to P.Č. over the period of 1981 to 1987.

References

Adams JW, Botinelly Th, Sharp WN, Robinson K (1974) Murataite, a new complex oxide from El Paso County, Colorado. Am Mineral 59:172–176

Adusumili MS, Kieft C, Burke EAJ (1974) Tantal-aeschynite, a new mineral of the aeschynite group from the Borborema group, north-eastern Brazil. Min Mag London 39:571–576

Amark K (1941) Minerals of the Varuträsk pegmatite XXIX. An X-ray study of stanniferous columbite from Varuträsk and of the related Finnish minerals ainalite and ixiolite. Geol Fören Stockholm Förh 63:295–299

Amichba TM, Dubakina LS (1976a) Wolframoixiolite in ores of tin-tungsten deposits of Yakutia. Sb Nauch Trud Moscow Br Vses Min Obschtch 1974, pp 11–18 (in Russian)

Amichba TM, Dubakina LS (1976b) Once again, wolframoixiolite. Sb Nauch Trud Moscow Br Vses Min Obschtsch 1974, pp 25–28 (in Russian)

Beddoe-Stephens B, Fortey NJ (1981) Columbite from the Carrock Fell tungsten deposit. Min Mag 44:217–223

Bjørlykke H (1937) Mineral parageneses of some growth pegmatites near Kragerø, southern Norway. Norsk Geol Tiddskr 17:1–16

Borisenko LF, Maximova NV, Kazakova MYe (1969) Scandian ixiolite, a new tantalo-niobate species with formula $(A,B)_n O_{2n}$. Dokl Acad Sci USSR 189:148–151 (in Russian)

Bowley H (1923) Annu Rep Mines W Austr Div 7 Sec 2, pp 117–120

Brøgger WC (1897) Über den Mossit und über den Krystallsystem des Tantalit (Skogbölit) aus Finnland. Vidensk Christiania Oslo Srk 1 Math Naturviss Kl 7:1–19

Bulakh AG, Kukharenko AA, Knipovich YuN, Kondratyeva VV, Baklanova KA, Baranova EN (1960) Some new minerals in cabonatites of the Kola Peninsula. Trudy Vses Nauch-Issl Geol Inst 1960 114–116 (in Russian)

Burke EAJ, Kieft C, Felius RO, Adusumili MS (1969) Staringite, a new Sn-Ta mineral from north-eastern Brazil. Min Mag 37:447–452

Buttgenbach H (1933) Minéraux du Congo belge. Ann Soc Geol Belg 56:327

Cabri LJ (1987) The mineralogy of precious metals: new developments and metallurgical implications. Can Mineral 25:1–7

Černý P, Ercit TS (1985) Some recent advances in the mineralogy and geochemistry of Nb and Ta in rare-element granitic pegmatites. Bull Mineral 10:499–532

Černý P, Turnock AC (1971) Niobium-tantalum minerals from granitic pegmatites at Greer Lake, south-eastern Manitoba. Can Mineral 10:755–772

Černý P, Hawthorne FC, Laflamme JHG, Hinthorne J (1979) Stibiobetafite, a new member of the pyrochlore group from Vezna, Czechoslovakia. Can Mineral 17:583–588

Černý P, Trueman DL, Ziehlke DV, Goad BE, Paul BJ (1981a) The Cat Lake-Winnipeg River and the Wekusko Lake pegmatite fields, Manitoba. Manitoba Dep Energ Mines Min Res Div Econ Geol Rep ER80-1, 234 pp

Černý P, Paul BJ, Hawthorne FC, Chapman R (1981b) A niobian rutile-disordered columbite intergrowth from the Huron Claim pegmatite, southeastern Manitoba. Can Mineral 19:541–548

Černý P, Ercit TS, Wise MA (1984) Compositional and structural variability in columbite-tantalite and tapiolite. Geol Soc Am Abstr Progr 16:466

Černý P, Roberts WL, Ercit TS, Chapman R (1985) Wodginite and associated minerals from the Peerless pegmatite, Pennington County, South Dakota. Am Mineral 70:1044–1049

Černý P, Goad BE, Hawthorne FC, Chapman R (1986) Fractionation trends of the Nb- and Ta-bearing oxide minerals in the Greer Lake pegmatitic granite and its pegmatite aureole, southeastern Manitoba. Am Mineral 71:501–517

Černý P, Fransolet A-M, Ercit TS, Chapman R (1988a) Foordite $SnNb_2O_6$, a new mineral, and the foordite-thoreaulite series. Can Mineral 26:889–900

Černý P, Ercit TS, Hawthorne FC, Chapman R (1988b) The Nb,Ta,Sn,Sb-bearing mineral assemblage from the Odd West pegmatite, southeastern Manitoba (in preparation)

Chao GY (1979) Changbaiite. Am Mineral 64: New Mineral Names Abstr 242

Clark AM (1974) A tantalum-rich variety of sphene. Min Mag 39:605–607

Clark AM, Fejer EE (1978) Tapiolite, its chemistry and cell dimensions. Min Mag 42:477–480

Clark AM, Fejer EE, Donaldson DJ, Silver J (1976) The ^{119}Sn Mössbauer spectra, cell dimensions, and minor element contents of some cassiterites. Min Mag 40:895–898

Cohen AJ, Adekeye JID, Hapke B, Partlow DP (1985) Interstitial tin (2+) ion in synthetic and natural cassiterite crystals. Phys Chem Min 12:363–369

Correia-Neves JM, Lopes Nunes JE, Sahama ThG, Lehtinen M (1974) Bismuth and antimony minerals in the granite pegmatites of northern Mozambique. Rev Cienc Geol 7 Univ Lourenço Marques, pp 1–37

Dunn PJ, Gaines RV, Wolfe CW, Barbosa CdoP (1978) Epitaxial wodginite and cassiterite from Lavra Jabuti, Baixio, Galilea, Minas Gerais, Brazil. Min Rec 9:14–18

Dunn PJ, Gaines RV, Kristiansen R (1979) Mossite discredited. Min Mag 43:553–554

Dunn PJ, Peacor DR, Sturman DB, Ramik RA, Roberts WL, Nelen JA (1986) Johnwalkite, the Mn-analogue of olmsteadite, from South Dakota. N Jahrb Mineral Monatsh 1986:115–120

Eid AS, Knorring O von (1976) Geochemical aspects of the tantalum mineral microlite from some African pegmatites. Annu Rep Inst Afr Geol Univ Leeds 1976:56–58

Ercit TS (1986) The simpsonite paragenesis: the crystal chemistry and geochemistry of extreme tantalum fractionation. PhD Thesis, Univ Manitoba, Winnipeg

Ercit TS (1989) Structural classification of niobium and tantalum oxide minerals (in preparation)

Ercit TS, Černý P (1982) The paragenesis of simpsonite. Progr Abstr GAC-MAC Annu Meet Winnipeg 7:48

Ercit TS, Černý P (1988) The crystal structure of foordite. Can Mineral 26:901–906

Ercit TS, Černý P, Hawthorne FC (1984) Wodginite crystal chemistry. Geol Soc Am Annu Meet Progr Abstr 16:502

Ercit TS, Hawthorne FC, Černý P (1985a) The crystal structure of synthetic natrotantite. Bull Min 108:541–549

Ercit TS, Černý P, Hawthorne FC (1985b) Normal and inverse pyrochlore group minerals. GAC-MAC Abstr Progr 10:A17

Ercit TS, Černý P, Hawthorne FC (1986a) Parabariomicrolite, a new relative of the pyrochlore group from Brazil. GAC-MAC Annu Meet Abstr Progr 11:67

Ercit TS, Hawthorne FC, Černý P (1986b) Parabariomicrolite, a new species, and its structural relationship to the pyrochlore group. Can Mineral 24:655–663

Ercit TS, Hawthorne FC, Černý P (1989) The crystal structure of alumotantite: its relation to simpsonite and (Al,Ga) $(Ta,Nb)O_4$. Can Min 27 (in press)

Ewing RC (1974) Spherulitic recrystallization of metamict polycrase. Science 184:561–562

Erwing RC (1975) Alteration of metamict, rare-earth, AB_2O_6-type Nb-Ta-Ti oxides. Geochim Cosmochim Acta 39:521–530

Ewing RC (1976a) Metamict columbite re-examined. Min Mag 40:898–899

Ewing RC (1976b) A numerical approach toward the classification of complex, orthorhombic, rare-earth, AB_2O_6-type Nb-Ta-Ti oxides. Can Mineral 14:111–118

Ewing RC, Ehlmann AJ (1975) Annealing study of metamict, orthorhombic, rare-earth, AB_2O_6-type Nb-Ta-Ti oxides. Can Mineral 13:1–7

Ewing RC, Snetsinger KG, Bunch TE (1977) Euxenite from Ampangabé, Madagascar. Can Mineral 15:92–96

Ferguson RB, Hawthorne FC, Grice JD (1976) The crystal structures of tantalite, ixiolite and wodginite from Bernic Lake, Manitoba. II. Wodginite. Can Mineral 14:550–560
Ferreira KJ (1984) The mineralogy and geochemistry of the lower Tanco pegmatite, Bernic Lake, Manitoba, Canada. M Sc Thesis, Univ Manitoba, Winnipeg
Fleischer M (1977) Wolframo-ixiolite. In: New mineral names. Am Mineral 62:1262
Fleischer M (1984) Unnamed monoclinic dimorph of columbite (?). In: New mineral names. Am Mineral 69:213
Fleischer M (1987) Glossary of mineral species, 5th edn. Min Rec Tucson, 227 pp
Foord EE (1981) Bismuthian stibiocolumbite-tantalite and other minerals from the Little Three Mine, Ramona, California, 4th Biennial MSA. FM Symp, Tucson, Abstr, pp 7–8
Foord EE (1982a) Bismuthian stibiocolumbite-tantalite and other minerals from the Little Three mine, Ramona, California (Abstr). Am Mineral 67:181
Foord EE (1982b) Minerals of tin, titanium, niobium and tantalum in granitic pegmatites. In: Černý P (ed) Granitic pegmatites in research and industry. Min Assoc Can Short Course Handb 8:187–238
Foord EE, Mrose ME (1978) Rynersonite, $Ca(Ta,Nb)_2O_6$, a new mineral from San Diego County, California. Am Mineral 63:709–714
Foord EE, Sharp WN, Adams JW (1984) Zinc- and Y-group-bearing senaite from St. Peters Dome, and new data on senaite from Dattas, Minas Gerais, Brazil. Mineral Mag London 48:97–106
Foord EE, Taggart JE, Jr., Gaines RV, Grubb PLC, Kristiansen R (1986) Zimbabweite, a new alkali-lead-arsenic tantalate from St. Anns mine, Karoi District, Zimbabwe. Bull Min 109:331–336
Gatehouse BM, Grey IE, Kelly PR (1979) The crystal structure of davidite. Am Mineral 64:1010–1017
Ginsburg AI, Gorzhevskaya SA, Sidorenko GA, Ukhina TA (1969) Wolframoixiolite, a new variety of ixiolite. Zapiski Vses Min Obschtsch 98:63–73 (in Russian)
Gorzhevskaya SA, Sidorenko GA, Ginsburg AI (1974) Titano-tantalo-niobates (properties, compositional characteristics and conditions of formation). Nedra, Moscow, 343 pp (in Russian)
Graeser S, Schwander H, Hänni H (1979) Vigezzite, (Ca,Ce) $(Nb,Ta,Ti)_2O_6$, a new aeschynite-type mineral from the Alps. Min Mag 43:459–462
Graham J, Thornber MR (1974) The crystal chemistry of complex niobium and tantalum oxides II. Composition and structure of wodginite. Am Mineral 59:1040–1044
Groat LA, Carter RT, Hawthorne FC (1985) Tantalian niobian titanite from the Irgon Claim, southeastern Manitoba. Can Mineral 23:569–571
Groat LA, Černý P, Ercit TS (1987) Re-instatement of stibiomicrolite as a valid species. Geol Fören Stockholm Förh 109:105–109
Groult D, Pannetier J, Raveau B (1982) Neutron diffraction study of the defect pyrochlores $TaWO_{5.5}$, $H_2Ta_2O_6$, and $HTaWO_6.H_2O_2$. J Solid State Chem 41:277–285
Grubb PLC, Hannaford P (1966) Magnetism in cassiterite. Min Dep 2:148–171
Hawthorne FC (1983) Graphical enumeration of polyhedral clusters. Acta Crystallogr A39:724–736
Hawthorne FC (1985) Towards a structural classification of minerals: The $^{vi}M^{vi}T\phi_n$ minerals. Am Mineral 70:455–473
Hawthorne FC (1986) Structural hierarchy in $^{vi}M_x{}^{iii}T_y\phi_z$ minerals. Can Mineral 24:625–642
Heinrich EWm (1962) Radioactive columbite. Am Mineral 47:1363–1379
Hogarth DD (1977) Classification and nomenclature of the pyrochlore group. Am Mineral 62:403–410
Izoret L, Marnier G, Dusausoy Y (1985) Caractérisation cristallochimique de la cassitérite des gisements d'étain et de tunsténe de Galice, Espagne. Can Mineral 23:221–231
Jambor JL, Sabina AP, Roberts AC, Bonardi M, Ramik RA, Sturman BD (1984) Franconite, a new hydrated Na-Nb oxide mineral from Montreal Island, Quebec. Can Mineral 22:239–243
Jambor JL, Sabina AP, Roberts AC, Bonardi M, Owens DR, Sturman BD (1986) Hochelagaite, a new calcium-niobium oxide mineral from Montreal, Quebec. Can Mineral 24:449–453
Khvostova VA, Arkhangelskaya VN (1970) A find of the manganese variety of tapiolite. Dokl Acad Sci USSR 194:677–678 (in Russian)
Khvostova VA, Pavlova NV, Alexandrov VB, Maximova NV (1966) First find of wodginite in USSR. Dokl Acad Sci USSR 167:1135–1138 (in Russian)
Khvostova VA, Lebedeva SI, Maksimova NV (1983) Tin-bearing tantaloniobates and their typomorphic features. Int Geol Rev:922–932
Knorring O von (1960) Some geochemical aspects of a columbite-bearing soda granite from south-east Uganda. Nature (London) 188:204–206
Knorring O von (1974) On tantalum mineralization in some pegmatites of equatorial and southern Africa. 20th Annu Rep Res Inst Afr Geol Univ Leeds, pp 34–38

Knorring O von, Fadipe A (1981) On the mineralogy and geochemistry of niobium and tantalum in some granite pegmatites and alkali granites of Africa. Bull Min 104:496–507
Knorring O von, Hornung G (1963) Simpsonite and stibiotantalite from Benson pegmatite mine, Mtoko, southern Rhodesia. Mineral Mag London 33:458–466
Knorring O von, Sahama ThG (1979) A note on rynersonite from Uganda. Schweiz Mineral Petrol Mitt 59:15–18
Knorring O von, Sahama ThG, Saari E (1966) A note on the properties of manganotantalite. C R Soc Geol Finl 38:47–50
Knorring O von, Sahama ThG, Lehtinen M (1969a) Ferroan wodginite from Ankole, south-west Uganda. C R Soc Bull Geol Soc Finl 41:65–69
Knorring O von, Sahama ThG, Lehtinen M (1969b) Scandian ixiolite from Mozambique and Madagascar. C R Soc Bull Geol Soc Finl 41:74–77
Knorring O von, Vorma A, Nixon PH (1969c) Rankamaite, a new tantalum mineral from Kivu, Congo. Bull Geol Soc Finl 41:47–56
Kobayashi A, Kawai T, Sakurai K (1970) On ferrosamarskite from the Ishikawa District, Fukushima Prefecture, Japan. Geosci Mag 21:413–415 (in Japanese)
Komkov AI (1965) Crystal structure and chemical constitution of samarskite. Dokl Acad Sci USSR Earth Sci Sec 160:127–129
Komkov AI (1970) Relationship between the X-ray constants of columbites and composition. Dokl Acad SciUSSR 194:434–436 (in Russian)
Komkov AI (1973) Polymorphic relationships in compounds of $TRNbTiO_6$ and $TRTaTiO_6$ types. Sov Phys Dokl 17:959–960
Komkov AI, Belopolskyi MP (1966) Isomorphism in the system $YNbTiO_6$-$CeNbTiO_6$. Dokl Acad Sci USSR Earth Sci Sec 170:163–164
Konovalenko SI, Voloshin AV, Pakhomovskyi YaA, Rossovskyi LN, Divnev SA (1982) Tungsten-bearing varieties of tantalates-niobates from the miarolitic pegmatites of southwestern Pamir. Min Zh 4:65–74 (in Russian)
Kornetova VA, Sidorenko GA, Kazakova ME, Troneva NV, Boyarskaya RV (1978) On iron-rich wodginite. Trudy Min Mus Acad Sci USSR 27:75–85
Kuzmenko MV, Eskova EM (1968) Tantalum and niobium; genetic types of deposits and geochemistry. Nauka, Moscow, 338 pp
Lahti SI (1987) Zoning in columbite-tantalite crystals from the granitic pegmatites of the Eräjärvi area, southern Finland. Geochim Cosmochim Acta 51:509–517
Lahti S, Johanson B, Virkkunen M (1983) Contributions to the chemistry of tapiolite – manganotapiolite, a new mineral. Bull Geol Soc Finl 55:101–109
Lebedeva SI (ed) (1968) Physical properties (density and microhardness) of minerals of the ilmeno-rutile-strüverite group, and their dependence on composition. In: Physical properties of rare-metal minerals and methods of their examination. Nauka, Moscow, pp 3–11 (in Russian)
Lindquist K, Rehtijärvi P (1979) Pyrochlore from the Sokli carbonatite complex, northern Finland. Bull Geol Soc Finl 51:81–93
Lugovskoi GP, Stolyarova TI (1969) Strüverite from metasomatically altered granites. Trudy Min Mus Acad Sci USSR 19:179–183 (in Russian)
Lumpkin GR, Chakoumakos BC, Ewing RC (1986) Mineralogy and radiation effects of microlite from the Harding pegmatite, Taos County, New Mexico. Am Mineral 71:569–588
MacGregor AM (1946) Simpsonite and other tantalates from Bikita, southern Rhodesia. Min Mag 26:157–165
Maksimyuk IE, Lebedeva SI (1968) A study of physical properties of cassiterite aimed at clarification of the mode of tantalum and niobium incorporation. In: Lebedeva SI (ed) Physical properties of rare-element minerals and methods of their examination. Nauka, Moscow, pp 31–43 (in Russian)
Mihálik P (1967) A chemical and mineralogical study of microlite from the Noumas pegmatite. M Sc Thesis, Univ Witwatersrand
Mitchell RS, Zulkiewicz RJ (1970) Virginia metamict minerals: pyrochlore-microlite series. SE Geol 11:173–185
Moore PB (1982) Pegmatite minerals of P(V) and B(III). MAC Short Course Handb 8:267–290
Moore PB, Araki T, Kampf AR, Steele IM (1976) Olmsteadite, $K_2Fe_2^{2+}[Fe_2^{2+}(Nb,Ta)_2^{5+}O_4(H_2O)_4(PO_4)_4]$, a new species, its crystal structure and relation to vauxite and montgomeryite. Am Mineral 61:5–11
Mrose ME, Rose HJ, Jr. (1961) Behierite, $(Ta,Nb)BO_4$, a new mineral from Manjaka, Madagascar. Min Soc Am Progr Abstr 111A

Mücke A, Strunz H (1978) Petscheckite and liandratite, two new pegmatite minerals from Madagascar. Am Mineral 63:941–946
Nekrasov IYa, Nekrasova RA, Tsepin AI, Sipavina LV, Remez VK (1982) Thoreaulite from rare-metal pegmatites of Siberia. Min Zh 4:11–20 (in Russian)
Nickel EH (1964) Latrappite – a proposed new name for the perovskite-type calcium niobate mineral from the Oka area of Quebec. Can Mineral 8:121–122
Nickel EH, Robinson BW (1985) Kimrobinsonite, a new tantalum mineral from western Australia, and its association with cesstibtantite. Can Mineral 23:573–576
Nickel EH, Rowland JF, McAdam RC (1963a) Wodginite – a new tin-manganese tantalate from Wodgina, Australia, and Bernic Lake, Manitoba. Can Mineral 7:390–402
Nickel EH, Rowland JF, McAdam RC (1963b) Ixiolite – a columbite substructure. Am Mineral 48:961–979
Ovchinnikov LN, Solodov NA (eds) (1980) Deposits of lithophile rare metals. Nedra, Moscow, 559 pp (in Russian)
Paul BJ, Černý P, Chapman R (1981) Niobian titanite from the Huron Claim pegmatite, southeastern Manitoba. Can Mineral 19:549–552
Petruk W, Owens DR (1975) Electron microprobe analyses for pyrochlores from Oka, Quebec. Can Mineral 13:282–285
Pryce MW (1971) Holtite: a new mineral allied to dumortierite. Min Mag 38:21–25
Pudovkina ZV, Dubakina LS, Lebedeva SI, Pyatenko YuA (1974) Study of Brazilian zirkelite. Zapiski Vses Min Obsch 103:368–372 (in Russian)
Ramdohr P (1961) Magnetische Cassiterite. C R Soc Geol Finl 33:473–480
Roth RS (1957) Classification of perovskite and other ABO_3-type compounds. J Res Natl Bur Stn 58:45–58
Ruck R, Babkine J, Nguyen C, Marnier G, Dusausoy Y (1986) Geochemical association of Fe and Nb in synthetic and natural cassiterites and rutiles. Int Symp Exp Min Geochem, Nancy, Abstr, pp 122–123
Rudakova, ZhN (1972) On the quantitative relationships of some elements in the structure of cassiterite. Zapiski Vses Min Obshtch 101:317–322
Saari E, Knorring O von, Sahama ThG (1968) Niobian wolframite from the Nuaparra pegmatite, Zambezia, Mozambique. Lithos 1:64–68
Sahama ThG (1978) Niobian rutile from Muiane, Mozambique. J Min Recife 7:115–118
Semenov EI (1957) Oxides and hydroxides of titanium and niobium in the Lovozero alkalic massif. Trudy IMGRE 1:41–59 (in Russian)
Stern LA, Brown GE, Jr., Bird KD, Jahns RH, Foord EE, Shigley JE, Spaulding LBJr (1986) Mineralogy and geochemical evolution of the Little Three pegmatite-aplite layered intrusive, Ramona, California. Am Mineral 71:406–427
Sturdivant JR (1930) The crystal structure of columbite. Zeits Krist 75:88–108
Sugitani Y, Suzuki Y, Nagashima K (1985) Polymorphism of samarskite and its relationship to other structurally related Nb-Ta oxides with the α-PbO structure. Am Mineral 70:856–866
Sveshnikova EV, Zhabin AG, Yakovlevskaya TA, Aleksandrov VB (1965) On titanium-bearing columbite from alkaline massives. Trudy Min Mus Acad Sci USSR 16:266–270
Taylor RG (1979) Geology of tin deposits. Dev Econ Geol 11. Elsevier, Amsterdam
Thoreau J (1949) Les phases de la pegmatite de Manono, Katanga. Bull Acad R Belg, Cl Sci 35:1008–1015
Turnock AC (1966) Synthetic wodginite, tapiolite, and tantalite. Can Mineral 8:461–470
van d Veen AH (1963) A study of pyrochlore. Nederl Geol Min Gen Verh Geol Ser 22:188
Vlasov KA (ed) (1964) Geochemistry and mineralogy of rare elements and genetic types of their deposits, vol 2: Mineralogy of rare elements. Moscow. Isr Progr Sci Transl, Jerusalem 1966, 945 pp
Voloshin AV, Pakhomovskyi YaA (1982) Study of chemical composition and heterogeneity of microlite. In: Balkov IV, Voloshin AV (eds) Minerals and mineral parageneses of rocks of the Kola Peninsula. Kola Br Acad Sci USSR, pp 3–9 (in Russian)
Voloshin AV, Gordiyenko VV, Gelman EM, Zorina ML, Elina MA, Kulchitskaya EA, Menshikov YuP, Polezhaeva LI, Ryzhova RI, Sokolov PB, Utochkina GI (1977) Holtite (first find in the USSR) and its relationships to other tantalum minerals in rare-element pegmatites. Zap Vses Min Obschtsch 106:337–346 (in Russian)
Voloshin AV, Menshikov YuP, Pakhomovskyi YaA (1981a) Alumotantite and natrotantite – new minerals of tantalum in granitic pegmatites. Zap Vses Min Obschtsch 110:338–345 (in Russian)

Voloshin AV, Menshikov YuP, Pakhomovskyi YaA, Polezhayeva LI (1981b) Cesstibtantite (Cs,Na) $SbTa_4O_{12}$ – new mineral from granitic pegmatites. Zap Vses Min Obschtsch 110:345–351 (in Russian)

Voloshin AV, Menshikov YuP, Pakhomovskyi YaA (1982a) Sosedkoite $(K,Na)_5Al_2(Ta,Nb,Sb)_{22}O_{60}$ – new mineral from granitic pegmatites. Dokl Acad Sci USSR 264:442–445 (in Russian)

Voloshin AV, Pakhomovskyi YaA, Tyusheva FN (1982b) Calciotantite $CaTa_4O_{11}$ – new mineral from granitic pegmatites of Kola Peninsula. Min Zh 4:75–79 (in Russian)

Voloshin AV, Pakhomovskyi YaA, Stepanov VI, Tyusheva FN (1983a) Lithiotantite $Li(Ta,Nb)_3O_8$ – new mineral from granitic pegmatites of eastern Kazakhstan. Min Zh 5:91–95 (in Russian)

Voloshin AV, Pakhomovskyi YaA, Perlina GA (1983b) Tantite Ta_2O_5 – new mineral from granitic pegmatites of Kola Peninsula. Min Zh 5:90–93 (in Russian)

Voloshin AV, Pakhomovskyi YaA, Stepanov VI, Tyusheva FN (1983c) Natrobistantite, (Na,Cs) $Bi(Ta,Nb,Sb)_4O_{12}$ – new mineral from granitic pegmatites. Min Zh 5:82–86 (in Russian)

Voloshin AV, Pakhomovskyi YaA, Bulgak LV, Perlina GA (1985a) Ungursaite – a new tantalate of calcium and sodium from granitic pegmatites. Min Zh 7:88–94 (in Russian)

Voloshin AV, Pakhomovskyi YaA, Bulgak LV, Perlina GA (1985b) Irtyshite – a new mineral from granitic pegmatites. Min Zh 7:83–87 (in Russian)

Voronina LB, Gaydukova VS, Dobrovolskaya NV, Korovushkin VV (1979) Forms of occurrence of iron in cassiterite. Geochem Int 10:128–143

Wambeke L van (1970) The alteration processes of the complex titano-niobotantalates and their consequences. Neues Jahrb Min Abh 112:117–149

Wang X-J, Zou T-R, Xu J-G, Yu X-Y, Qiu Y-Zh (1981) Study of pegmatite minerals of the Altai region. Sci Publ House, Beijing (in Chinese)

Weitzel H (1976) Kristallstrukturverfeinerung von Wolframiten und Columbiten. Zeits Krist 144:238–258

Williams FA, Meehan JA, Paulo KL, John TU, Rushton HG (1956) Economic geology of the decomposed columbite-bearing granites, Jos Plateau, Nígeria. Econ Geol 51:303–332

Wise MA (1987) Geochemistry and crystal chemistry of Nb,Ta and Sn minerals from the Yellowknife pegmatite field, N.W.T. Unpubl Ph D Thesis, Univ Manitoba, Winnipeg, 368 pp

Wise MA, Černý P (1986) The status of ixiolite. Int Min Assoc XIV. Meet Stanford Progr Abstr 14:265

Wise MA, Turnock AC, Černý P (1985) Improved unit cell dimensions for ordered columbite-tantalite end-members. N Jahrb Mineral Monatsh 1985:372–378

Zayachkivsky B (1985) Granitoids and rare-element pegmatites of the Georgia Lake area, northwestern Ontario. Unpubl M Sc Thesis, Lakehead Univ, 234 pp

Zdorik TB, Sidorenko GA, Bykova AV (1961) A new calcium titanozirconate-calzirtite. Dokl Acad Nauk USSR 137:681–684 (in Russian)

Zhang R-B, Tian H-X, Peng Z-H, Ma Z-S, Han F-M, Jing Z-B (1980) A new mineral – ashanite $(Nb,Ta,U,Fe,Mn)_4O_8$. Kexue Tongbao 25:510–514

Note added in press: Continuing examination of Ta-rich wodginites has revealed the presence of significant quantities of Li, suggestive of a $Li_4Ta_{12}O_{32}$ component known as a synthetic wodginite-type phase (Gatehouse BM, Leverett P 1972, Lithium triniobate (V), $LiNb_3O_8$; Cryst Struct Comm 1:83–86). This development will require modification of the wodginite systematics presented here.

Pyrochlore Group Minerals from the Qaqarssuk Carbonatite Complex

C. KNUDSEN

Abstract

The pyrochlore group minerals (PyGM) found in the Qaqarssuk carbonatite complex are located either in late carbonatite or in glimmerite and fenite adjacent to late stage lanthanide-rich carbonatite, often together with apatite, magnetite and zircon. No PyGM are found in carbonatites of the main intrusive phase. This indicates that Nb and Ta have been concentrated in, and precipitated from, the residual fluids.

Elements like Nb, Ta, U, Th and lanthanides are likely to have been soluble in the carbonatite magma as complexes with ligands like PO_4^{3-}, F^- and CO_3^{2-} possibly also influenced by Na^+ and K^+. The PyGM were precipitated where there was a decrease in concentration of the complexing ligands (1) simultaneously with precipitation of apatite, (2) in contact with or in the metasomatized wall rock (brown glimmerite and fenite), where there has been a high gradient of the complexing ligands and loss of alkalis during the metasomatic alteration of the wall rock.

The PyGM in the Qaqarssuk carbonatite complex show a wide compositional variation. Pyrochlore with a composition close to the end member $NaCaNb_2O_6$ (O,OH,F) is the most common PyGM species and species like bariopyrochlore, uranpyrochlore and betafite are present. A new mineral species, a Ba-betafite analogue, with Ba as the dominating A-cation and Ba exceeding 20% of total A-cations, has been found.

The composition of the PyGM varies with the rock type in which the minerals are found. U- and Ta-rich PyGM are found in fenite, PyGM with high Th and Ce are found in glimmerite and "pure" pyrochlore or zoned pyrochlore with a "pure" rim is found in carbonatite. The compositional variation in the PyGM with rock type is interpreted as precipitation of the most mobile elements in the most distal environment, the most mobile being those with the largest capacity for retaining the anions with which they form soluble complexes. With decreasing concentration of e.g. PO_4^{3-}, F^- and CO_3^{2-} away from the intruding carbonatite, the cations as U, Th and lanthanides will be precipitated in PyGM in a sequence of increasing field strength.

A similar zonation can be observed in compositionally zoned pyrochlore grains with a U- and Ta-rich core, a Th-, Ce- or Ba-rich intermediate zone and a pure pyrochlore rim. The zoning of PyGM grains is interpreted as the result of a more or

Geological Survey of Greenland, Østervoldgade 10, DK-1350 København K

Lanthanides, Tantalum and Niobium
P. Möller, P. Černý and F. Saupé (Eds.)

less continuous precipitation of pyrochlore with a composition in equilibrium with a fluid with increasing concentration of the complexing ligands, during the intrusion of carbonatite.

1 Introduction

Pyrochlore is the primary and most important Nb mineral in carbonatites (Deans 1966). A general formula $A_{2-m}B_2O_6(O,OH,F)_{1-n}\cdot pH_2O$ has been established, and a classification and nomenclature recommended (Hogarth 1977) (Table 1). The most common elements in the A position are Na and Ca, in the B position Nb, and most pyrochlore has a composition close to $NaCaNb_2O_6(OH,F)$. Pyrochlore containing appreciable amounts of elements like U, Th, REE, Ba, K in the A position and Ta, Ti, Fe^{III}, Zr in the B position is less common. This type of PyGM is often located in metasomatized wall rock (Kapustin 1980). The substitution of Na and Ca by elements with larger size or higher charge in the A position generally leads to site vacancies (m in the general formula), either primary due to substitution during initial growth of the mineral or secondary due to alteration processes (Borodin and Nazarenko 1959). The number of atoms in the B position is very constant, whereas the amount of water in the mineral varies with the amount of A-site vacancies (Borodin and Nazarenko 1959).

The Qaqarssuk carbonatite complex is situated 60 km east of Sukkertoppen, West Greenland. The complex was found by Kryolitselskabet Øresund A/S in 1962, and the company worked in the area until 1976 mainly with Nb exploration, but no economic deposits were located. The complex was mapped by the author during the summers of 1984 and 1985, as a part of a phosphate exploration programme (Knudsen 1985, 1986). The aim of the present work is to study the geological environment in which pyrochlore is deposited, and its compositional variation with geological environment.

2 Geological Setting

The Qaqarssuk carbonatite complex was intruded into the Archaean basement 173 m.y. ago (Larsen et al. 1983) and covers an area of 15 km^2 (Fig. 1). The complex consists mainly of calcite-dominated carbonatite (søvite) and dolomite-dominated carbonatite (rauhaugite) ring dykes. These ring dykes dip outwards, with an increasing dip, decreasing grain size and increasing iron-content in the carbonate, from the core of the complex to the margin, where a fine-grained ferrocarbonatite is found. The carbonatite is surrounded by metasomatically altered basement rocks, mainly Na-fenites and ultramafic rocks consisting of phlogopite, hornblende, diopside and calcite. No pyrochlore has been found in søvite and rauhaugite belonging to the main phase of intrusion, and the concentration of Nb varies between 2 to 70 ppm (mean of 61 samples is 26 ppm) and 4 to 150 ppm (mean of 15

Table 1. See page 81 for declaration

Subgroups defined by B atoms viz Nb, Ta, Ti			Pyrochlore $Nb+Ta>2Ti$; $Nb>Ti$	Microlite $Nb+Ta>2Ti$; $Ta>Nb$	Betafite $2Ti>Nb+Ta$
	Na + Ca, but no other A-atoms > 20% of total A-atoms		**pyrochlore**	microlite	
Species defined by A-atoms viz K, Sn, Ba, REE, Pb, Bi, U	One or more A-atoms, other than Na and Ca, > 20% total A-atoms.	K	kalipyrochlore		
		Sn		stannomicrolite	
	Species named by most abundant A-atom other than Na or Ca.	Ba	**bariopyrochlore**	bariomicrolite	**n.n.**
		REE	yttropyrochlore ceriopyrochlore		yttrobetafite
		Pb	plumbopyrochlore	plumbomicrolite	plumbobetafite
		Bi		bismutomicrolite	
		U	**uranpyrochlore**	uranmicrolite	**betafite**

After Hogarth (1977)
bold face = minerals found in the Qaqarssuk carbonatite complex

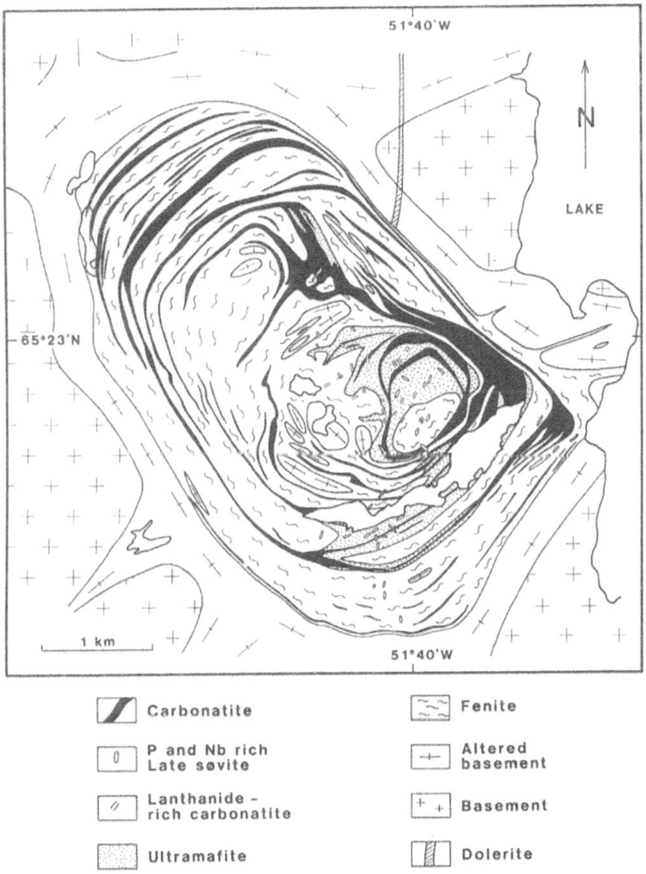

Fig. 1. Geological map of the Qaqarssuk carbonatite complex

samples is 50 ppm) respectively. The søvite and rauhaugite ring dykes are cut by a late generation of coarse søvite veins, lanthanide-rich carbonatite veins, beforsite dykes and lamprophyre dykes.

The PyGM is generally found in, or associated with, these late carbonatites, mainly located in the central parts of the complex. The pyrochlore-bearing rock types are:

a) Late stage søvite;
b) Fine-grained ferrocarbonatite;
c) Brown glimmerite;
d) Fenite.

3 The Pyrochlore-Bearing Rock Types

3.1 Late Stage Søvite

The 0.1- to 10-m-wide late stage søvite veins are characterized by large platy calcite crystals rich in deformation lamellae. Other minerals are phlogopite, alkali amphibole, apatite and magnetite. The Nb content in the late stage søvite varies between 5 and 12 000 ppm (mean of 39 samples is 871 ppm). PyGM is often found in, or adjacent to schlieren of acicular apatite, alkali amphibole and magnetite. These schlieren are locally accumulated to a magnetite-apatite-alkali amphibole-pyrochlore rock, forming up to 10-m-wide sheets with up to 15 wt.% Nb_2O_5, but typically 1 wt.% Nb_2O_5. About 60% of the pyrochlore yet located in the complex is found in late stage søvite or magnetite-apatite-alkali amphibole-pyrochlore rock associated with the late stage søvite.

The pyrochlore in the late stage søvite generally occurs as euhedral, pale brown, transparent crystals often with radiating amphibole or apatite inclusions (Fig. 2a).

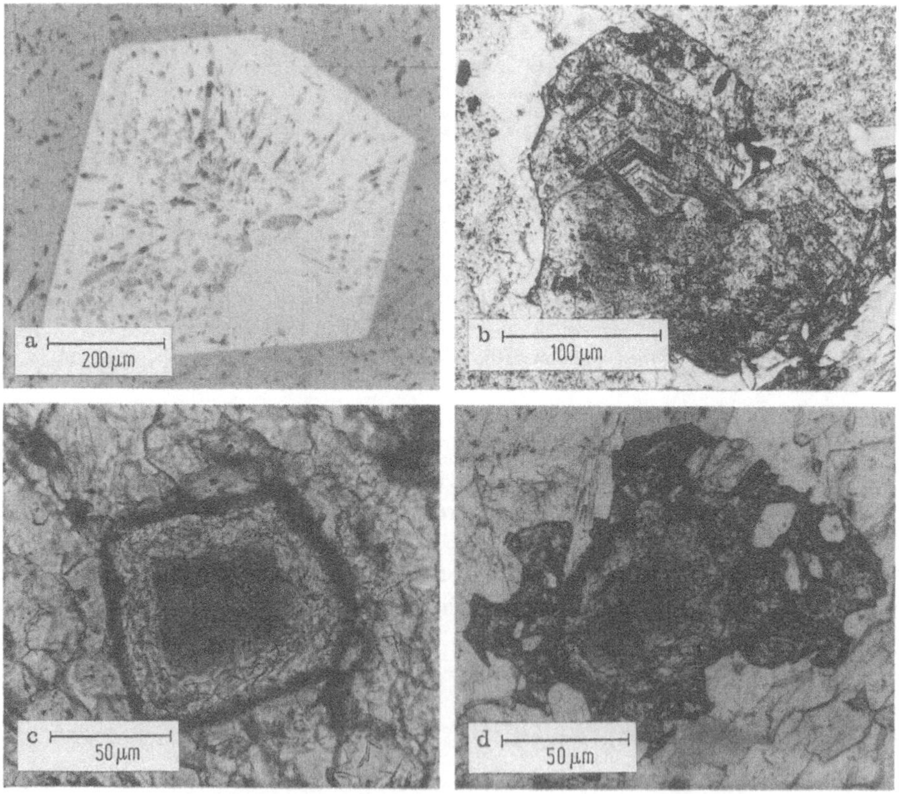

Fig. 2. Microphotographs of pyrochlore. **a** Euhedral pyrochlore with radiating richterite crystals and zones rich in inclusions. Reflected light; GGU 249831; late søvite. **b** Zoned pyrochlore overgrown by unzoned anhedral pyrochlore. Plane polarized light; GGU 320023; glimmerite. **c** Metamict betafite with an orange-yellow core and light yellow rim. Plane polarized light; GGU 266091; ferrocarbonatite. **d** Metamict orange to yellow betafite overgrown by light brown, anhedral pyrochlore. Plane polarized light; GGU 320407; late søvite

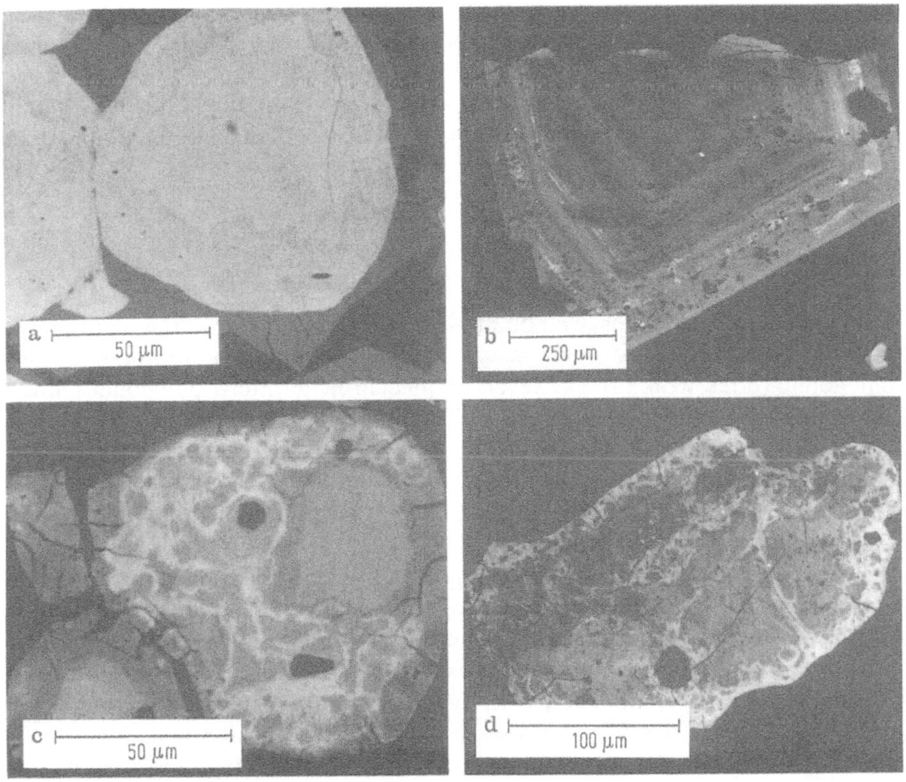

Fig. 3. Electron microprobe backscatter images. **a** Zoned pyrochlore, GGU 252509; late søvite. **b** Zoned pyrochlore, GGU 320154; glimmerite. **c** Zoned pyrochlore, GGU 266078; vein in fenite. **d** Zoned and veined pyrochlore, GGU 320056; disseminated in fenite

Zoning defined by varying amounts of inclusions, as well as zoning defined by alternating opaque and pale brown transparent pyrochlore, is seen. The grain size varies from 0.1 to 5 mm. Locally the pyrochlore contains a metamict, an orange-brown core (Fig. 2d), a transitional yellow zone and a colourless to light brown euhedral rim locally rich in inclusions. The core is homogeneous and has a jagged contact to the surrounding intermediate zone. The intermediate zone has a complex zonation (Fig. 3a), often cracks and the intermediate zone is slightly darker than the core on electron backscatter image. The contact of the intermediate zone to the surrounding pyrochlore is rounded. The euhedral pyrochlore rim is dark on the backscatter image. (Fig. 3a).

3.2 Ferrocarbonatite

The pyrochlore-bearing ferrocarbonatite occurs as 5- to 100-cm-wide sheets or veins in the marginal part of the complex. This ferrocarbonatite is a fine-grained rock consisting of iron-rich dolomite or ankerite, partly digested fenite inclusions, mica and acicular apatite. Pyrochlore in the fine-grained ferrocarbonatite occurs as fine-to medium-grained, euhedral to subhedral, homogeneous or zoned crystals, or metamict grains (Fig. 2c). The core is homogeneous and is surrounded by a complexly zoned intermediate part and locally a homogeneous rim. Concentrations of Nb vary between 100 and 1000 ppm (mean of 18 samples is 580 ppm) and the total amount of pyrochlore in ferrocarbonatite is estimated to be less than 5% of that found in the complex.

3.3 Brown Glimmerite

The pyrochlore-bearing brown glimmerite occurs adjacent to lanthanide-rich carbonatite veins. The brown glimmerite consists of medium- to coarse-grained phlogopite (often with a rim of reversely pleochroic tetraferri-phlogopite), minor alkali amphibole and interstitial calcite. The pyrochlore often has a pleochroic halo in the surrounding mica. The pyrochlore often occurs together with apatite, magnetite and zircon. Where the glimmerite is cut by minor lanthanide-rich carbonatite veins, the pyrochlore is often located at the carbonatite-glimmerite interface together with apatite and zircon.

The pyrochlore in the glimmerite is generally more fine-grained (0.05–0.5 mm) and slightly darker brown than pyrochlore in the late stage søvite. The pyrochlore is euhedral with a zonation reflecting euhedral growth (Figs. 2b and 3b). The Nb content in brown glimmerite varies between 50 and 20 000 ppm (mean of 22 samples is 2366 ppm). The pyrochlore-rich glimmerite occurs as up to 10-m-wide zones, which typically contain 0.8 wt.% Nb_2O_5, and the pyrochlore in the glimmerite accounts for approximately 35% of the pyrochlore observed within the complex.

3.4 Fenite

The pyrochlore-bearing fenite is a medium- to coarse-grained rock consisting mainly of albite, hornblende and aegirine augite, with interstitial carbonate or carbonate veinlets.

The pyrochlore may either be found in thin (1–5 mm) veinlets in the fenite or disseminated in the fenite adjacent to thin, lanthanide-rich carbonatite veinlets. The veins consist of pyrochlore, anhedral apatite, calcite, lanthanide carbonates and phlogopite. The pyrochlore is seen as fine-grained (0.1–0.5 mm), metamict, rounded, zoned brown to yellow grains. The pyrochlore has a homogeneous core with a jagged outline (Fig. 3c) surrounded by a complexly zoned rim with cracks post-dating the zonation. The pyrochlore disseminated in the fenite occurs together with anhedral apatite and locally an unidentified lanthanide silicate. This pyrochlore is seen as rounded, metamict, zoned grains with a brownish-orange core and yellow rim, and often with a pleochroic halo in surrounding mica or amphibole. On the backscatter image (Fig. 3d), a bright rim and bright veins can be seen as well as cracks post-dating these features. The Nb content in fenite varies from 16 to 5000 ppm (mean of 32 samples is 343 ppm). The Nb_2O_5 concentration may be as high as 10 wt.% in mineralized veinlets, but the total amount of pyrochlore in the fenite is considered insignificant.

4 Mineral Chemistry

4.1 Experimental Technique

The pyrochlore was analyzed with a JEOL 733 Superprobe at the Institute of Mineralogy, University of Copenhagen, using four wavelength dispersive X-ray spectrometers (15 kV, 30 nA, focussed beam) and link system energy dispersive equipment (15 kV, 1 nA, focussed beam, 200-s lifetime). Raw data were ZAF corrected on line. The results were calibrated relative to natural and synthetic standards and spectral interference was computed from the analyses of the standards. Good reproducibility between WDX and EDX was achieved, except for SiO_2, and the EDX analyses for SiO_2 were calibrated versus WDX analyses of pyrochlore.

The compositions of typical pyrochlores from the Qaqarssuk carbonatite complex are listed in Table 2. In the calculation of atomic contents, the B-cations are assumed to be constant (=2) (Borodin and Nazarenko 1959). The compositional variation in zoned grains is illustrated in Figs. 4, 5, 6 and 7. The fact that the sum of the A-cations locally exceeds 2, may be caused by Si in the B position, and not the A position assumed in the calculation of the formula.

4.2 Substitutions for Na and Ca

4.2.1 Late Stage Søvite

Most of the pyrochlore in the late stage søvite has a composition close to the end member $NaCaNb_2O_6(OH,F)$ corresponding to the composition of the rim of zoned grains. In zoned grains (GGU 252509 and 320407 in Fig. 4) the homogeneous core has a high UO_2 content of about 20 wt.%, Na_2O is about 6 wt.%, CaO is about 10 wt.% and contents of Th, lanthanides, Ba, Sr and Si are relatively low. In the intermediate zone, the U content is constant or decreasing, Na is very low and CaO decreases to 2 to 8 wt.%. When large ions such as Ba^{2+} or multivalent ions such as U^{4+} are introduced into the pyrochlore, Na is lost in preference to Ca (Fig. 4). Sometimes U accounts for more than 20% of the total A-cations and the PyMG must then be termed uranpyrochlore. The amount of vacancies in U-rich pyrochlore is low (Table 2, 252509-core and 320407-core) compared to what could be expected (Borodin and Nazarenko 1959). The content of Ba, Sr, Th and lanthanides increases in the intermediate zone as well as the content of water and possibly F (low sum). The introduction of large divalent ions such as Ba and Sr creates A-site vacancies (Table 2). Locally the number of Ba cations exceeds the number of U cations in the A position and constitutes more than 20% of the total A-cations; the mineral is then bariopyrochlore. In GGU 320407 no sharp boundary between the core and the intermediate zone can be seen, and the continuous decrease in U and Ta content indicates a more continuous precipitation history than e.g. in GGU 252509. There is a slight increase in Si content in the intermediate zone. In the rim, elements like U, Ba and Th disappear, the Na_2O content increases to 8 wt.% and CaO to 16 wt.%, and the mineral approaches pure pyrochlore in composition.

Table 2. Representative pyrochlore analyses (Basis of calculation: $A_{2-x}\ B_2O_6(OH,F)$, B tot. = 2.00)

	320410[a]	320438[1]	320543[a]	320516[b]	252509[c] core	252509[c] interm.	252509[c] rim	320407[c] core	320407[c] interm.	320407[c] rim	320488[b] core	320488[b] rim
Nb_2O_5	64.15	65.80	64.80	46.09	34.27	30.07	67.14	37.24	45.12	66.87	51.93	63.62
Ta_2O_5	0.02	0.03	0.28	7.86	20.47	23.07	0.67	12.99	8.80	0.41	3.36	0.32
TiO_2	3.69	3.51	3.42	5.98	3.94	3.88	4.03	9.26	8.08	3.90	4.08	2.91
ZrO_2	0.07	0.17	0.20	0.74	0.41	0.46	0.85	0.07	b.d.	0.26	0.29	b.d.
Fe_2O_3	0.49	0.08	0.06	0.18	1.78	1.72	b.d	0.25	1.27	0.10	2.26	0.10
Na_2O	8.19	8.03	7.08	6.92	6.70	0.32	7.57	6.16	0.57	8.53	2.00	7.95
CaO	14.31	15.82	12.26	10.47	8.08	5.56	16.16	9.92	5.37	15.21	6.95	13.11
UO_2	0.35	0.54	0.10	15.68	20.86	20.32	b.d.	17.65	5.71	b.d.	2.81	0.31
ThO_2	0.24	0.20	0.10	0.37	0.02	0.07	0.20	0.89	3.70	0.24	0.26	0.18
La_2O_3	0.37	0.14	0.25	0.04	0.43	0.44	0.20	0.43	0.77	0.19	0.35	0.25
Ce_2O_3	1.41	0.64	0.82	0.26	0.09	0.73	0.45	1.17	2.67	0.85	1.24	0.39
Nd_2O_3	0.35	0.15	0.11	0.17	0.13	0.02	0.10	0.11	0.67	b.d.	0.13	0.04
SiO_2	2.49	0.42	0.35	0.55	0.43	0.81	b.d	0.65	1.48	0.61	7.21	0.29
BaO	n.a.	n.a.	n.a.	n.a.	0.36	2.05	0.20	b.d.	2.64	b.d.	n.a.	n.a.
SrO	n.a.	n.a.	n.a.	n.a.	0.54	1.16	1.29	0.30	b.d.	0.65	n.a.	n.a.
Total	96.13	95.53	89.83	95.31	98.51	90.68	98.68	97.09	86.85	97.82	82.87	89.47
Atomic contents												
B – cations:												
Nb	1.802	1.828	1.826	1.490	1.212	1.119	1.787	1.222	1.368	1.806	1.603	1.849
Ta		0.001	0.005	0.153	0.436	0.516	0.011	0.256	0.160	0.007	0.062	0.006
Ti	0.172	0.162	0.160	0.322	0.232	0.240	0.178	0.505	0.407	0.175	0.209	0.141
Zr	0.002	0.005	0.006	0.026	0.016	0.018	0.024	0.002		0.008	0.010	
Fe	0.023	0.004	0.003	0.010	0.105	0.106		0.014	0.064	0.004	0.116	0.005
Tot.	2.000	2.000	2.000	2.000	2.000	2.000	2.000	2.000	2.000	2.000	2.000	2.000
A – cations:												
Na	0.987	0.957	0.856	0.959	1.016	0.051	0.864	0.867	0.074	0.988	0.265	0.991
Ca	0.953	1.042	0.819	0.802	0.677	0.490	1.019	0.771	0.386	0.974	0.508	0.903
U	0.005	0.007	0.001	0.250	0.363	0.372		0.285	0.085		0.043	0.004

Th	0.003	0.003	0.001	0.006		0.001	0.003	0.015	0.056	0.003	0.004	0.003
La	0.008	0.003	0.006	0.001	0.012	0.013	0.004	0.012	0.019	0.004	0.009	0.006
Ce	0.032	0.014	0.019	0.007	0.003	0.022	0.010	0.031	0.066	0.019	0.031	0.009
Nd	0.008	0.003	0.002	0.004	0.004	0.001	0.002	0.003	0.016		0.003	0.001
Si	0.155	0.026	0.022	0.039	0.034	0.067		0.047	0.099	0.036	0.492	0.019
Ba					0.011	0.066			0.069			
Sr					0.025	0.055	0.044	0.013		0.023		
Tot.	2.151	2.056	1.726	2.069	2.145	1.139	1.946	2.043	0.871	2.047	1.355	1.936
O_6	6.000	6.000	6.000	6.000	6.000	6.000	6.000	6.000	6.000	6.000	6.000	6.000
O in (OH,F)	1.041	0.968	0.640	0.885	0.917	−0.097	0.845	0.776	−0.397	0.951	0.129	0.860
A/B	1.075	1.028	0.863	1.034	1.073	0.569	0.973	1.022	0.435	1.023	0.677	0.968
Name	Pyrochlore (pc.)	pc.	pc.	pc.	pc.	Uran-pyrochlore	pc.	pc.	pc.	pc.	pc.	pc.
	266091[c] core	266091[c] rim	320159[a]	320154[a]	320023[c]	320113[b]	266078[c] core	266078[c] interm.	266078[c] rim	320056[c] core	320056[c] interm.	320056[c] rim
Nb_2O_5	42.51	47.19	69.13	65.11	68.95	62.84	52.48	49.56	49.41	31.42	28.67	28.24
Ta_2O_5	9.00	4.55	0.23	0.40	0.53	1.30	10.19	8.30	10.18	9.46	9.47	8.18
TiO_2	5.25	3.89	3.94	3.36	2.00	3.11	3.55	4.10	3.30	11.26	10.35	12.24
ZrO_2	1.03	0.32	0.29	0.77	0.60	b.d.	n.a.	n.a.	n.a.	n.a.	n.a.	n.a.
Fe_2O_3	0.32	1.00	0.29	0.22	0.13	0.22	1.06	0.58	0.69	0.92	1.18	1.01
Na_2O	6.03	2.73	8.00	7.26	8.49	8.11	6.49	0.24	1.09	2.01	0.54	0.46
CaO	10.40	6.15	16.91	12.35	10.44	13.42	12.33	2.80	11.44	11.42	7.15	1.24
UO_2	13.40	5.34	0.13	0.39	0.63	1.51	8.60	7.84	9.01	22.37	22.50	20.20
ThO_2	0.34	0.27	0.06	0.65	2.14	2.93	0.09	0.44	0.67	0.25	0.29	0.30
La_2O_3	b.d.	b.d.	0.04	0.13	0.68	0.57	0.37	b.d.	0.40	b.d.	0.08	b.d.
Ce_2O_3	0.26	0.03	0.31	0.65	2.13	1.80	0.50	0.64	0.75	1.19	0.73	1.55
Nd_2O_3	b.d.	0.26	0.09	0.18	0.45	0.59	0.33	0.39	0.30	b.d.	0.10	0.45
SiO_2	0.43	2.08	0.25	0.33	0.38	0.19	0.51	0.77	0.66	0.78	0.96	1.18
BaO	b.d.	9.39	n.a.	n.a.	b.d.	n.a.	0.29	14.56	0.39	0.04	7.30	13.49
SrO	0.57	1.91	n.a.	n.a.	0.65	n.a.	0.40	1.54	0.20	0.01	0.76	0.62
Total	89.54	85.11	99.67	91.80	98.20	96.59	97.19	91.76	88.49	91.13	90.08	89.16

Table 2. *Cont.*

	320410[a]	320438[1]	320543[a]	320516[b]	252509[c] core	252509[c] interm.	252509[c] rim	320407[c] core	320407[c] interm.	320407[c]	320488[b]	320488[b]
Atomic contents:												
B – cations:												
Nb	1.458	1.616		1.805	1.877	1.817	1.584	1.590	1.590	1.095	1.071	1.023
Ta	0.186	0.094		0.007	0.009	0.023	0.185	0.160	0.197	0.198	0.213	0.178
Ti	0.300	0.222		0.155	0.091	0.150	0.178	0.219	0.177	0.653	0.643	0.738
Zr	0.038	0.012		0.023	0.018							
Fe	0.018	0.057		0.010	0.006	0.011	0.053	0.031	0.037	0.053	0.073	0.061
Tot.	2.000	2.000		2.000	2.000	2.000	2.000	2.000	2.000	2.000	2.000	2.000
A. – cations:												
Na	0.887	0.401		0.863	0.991	1.006	0.840	0.033	0.150	0.301	0.087	0.071
Ca	0.846	0.499		0.811	0.674	0.920	0.882	0.213	0.872	0.944	0.633	0.106
U	0.226	0.090		0.005	0.008	0.021	0.128	0.124	0.143	0.384	0.414	0.360
Th	0.006	0.005		0.009	0.029	0.043	0.001	0.007	0.011	0.004	0.005	0.005
La				0.003	0.015	0.013	0.009		0.010		0.002	
Ce	0.007	0.001		0.015	0.047	0.042	0.012	0.017	0.020	0.034	0.022	0.045
Nd		0.007		0.004	0.01	0.013	0.008	0.010	0.008		0.003	0.013
Si	0.033	0.158		0.020	0.023	0.012	0.034	0.055	0.047	0.060	0.079	0.095
Ba		0.279					0.008	0.405	0.011	0.001	0.236	0.424
Sr	0.025	0.084			0.023		0.015	0.063	0.008		0.036	0.029
Tot.	2.03	1.523		1.731	1.820	2.071	1.937	0.926	1.280	1.728	1.518	1.149
O_6	6.000	6.000		6.000	6.000	6.000	6.000	6.000	6.000	6.000	6.000	6.000
O in (OH,F)	0.843	0.349		0.632	0.760	0.986	0.795	-0.214	0.155	0.348	0.123	-0.281
A/B	1.015	0.761		0.865	0.910	1.035	0.969	0.463	0.640	0.864	0.759	0.575
Name	pc.	pc.		pc.	pc.	pc.	pc.	Bario-pyrochlore	pc.	Uran-pyrochlore	Betafite	No name

Søvite : 252509, 320407, 320410, 320516, 320543
Ferrocarbonatite : 266091, 320488
Glimmerite : 320023, 320113, 320154, 320159
Vein in fenite: 266078
Disseminated in fenite : 320056

[a] Avg. of analyses in one homogeneous grain (spectrometer).
[b] Point analyses (spectrometer).
[c] Point analyses (EDX.)

n.a. = Not analyzed.
b.d. = Below detection limit.

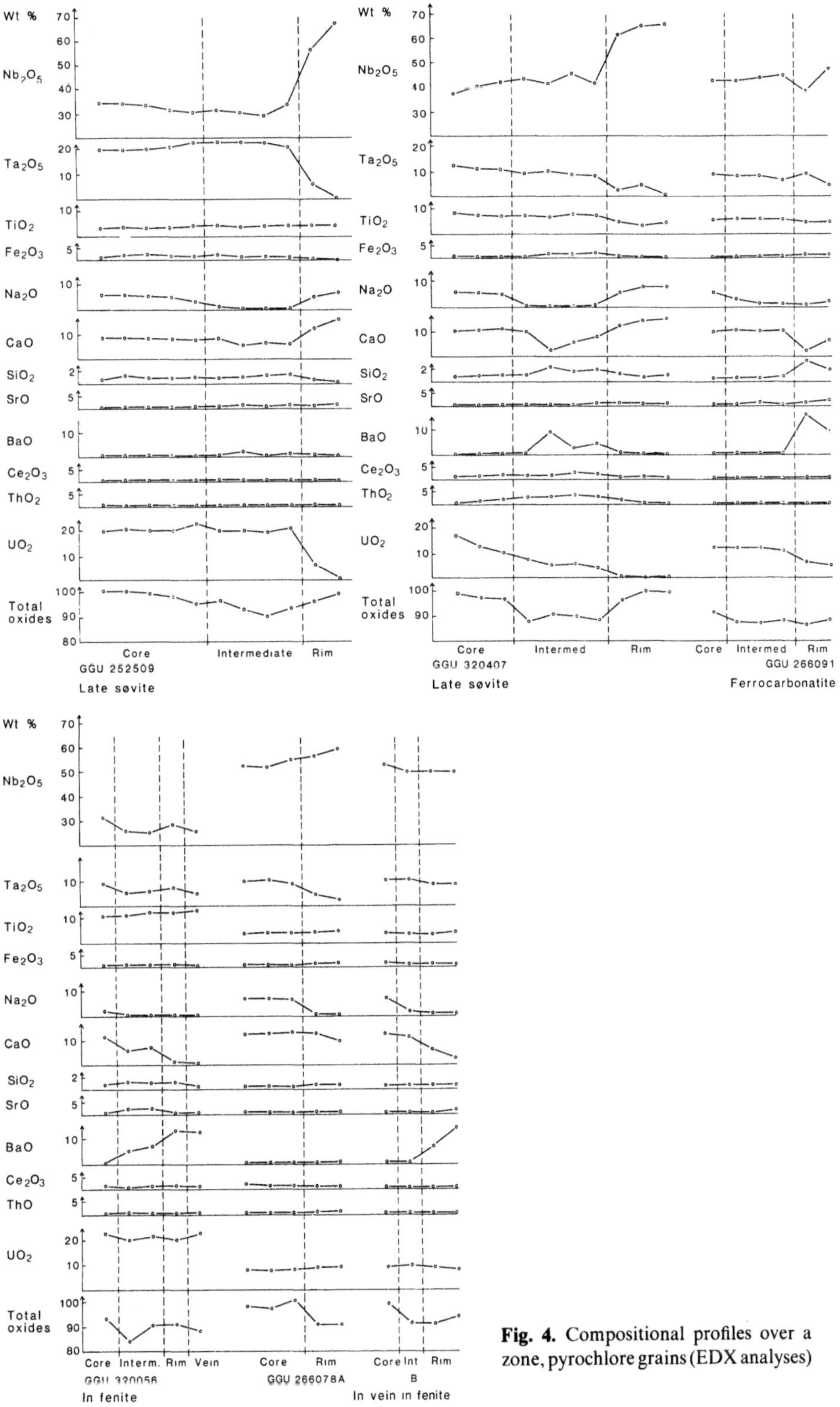

Fig. 4. Compositional profiles over a zone, pyrochlore grains (EDX analyses)

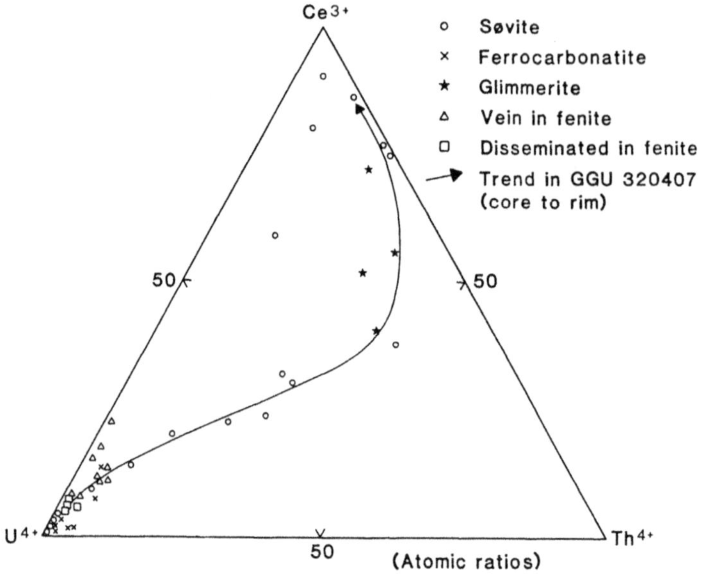

Fig. 5. Ternary plot of UO_2–ThO_2–Ce_2O_3 in pyrochlore

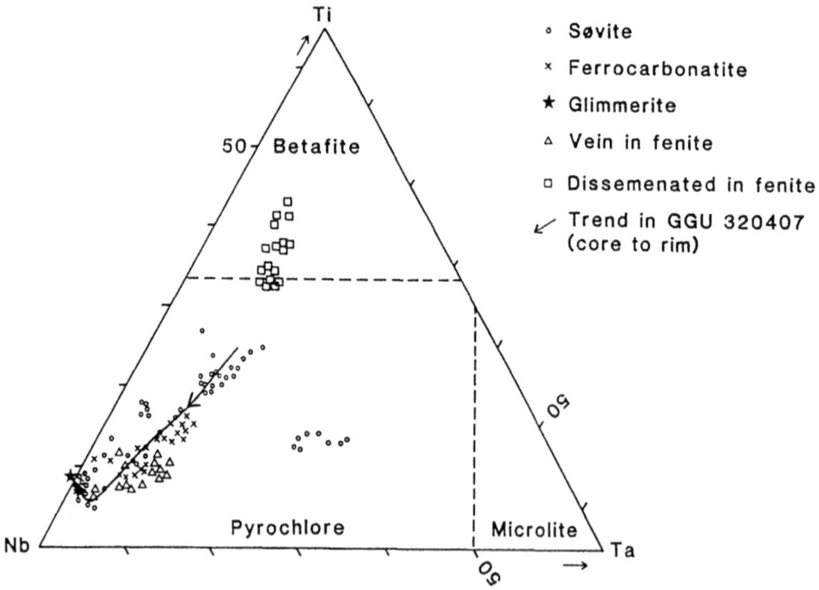

Fig. 6. Ternary plot of elements in the B position in the pyrochlore formula

Fig. 7. Plot of cations in the pyrochlore formula A_{2-m} $B_2O_6(OH,F)_{1-n}$. *From top to bottom* $U^{4+}+Th^{4+}+La^{3+}+Ce^{3+}+Nd^{3+}$ vs Ta^{5+}. $U^{4+}+Th^{4+}+La^{3+}+Ce^{3+}+Nd^{3+}$ vs Ti^{4+}. $U^{4+}+Th^{4+}+La^{3+}+Ce^{3+}+Nd^{3+}$ vs $Ta^{5+}+Ti^{4+}$

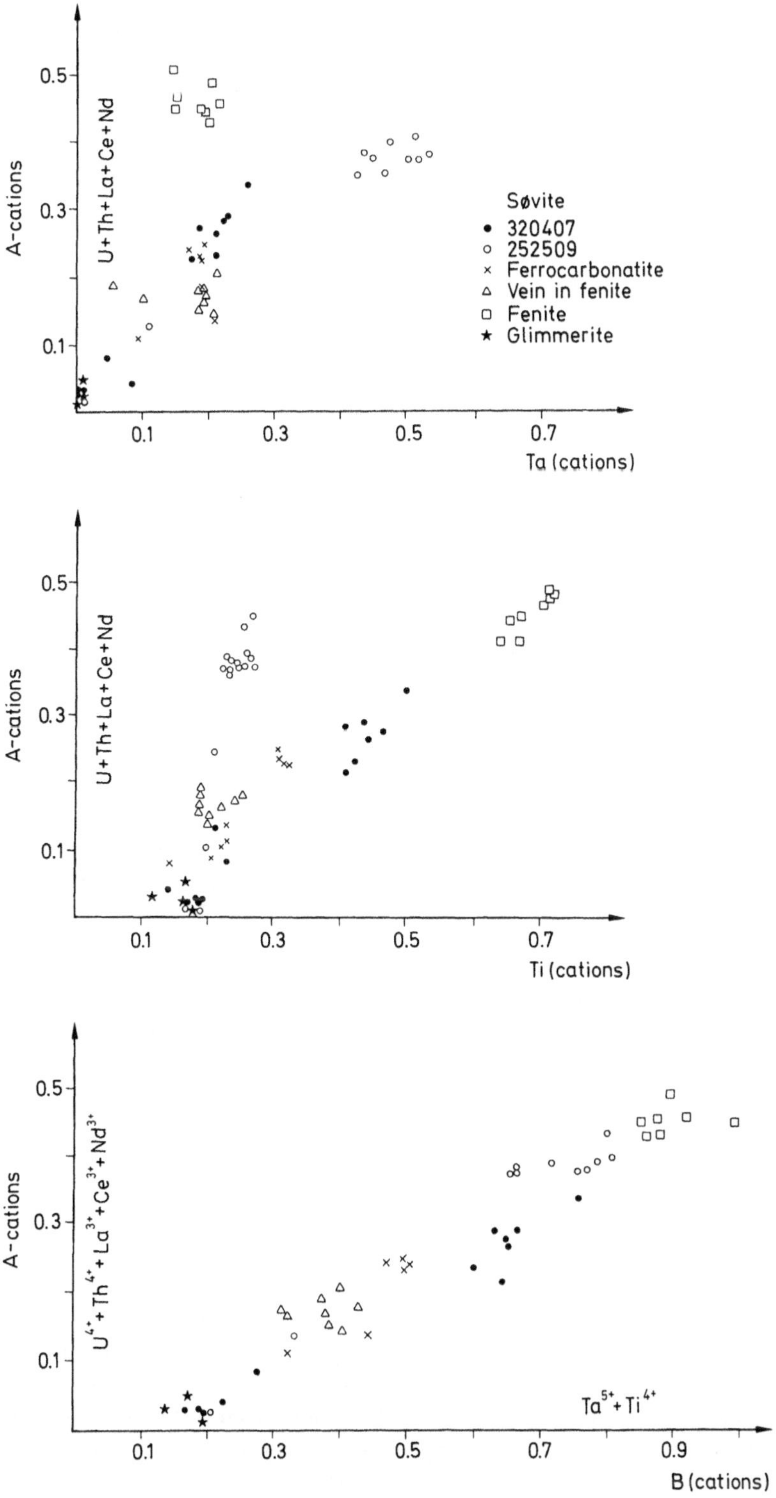
A-cations
U+Th+La+Ce+Nd
Søvite
320407
252509
Ferrocarbonatite
Vein in fenite
Fenite
Glimmerite
0.1
0.3
0.5
0.7
Ta (cations)
A-cations
U+Th+La+Ce+Nd
Ti (cations)
A-cations
U4+ + Th4+ + La3+ + Ce3+ + Nd3+
Ta5+ + Ti4+
0.9
B (cations)

4.2.2 Ferrocarbonatite

The compositional zoning in the pyrochlore in the ferrocarbonatite is similar to that seen in the late stage søvite.

4.2.3 Glimmerite

The composition of the pyrochlore in glimmerite shows only small variations and is close to the end member $NaCaNb_2O_6(OH,F)$. The content of Th and lanthanides is slightly higher than in the major part of pyrochlore in late stage søvite. The zoning, indicating euhedral growth (Figs. 2b, 3b), is compositionally reflected in variable totals (content of water or F).

4.2.4 Fenite

Pyrochlore in veins in the fenite has a homogeneous core of U- and Ta-rich pyrochlore and betafite surrounded by a complexly zoned rim. The U content is approximately 8 wt.% UO_2 in both core and rim, whereas Na almost disappears and Ca decreases in the rim relative to the core. The complex zonation seen in Fig. 3c illustrates the variable Ba content, with the bright layers having as much as 14 wt.% BaO (bariopyrochlore).

The pyrochlore group minerals disseminated in the fenite have a high content of Ti (Figs. 6 and 7) and belong mainly to the betafite subgroup. The UO_2 content is high, between 20 and 24 wt.%. The betafite is zoned and cut by late betafite veins (Fig. 3d). The orange-brown core is characterized by 1–3 wt.% Na_2O, 10 wt.% CaO and only small amounts of Ba and Sr. This core is surrounded by an orange zone with decreasing contents of Na and Ca and increasing Ba. The bright rim and veins in Fig. 3d are rich in BaO (up to 14 wt.%), and Ba is the dominating A-cation (GGU 320056, Table 2), constituting more than 20% of the A-cations. The mineral is a member of the betafite subgroup, and must be a new pyrochlore group mineral. Because the mineral only occurs as narrow rims on, and veins in, the betafite (Fig. 3d), it has not been possible to extract sufficient amounts for X-ray diffraction studies, and a new mineral species has not been established. The core of the grain falls in the pyrochlore field (Fig. 6) and must be termed uranpyrochlore.

4.3 Substitutions for Nb

There is variation in the amount of Ta and Ti substituting Nb in the B position in pyrochlore, with a rock type, similar to that observed concerning elements substituting Na and Ca in the A position. Ta- and Ti-rich pyrochlore group minerals are located in fenite, and form cores of zoned grains in carbonatite. Ta_2O_5 content of pyrochlore in glimmerite varies from 0.1 to 1.3 wt.%, which is slightly higher than that found in unzoned pyrochlore in late stage søvite and ferrocarbonatite (and in the rim of zoned crystals in the late stage søvite and ferrocarbonatite), with 0.0 to 0.5 wt.%

Ta_2O_5. The TiO_2 content varies from 12 wt.% in fenite to about 3 wt.% in glimmerite and søvite.

The pyrochlore in søvite and ferrocarbonatite with a compositional zonation with respect to the A-cations has a compositional zoning of the B-cations as well. The concentration of Ti and Ta is high in the core and intermediate zones of grains also enriched in U, Th, Ba and lanthanides (Fig. 4). The TiO_2 content may reach 12 wt.% and the Ta_2O_5 content 23 wt.% and, as observed regarding substitutions in the A position, there is a coincidence between (1) the trend from fenite to carbonatite and (2) the trend from core to rim in zoned grains (Fig. 6). The highest Ta contents are found in cores of pyrochlore from the magnetite-rich late stage søvite GGU 252509. This pyrochlore has low Ti contents compared to other U-rich pyrochlores (Fig. 7b), but the high Ta contents (Fig. 7a) compensate for this so that the analyses still fall "on the line" in Fig. 7c. The Ta and Ti concentrations decrease in the rim, parallel to the evolution in the concentrations of elements (U, Th, Ba and lanthanides) in the A position (Fig. 7c).

5 Discussion

The following is based on the described field relations, petrography and mineral chemistry. No attempts have been made to model the behaviour of the elements and phases during the evolution of the complex; thus, the present discussion must be regarded as qualitative.

5.1 Factors Controlling the Location of Pyrochlore Mineralization

The concentration of Nb and Ta in the carbonatites of the main intrusive phase is relatively low, and pyrochlore has not been found in these carbonatites, but only in, or associated with, late stage carbonatites. This indicates that Nb and Ta are concentrated in, and have been precipitated from, the residual fluids.

In the carbonatite magma Nb and Ta are soluble as, e.g., carbonate, fluorine, phosphate and sulphate complexes (Aleksandrov 1967; Aleksandrov et al. 1985). Constituents like U and Th are soluble as fluorine and sulphate complexes under acid conditions, as phosphate complexes under neutral conditions and as carbonate and hydroxide complexes under alkaline conditions (Boyle 1982). Lanthanides form soluble complexes with halides (Flynn and Burnham 1978) and carbonate (Kosterin 1959; Aleksandrov 1967). Alkalis in the solution are likely to enhance the solubility of Th, U and lanthanides (Kosterin 1959; Boyle 1982). Because the concentrations of the complexing ligands such as phosphate, halides and carbonate are high in the carbonatite, the Nb, Ta, U, Th and lanthanides are likely to be soluble and hence not precipitated in pyrochlore. As the concentrations of phosphate, halides and carbonate decrease, the heavy ions kept in solution by these will be precipitated. A drop in concentration of the complexing anions will occur: (1) from the carbonatite to the wall rock; (2) where these anions are lost, e.g. phosphate and fluorine by precipitation of apatite; (3) when pressure is released.

The fact that pyrochlore in Qaqarssuk is found (1) at the carbonatite/wall rock boundary, (2) together with large amounts of apatite, indicates that a drop in concentration of the complexing ligands actually was the process responsible for the precipitation of pyrochlore. Pyrochlore often occurs together with magnetite and zircon which indicates that the precipitation of these minerals is controlled by similar processes.

5.2 Factors Controlling the Composition of the Pyrochlore

5.2.1 A-Cations

U-rich pyrochlore is found in the fenite, whereas the pyrochlore in the carbonatite is generally pure. This can be explained as follows.

The sequence of precipitation of the heavy ions is likely to be dependent on the ability of the heavy cations to retain the complexing anions and stay in solution, i.e. it is dependent on their field strength (charge/radius). For U^{4+}, Th^{4+} and Ce^{3+} the field strengths are 3.70, 3.57 and 2.46 respectively (cubic coordination and ionic radii from Whittaker and Muntus, 1970). That this is the sequence precipitated in pyrochlore is seen e.g. in the compositionally zoned grain in GGU 320407 (Fig. 5). Further, the most distal pyrochlores relative to the carbonatite (in fenite) are the most U-rich, the intermediate (in glimmerite) are Th-rich and the most proximal and those in the carbonatite are lanthanide-dominated or have very low contents of heavy ions. The zonation in pyrochlore with a betafite core and "pure" rim as e.g. GGU 320407 (late stage søvite) (Figs. 2d, 4–6) can be generated as follows. The core of grain was formed in a small fracture in fenite, at low pressures of complexing ions, at an early stage of the carbonatite intrusion. The sequence U –> Th –> lanthanides –> pure pyrochlore is an expression of increasing concentration of complexing anions during the intrusion of the carbonatite (Fig. 8). The pure pyrochlore in the rim is consistent with the present location in søvite.

The light lanthanides, especially Ce, are dominant over the heavy ones in the pyrochlore.

The complexly zoned, Na-poor, Ba- and water-rich intermediate zone observed in most zoned pyrochlore grains is very similar to that described as leached pyrochlore (Borodin and Nazarenko 1959; Van Wambeke 1970). The almost constant U, Ta and Ti contents over the boundary from core to intermediate zone in most zoned pyrochlores together with the very jagged outline of the core (Fig. 3a) support a solid state leaching. The botryoidal shapes often seen in the intermediate zone (Fig. 3d) can be interpreted as being due to precipitation from a colloidal medium, and the (syneresis) cracks (Fig. 3c and d) as being generated during dehydration of a gel. In contrast, the continuous evolution seen in GGU 320407 indicates that the Na- and Ca-poor, Ba- and water-rich pyrochlore could have been formed during continuous precipitation. The high water content in this zone indicates that water was abundant during the formation of the zone.

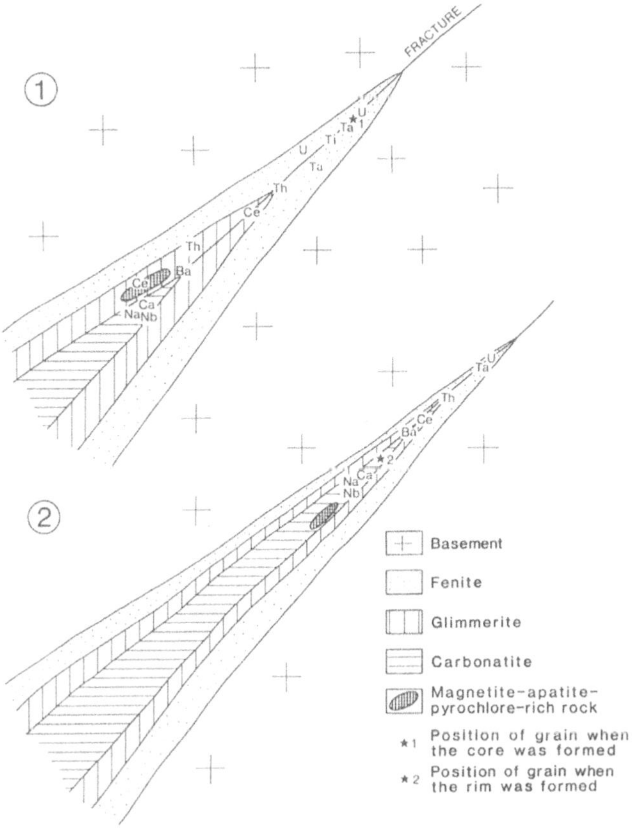

Fig. 8. Schematic stages in the development of zonation with respect to pyrochlore composition in the complex, and in individual grains

5.2.2 B-Cations

Ta- and Ti-rich pyrochlore is found in the fenite, whereas the pyrochlore in the carbonatite generally has low contents of these elements. The solubility of Nb and Ta is pH-dependent, and Ta is relatively more soluble under acid conditions relative to Nb, which is the most soluble under neutral and alkaline conditions (Aleksandrov et al. 1985). The pH was probably decreasing away from the strongly alkaline carbonatite, and hence Ta is relatively more soluble and mobile in the distal environment. There is a relatively good correlation between both Ta and Ti content in the pyrochlore and the content of tri- and tetravalent cations at the A position (mainly U) (Fig. 7a and b). When plotted together the correlation is even better (Fig. 7c). The line along which the analyses plot corresponds to two substitutions of Ta and Ti for one substitution of one tri- or tetravalent cation. If a Ta valency of 4 is assumed, the introduction of Ta and Ti compensates the substitutions on the A site, which however is not very likely. The relation between Ta and U is best explained as these elements, being the most mobile, are deposited in a distal environment and overgrown by less U- and Ta-rich pyrochlore.

6 Conclusions

1. Precipitation of pyrochlore has not taken place in the main phase of carbonatite intrusion, but is related to late stage carbonatite.
2. Pyrochlore is mainly located in late stage søvite and glimmerite adjacent to late lanthanide-rich carbonatite. Small amounts are located in ferrocarbonatite and fenite.
3. Pyrochlore has been precipitated (1) simultaneously with precipitation of apatite, (2) at or in the wall rock adjacent to late carbonatites. In this environment there has been a rapid decrease in the concentration of the complexing ligands responsible for keeping the elements Nb, Ta, U, Th and lanthanides in solution.
4. Pyrochlore with a composition close to $NaCaNb_2O_6(OH,F)$ is the most abundant pyrochlore group mineral in the Qaqarssuk carbonatite complex, but a wide compositional variation is observed and pyrochlore enriched in Ta, Ti, U, Th, lanthanides, Ba, and Sr is found.
5. A betafite subgroup mineral with Ba as the dominant A-cation has been found. This composition has not been described before.
6. The most U1, Ti- and Ta-rich pyrochlore is formed in a distal environment relative to the carbonatite, and pure pyrochlore is formed close to or in the carbonatite.
7. The contents of Ta and Ti in the pyrochlore can be correlated with the contents of tri- and tetravalent ions in the A position.

Acknowledgements. Thanks are due to J. Rønsbo for guiding the microprobe analyses, T. Tukiainen, L.M. Larsen, A.K. Petersen and K. Secher for valuable discussions and comments on the manuscript. The field work was supported by the European Economic Community through contract number MSM119DK. The author is funded by the Danish Natural Science Research Council, contract number 11-5835. Publication is authorized by the Director of the Geological Survey of Greenland.

References

Aleksandrov IV (1967) Niobium in the carbonate solutions and some considerations on the migration of rare elements under hydrothermal conditions. Geochem Int 4:558–566

Aleksandrov IV, Krasov AM, Kochnova LN (1985) The effects of potassium, sodium and fluorine on rock-forming mineral assemblages and the formation of tantaloniobate mineralization in rare-element granite pegmatites. Geochem Int 22 8:85–94

Borodin LS, Nazarenko II (1959) Chemical composition of pyrochlore and diadochic substitutions in the $A_2B_2X_7$ molecule. Geochemistry 4:330–349

Boyle RW (1982) Geochemical prospecting for thorium and uranium deposits. Elsevier, Amsterdam

Deans T (1966) Economic mineralogy of African carbonatites. In: Tuttle OF, Gittins J (eds) Carbonatites. Wiley & Sons, New York, pp 385–413

Flynn RT, Burnham CW (1978) An experimental determination of rare earth partition coefficients between a chloride containing vapour phase and silicate melts. Geochim Cosmochim Acta 422:685–701

Hogarth DD (1977) Classification and nomenclature of the pyrochlore group. Am Miner 62:403–410

Jäger E, Niggli E, van d Veen AH (1959) A hydrated barium-strontium pyrochlore in a biotite rock from Panda Hill, Tanganyika. Min Mag 32:10–25

Kapustin YuL (1980) Mineralogy of carbonatites. Amerind, New Delhi

Knudsen C (1985) Investigations of the Qaqarssuk carbonatite complex, southern West Greenland. Rapp Grønlands Geol Unders 125:34–40

Knudsen C (1986) Apatite mineralisation in carbonatite and ultramafic intrusions in Greenland. EEC contract MSM 119 DK. Final Rep Geol Surv Greenland, Copenhagen
Kosterin AV (1959) The possible modes of transport of the rare earth by hydrothermal solutions. Geochemistry 4:381–387
Larsen LM, Rex DC, Secher K (1983) The age of carbonatites, kimberlites and lamprophyres from southern West Greenland: recurrent alkaline magmatism during 2500 mill. years. Lithos 16:215–221
Veen AH van d. (1963) A study of pyrochlore. Ned Geol Minbouw Gen Geol Ser 22
Wambeke L van (1970) The alteration processes of the complex titano-niobotantaeates and their consequences. N Jahrb Miner Abh 112:117–149
Whittaker EJW, Muntus R (1970) Ionic radii for use in geochemistry. Geochim Cosmochim Acta 34:945–956

Part II Geochemistry

REE(Y), Nb, and Ta Enrichment in Pegmatites and Carbonatite-Alkalic Rock Complexes

P. MÖLLER

1 Introduction

Under favourable conditions REE(Y) and Ta(Nb) are enriched in complex granitic pegmatites but not both in the same body. The two groups of elements follow different fractionation trends. In alkalic rock-carbonatite complexes, however, REE(Y) and Nb(Ta) are always concentrated together. This geochemical coherence is remarkable because REE(Y) are expected to act as cations, whereas Ta and Nb form nuclei of complex anions. Both element suites are considered as "incompatible", but their degree of incompatability differs. REE(Y) substitute for Ca, Fe and to some degree Mg in common rock-forming minerals and, as a result, are widely distributed. The anionic species of Ta and Nb do not substitute for SiO_4^{4-} in silicates. Ta and Nb commonly substitute for Ti, Zr, W and Sn. The enrichment of REE(Y), Nb and Ta is closely linked with the petrogenesis of pegmatites, carbonatites and alkalic rocks.

Late and post-tectonic granitic pegmatites from orogenic belts are considered as crustal differentiates, whereas carbonatites undoubtedly originate deep in the earth's mantle. The source of alkalic rocks associated with carbonatites is not as clear. The different sources of melts lead to specific patterns of occurrence: granitic pegmatites are mostly emplaced along contact zones and axial faults in greenschist belts or higher-grade metamorphic terrains. Carbonatites are exclusively bound to extensional structures. Different sources of melts of the pegmatites, alkalic rocks and carbonatites account for differences in (1) chemical composition, (2) P-T regime, (3) contents of volatiles such as H_2O and CO_2.

This chapter discusses the prerequisite conditions for the formation of rare-element pegmatites and carbonatite-alkalic rock complexes based on (1) field observations, (2) results of laboratory experiments and (3) theoretical considerations based on both phase relationships and chemical behaviour of the rare metals under respective P-T-X conditions known so far. Only a combination of the different approaches will yield satisfactory models for the rare element hosting complex pegmatites and alkalic rock-carbonatite complexes. Additional insight may be gained from a comparison of the processes which lead to REE(Y) and Ta,Nb enrichment in these bodies and an attempt will be made to outline the chemical constraints for rare-metal fractionation.

Hahn-Meitner-Institut Berlin GmbH, AG Geochemie, Postfach 390128, D-1000 Berlin 30

Lanthanides, Tantalum and Niobium
P. Möller, P. Černý and F. Saupé (Eds.)

2 Pegmatites

2.1 Petrologic and Geochemical Framework

Common pegmatites consist predominantly of the rock-forming elements that make up granites. Complex pegmatites host a number of rare elements such as Li, Rb, Cs, Ta > Nb, B, P, F and occasionally Nb > Ta and REE(Y). The economically interesting rare-element pegmatites are mostly Precambrian, and are located in high-T, low-P greenschist belts, predominantly the andalusite-cordierite-muscovite subfacies (Černý 1982). Pegmatites are generally described as younger than the regional metamorphic environment (Register 1979; Gaupp et al. 1984). Ginsburg et al. (1979) approached the highly controversial genetic question of anatexis of high-grade metamorphic rocks vs igneous fractionation from granitic intrusions by comparing the geological setting of the various types of pegmatites with appropriate genetic types in specific crustal situations (Table 1). According to their analysis, REE(Y)- and Ta(Nb)-bearing pegmatites are formed at intermediate geobarometric levels of the cordierite-amphibolite facies.

Rare-element pegmatites are either spatially associated with peraluminous granites or a granitic pluton is thought to occur at depth. The relationship between rare-metal granitic pegmatites and specialized granites is demonstrated by the pre-enrichment of rare-element parental granites. Examples are (many more could be quoted if other than REE-, Nb-, Ta-rich bodies are considered):

1. REE(Y)-rich pegmatites in the vicinity of the REE(Y)-enriched Silver Plume granite in the Jamestown District, Boulder County, Colorado (Bray 1942), or in the REE(Y)-enriched Pikes Peak batholith, South Platte District, Colorado (Simmons et al. 1987);
2. Ta-bearing pegmatites in the Finnish Precambrian igneous complex with Ta-enriched young granites (Rankama 1944);
3. The Harding pegmatite, New Mexico, which is related to the adjacent Ta-rich Dixon granite (Montgomery 1950);

Table 1. Classification of pegmatites (after Ginsburg et al. 1979)

Mechanism of formation	Type of pegmatite
Crystallization from fluids in apical parts of granite intrusions	Miarolitic pegmatites; quartz, fluorite, topaz, beryl
Fractionation of differentiated allochthonous granites	Rare-element pegmatites; Li,Rb,Cs,Be,Ta,Nb,Sn,REE(Y)
Products of anatexis, more or less autochthonous granites	Mica, scarce rare elements
Migmatites in granulite-facies terrains	No economic mineralization

4. The rare-element pegmatites surrounding the Harney Peak Granite in the Black Hills, South Dakota (Shearer et al. 1987);

5. The Greer Lake pegmatite granite, southeastern Manitoba (Černý et al. 1986); and

6. Nb-rich pegmatites of SE Uganda which are located in a columbite-bearing granite province (von Knorring 1960). The eventual stability of many of the rare-element-bearing accessory minerals such as fluorite(REE), apatite(REE), columbite, tantalite and others (Shearer et al. 1987) is responsible for the enrichment of the incompatible elements (e.g. F, P, Ta, Nb, REE) in the parental granites during solidification.

2.2 Differentiation of Parental Melt

During crystallization of the specialized leucogranitic melts, the trace and minor elements are fractionated between rock-forming and accessory minerals and melt, the latter changing continuously in chemical composition. This fractionation is controlled by respective partition coefficients. These partition coefficients cannot be considered to be constant due to changes of the volatile contents of the melt which affect both its structure (Manning et al. 1980; Mysen 1986; London 1986a,b; Jolliff et al. 1987a,b) and the amount of complexing agents. The incompatible elements become enriched in the fluid phase towards the final stages of solidification. Since Rb^+ (1.75 Å) and Cs^+ (1.88 Å) are larger than K^+ (1.65 Å), it is convenient to describe the increasing fractionation qualitatively by both increasing Rb, or Cs contents or by Rb/K, Cs/K ratios. These parameters increase in the residual melt and the feldspars and micas crystallizing from it.

Vesiculation and thermal gravitational convection-diffusion (Shaw et al. 1976; Hildreth 1979) during fractional crystallization in the parental melt also concentrate incompatible elements that partition into the low density phase which accumulates in the cupola of the solidifying granites. Incompatible elements are, therefore, enriched at the top of the solidifying parental melts. This chemical evolution of melt leads to alkali-rich silicic hydrosaline fluids (Shaw et al. 1976; Hildreth 1979). Such highly rare-element-enriched fluids have been considered as the source of pegmatites (Simmons and Heinrich 1980). Along similar lines Simmons et al. (1987) argued that the South Platte pegmatite fluids separated from granitic and quartz-monzonitic magma during fractional crystallization, giving rise to polyzonal quartz-core pegmatites with abundant heavy REE(Y) minerals and bizonal composite-core pegmatites largely containing light REE minerals. The REE(Y) fractionation is explained by roofward diffusion of volatiles and fluorine-complexed heavy REE(Y) leading to depletion of heavy REE at the bottom of a stratified magma chamber with granitic magma near the top and quartz-monzonitic magma near the base.

The effectiveness of the accumulation process of "incompatible" compounds is demonstrated by increasing amounts of dissolved gases. The content of included gases in pegmatitic quartz (about 120 cm^3 gas/kg) is considerably higher than in minerals from granites (about 15 cm^3 gas/kg) (Zakharchenko 1964).

2.3 Fractionation of Chemically Coherent Elements

In contrast to their behaviour in common magmatic rocks, the ratios of pairs of chemically coherent elements such as Ga/Al, Hf/Zr, Ta/Nb and heavy REE/light REE are found to increase in pegmatite minerals. Extreme differentiation of melts is required to achieve changes of these ratios because of the nearly identical chemical behaviour of the related elements which is largely the result of their correspondence in ionic radii and valence. Some fractionation occurs between heavy and light REE(Y) due to a decrease in ionic radii from La to Lu. Although the behaviour of Ga and Hf is not the subject of this chapter, it has to be mentioned here because it characterizes the fractionation that leads to pegmatites.

2.3.1 Ga/Al Ratio

The Ga/Al ratio is nearly invariable in common magmatic, metamorphic and sedimentary rocks (Table 2). In highly specialized granites and more so in pegmatites, the Ga/Al ratio slightly increases from granites to pegmatites with increasing Rb/K ratio as demonstrated in Fig. 1 (von Knorring 1960; Vlasov 1966; Černý et al. 1981; 1985). Notable enrichment in Ga is encountered in leucogranites and pegmatitic granites being parental to fields of rare-element pegmatites. The Tanco deposit, Canada, for instance, seems to be the most Ga-enriched pegmatite

Table 2. Ga/Al and Hf/Zr weight ratios in various types of igneous, metamorphic and sedimentary rocks (after Burton and Culkin 1969; Erlank et al. 1969)

Rock type	Ga/Al	Hf/Zr
Igneous:		
granites	$2\ 10^{-4}$	1:39
basalts	$2\ 10^{-4}$	1:43
Metamorphic	$2\ 10^{-4}$	1:35
Sedimentary		
arenaceous	$2\ 10^{-4}$	1:55
argillaceous	$2\ 10^{-4}$	1:40
carbonates	$2\ 10^{-4}$	1:60
Average	$2\ 10^{-4}$	1:43

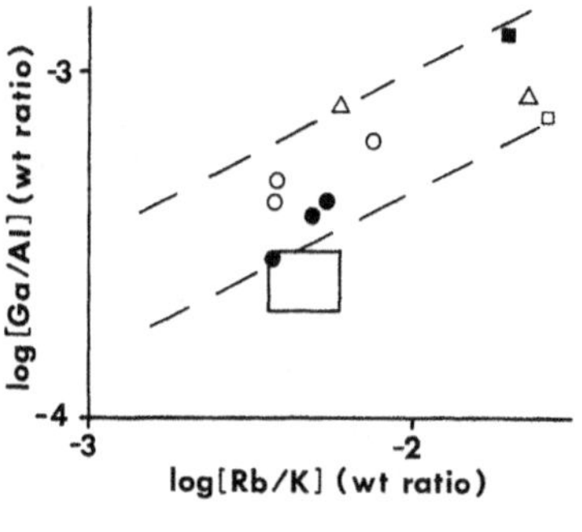

Fig. 1. Ga/Al weight ratio vs Rb/K weight ratio in leucogranites, pegmatitic granites and pegmatites (analyses from Černý et al. 1981). *Filled circles*: biotite granite; *open circles*: fine-grained leucogranite; *triangles*: pegmatitic leucogranite; *open square*: sodic aplite; *solid square*: potassic pegmatite; *rectangle*: mean of granites (data from Burton and Culkin 1969; Heier and Billings 1969)

(Černý et al. 1985) with Ga/Al ratios in mica of about × 10^{-2}, whereas the K-feldspars are not significantly more enriched than the leucogranites and pegmatitic granites of that district. This demonstrates that most of the Ga/Al fractionation is performed during the differentiation of the granites. During the crystallization of pegmatites, Ga/Al fractionation largely continues between the various minerals with mica showing the largest spread of Ga/Al ratios and the highest concentrations (Černý et al. 1985).

2.3.2 Hf/Zr Ratio

Hf/Zr ratios are found to increase in the sequence: granite, pegmatitic granite, aplite and potash pegmatite (Vlasov 1966; Černý et al. 1981, 1985) (Fig. 2). Extreme fractionation of Hf and Zr is observed in cassiterites from pegmatites and alkali granites or their greisens with ratios of 4 ± 2 (Möller and Dulski 1983). In contrast, SnO_2 samples from hydrothermal deposits show Zr/Hf ratios of 28 ± 10. Both these ratios should be compared with 43 ± 8 in magmatic rocks (Fig. 3). Assuming a simple fractionation mechanism with a constant difference of solid-melt partition coefficients for Zr and Hf (Fig. 4), the extreme Hf/Zr ratios of cassiterites from pegmatites and alkali granites indicate that SnO_2 crystallized from residual fluids which represent less than 10^{-4} of the initial parental mass (Möller and Dulski 1983). The conspicuous difference in Hf/Zr ratios of pegmatitic and hydrothermal cassiterite might be related to the different emplacement levels of the parental granites: Sn, W granites with hydrothermal vein mineralizations are largely shallower-seated than the granites with pegmatite aureoles. In the presence of sufficient water, fluid phases should separate at an earlier stage of granite solidification and carry with them a less fractionated Hf/Zr ratio. In contrast, hydrous pegmatite melts are released at greater depth and at much later stages of parental melt differentiation.

A similar range of Zr/Hf ratios is displayed by zircons from pegmatites (Černý et al. 1985).

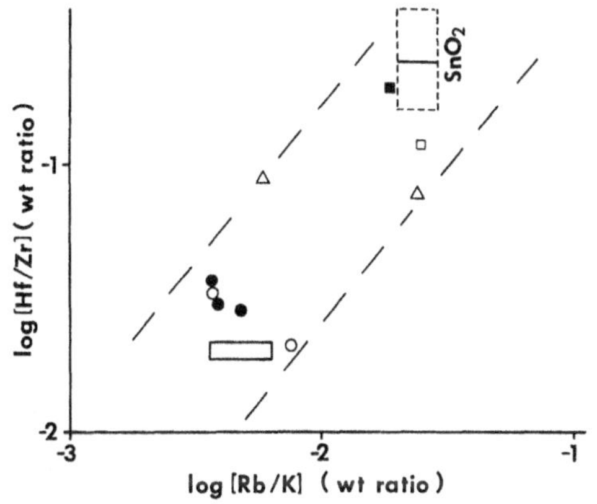

Fig. 2. Hf/Zr wt. ratio vs Rb/K wt. ratio in leucogranites, pegmatitic granites and pegmatites (analyses from Černý et al. 1981). *Solid circles:* biotite granite; *open circles*: fine-grained leucogranite; *triangles*: pegmatitic leucogranite; *open square*: sodic aplite; *solid square*: potassic pegmatite; *rectangle*: mean of granites (data from Erlank et al. 1969; Heier and Billings 1969)

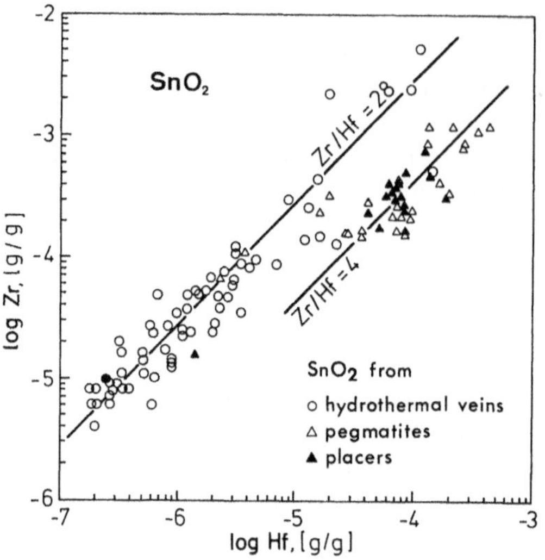

Fig. 3. Zr vs Hf contents in cassiterites from hydrothermal veins, pegmatites and placers. The average Zr/Hf ratio in cassiterite from placers and pegmatites is very different from that in hydrothermal veins (data from Möller and Dulski 1983)

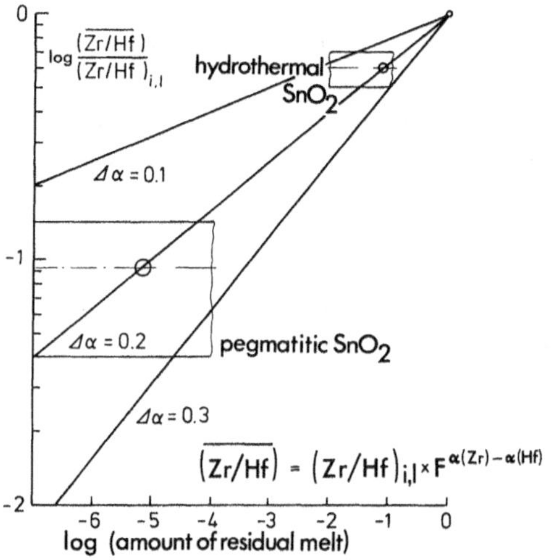

Fig. 4. Fractionation of Zr and Hf assuming individual fractionation coefficients α. y-axis is given by mean Zr/Hf ratio of hydrothermal and pegmatitic cassiterites normalized by those of granitic melts. The difference of $\alpha(Zr)-\alpha(Hf)=\alpha$ is given as a parameter. Starting from magmatic melts (*upper right corner*), it is assumed that hydrothermal fluids with a ratio of 28/48 (data from Fig. 3) are generated in the range of 90 to 99% solidification of the granite. This fixes α to be in the range of 0.1 to 0.3. For the normalized mean pegmatite Zr/Hf ratio of 4/48 the solidification equivalent of the parental melt amounts to $>99.99\%$, which indicates that the pegmatites represent less than 10^{-4} of the parental melt (Möller and Dulski 1983)

2.3.3 Tantalum and Niobium

Ta-Nb fractionation can be related to the Ti contents of the parental magmas. Both elements are hosted by Ti phases such as rutile, ilmenorutile, titanite, perovskite and ilmenite (Vlasov 1966; Wedepohl 1969; Heinrich 1969; Nekrasov 1970; Haapala 1977). Most probably $(Fe,Mn)(Ta,Nb)_2O_6$ substitutes Ti_3O_6 in rutile.

Ta/Nb ratios show a gradation from late orogenic peraluminous (1/10) to anorogenic alkaline (1/15) granites. This difference could be related to decreasing TiO_2 contents (0.37 and 0.2% respectively; data are from Wedepohl 1969; Heinrich 1969; Correns 1969). Nb content and the Nb/Ti ratio increase, whereas the Ta/Nb ratios decrease during differentiation of granitic melts (Vlasov 1966). The soda granites associated with Nb-rich pegmatites from SE Uganda, for instance, show exceptionally low TiO_2, but high columbite contents in the range of 0.05 to 0.2 wt.% (von Knorring 1960). In the Dixon granite, to which the Harding pegmatite is related, Ta is hosted by titanite (Montgomery 1950). Znamenskii et al. (1962) reported that Ta preferably partitions into titanite and ilmenite, whereas Nb is accumulated in cogenetic biotites. Recently, Green and Pearson (1987) demonstrated experimentally that:

1. Ta and Nb partition strongly into titanite, rutile and moderately into ilmenite and titanomagnetite;
2. Partition values for titanite increase with decreasing temperature and increasing SiO_2 content of co-existing liquids;
3. D_{Ta}/D_{Nb} decreases in the series titanite (~3), rutile (~1.7), ilmenite (~1.3) and Ti-magnetite (~1.2). Whenever titanite is involved in fractional crystallization, the Ta/Nb ratio of any derivative liquid will show a lower ratio.

Observations in drill cores from the Beauvoir granite/France (Cuney et al. 1985) reveal that albite formation and increase of Ta contents are related to fluorine increase in granitic melts. High F and H_2O contents shift the minimum melt composition towards the more albitic composition and strongly decrease the solidus temperature (Manning 1981). This agrees with results reported by Aleksandrov et al. (1985): (1) Ta and Nb are enriched in pegmatite fluids by CO_2 and F_2; (2) temperature decrease and Na increase (decrease of K/Na ratio) favour albite rock formation and accumulation of tantaloniobates; (3) Ta and Nb deposition is favoured by pressure reduction; (4) tantaloniobates form in weakly acid fluids and (5) the Ta/Nb ratio increases with decreasing pH. Černý and Ercit (1985) and Černý et al. (1986) reported that Ta-rich minerals are often associated with lepidolite-bearing pegmatites.

Highest Nb contents are observed in the outer zones of individual pegmatites, not in the centre (Vlasov 1966). In general, Ta/(Ta + Nb) increases from the wall zone to the core, similar to the Cs/Rb ratio in co-existing micas (Jolliff et al. 1987a) or blocky core-margin K-feldspars (Černý et al. 1985).

In primary pegmatite white mica, Cs and Ta increase relative to their concentrations in low Ca granites by about three and two orders of magnitude respectively (Fig. 5). White mica from each individual pegmatite covers a limited range of Ta and Cs contents (Fig. 5). For each pegmatite body, the range of "primary white mica" (Gaupp et al. 1984) covers no more than a factor of five. This is empirically independent of the total number of analyses performed. This limited variability is attributed to the degree of fractionation during pegmatite crystallization. The arrangement of pegmatite data fields in Fig. 5, however, demonstrates that initial pegmatitic fluids were already fractionated to different degrees. This type of fractionation is inherited from the parental leucogranites. The inherited fractionation exceeds the in situ fractionation and varies by factors of 50 or 100 for Ta and Cs respectively. The low-Ca granites also fall on the same Ta and Cs fractionation trend line. The observed trend indicates that the average partition coefficients of Ta are less than unity, but larger than that of Cs and Rb during solidification of the pegmatite.

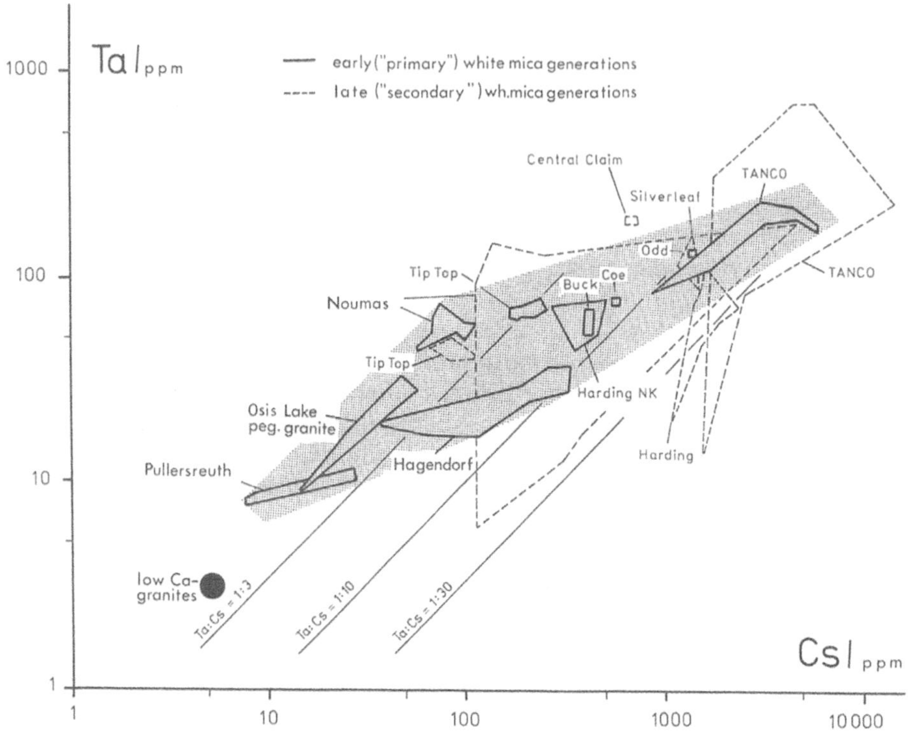

Fig. 5. Ta-Cs relationship in white mica from pegmatites (Gaupp et al. 1984; Morteani and Gaupp, this Vol.). For comparison the *solid circle* represents the respective values in low Ca granites

2.3.4 Rare Earth Elements

In granites REE(Y) are largely contained by accessory minerals, and the remaining REE in rock-forming minerals are preferentially hosted by mica and only to a lesser extent by feldspars. In cases when partition coefficients exceed unity [e.g. REE(Y) in biotite and apatite; Eu in feldspars], the negative Eu anomalies develop in the evolved melts (e.g. leucogranites in Fig. 6) In highly polymerized melts silicate-Eu^{2+} complexes are more stable than the Eu^{3+} equivalents and the other trivalent REE(Y) complexes (Möller and Muecke 1984). The Eu^{2+}-silicate complex, with its assumed double four-ring structure, is preferentially incorporated in feldspars, because it can be considered as a subunit of the feldspar structure. The uptake of Eu^{2+} complexes by crystallizing feldspars leads to the negative Eu anomaly in evolved granitic melts and the pegmatites derived from them.

Comparing REE(Y) distribution patterns of rare-element pegmatites with those of the parental granites, three features are conspicuous:

1. Pegmatites (excluding zones of REE mineralization) are lower in REE(Y) contents than the related granites (Simmons et al. 1987);

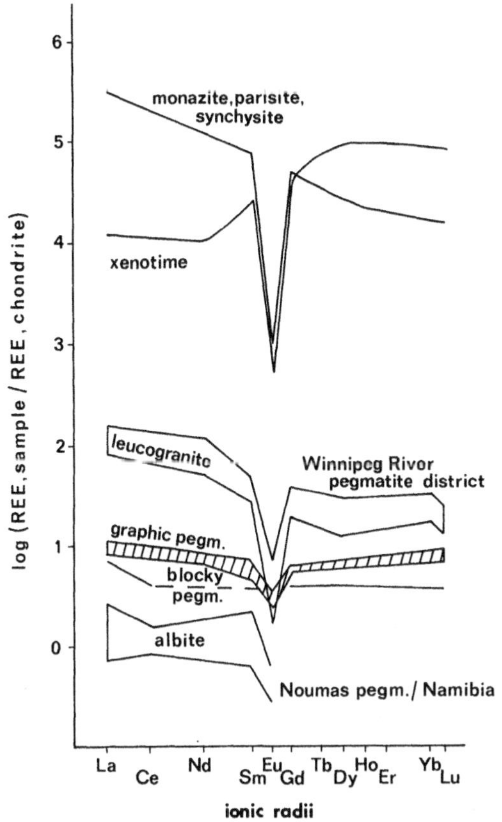

Fig. 6. REE distribution patterns of leucogranites thought to be parental to rare-element pegmatites, blocky and graphic pegmatites (Černý 1982), albites from the Noumas pegmatite (Gaupp et al. 1984) and of REE minerals (means) from REE-rich pegmatites (unpubl. data). The REE patterns of the albites are incomplete because the heavy REE are below detection limits

2. REE(Y) patterns of pegmatites are flatter than those of the granites (Černý 1982; Simmons et al. 1987); and
3. Both show commonly negative Eu anomalies, but they are sometimes less developed in pegmatites than in the parental granitoids (Černý 1982).

The rather low REE(Y) contents and the REE(Y) distribution patterns in most of the complex pegmatites indicate crystallization from fluids that originated from extremely late stages of solidification of a parental granitic melt.

In contrast, REE(Y) fractionation during metamorphic processes leads to considerable enrichment of the heavy REE(Y) in minerals from open veins (Fig. 7). Depletion of light REE during remobilization of REE(Y)-bearing minerals is common due to increasing complex stability of the medium to heavy REE(Y) in aqueous solutions (Möller 1983). Such trends and the absence of negative Eu anomalies are never observed in distribution patterns of REE(Y) minerals from pegmatites. Therefore, formation of rare-element pegmatites can hardly be understood as a product of metamorphism.

The REE(Y) distribution patterns of anatectic melts are largely indistinguishable from those of the parental rocks if melting exceeds 10% (Henderson 1984). Leucosomes of migmatites show no or positive Eu anomalies (unpubl.). Pegmatite

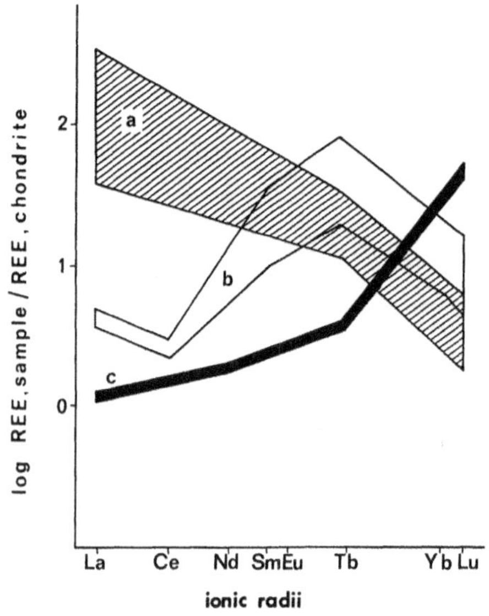

Fig. 7. Typical REE distribution patterns of (**a**) primary hydrothermal and (**b**) remobilized fluorites (Möller et al. 1982) and (**c**) calcites from open Alpine fissures (Lausch et al. 1974)

feldspars crystallizing from partial melts, regardless of how small the melt fraction may have been, should show positive Eu anomalies. This signature, however, is in no way typical of REE(Y) distribution patterns in feldspars from rare-element pegmatites [The only exceptions known to the author are the reported feldspars of the wall and first intermediate zone from the Tin Mountain pegmatite, Black Hills, South Dakota (Walker et al. 1986).] In any case, anatexis cannot explain the formation of rare-element pegmatites.

In solidifying granitoids Eu contents decrease from early to late phases of crystallization. If fluids exsolve from late stage water-enriched melts, this vapour phase will slowly escape through fractures and the intergranular space of the almost solidified outer parts, whereby they come into contact with feldspars being considerably richer in Eu than those that formed from late melt fractions. Recrystallization should result in redistribution of Eu. As a result of this, the escaping fluids will lose some of their strong negative Eu anomalies. Such a process will also lead to re-equilibration of other ions, an effect that has not been studied, to the best of the author's knowledge.

In aqueous solutions REE(Y) are strongly chemically complexed by F^-, PO_4^{3-} and CO_3^{2-}, but only slightly by Cl^- (Bilal and Koβ 1982; Taylor and Fryer 1983; Möller 1983; Becker and Bilal 1985; Bilal and Langer 1988). If the ligands are present in parental melts in sufficient quantities, REE(Y) become enriched in the late hydrous fluids. In such cases, REE(Y) predominantly accumulate in topaz, fluorite, apatite etc. which form in fractures, pores and intergranular space and fractures of the parental granite (Simmons and Heinrich 1980). If such late fluids are channelled into pegmatites, then REE-bearing minerals (e.g. fluorite, apatite, monazite, xenotime and others) are formed.

In the South Platte district (South Dakota, USA), REE contents in whole rock drop from the hosting granites to the border and wall zones by factors as large as 6, e.g. for La and Ce, the LREE being more depleted than the HREE. It has further been estimated that the REE abundance in the Oregon No. 3 pegmatite, South Platte, is between 1 to 2%. This is more than ten times the concentration of the hosting granite, although even the granite is super-enriched in REE. Most of the REE are accumulated in accessories with Eu largely depleted (Simmons and Heinrich 1980; Simmons et al. 1987). In order to explain the REE mineralization at the quartz core and intermediate zone interface, Simmons and Heinrich (1980) supposed that a "second hydrous phase apparently had a larger capacity for these elements than the remaining silicate melt and consequently additional quantities were fractionated into these fluids. Once in the pegmatitic fluids, the REE were strongly partitioned into the progressively more hydrous-rich phases as crystallization continued. Only small quantities precipitated in the wall-zone rocks and virtually none in the intermediate microcline, core-margin biotite, core quartz, or core feldspar".

Similarly, Walker et al. (1986) reported virtually no REE fractionation by crystallization of feldspars from the wall zone through three intermediate zones and the core. With the exception of the wall zone and first intermediate zone, which show positive Eu anomalies, even the negative Eu anomaly stays unchanged, although the REE in the fluid must have increased as a consequence of solidification. The conspicuous change of Eu anomalies between the first and the second intermediate zone has been interpreted by the authors to indicate either a change in melt structure (Möller and Muecke 1984) or in fo_2 (Drake and Weill 1975).

In the Bob Ingersoll pegmatite, South Dakota, Jolliff et al. (1987b) found that the REE decrease from the border zone (depleted in light REE) to the core. This is explained by early crystallization of apatite and other "REE-sink" minerals. As with Walker et al. (1986), the border zone tourmalines (as well as individual samples from other zones) show no Eu anomaly, but a tourmaline sample from the core exhibits a positive one. The Eu anomalies seem to be related to the Ca content of tourmaline. Positive Eu anomalies are considered to represent "a late stage 'mini-flood' of Ca^{2+} (Jahns 1982) that resulted from subsolidus exsolution of an early calcic plagioclase component in perthite to form more sodic albite" (Jolliff et al. 1987a,b).

The REE(Y)-rich late fluids finally form minerals such as fluorite, monazite, xenotime, allanite, gadolinite, samarskite and others, mostly by replacing other minerals. These replacement reactions seem to be caused by the low solubility of REE(Y)-fluorides, -phosphates and -silicates. All REE(Y) minerals from pegmatites are characterized by strong negative Eu anomalies in their REE(Y) distribution patterns (Fig. 6). The anomalies are inherited from the Eu-depleted fluids which originate from evolved granites.

2.4 Isotope Signatures in Pegmatites

Rb/Sr mineral age determinations on pegmatites aimed at establishing the cooling history are extremely ambiguous because of the uncertainty of the respective Sr-initial ratios (Clark 1982). Riley (1970), Clark (1982) and Clark and Černý (1987) explained the highly variable Sr-initial ratios by high ^{87}Sr mobility. Since the Rb-poor minerals are significantly enriched in radiogenic Sr, these authors claim that the mobility of radiogenic ^{87}Sr is confined to the pegmatite bodies internally, excluding any need of a fluid from outside. If a pegmatite, however, is fed at least by fluids that originated from different domains of element fractionation in the parental source, and if these fluids are mixed in varying proportions in the pegmatite fluid, then there should be no uniform Sr-initial ratios for dating of minerals from pegmatites. Due to fractionation in the parental granite the last fluids should be highest in Rb and extremely low in Sr, and therefore high $^{87}Sr/^{86}Sr$ ratios should result. The increased Sr-initial ratio depends on the Rb/Sr fractionation, the amount of selectively mobilized radiogenic ^{87}Sr, and the time elapsed between the events of Rb/Sr fractionation in the parental granite.

Figure 8 shows the trends of Rb and Sr in K-feldspars across the symmetrically zoned Tsaobismund pegmatite/Namibia (Bergholz 1978 and unpubl. data) and a half-profile from the zoned Vezna pegmatite/CSSR (Černý et al. 1984). Rb, Sr contents and Rb/Sr ratios vary irregularly from the border zone towards the centre and decrease towards the central core. It may be argued that the number of data are not statistically sufficient, but the irregular trends of Sr and Rb demonstrate that the fluids from which the K-feldspars crystallized were not fractionated systematically. Similarly, Walker et al. (1986) tried to fit their data to a Rayleigh fractionation model, but with poor success. The scatter of data is too large to be explained by a straightforward, simple single-stage fractionation process. The resulting variable Rb/Sr ratio could be explained by superimposing three effects:

1. Fractionation of Rb and Sr during solidification of the parental granite;
2. Fractionation during crystallization of the pegmatite; and

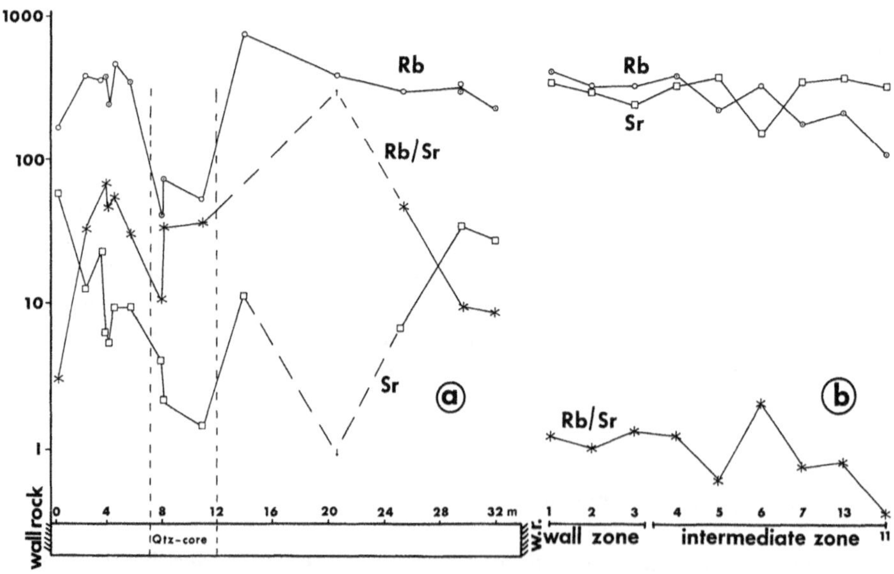

Fig. 8. Rb, Sr contents (ppm) and Rb/Sr wt. ratios across (**a**) the bilateral beryl-bearing Tsaobismund pegmatite/Namibia (Bergholz 1978), with the exception of the core quartz sample (unpubl. data); (**b**) zoned Vezna pegmatite/CSSR (Černý et al. 1984). Samples *1–3*: granular K-feldspar from the wall zone; samples *4–11*: graphic K-feldspar from the intermediate. All sample numbers refer to the KVZ numbers of Černý et al. (1984). The blocky feldspars from the intermediate zone have considerably higher Rb/Sr ratios because the Sr content is below the detection limit

3. Inhomogeneous pegmatite fluid due to mixing of liquids with greatly differing Rb/Sr ratios.

The initial melt of any pegmatite is not as strongly enriched in Rb as those of the final stages. With directed inward growth, e.g. of the feldspars, the Rb content is expected to increase. This increase of Rb (Jahns 1982) and decrease of Sr is not as continuous as it should be expected for a continuous in-situ fractionation in a closed system (Fig. 8).

Mixing of differently fractionated fluids could have been the reason for the observed non-systematic distribution of Rb and Sr. Mixing of the evolving pegmatite melt with late hydrous rare-element-rich fluids from the intergranular space of the parental granite would offer a simple model to account for the observations. If Rb/Sr variability is predominantly the effect of fluid mixing, then this would explain the increase of initial Sr ratios. The large changes of Rb and Sr contents in feldspars (Fig. 8a) indicate that the second fluid entering the pegmatite shortly after starting solidification was enriched in Rb, but greatly impoverished in Sr: less than 10% of pegmatite crystallization can only increase the Rb content in the residual pegmatite by 10% at maximum. However, Rb increases by a factor of 2 and Sr decreases by a factor of 5. Explaining the observed variability by "dirty effects" means to question the usefulness of partition coefficients in magmatic systems.

$\delta^{18}O$ studies largely reveal changes of the pegmatite fluid during growth. For instance, in quartz $\delta^{18}O$ increases from base to top in a single crystal or in crystal overgrowths (Taylor et al. 1979). In feldspars re-equilibration seems to be common (Longstaffe 1982). In general, pegmatites show higher $\delta^{18}O$ compared to related granites (Fig. 9) (Taylor and Epstein 1962). Longstaffe (1982) discussed this finding

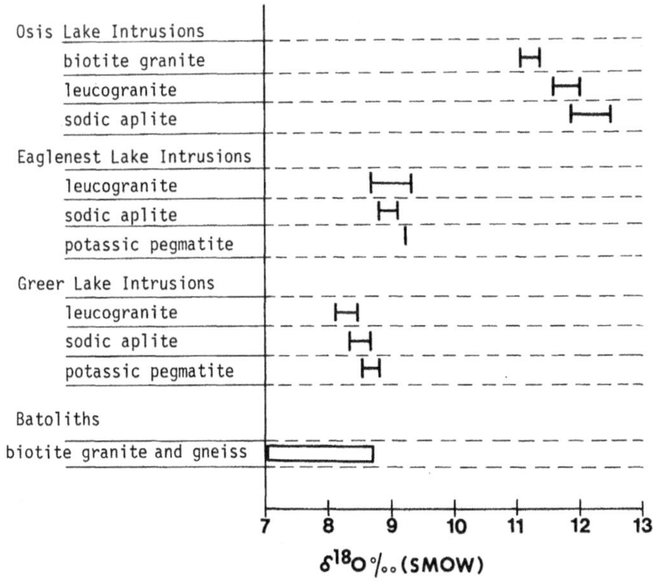

Fig. 9. $\delta^{18}O$‰ (SMOW) of granites, sodic aplites and potassic pegmatites from the Winnipeg River pegmatite district (after Černý et al. 1981)

as a result of (1) differentiation from an igneous source followed by interaction with host rocks, or (2) anatexis of $\delta^{18}O$-rich supracrustal rocks. Longstaffe (1982) favoured (1). The examples given by him show a systematic increase of $\delta^{18}O$ with the fractionation sequence, from pegmatitic leucogranites to sodic aplites and potassic pegmatites. Since late magmatic fluids have always high $\delta^{18}O$ (Hoefs 1980), an increase of $\delta^{18}O$ is to be expected when such fluids feed pegmatites during their growth. The open-system oxygen-isotope behaviour limits the usefulness of whole-rock $\delta^{18}O$ values for many petrogenetic studies, but provides a very powerful tool for identification of the type of fluids present in pegmatite systems (Longstaffe 1982).

Water-muscovite, hydrogen-isotope fractionation increases with decreasing temperature. As a result, δD increases from the pegmatite-aplite zone towards the rare-element mineralized pocket zone (Taylor et al. 1979). Involvement of meteoric water is ruled out by these authors.

2.5 Physico-Chemical Aspects of Pegmatite Formation

High H_2O contents in pegmatite fluids is suggested by the giant growth of crystals. This is atypical for minerals crystallizing from melts and only possible in fluids that are not oversaturated. Such crystals are in marked contrast to those of the aplitic rim with its small grain size. The aplitic minerals form under conditions of oversaturation due to some cooling of the injected hydrous silicic melt. Initially, the rate of nucleation is very high due to oversaturation (Jahns and Burnham 1969). Later on, mineral growth is controlled by secondary (surface) nucleation only. Mobility of ions is high enough to maintain these conditions during solidification of the intermediate and inner zones of pegmatites.

There are two contrasting ideas in explaining the nature of the pegmatitic fluids, the textures and the zonation of pegmatites:

1. Open system behaviour demanding some supply of fluids from the parental granite, and
2. Closed system behaviour calling for internal fractionation only.

Roedder (1984) described pegmatite bodies as "open system fluid inclusions" in which the various giant minerals grow from a fluid that could not have been a silicate melt. "Geologic evidence clearly indicates that after the common pegmatites formed, subsequent fluids (in some localities only) carried and deposited relatively large quantities of the rather rare metals. In many areas, there is good evidence that these deposits ... formed by alteration and replacement of the preexisting minerals"; or "rare-element pegmatites represent environments in which the residual fluid from a larger mass of rock has been channelled into the structure and has precipitated some part of its complement of rare metals. If such movement does not take place, the residual fluid will simply remain between the crystals [of the parental body, author] yielding miarolitic cavities lined with crystals like those found in pegmatites".

Roedder's view may be outdated, at least in part, by more recent results on Li- and B-rich pegmatites that have been found to be exclusively fractionated in a closed system. The understanding of Li-, F-, P- and B-rich pegmatites has been advanced by London (1986a,b, 1987) and Shearer et al. (1987). B_2O_3 is and HF are considered to depolymerize silicate melts, the former by the synergistic effect with H_2O, resulting in low viscosity of the fluids, enabling efficient convection of fluids and diffusion of components. Increased Li_2O, P_2O_5 and HF contents depress the liquidus of the granitic system. For example, the "actual bulk composition" of the pegmatites associated with the Harney Peak granite are close to the composition of minima in the experimental systems albite-quartz-eucryptite-H_2O (Stewart 1978) and albite-orthoclase-quartz-F-H_2O (Manning and Henderson 1981). Additionally, F- and PO_4^{3-} form metal complexes (Watson and Green 1981; Mysen and Virgo 1985).

Pegmatites behave as semipermeable systems which allow water and some of the highly soluble alkalis to escape into the country rocks, but retain the more hydrolyzing and therefore less soluble compounds. The exsolved hydrous fluid escapes from the solidifying pegmatite under its own overpressure. Around pegmatites, the released alkalis form well-known haloes that can be indicative of rare-metal mineralization (Shearer et al. 1986).

Multiphase fluid inclusions yield some insight into the salinity and P-T conditions of mineral formation. Solid-gas inclusions in aplitic and micrographic zones of common pegmatites homogenize to melts when heated to the range of 700° to 800°C. If the melts are very water-rich, this temperature is in the range of 540° to 690°C (Kosukhin 1978; Roedder 1984).

London (1986a) studied the evolution of the Tanco pegmatite and found that primary petalite, microcline units and footwall albitites with disordered Nb-Ta-Sn oxides crystallized in the range of 600° to 500°C and 320 to 290 MPa in the presence of a hydrous borosilicate fluid. Next to that, primary spodumene crystallized at 475° to 500°C and 270 to 290 MPa. Tourmaline crystallization caused formation of rare-element-rich albitites (ordered Ta-Sn oxides) and generation of a low-salinity CO_2-bearing aqueous fluid in the range of 420° to 475°C and 260 to 270 MPa. The late fluids started tourmalinization of the wall rock, sericitization of microcline and later propylitic alteration of the wall rock. During these alteration processes, the temperature and pressure decreased to 280° to 390°C and 180 to 250 MPa. Late stages of aqueous fluids may be responsible for pseudomorphs, metasomatism of earlier formed minerals and for alteration of pegmatite host rocks (London and Burt 1982).

Phase separation or immiscibility of fluids have been considered as an important feature in pegmatite formation. Koster Van Groos and Wyllie (1969) described immiscibility of silicate melt and NaCl solution in the simple system $NaAlSi_3O_8$-NaCl-H_2O. With dropping pressure, this system passes through a variety of different equilibria. However, the disadvantages of this well-known system is that it differs too much from the bulk composition of rare-element pegmatites; it is too high in NaCl and lacks the presence of the network modifiers B_2O_3, P_2O_5 and other fluxes.

More recent studies revealed that the Ab-NaCl-H_2O system does not meet the demand of Li- and B-rich rare-element pegmatites. Cl^- is not a major component. This should not be confused with the high salinity as determined by microthermometry in multiphase fluid inclusions. The total Cl^--contents of the bulk pegmatitic magma is estimated to be probably in the range of X0 to X00 ppm, whereas the determined NaCl equivalents are in the range of 45 to 80 wt.% caused by dissolved alkali silicates and -borates (Roedder 1963; Foord 1976; London 1986a,b, 1987). Such dense fluids were termed "solution melts" by Zakharchenko (1973). The composition and structure of the initial hydrous melt with its high solute load of alkali silicates is definitely different from that of the granitic melt. Densities up to 2 g/cm^3 have been estimated in fluid inclusions of aplites and such values are neither characteristic for a gaseous "pneumatolytic" fluid (London et al. 1982) nor typical for melts. This intermediate density must be caused by synergistic effects which allow dissolution of increased amounts of water in the presence of Li_2O, B_2O_3, P_2O_5 and HF.

Solute-poor aqueous vapours as manifested by secondary inclusions in pegmatite quartz only exist under subsolidus conditions (London 1986a).

Although the phase diagrams of the systems granite-H_2O or Ab-H_2O are known, it is extremely difficult to extrapolate to more complex systems that include additional compounds such as Li_2O, HF, P_2O_5, B_2O_3 etc, which are typically present in rare-element pegmatites. Some of these compounds increase the solubility of H_2O in silicate melts (Wyllie and Tuttle 1964; Pichavant 1981; Manning 1981), others increase the solubility of silicates in solutions even when present as "minor" constituents only.

Discussing pegmatite fluids in a P-T diagram may help to understand their relation to late granitic melts. Figure 10 shows the hydrous granitic minimum melting curve and the critical curve of undersaturated NaCl solutions for orientation. In addition, andalusite-cordierite, andalusite-sillimanite and muscovite-feldspar mineral equilibria are given. It is evident from the studies of Wyllie and Tuttle (1964), Manning (1981), Manning and Henderson (1984) and Pichavant (1983) that even small amounts of dissolved Li_2O, HF, or P_2O_5 shift many equilibria towards lower temperatures and pressures; i.e. the NaCl-dominated system will not be able to describe the pegmatite-granite system correctly. For example, the most probable path of crystallization of the Li-rich Tanco pegmatite (London 1986a) and some B-rich miarolitic pegmatites (London 1986b) are given. The solidus temperatures of Li-rich pegmatites are obviously shifted by more than 100°C towards lower values compared with that of hydrous granitic melts. Figure 10 encompasses the range of temperatures that is indicated by homogenization of multiphase fluid inclusions (compiled by Roedder 1984; London 1986a,b). Inclusions that homogenize to melts represent real melt inclusions and indicate P-T conditions in the field of the appropriate "granite solidus curves" which is caused by the presence of components such as B_2O_3, P_2O_5 and Li_2O. The influence of these compounds is demonstrated by the internal evolution of the Tanco pegmatite (London 1986a), and the San Diego (USA) and the Kulam (Afghanistan) pegmatites (London 1986b).

Gas-liquid inclusions result when fluids are trapped on the left-hand side of the compositionally controlled "granite solidus curve". The respective vapour-pressure curve is determined by the sum of solute contents of the liquids. With P-T parameters well above the critical curve of undersaturated NaCl solution, the system would be in the supercritical monophase if salinity of pegmatite fluids is caused by NaCl. Unfortunately, the critical curve for undersaturated alkali borate and -silicate fluids is not known. The observations of London (1986a), however, signalize that supercritical phases do not occur; i.e. the critical curve of pegmatitic fluids must be much steeper than that of undersaturated NaCl solutions.

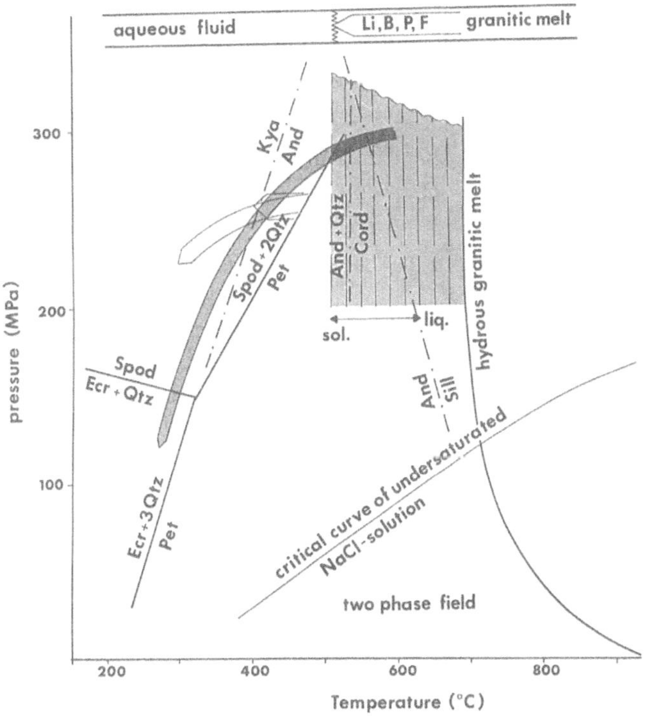

Fig. 10. P-T diagram showing the evolution of pegmatite melts. The *shaded area* indicates shifting of the granitic melt curve to the left when Li_2O, P_2O_5, HF, B_2O_3 is present in increasing concentrations. The Al-silicate equilibria characterize the metamorphic environment of rare-element pegmatites: _._._._ after Holdaway (1971) and Winkler (1976) (*Kya* = Kyanite; *And* = andalusite; *Sill* = sillimanite; *Qtz* = quartz; *Cord* = cordierite). The Li-mineral equilibria describe the P-T conditions of crystallization of rare-element pegmatites (London 1984) (*Ecr* = eucryptite; *Spod* = spodumene; *Pet* = petalite). The *arrows* demonstrate the evolution of pegmatite melts: *stippled*: Tanco deposit, Canada; *open*: Kulam pegmatite (Afghanistan) and San Diego pegmatite (USA). The range *"sol-liq"* indicates solidi and liquidi in F-, B-, P-rich pegmatitic melts (London 1986a,b)

An important question is whether the composition of the pegmatite-forming fluids change continuously between melt and solution or whether this change occurs suddenly because of immiscibility or exsolution. As far as the highly fractionated rare-element pegmatites are concerned, continuous development seems to dominate (London 1986a). London did not find any indication that an exsolved aqueous fluid existed in the early stage of pegmatite consolidation in the highly fractionated Tanco pegmatite. Foord (1976) and London (1986a) raised the question of the efficacy of H_2O in promoting chemical segregation in pegmatites if primary fluids do not contain appreciably halide salts. It has been suggested that alkali borate acts as a promotor of the H_2O solubility in the silicate-rich fluids, which depolymerizes the silicate network (Manning et al. 1980; Pichavant 1981; Tingle and Fenn 1984; Candela and Holland 1986).

The behaviour of REE in pegmatite fluids is significantly different when compared with that in related granites. For instance, Simmons et al. (1987) have observed that the wall zone is depleted in REE when compared with the hosting granite. In the pegmatite REE, together with Th and Zr, are concentrated into minerals crystallizing from the very last fluid.

After crystallization of minerals that largely consumed the network modifiers in the melt, a CO_2-bearing low-salinity homogeneous fluid is generated (London 1986a).

2.6 Synopsis of a Genetic Model

Rare-element pegmatites are related to granites enriched in rare elements. Under P-T conditions of the plutonic emplacement, when internal pressures

$$P_{\text{granite melt}} > P_{\text{metamorphites}},$$

small amounts of late water-enriched pegmatitic melts escape from the almost solidified granites along fractures and forcibly open weak zones in the surrounding metamorphic rocks (Fig. 11). The behaviour of trace and minor element distributions suggests that the pegmatitic melt is different in chemical composition and structure; for instance, across the granite-pegmatite boundary the REE concentration of whole rocks changes drastically (Simmons and Heinrich 1980). After the

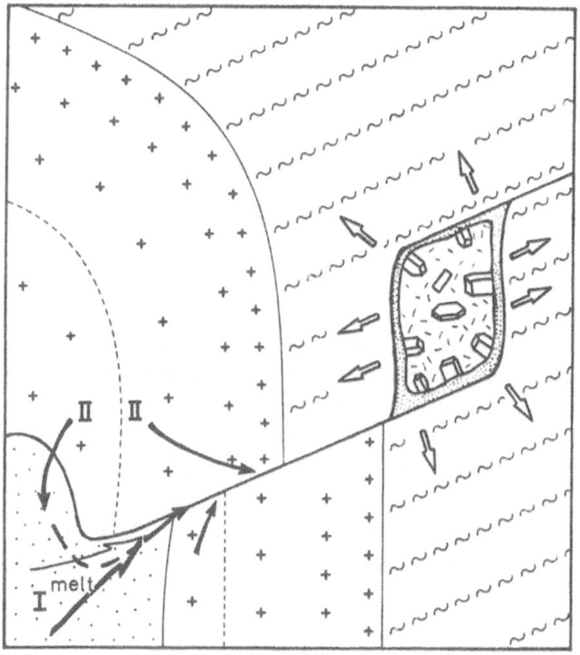

Fig. 11. A simplified model showing the discussed relationship of pegmatites to their parental granites and the different sources of fluids. *Fluid I* is a late residual hydrous granitic melt, *fluid II* represents the intergranular liquid which mixes continuously with the former. The pegmatite is considered as an open system at least for all non-hydrolyzing ions

release of the pegmatitic melt from the granite the residual fluid in the intergranular space will exsolve volatiles and overpressure forces these fluid to coalesce towards the still liquid centre of the granite or the pathway of the pegmatitic melt. The intergranular fluid is present only in small quantities, probably less than 10^{-4} of the original melt volume, but with extreme fractionation of coherent elements and extreme enrichment of incompatible elements. When both liquids mix, the concentration of rare elements increases in the pegmatite melt. The latest fluids move towards the pegmatite at a rate controlled by (1) the escape of volatiles from the pegmatite into the metamorphic host rocks (forming well-known haloes of alkalis) and (2) the maintenance of channels between the source of liquids and the pegmatite.

Evidence for an at least partly open system is represented by the alteration halo of individual elements, and the non-unidirectional variation of trace element concentrations in a given mineral, e.g. Rb and Sr in Fig. 5.

Under different aspects, a second fluid has already been assumed in order to interpret analytical results (Simmons and Heinrich 1980; Longstaffe 1982; Roedder 1984; Jolliff et al. 1987b). Although the need for its assumption was different, it may indicate that at least some of the rare-element pegmatites have been open systems for at least some time. Furthermore, this idea would not be in conflict with the model of pegmatite genesis described by Simmons et al. (1987).

Rare elements are enriched in at least two steps:

1. During solidification of the parental granites in the parent melts of the pegmatites and intergranular fluids; and
2. The fractional crystallization in the pegmatites leading to the distinct (zonal) distribution of minerals.

Fractionation in the parental melt is superior to fractionation in pegmatites (see discussion of Ga/Al, Fig. 1 and Ta/Cs, Fig. 5, fractionation). The latter enrichment has been described by Jahns (1982) to be controlled largely by exsolution of a vapour phase under the P-T conditions of the emplacement and the cooling history. Later, London (1986a,b, 1987) pleaded for a continuous change from melt to aqueous fluid. Depending on the preference of elements to either alkali borate-silicate melts or aqueous solution explains the distribution of the rare elements within pegmatite bodies.

3 Carbonatite-Alkalic Rock Complexes

Carbonatites are rocks which contain at least 50% carbonate minerals and are divided into calcic (sövite), dolomitic (beforsite), ferro- and natro-carbonatite (Bowden 1985). Carbonatites always occur together with alkalic rocks of greatly varying mineralogical and chemical composition, but alkalic rocks are not always associated with carbonatites. Many of the alkalic rocks are silica undersaturated with molar proportions of alkalis to silicon in excess of 1:3. They contain feldspathoids or alkali minerals such as acmite. Alkali feldspars are the predominant feldspar (Sörensen 1979, 1986). The association of carbonatites and alkalic rocks is by far the most important primary source of Nb and REE(Y).

3.1 Element Fractionation

3.1.1 Partition of Elements Between Alkalic Rocks and Carbonatites

Element distribution patterns are a useful tool in the discussion of element fractionation. In these patterns the elements are arranged according to their valencies and within each group of valencies with respect to ionic radii. The actual element contents are normalized to those of the primitive mantle. The characteristic differences in element distribution between alkalic rocks which are associated with carbonatites and those which are not are demonstrated by the two complexes of Grönnedal-Ika and Motzfeldt Centre/southern Greenland, respectively.

At the Motzfeldt Centre most alkalic rocks of the sequence syenite, nepheline syenite and amphibole-bearing syenite show conspicuous depletion of Ca, Sr, Ba and Eu (Fig. 12) as a function of magmatic differentiation. The segregation of Ca-rich cumulates (rich in plagioclase) from the alkalic melts accounts for Ca, Sr, Ba and Eu depletion in evolved melts; additional depletions of Ti, Sc and Fe suggest the crystallization of titanite and ilmenite. From top to bottom the graphs in Fig. 12 represent increasing element fractionation. During the first stages of differentiation, the incompatible elements Rb, Cs, Sr, Ba, Sc, REE, Hf, Nb, Ta, Th and U are only slightly enriched (less than one order of magnitude), partly because they are taken up by the precipitating feldspar (Ba, Sr, Eu), and partly because the volume reduction of the melt by feldspar segregation cannot have exceeded a few tens of percent. A considerable enrichment of incompatible elements in suites of syenitic rocks is observed when SiO_2-undersaturated melts become saturated (Fig. 12). This can be explained by the reaction melting of intruded undersaturated liquids with their roof rocks. A possible reaction would be

$$Ca^{2+}(m) + 2\ KAlSi_3O_3(s) = CaAl_2SiO_8(s) + 2\ K^{+}(m) + 4\ SiO_2(m)$$
$$(m) = \text{melt};\ (s) = \text{solid}.$$

The released SiO_2 effectively saturates or even oversaturates the melt. Simultaneously, the alkali content of the melt increases. In the course of this process other minerals of the country rocks may be selectively dissolved in the melt causing the observed increase in Th, U, Nb.

Alkalic rocks associated with carbonatites (Fig. 13) show some features different from those in Fig. 12. A comparison of the syenites, nepheline syenites and carbonate-bearing basalts with the carbonatites demonstrates that

1. Carbonatites are enriched in Sr;
2. REE(Y) show rather straight line distribution patterns in alkalic rocks compared with the convex curvature of the same in carbonatites and carbonate-bearing basalts;
3. Eu anomalies are not developed, although Ca is extremely low in some of the samples;
4. Normalized Ta/Nb ratio $\ll$ unity are typically found in carbonatites and carbonate-rich alkalic rocks.

The difference in behaviour of Eu in nepheline syenites from the Motzfeldt Centre (Fig. 12) versus Grönnedal-Ika (Fig. 13) can be explained by plagioclase fractionation in the former, and segregation of carbonate-rich phases in the latter.

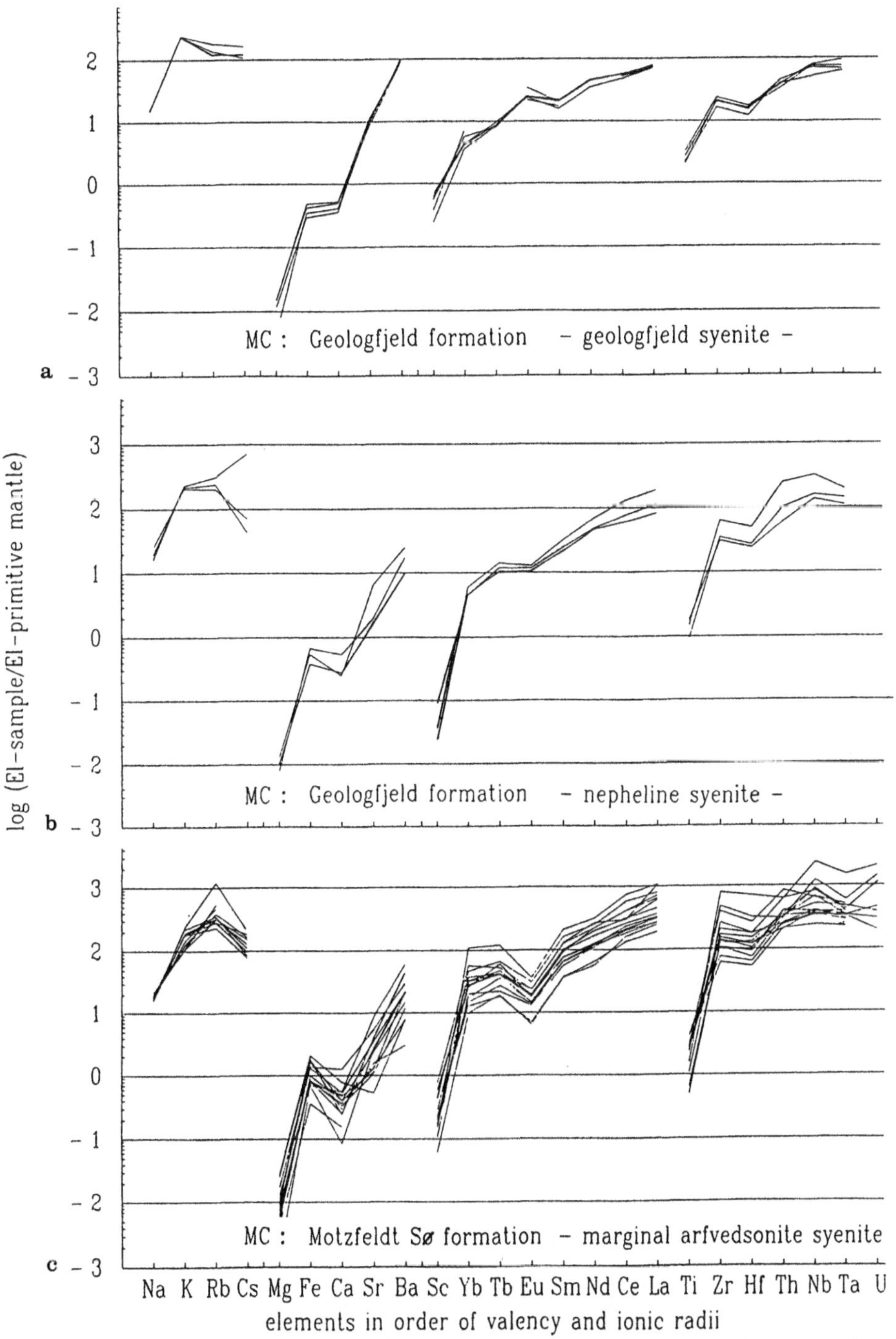

Fig. 12. Element distribution patterns of alkalic rocks of Motzfeldt Centre, Greenland. **a** syenites considered to be the least differentiated among the sequence of alkalic rocks; **b** nepheline syenites thought to be differentiation products of the foregoing syenite; **c** amphibole-bearing syenites considered to be the result of reactions of undersaturated melts with alkali-feldspars from the lower crust. Note that the normalized Nb/Ta ratios are near unity in all alkalic rocks. For normalization a set of data has been derived from Wood (1979), Sun and Nesbitt (1977) and Sun (1982)

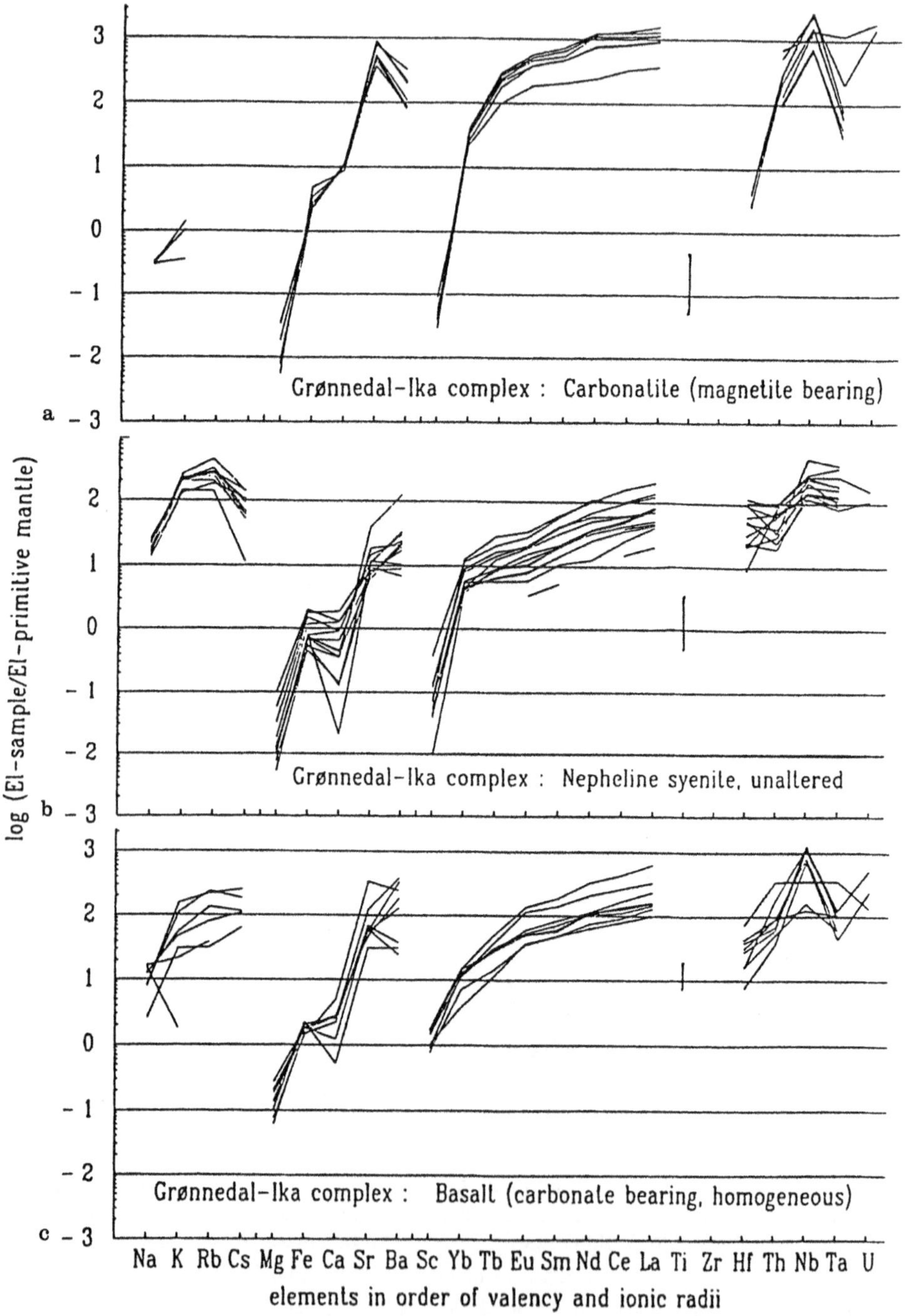

Fig. 13. Element distribution patterns of carbonatite and associated alkalic rocks, Grönnedal-Ika, Greenland. **a** magnetite-bearing carbonatites with REE distribution patterns and normalized Nb/Ta ratios distinctly different from those of the nepheline syenites in **b**; **c** late basalts which interacted with a carbonatite and acquired carbonatitic characteristics. For normalization, see legend of Fig. 12

Alkali silicate melts that separated from carbonate melts must have inherited both the latter's REE(Y) distribution patterns and Ta/Nb ratios. Ta/Nb and heavy REE(Y)/light REE fractionation should be a reliable tool to distinguish between consanguinous and co-existing rocks. Incompatible elements are accumulated in carbonatite liquids and alkali silicate liquids to different degrees: carbonatites are preferentially enriched in Sr, Ba and light REE, while alkalic melts are enriched in Ta, Nb, U, Th and heavy REE.

3.1.2 Rare Earth Elements

REE(Y) are very common in carbonatites and alkalic rocks. The undifferentiated mantle is characterized by a flat REE(Y) distribution pattern with normalized values of unity. Under high temperature conditions the light REE carbonates should be more resistant to thermal decomposition than the heavy REE which are less polarizable in their compounds. This would explain the typical REE(Y) pattern of both carbonatites and alkalic rocks in which the ratios between the elements decrease slightly from La to Lu (Fig. 14). Similar trends of REE(Y) distribution patterns of both carbonatites and alkalic rocks are typical for partial melts from the earth's mantle. The REE(Y) distribution patterns characteristically show strong enrichment of the light REE with the exception of samples representing melts from a depleted upper mantle (e.g. MOR basalts).

Negative Eu anomalies in carbonatites are almost unknown. Anomalies have only been reported for individual minerals such as apatite from Oka (Eby 1975) and

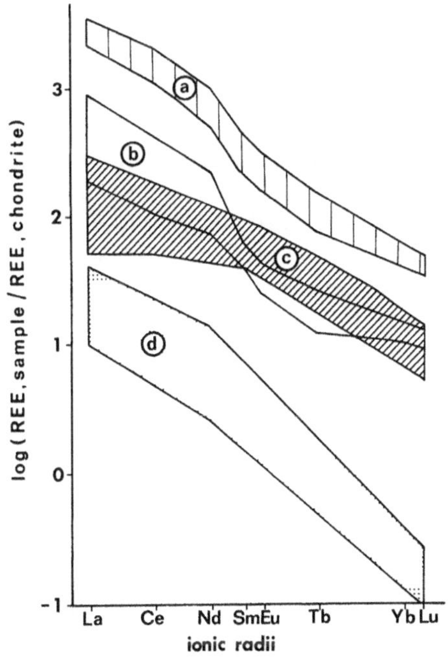

Fig. 14. REE distribution patterns of sövites (**a**) and associated alkalic rocks (**b**) from the Kaiserstuhl/Germany, alkali basalts (**c**) and K-rich peridotites (**d**). All patterns show similar trends which are typical for partial melting

have been observed in various alkalic rocks from southern Greenland not associated with carbonatites (Figs. 12 and 13).

Experimental partition coefficients of REE(Y) between silicate and carbonate melt (Wendlandt and Harrison 1979; Bedson 1983; Hamilton et al. in press) demonstrate that light REE are enriched in carbonate melts. In carbonate-phlogopite association the silicate-carbonate distribution coefficients, $D^{S/C}$, of REE(Y) are in the range of 0.2 to 3. The ratios increase from La to Lu, and decrease with an increase of pressure and with falling temperature; P_2O_5 may further decrease $D^{S/C}$. At low temperature and high pressure the light REE are preferentially accumulated in carbonatites and the enrichment is further increased with falling temperature and pressure (Bedson 1983; Hamilton and Bedson 1986).

3.1.3 Niobium and Tantalum

Ta is not concentrated into carbonate fluids to the same extent as Nb. Nb solubility is always greater than Ta in fluids rich in carbonates, fluorides and phosphates (Aleksandrov et al. 1985); e.g. 3–4% pyrochlore is soluble in fluorite-calcite-portlandite melts. Alkalis seem to be unimportant for pyrochlore crystallization, but early crystallization of apatite removes P_2O_5 and F^- which lowers the Nb_2O_5 solubility and causes pyrochlore to cocrystallize (Gittins 1986). Aleksandrov (1967) reported considerable solubility of polynuclear carbonato-complexes of Nb in alkali-carbonate melts to exist in the range of 50° to 450°C and pressures up to 1 kbar. Stable fluoro-complexes are known at high F^- contents. F^--rich carbonatites are therefore notoriously poor in Nb compounds. The solubility behaviour may also be related to the reactivity of the pentoxides with $CaCO_3$ which is less for Ta_2O_5 than for Nb_2O_5 (Gmelin 1971). Ta is found to be strongly enriched in the silicate liquid during separation of immiscible carbonate- and alkaline silicate fluids (Hamilton et al. in press). Nb-rich carbonatites are observed also to be rich in Fe (Figs. 12 and 13), which points to digestion of specific Fe-Nb compounds during the ascent of the parental magma.

In cases where an independently generated carbonate melt comes into contact and partly mixes with an alkaline silicate melt, or its rock precursor, Nb and Ta are partitioned between the two. Thereby alkalic rocks, formed at higher levels of the upper mantle than the carbonatites, may gain some Nb from the carbonate melt by exchange. This characteristically leads to a decrease in the Ta/Nb ratio, which is about 1/10 and 1/15 for crustal and upper mantle rocks respectively (Wedepohl 1969; Heinrich 1969). Alkalic rocks not associated with carbonatites show nearly normal crustal or mantle ratios (Table 2), although the contents of Nb and Ta vary considerably. Carbonatites always show lowered Ta/Nb ratios, often 1/100 only.

3.2 Isotope Chemistry

The low Sr-initial ratios of alkalic rocks and carbonatites indicate a source with a low Rb/Sr ratio. Sr-initial ratios in carbonatites and most of the alkalic rocks are consistent with Rb/Sr = 0.03 for the primitive upper mantle (Fig. 15). From this

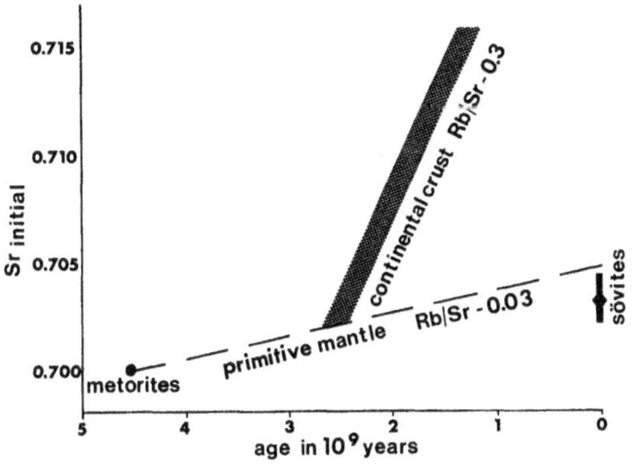

Fig. 15. Systematics of initial Sr-isotope ratios. The *broken line* demonstrates the development of Sr-initial ratios of the primitive mantle. Since carbonatites are Rb-poor, their actual Sr ratios correspond to Sr-initial ratios. The development in alkalic rocks is determined by the amount of assimilated Rb. Sr-isotope ratios of alkali cover the range from 0.703 to 0.72

mantle the alkalic rocks and carbonatite are derived with Rb/Sr < 0.2 and Rb/Sr < 0.01 respectively (Faure and Powell 1972; Sörensen 1979). The Rb/Sr ratios of some alkalic rocks exceed 0.03. These rocks either assimilated phlogopite in the upper mantle or lower crustal material during ascent. The above Rb/Sr ratios in carbonatites and alkalic rocks do not really represent ratios in the source area because of the differences in distribution coefficients of Rb and Sr, mineralogy and degree of partial melting.

Positive Nd and negative Sr values in the carbonatites of the Canadian Shield and East African Rift indicate a large ion lithophile depleted subcontinental mantle source region (Kwon and Tilton 1986; Bell and Blenkinsop 1987). At Palabora the Nd and Sr indicate mixing of magmas from two distinct mantle sources, one similar to basaltic kimberlites, the other more enriched than micaceous kimberlites (Allsopp 1986). An additional crustal source for U, Pb and C is suggested at Palabora, since $\mu = 11$ and $\delta\ ^{13}C$ in the range of -3 to -5‰ (PDB) have been found.

Pb isotope ratios of the Canadian carbonatites plot below the Stacey-Kramers line and point to a mantle source similar to that which yields ocean island basalts. Mixing of depleted and undepleted melts from the mantle source is therefore suggested by Bell and Blenkinsop (1987) for the Canadian carbonatites. They advocate a closed mantle reservoir from which the parental magmas are derived. At least up to the time of magma formation the mantle was rigidly coupled to the crust. Similarly, Kwon and Tilton (1986) use Nd, Sr, Sr-initial ratios, and Pb-isotope ratios to deduce that carbonatites had geochemical fractionation histories similar to oceanic island basalts.

The probable field of primary igneous carbonatites (Taylor et al. 1967) is given by:

$-8 < \delta^{13}C < -5‰$ (PDB)

$+6 < \delta^{18}O < +9‰$ (SMOW).

However, higher values are also reported, these are considered indicative of differentiation (increasing $\delta^{13}C$ and $\delta^{18}O$) or alteration (increase of $\delta^{18}O$ alone).

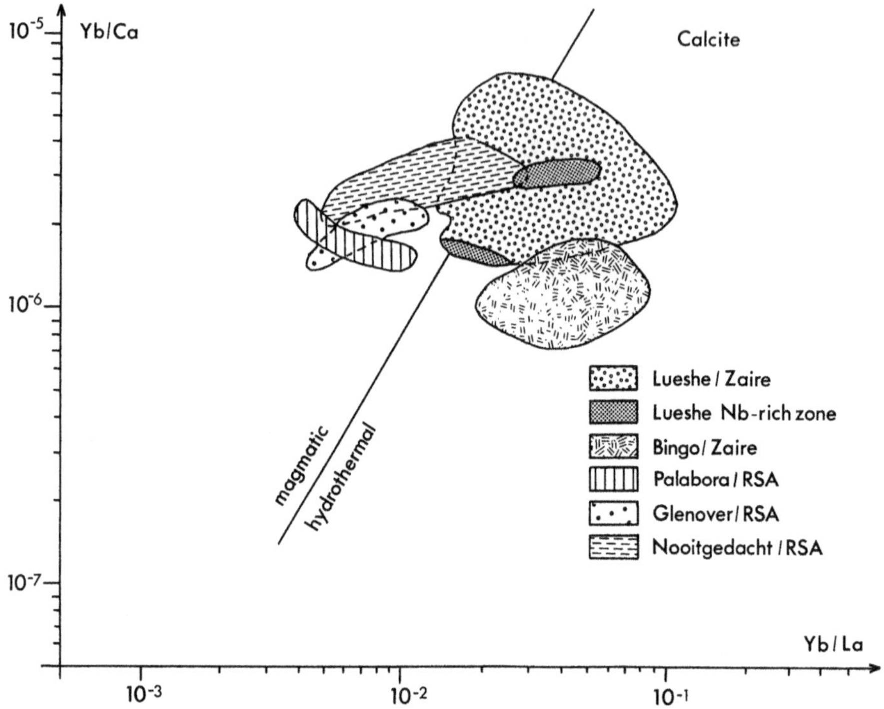

Fig. 16. Yb/Ca vs Yb/La diagram for calcite. The *borderline* that separates hydrothermal from magmatic calcites is taken from Möller (1983). Many of the calcite samples plot in the field of hydrothermal calcite; such samples indicate post-magmatic alteration processes. In particular, the Nb-rich zones of Fen/Norway and Lueshe/Zaire fall into the hydrothermal field. St. Honore/Canada and Oka/Canada have been mined, Lueshe/Zaire is actively mined for pyrochlore and Bingo/Zaire and Mbeya Hill/Tanzania have considerable reserves. Kaiserstuhl/Germany, Alnö/Sweden, Glenover/RSA, and Palabora/RSA are barren of Nb mineralization (unpubl. data)

3.3 Alteration Processes

Post-magmatic alteration severely affects the distribution of incompatible elements, e.g. REE(Y) are strongly complexed by carbonate-rich fluids. More importantly, they are adsorbed on carbonate surfaces where REE(Y) start to form their own minerals. To some degree they are also leached from the carbonates and co-precipitated in hydrothermal vein mineralizations of the country rocks. The effect of leaching can be seen in the increased Yb/La ratio (Fig. 16) and simultaneous loss of Ba by up to two orders of magnitude (Fig. 17). During post-magmatic alteration, crystallization of Nb minerals is favoured, probably because Nb species are mobilized in alkaline fluids. Movement of fluids along contact zones, fractures and faults, however, can result in considerable local Nb enrichment.

Weathering of carbonatites enriches Nb minerals in the sediment cover. This is a very important physical enrichment process. Araxa is the most important producer of pyrochlore which is recovered from deeply weathered carbonatites. Finely dispersed REE(Y) in $CaCO_3$ tend to form their own minerals during weathering.

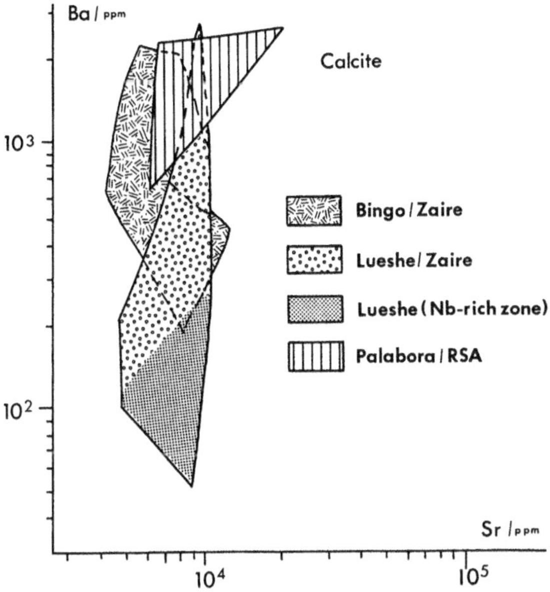

Fig. 17. Ba-Sr correlation in calcite from carbonatites. Ba is often observed to be highly variable, whereas Sr stays rather constant. The decrease of Ba in Lueshe is related to weathering processes during which the calcite recrystallized (unpubl. data)

This is considerably favoured if PO_4^{3-} and F^- are present. In the presence of oxygen, florencite, REE(Y)$Al_3[(OH)_6(PO_4)_2]$, (showing a negative Ce anomaly) and cerianite), $(Ce, Th)O_2$, are formed. Synchysite, $CaCe[F/(CO_3)_2]$ is a secondary mineral in the carbonatites of Glenover/RSA and Mountain Pass/Cal., USA.

The excess of alkalis induces fenitization (alkali and ferric iron metasomatism) of the wall rocks. Fe^{3+} ions substitute for Al^{3+} in feldspars. These fenites are also often enriched in Nb (Vlasov 1966; von Maravic 1983), indicating the tendency of Nb compounds to stay in highly alkaline solutions.

3.4 Melt Relationships

A much debated question is whether carbonatites originate from partial melts of the deep upper mantle by

1. Fractional crystallization of silicates from a CO_2-rich alkaline magma, or
2. Separation of an immiscible carbonate melt from an initially homogeneous CO_2-rich alkaline silicate magma.

Carbonate melts have low viscosities (~ 0.05 poise), low density (~ 2.2 g/cm³) and lower heats of fusion (~ 300 J/g) (Treiman 1986) than common silicates (~ 500 J/g). Due to these features coarse carbonate crystals would settle quickly (Bedson 1984; Treiman 1986). If carbonate fluids form as primary melts they should form earlier and ascend faster than associated silicate magmas. This has been used as an argument for melt separation during ascent within the earth's crust shortly before intrusion or extrusion. In contrast, fractional crystallization could have occurred during 10^3 to 10^5 years before intrusion or eruption (Barker 1986). In order to produce the coarse, gneissic texture of many carbonatites, subsolidus plastic deformation and sald-domelike emplacement of almost solid carbonatite has to be assumed as the very last act. Carbonate melts intrude most or all affiliated silicate rocks. Exsolution of alkali-rich fluids may cause carbonate crystallization (Barker 1986).

Fluid inclusions in minerals from carbonatites reveal a wide range of temperatures and compositions. They include carbonate melts, various carbonate-silicate-salt-H_2O mixtures and aqueous solutions. Melt inclusions that have been studied in silicates and carbonatites show lower homogenization temperatures in the latter (Roedder 1984). Haapala (1980) reported primary polyphase aqueous inclusions in apatite from Sokli carbonatite, Finland, with >60 vol% daughter crystals. The vapour phase homogenized between 250° and 350° C. Daughter crystals started to dissolve below 500° C, but most of them melted between 500° and 550° C. Total homogenization was not reached. Rankin and Le Bas (1974) observed directly liquid carbonate melt immiscibility during heating of fluid inclusions in apatite from carbonatites. The inclusions homogenized at higher temperatures and unmixed on cooling.

In carbonate-silicate systems large solvus areas exist at about 1 bar. At 30 kbar, solvi are observed neither in Ca^{2+} nor Mg^{2+} carbonate-silicate systems (Huang and Wyllie 1974). In the system Na_2O-K_2O-CaO-SiO_2-Al_2O_3-CO_2 a solvus is known between SiO_2-Al_2O_3 melt and alkali carbonate liquid which closes towards $CaCO_3$ (Hamilton et al. 1979; Freestone and Hamilton 1980). CaO-rich carbonate and silicate melts do not show an immiscibility gap at pressures of 1 to 8 kbar. Therefore, sövite-type carbonatites either do not owe their origin to immiscibility or have lost their alkalis before emplacement (Hamilton and Bedson 1986). Recently, Kjarsgaard and Hamilton (in press) have experimentally demonstrated the direct generation of sövite-type melt by liquid immiscibility in alkali-poor systems. Le Bas and Handley (1979) considered sövites and ijolites as crystalline products of conjugate immiscible liquids.

Koster Van Groos and Wyllie (1966, 1973) and Koster Van Groos (1968) outlined a liquid immiscibility field intersected by joins through the system Ab-An-Cc-H_2O at 1 kbar, and discussed its petrologic application involving the coexistence of a nepheline-normative silicate, a carbonate-rich liquid and a vapour phase. The extension of these studies to 10 kbar by Koster Van Groos (1975) confirms that in the presence of CO_2 immiscible carbonate liquids separate from a wide range of igneous rock compositions. The carbonate liquid is always enriched in alkalis. Although the quantities of alkalis in peridotites are small, their presence during magmatic processes in the mantle could be of great significance (Huang and Wyllie 1974).

Immiscibility of fluids in carbonate-silicate systems is often considered to be of fundamental importance in the formation of carbonatite-alkalic rock complexes (Middlemost 1974; Mitchell and Brunfelt 1975; Koster Van Groos 1975; Brey and Green 1976; Le Bas 1977; Verwoerd 1978; Freestone and Hamilton 1980). Gittins

(1986), however, finds this "importance" exaggerated. To him the phenomenon of liquid immiscibility is of minor importance in enrichment processes because immiscibility will only occur at low pressures, i.e. towards the final stage of intrusion.

3.5 Source of CO_2

The groundmass of kimberlites contains carbonates which unquestionably originate from the earth's mantle. Deep source ocean island basalts are notoriously gas-rich, with CO_2 predominating (Thompson et al. 1984). Fluid inclusions in lherzolite xenoliths from kimberlites often contain only dense CO_2 which seems to be the most important volatile component in the mantle (Roedder 1965; Jackson and Wright 1970; Green and Radcliffe 1975). There is little doubt that volatiles like CO_2 or H_2O play an important role in the formation of the asthenosphere. The controversy is whether melting of carbonates (Wyllie and Huang 1975) or separation of liquid CO_2 from mantle silicates (Green 1972) support the formation of the asthenosphere.

H_2O and CO_2 are considered to be critical compounds in the melting process of mantle material. For example, H_2O depolymerizes silicate melts

$$\text{-Si-O-Si} + H_2O = 2\ \text{-Si-OH},$$

whereas CO_2 enhances their degree of polymerization

$$2\ \text{-Si-O-M} + CO_2 = \text{-Si-O-Si-} + M_2CO_3$$
$$M = \text{monovalent metal ion.}$$

The resulting lowered viscosity of hydrous relative to anhydrous melts may add to the reduction of seismic velocities of the asthenosphere. In contrast, higher seismic velocities may be expected for CO_2-bearing silicate melts at the same pressure and temperature (Mysen 1977). These effects may be overshadowed by the degree of partial melting.

3.6 Evolution of Silicate-Saturated Carbonate Melts

The subcontinental mantle is considered to be lighter and more refractory than the oceanic because of its early depletion during formation of the crust, but seems to be still sufficiently fertile to produce alkaline magmas. This apparent paradox is solved if one assumes that parts of the subcontinental mantle are metasomatized by fluids from greater depth that approximate alkali basalts in composition, i.e. fluids that are rich in incompatible elements (Jordan 1978). The generation of these metasomatizing fluids may be triggered by volatiles from the deep mantle which is assumed to exude carbonate-rich fluids. They are though to dissolve accessory mineral components such as phosphates, niobates, tantalates and, of course, silicates. This silicate-saturated carbonate fluid preferentially leaches compounds of incompatible ions that are not hosted by rock-forming minerals. Cation such as REE(Y) U, Th, Zr etc. are concentrated by leaching as chemical complexes in the fluids. Frey and Green (1974) questioned whether such small melt volumes of strongly alkaline composition could penetrate the relative cold and rigid overlying lithosphere. They are expected to freeze in the lower lithosphere and give rise to a "veined" mantle. During enforced crystallization hydrous CO_2-rich fluids are expelled which initiate metasomatic reactions in the surroundings and some of the volatiles emanate into the crust. Metasomatic mineral reactions in the mantle under continents may be considered as an initial enrichment process of incompatible elements. Mantle metasomatism is promoted by escaping volatiles which form an aureole of reaction products around the frozen magma. For instance, in the reaction

$$CO_2 + CaMgSi_2O_6 = CaCO_3 + MgSiO_3 + SiO_2$$

the light REE are released because the heavy REE are preferentially incorporated into pyroxenes such as enstatite and bronzite (Möller 1988).

The most common carbonate-silicate magma is of olivine-sövite composition and is enriched in Na_2O and K_2O (Gittins 1986). Rare-element-enriched carbonate magmas are thought to be generated by fractional melting of carbonate-bearing peridotites (Watkinson and Wyllie 1971; Eby 1975) at pressures > 27 kbar in regions where CO_2, Nb, REE(Y), P and F are (pre-)concentrated. These conditions, however, may not be very common (Gittins 1986).

3.7 Source of Alkaline Melts

Experimental studies by Eggler (1974) demonstrated the role of the volatiles H_2O and CO_2 during partial melting in determining basaltic versus nephelinitic composition of melts. Sr and Nd isotope data indicate that most of the evolved alkalic rocks are formed by partial melting of deep-seated basaltic rocks or from melts from the upper mantle (Neuman 1980; Blaxland et al. 1978). Other authors assume that alkalic magmas are derived from a deeper level of the mantle or represent smaller degrees of partial melting than tholeiitic magmas (Gast 1968; Oxburgh and Turcotte 1968). Potassium-rich melts should form at deeper levels in the mantle than sodium-rich ones (Ryabchikov and MacKenzie 1984). From mantle nodules, trace element data and isotope data Lloyd and Bailey (1975) deduced that such melts formed from a metasomatically altered mantle. In contrast to these concepts, Clarke et al. (1983) provided evidence for the formation of alkalic melts by extreme differentiation of tholeiitic melts trapped in accreting oceanic plates.

The majority of petrologists would agree that the alkalic rocks are derived from mantle material (Sörensen 1986). Most of the characteristic features of alkalic rocks can be explained by their mantle origin. Crustal origin is considered to be less reasonable. If a contribution of the lower crust was dominant, a much wider scatter of isotope signatures and trace element distributions would be expected. In the following discussion a mantle origin of alkalic rocks will therefore be assumed.

It has been argued that the subcontinental mantle is partly metasomatized by fluids that are rich in hygromagmatophile elements (Jordan 1978). Evidence for a metasomatized mantle below continents is given by Frey and Green (1974), Sun and Hanson (1975), Lloyd and Bailey (1975), Frey et al. (1978) and Tarney et al. (1980). It has further been suggested that the heterogeneously metasomatized mantle is more likely to be conserved beneath stable continents than the ocean crust because with time the convective cycling in the oceanic mantle with effectively disperse any significant chemical and mineralogical inhomogeneities.

Saunders et al. (1980) assumed that alkalic magmas formed under conditions of low PH_2O. Under such "rather dry" conditions the metasomatic mantle hosts mineral suites comparable to the "MARID" suite known from kimberlites (Gurney and Harte 1980), i.e.

*m*ica[REE(Y)], *a*mphiboles(REE), *r*utile(Nb,Ta),
*i*lmenite(Nb,Ta) and *d*iopside(REE).

These pre-enriched phases will be contributors of REE(Y), Nb and Ta to the magmas.

Mantle upwelling initiates doming of the lithosphere and the resultant pressure release accelerates degassing. Deep faults may act as foci for volatiles to start metasomatism. Fracturing of these domes triggers pressure release leading to partial melting of the metasomatized lithosphere resulting in alkaline magmatism (Woolley 1986).

Compared with the composition of the lithosphere, carbonatites are enriched in Ca, Ti, P, Zr, Nb>Ta and REE(Y). The occurrence of perovskite, pyrochlore, ilmenite and magnetite in carbonatites is mineralogically salient (Haggerty 1986) which is supposed to originate from "flash" melting of mantle metasomatites which include titanates, pyrochlore, amphiboles and carbonates. Carbonatites are super-enriched with Ti, Zr, Nb and REE(Y) compared with kimberlites. Although the ultimate sources of carbonatites and kimberlites are similar, the enrichment is quite different. It seems that the distinctions are primarily the result of time-integrated metasomatism which is short for kimberlites and rather long for carbonatites (Haggerty 1986).

Loss of alkalis from the carbonate melt causes it to separate into a silicate melt and a carbonate-rich fluid (Hamilton et al. 1979). At that time both fluid phases are saturated with respect to the components of the other. During cooling and crystallization each in turn may continue to suffer liquid fractionation.

The combined evidence from isotope and REE(Y) distribution patterns of alkalic rocks provides an overwhelming case for rock assimilation by silicate fluids on their way into the crust. This melt-rock interaction differs from place to place and may be related to the rate of diapiric ascent. There is no direct evidence for carbonate melt-country rock interaction, except through late reaction or mixing, as shown in Fig. 13 for carbonate-bearing basalts. Magmas derived from the deep mantle, rich in incompatible elements, locally metasomatized the earlier depleted upper mantle (Fig. 18,a) or accumulated in advancing magma chambers where differentiation continued to take place (Fig. 18b,c).

3.8 Cogenetics vs Coexisting Melts

Widely different models have been proposed to explain carbonatite-alkalic rock associations: differentiation of a carbonated silicate melts (cogenetic) or independent sources for the different melts in the mantle (coexisting). These diverging views are still debated: Middlemost (1974); Koster Van Groos (1975); Le Bas (1977); Freestone and Hamilton (1980); Wyllie (1987); Kjarsgaard and Hamilton (in press).

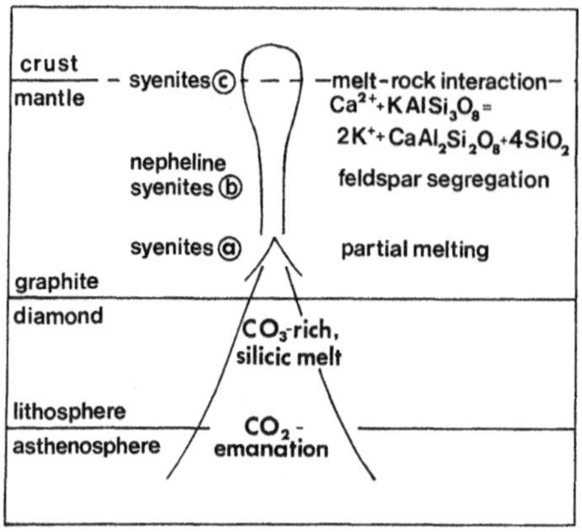

Fig. 18. Outline of the different origins of alkaline silicate and carbonate melts. CO_2 probably has its source near the lithosphere-asthenosphere boundary and starts partial melting of material below the graphite-diamond transition boundary. Alkalic rocks largely originate from higher levels in the metasomatized mantle (**a**). Generation of alkalic silicate melts may be triggered by volatiles from below. Evolution of alkalic silicate melts during ascent leads to SiO_2-undersaturated melts (**b**) by feldspar segregation, or by melt-rock interaction to SiO_2-oversaturated syenites (**c**). This model explains why all carbonatites are associated with alkalic rocks, whereas alkalic rocks are not necessarily accompanied by carbonatites. Probably the carbonate melts use the pathways of the preceding alkalic silicate melts

Mitchell and Brunfelt (1975) and Cullers and Medaris (1977) assumed that initial alkaline melts which form by less than 5% partial melting of upper mantle material have lower REE(Y) contents than ijolites. The presumed primary magma could have undergone fractional crystallization of phases such as clinopyroxene, olivine and garnet to produce residual melts with REE(Y) contents typical of ijolites. The carbonatite is assumed to have formed from an immiscible carbonate-rich liquid which separated from the "parent" ijolite liquid. Any residual liquid after ijolite formation is too low in REE(Y) contents to produce the higher REE(Y) contents of carbonatites. Therefore, extensive fractional crystallization of ijolite to produce residual carbonate melt does not appear to be a viable model to explain the high REE(Y) content in carbonatites. Kjarsgaard and Hamilton (in press) propose that exsolution of a first carbonatite liquid occurred at about 20 km depth. Only then will the carbonate contain, among other incompatible elements, high amounts of REE and Nb. Immiscibility at higher levels should lead to barren carbonatites. Probably there will always be a miscibility gap between silicate and carbonate melts in the anorthite-$CaCO_3$ system (Hamilton, personal communication).

Wyllie and Huang (1975) postulated from experimental results that partial melting of CO_2-bearing peridotite at depths > 80 km would produce carbonatite. With progressive fusion, kimberlitic and basaltic magma would form.

It also seems reasonable that independent carbonate and silicate magmas are produced separately at different levels. Such melts could maintain contact with one another as they were forcefully injected to shallow depth. REE(Y) are concentrated in the carbonatite melt relative to the alkaline silicate melt. Alkalic rocks with significant negative Eu anomalies and element distributions very different from those of carbonatites must be taken as proof that the two melts evolved independently from each other. It follows that the carbonate liquid originates from a deeper, CO_2-bearing level in the mantle than the alkalic rocks. Thus, in the Motzfeldt Centre (Fig. 12) only late carbonatitic dykes are found and the various alkalic rocks show systematically changing degrees of negative Eu anomalies. Late SiO_2-oversaturated syenites from Grönnedal-Ika show the same features. These late syenites cannot have evolved by fractional crystallization or immiscibility from a common carbonate-silicate melt because carbonatites with negative Eu anomalies are unknown. On similar grounds nepheline syenites showing Eu anomalies cannot directly be related to carbonatites.

If associated alkalic rocks and the carbonatites are derived from a common parental melt, both rock types should show sympathetic element distribution patterns. If alkaline silicate melts and carbonate magma are immiscible pairs, then separation must have occurred prior to feldspar segregation. Evolution of syenites and nepheline syenites with negative Eu anomalies must have taken place later because carbonatites do not exhibit REE anomalies. In this case the carbonatites should have intruded first because of their low density and viscosity. The observed intrusion sequence contradicts this assumption. For instance, the syenites and nepheline syenites of the Motzfeldt Centre are marked by negative Eu anomalies and cannot be the direct silicate partner of a parental alkaline silicate-carbonate magma. At this locality the alkaline silicate melts were generated independently from those of the carbonatites. Independent sources for carbonate and alkali silicate melts have already been quoted by von Maravic (1983) and have been discussed by Cullers and Medaris (1977) for individual bodies.

If the coexisting alkalic rocks inherited their REE(Y) patterns from carbonate melts, both REE(Y) patterns may resemble each other depending on the degree of equilibration. Subparallelism alone does not prove that alkaline silicate melts had the same parent; they could also be generated independently and equilibrated later. In the latter case they could have mixed with or have only contacted each other during diapirism. Admixing of carbonate melts with silicic melts would overprint typical features of element distribution patterns to produce hybrid patterns. These would be characterized by (Fig. 13):

1. Curvature of REE(Y) patterns;
2. $Ta/Nb_{normalized} < 1$;
3. $Sr_{normalized} > Ba_{normalized}$.

The ratio between the means of individual REE(Y) in carbonatites and alkalic rocks cannot be expected to be constant because $D^{S/C}$ depends strongly on temperature and pressure (Bedson 1984; Hamilton et al. in press) and probably on the presence of variable amounts of complexing ions and volatiles.

3.9 Synthesis of a Genetic Model

Figure 18 comprises the various aspects of petrogenesis of carbonatite-alkalic rock complexes.

Mantle convection cause doming of the lithosphere, followed by tectonic events with subsequent partial melting of the metasomatic mantle. Olivine nephelinites

(Fig. 18,a), syenites and nepheline syenites (Fig. 18,b) indicate the evolution of melts by fractional crystallization. If the SiO_2-undersaturated parental syenite melts are trapped in an environment in which they are completely out of equilibrium, saturated to oversaturated melts can result from interaction with surrounding rocks. Such a situation exists in the lower crust where the roof rocks are of granulitic composition (Fig. 18,c). Depending on the amount of reaction melting, the relative concentration of Ca and incompatible elements vary considerably. Heavy REE(Y) distribution patterns flatten and isotope signatures become more like those of the crust. Plagioclase segregation leads to formation of amphibole-bearing rocks which are derived from highest level magma sources during the final stages of magma evolution (Fig. 12). These alkalic rocks are not associated with carbonatites and probably gain REE(Y), Nb and Th by piecemeal stoping of roof rocks. The high chemical variability of these rocks indicates that their melts were not generated by magmatic differentiation. It is, therefore, better described by the term "melt evolution" (O'Hara and Mathews 1981). Magma-roof rock interaction probably takes some time and requires a quiescent period in the intrusive cycle. The maximum enrichment of the incompatible elements by about a factor of ten at this stage of magma evolution is therefore not the effect of volume reduction of the parental liquid by fractional crystallization, but reflects processes such as "leaching", extraction", "assimilation" or "exchange". The relative increase of Fe over Ca in association with the incompatible ions is observed to be systematic in late alkalic rocks.

If mantle fracturing reached deeper, emanation of volatiles from the asthenosphere triggered the generation of alkali-olivine-sövitic melts. On their way up immiscibility occurs, giving the light carbonate melt a chance to intrude the earlier intruded and already solidified alkalic rocks. The latter are chemically quite similar to those which would be expected to follow the carbonatites because all alkalic rocks had their ultimate source in the mantle. The spatial association of alkalic rocks and carbonatites is the result of deep-rooted fractures which serve as pathways in the ascent of both liquids. The assumption of a stratified upper mantle, as outlined in Fig. 18, offers plausible explanations for the intrusion sequence in any carbonatite-alkalic rock complex. The process starts with alkalic magmas that originate from high levels in the metasomatized upper mantle. Alkalic rocks without negative Eu anomalies suggest a rapid ascent without a chance for feldspar segregation. Due to lower densities and viscosities, the carbonate melts are expected to ascend faster than the related (immiscible) silicate melts. Therefore, the observation that carbonatites intrude alkalic rocks has to be interpreted as an intrusion into older alkalic rocks that were emplaced earlier.

Loss of alkalis is considered to lead to a decrease in solubility of apatite (REE) and pyrochlore($Nb > Ta$). The carbonatites represent the residue which remains after the release of an alkali-rich fluid (Gittins and McKie 1980). If alkali-rich carbonate fluids overprint alkaline melts or rocks (Fig. 13), the element distribution patterns would feign a cogenetic relationship.

In the proposed model (Fig. 18) immiscibility of carbonate and silicate melts is important, but not in a genetic sense with respect to surrounding alkalic rocks. Some of the alkalic rocks have a hybrid source, others may be consanguinous. The former originate from a higher level in the mantle than the carbonatites.

4 Conclusions

Pegmatites and alkalic rock-carbonatite complexes are considered to mark the end of long-lasting fractionation processes of large dimensions in the crust and deep upper mantle respectively. This explains (1) their scarcity on earth, particularly when considered in terms of volume and (2) the considerable enrichment of incompatible elements.

Enrichment of REE(Y), Ta and Nb in pegmatites and carbonatite-alkalic rocks is largely the result of ion fractionation during magmatic differentiation and partial melting respectively. The generalized result of fractionation of chemically coherent elements are shown in Table 3. The enriched species are italicized. Comparing the behaviour of coherent element pairs in the three rock types, it will be evident that in pegmatites and carbonatites always the higher and lower chemical homologs are enriched respectively. In alkalic rocks light REE dominate over the heavy and the Hf/Zr, Ta/Nb and Ga/Al almost occur in the primitive earth mantle ratio.

The Eu anomalies in alkalic rocks (not associated with carbonatites) and pegmatites are almost always negative. Carbonatites do not exhibit Eu anomalies. This feature seems to be related to differences in segregation of feldspars during magmatic differentiation.

Alkalic rock-carbonatite complexes represent differentiates of fractional melts of the asthenosphere (carbonatites) and the metasomatized mantle (alkalic rocks). The latter can be considered to be pre-enriched with respect to light REE, Nb and Ta as known from minerals from mantle xenoliths (e.g. the MARID suite). The contents of the elements is controlled by the degree of melting or "leaching" of mantle material. These melt fractions are highly enriched in incompatible elements such as REE, Ta and Nb. CO_3^{2-}-rich melts preferentially concentrate Nb, whereas alkali silicate melts have Nb and Ta in their primitive ratio.

In contrast, pegmatites represent crustal granitic differentiates. During differentiation, granitic melts only concentrate their REE(Y), Ta and Nb contents in the melt if complex-forming species are present. For example, F^-, PO_4^{3-} and CO_3^{2-} effectively complex REE(Y) and Nb > Ta, and enable the lanthanides to concentrate in accessory minerals such as fluorite, topaz, monazite, apatite etc. Ta and Nb are easily taken up by titaniferous minerals from which they are released during pegmatite-forming processes. The fact that REE(Y) and Ta,Nb are controlled by different minerals leads to the separation of the two groups of elements.

Table 3. Comparison of pairs of coherent elements in pegmatites, carbonatites and alkalic rocks (p.m. = primordial mantle ratio)

Pegmatites	Carbonatites/ alkalic rocks	Alkalic rocks not associated with carbonatites
Ga/Al	p.m.	Slightly increased
Hf/Zr	p.m.	p.m.
HREE/LREE	HREE/*LREE*	HREE/*LREE*
Ta/Nb	Ta/*Nb*	± p.m.
Eu negative	Normal	Eu negative

Pegmatites, alkalic rocks and carbonatites each enrich characteristic suites of incompatible rare elements and volatile components from pre-enriched environments:

pegmatites: Li, Rb, Cs, Be, B, Sn, REE(Y), Ta>Nb, F, B, P;
carbonatites: Sr, Ba, REE(Y), Ta(Nb), P, F;
alkalic rocks: Rb, Cs, Ba, Sr, REE(Y), Zr, Nb>Ta.

The rare elements are collected by mobile, volatile-rich intergranular fluids that have coalesced. In the case of pegmatites, the extremely fractionated intergranular fluids are mixed with the late melt fractions of specialized granites, whereas in case of the carbonatites the initial fluid is thought to trigger partial melting in the metasomatic mantle. In both cases the incompatible elements are mobilized after pre-enrichment. High enrichment of REE(Y), Nb and Ta is the result of extreme fractionation in granites (e.g. pegmatites) or differentiation of partial melts combined with the phenomenon of immiscibility of melts (e.g. carbonatite-alkalic rocks).

REE, Ta>Nb enrichment in pegmatites depends largely on the degree of pre-enrichment in parental leucogranites and less on fractionation during crystallization of the pegmatite itself, although the latter is important. For instance, REE minerals crystallize from the latest fluids in pegmatites. The Nb(Ta) content of carbonatites or alkalic rocks largely depends on the amount of assimilation of metasomatically enriched mantle. The pegmatites vary much more in concentration of incompatible elements than the carbonatites because they represent differently composed late phases of chemical fractionation, in contrast to the early partial melts from which carbonatites and alkalic rocks are derived.

Ca, Sr, Ba and Eu are greatly depleted by segregation of feldspars prior to formation of pegmatites and alkalic rocks not associated with carbonatites. Both rock types are therefore devoid of the large lithophile ions such as Ba^{2+}, Sr^{2+}, Eu^{2+} and Ca^{2+} (about 1/10 to 1/100 from the enrichment of other elements). By contrast, carbonatites and associated alkalic rocks are considerably enriched in these elements and demonstrate that feldspar crystallization was not essential in the development of their melts.

Individual models have been proposed in order to explain the formation of pegmatites and carbonatites and characteristic lines of enrichment of rare elements in these bodies. From a physico-chemical point of view both models have in common that the rare elements are collected into extremely small volumes by highly mobile, volatile-rich melts. Although one is tempted to generalize pegmatite and carbonatite genesis, different local conditions will lead to a variety of different features in individual geological bodies. Phase equilibria in these systems are very sensitive to the contents of minor components. Loss of volatiles at different stages of the chemical evolution of the fluids may lead to the observed great variety in mineralogy and rare element composition, particularly in pegmatites. The proposed models can therefore be considered only to serve as general guidelines. Nature chooses various paths, some are concomitant with the models, others may be still obscure.

Acknowledgements. This chapter includes short contributions by G.K. Muecke, M. Cuney, Ch. Knudsen and P. Černý which were given during the meeting. The author is particularly indebted to P. Černý (pegmatite part) and D.L. Hamilton (carbonatite part) for their critical comments; their comments significantly improved the manuscript. The assistance of G. Morteani and B. Trumbull in preparing the manuscript is gratefully acknowledged.

References

Aleksandrov IV (1967) Niobium in carbon dioxide solutions and considerations concerning migration of rare elements under hydrothermal conditions. Geokhimiya: 684–693

Aleksandrov IV, Krasov AM, Kochnova LM (1985) Effect of potassium, sodium and fluorine on association of rock-forming minerals and formation of tantalum-niobium ore mineralization in rare metal granite pegmatites. Geokhimiya: 620–629

Allsopp HL (1986) The Palaborwa complex: isotopic evidence for ancient lithospheric enrichment. Joint Annu Meet GAC, MAC, CGU, Ottawa 11:40

Barker DS (1986) Carbonatite emplacement mechanisms: a review. Geol Assoc Can Joint Annu Meet, Ottawa 11:43

Becker P, Bilal BA (1985) Lanthanide-fluoride ion association in aqueous sodium chloride solutions at 25°C. J Solut Chem 14:407–415

Bedson P (1983) The origin of the carbonatites and their relation to other rocks by liquid immiscibility. Thesis, Univ Manchester, UK

Bedson P (1984) Rare earth element distribution between immiscible silicate and carbonate liquids. Nat Environ Res Council Publ Ser D 25:12–19

Bell K, Blenkinsop J (1987) Archean depleted mantle: evidence from Nd and Sr initial isotopic ratios of carbonatites. Geochim Cosmochim Acta 51:291–298

Bergholz J (1978) Spurenanalysen an Pegmatiten SW-Afrikas. Thesis, Tech Univ Berlin, D83, 157pp

Bilal BA, Koß V (1982) Complex formation of trace elements in geochemical systems-IV. Polyhedron 1:239–241

Bilal BA, Langer P (1987) Complex formation of trace elements in geochemical systems: Stability constants of fluoro-lanthanide complexes in 1 m NaCl aqueous medium up to 200°C and 1000 bars. Inorg Chim Acta 140:297–298

Blaxland AB, Bremen O v., Emeleus CH, Anderson JG (1978) Age and origin of the major syenite centers in the Gardar province of south Greenland: Rb-Sr studies. Geol Soc Am Bull 89:231–244

Bowden P (1985) The geochemistry and mineralization of alkaline ring complexes in Africa (a review). J Afr Earth Sci 3:17–39

Bray JM (1942) Spectroscopic distribution of minor elements in igneous rocks from Jamestown, Colorado. Geol Soc Am Bull 53:765–814

Brey GP, Green DH (1976) Solubility of CO_2 in olivine melilitite at high pressures and role of CO_2 in the earth's upper mantle. Contrib Mineral Petrol 55:217–230

Burton JD, Culkin F (1969) Gallium. In: Wedepohl KH (ed) Handbook of geochemistry. Springer, Berlin Heidelberg New York, II: 31 B-O

Candela PA, Holland HD (1986) A mass transfer model for copper and molybdenum in magmatic hydrothermal systems: the origin of porphyry-type ore deposits. Econ Geol 81:1–19

Černý P (ed) (1982a) Anatomy and classification of granitic pegmatites. In: Granitic pegmatites in science and technology. Mineral Assoc Can, Winnipeg 8:1–40

Černý P, Ercit TS (1985) Some recent advances in the mineralogy and geochemistry of Nb and Ta in rare-element granitic pegmatites. Bull Mineral 108:499–532

Černý P, Trueman DL, Ziehlke DV, Goad BE, Paul BJ (1981) The Cat Lake-Winnipeg River and the Wekusko Lake Pegmatite Fields, Manitoba. Manitoba Dep Energ Min, 216 pp

Černý P, Smith JV, Mason RA, Delaney JS (1984) Geochemistry and petrology of feldspar crystallization in the Veszna pegmatite, Czechoslovakia. Can Mineral 22:631–651

Černý P, Meintzer RE, Anderson AJ (1985) Extreme fractionation in rare-element granitic pegmatites: selected examples of data and mechanisms. Can Mineral 23:381–421

Černý P, Goad BE, Hawthorne FC, Chapman R (1986) Fractionation of the Nb- and Ta-bearing oxide minerals in the Greer Lake pegmatitic granite and its pegmatite aureole, southeastern Manitoba. Am Mineral 71:501–517

Clark GS (1982) Rubidium-strontium isotope systematics of complex granitic pegmatites. In: Černý P (ed) Granitic pegmatites in science and technology. Mineral Assoc Can, Winnipeg 8:347–372

Clark GS, Černý P (1987) Radiogenic ^{87}Sr, its mobility, and the interpretation of Rb-Sr fractionation trends in rare-element granitic pegmatites. Geochim Cosmochim Acta 51:1011–1018

Clarke DB, Muecke GK, Pe-Piper G (1983) The lamprophyres of Ubekendt Ejland, West Greenland: products of renewed partial melting or extreme differentiation? Contrib Mineral Petrol 83:117–127

Correns CW (1969) Titanium. In: Wedepohl KH (ed) Handbook of geochemistry. Springer, Berlin Heidelberg New York II: 22 B-O

Cullers RL, Medaris G (1977) Rare earth elements in carbonatite and cogenetic alkaline rocks: examples from Seabrook Lake and Callander Bay, Ontario. Contrib Mineral Petrol 65:143–153

Cuney M, Autran A, Burnol L (1985) Premiers résultats sur le sondage G.P.F. de 900 m réalisé sur le granite sodolithique et fluoré à minéralisation disséminée de Beaurai. Chron Rech Min 481:59–63

Drake MJ, Weill DF (1975) Partition of Sr, Ba, Ca, Y, Eu^{2+}, Eu^{3+}, and other REE between plagioclase feldspar and magmatic liquid: an experimental study. Geochim Cosmochim Acta 39:689–712

Eby GN (1975) Abundance and distribution of the rare-earth elements and yttrium in the rocks and minerals of the Oka carbonatite complex, Quebec. Geochim Cosmochim Acta 39:597–620

Eggler DH (1974) Effect of CO_2 on the melting of peridotite. Annu Rep Div Geophys Lab Washington 73:215–224

Erlank AJ, Smith HS, Marchant JW, Cardoso MP, Ahrens LH (1969) Hafnium. In: Wedepohl KH (ed) Handbook of geochemistry. Springer, Berlin Heidelberg New York, II: 72 B-O

Faure G, Powell JL (1972) Strontium isotope geology. Springer, Berlin Heidelberg New York, 182 pp

Foord EE (1976) Mineralogy and petrogenesis of layered pegmatite-aplite dikes in the Mesa Grande district, San Diego County, California. PhD Thesis, Stanford Univ

Freestone IC, Hamilton DL (1980) The role of liquid immiscibility in the genesis of carbonatites – an experimental study. Contrib Mineral Petrol 73:103–117

Frey FA, Green DH (1974) The mineralogy, geochemistry and origin of lherzolite inclusions in Victorian basanites. Geochim Cosmochim Acta 38:1023–1059

Frey FA, Green DH, Roy SD (1978) Integrated models of basalt petrogenesis. J Petrol 19:463–513

Gast PW (1968) Trace element fractionation and the origin of tholeiitic and alkaline magma types. Geochim Cosmochim Acta 32:1057–1086

Gaupp R, Möller P, Morteani G (1984) Tantal – Pegmatite. Geologische, petrologische und geochemische Untersuchungen. Monogr Ser Mineral Deposits 23:124 pp

Ginsburg AE, Timofeyev IN, Feldman LG (1979) Principle of geology of the granitic pegmatites. Nedra, Moscow, 296 pp

Gittins J (1986) Genesis and evolution of carbonatite magma. Geol Assoc Can Joint Annu Meet, Ottawa 1986 11:73

Gittins J, McKie D (1980) Alkali carbonatite magmas: Oldoinyo Lengai and its wider applicability. Lithos 13:213–215

Gmelin's Handbook of inorganic chemistry (1971) Tantal, B2: 383 pp

Green HW (1972) A CO_2-charged asthenosphere. Nat Phys Sci 238:2–5

Green HW, Radcliffe SV (1975) Fluid precipitates in rocks from the earth's mantle. Geol Soc Am Bull 86:846–852

Green TH, Pearson NJ (1987) An experimental study of Nb and Ta partitioning between Ti-rich minerals and silicate liquids at high pressure and temperature. Geochim Cosmochim Acta 51:55–62

Gurney JJ, Harte B (1980) Chemical variations in upper mantle nodules from southern African kimberlites. Philos Trans R Soc London Ser A 297:273–293

Haapala I (1977) Petrography and geochemistry of the Eurajoki stock, a rapakivi-granite complex with greisen-type mineralization in southwestern Finland. Bull Geol Surv Finl 286:128 pp

Haapala I (1980) Fluid inclusions in the apatite of the Sokli carbonatite, Finland – a preliminary report. Geologi 32:83–87

Haggerty SE (1986) Kimberlite-carbonatite relations: brethren or distant cousins? Joint Annu Meet GAC, MAC, CGU, Ottawa, 11:76

Hamilton DL, Bedson P (1986) Carbonatites by liquid immiscibility. Joint Annu Meet GAC, MAC, CGU, Ottawa 11:77

Hamilton DL, Freestone IC, Dawson JB, Donaldson CH (1979) Origin of carbonatites by liquid immiscibility. Nature (London) 279:408–410

Hamilton DL, Bedson P, Essen J (in press) The behaviour of trace elements in the evolution of carbonatites. In: Bell K (ed) Carbonatites

Heier KS, Billings GK (1969) Potassium. In: Wedepohl KH (ed) Handbook of geochemistry. Springer, Berlin Heidelberg New York, II:19 B-N

Heinrich EW (1969) Niobium. In: Wedepohl KH (ed) Handbook of geochemistry, Springer, Berlin Heidelberg New York, II:41 D

Henderson P (1984) Rare earth element geochemistry. Elsevier, Amsterdam, Oxford, New York, Tokyo, 510 pp

Hildreth W (1979) The Bishop tuff: evidence for the origin of compositional zonation in silicic magma chambers. Geol Soc Am Spec Pap 180:43–75

Hoefs J (1980) Stable isotope geochemistry, 2nd edn. Springer, Berlin Heidelberg, New York, 208 pp

Holdaway MJ (1971) Stability of andalusite and the aluminum silicate diagram. Am J Sci 271:97–131

Huang WL, Wyllie PJ (1974) Melting relations of muscovite with quartz and sanidine in the K_2O-Al_2O_3-SiO_2-H_2O system at 30 kbars and an outline of paragenetic melting relations. Am J Sci 274:378

Jackson ED, Wright TL (1970) Xenoliths in the Honolulu volcanic series, Hawaii. J Petrol 11:405–430

Jahns RH (1982) Internal evolution of pegmatite bodies. In: Černý P (ed) Granitic pegmatites in science and technology. Mineral Assoc Canada, Winnipeg, 8:293–327

Jahns RH, Burnham CW (1969) Experimental studies of pegmatite genesis. I. A model for the derivation and crystallization of granitic pegmatites. Econ Geol 64:843–864

Jolliff BL, Papike JJ, Shearer CD (1987a) Fractionation trends in mica and tourmaline as indicators of pegmatite internal evolution: Bob Ingersoll pegmatite, Black Hills, South Dakota. Geochim Cosmochim Acta 51:519–534

Jolliff BL, Papike JJ, Laul JC (1987b) Mineral recorders of pegmatite internal evolution: REE contents of tourmaline from the Bob Ingersoll pegmatite, South Dakota. Geochim Cosmochim Acta 51:2225–2232

Jordan TH (1978) Composition and development of the continental tectosphere. Nature (London) 274:544–548

Kjarsgaard BA, Hamilton DL (in press) Liquid immiscibility and the origin of alkali-poor carbonatites. Mineral Mag

Knorring O von (1960) Some geochemical aspects of columbite-bearing soda granite from south-east Uganda. Nature (London) 188:204–206

Koster Van Groos AF (1968) Liquid immiscibility in the join $NaAlSi_3O_8$-Na_2CO_3-H_2O and its bearing on the genesis of carbonatites. Am J Sci 266:932–967

Koster Van Groos AF (1975) The effect of high CO_2-pressures on alkalic rocks and its bearing on the formation of alkalic ultrabasic rocks and the associated carbonatites. Am J Sci 275:163–185

Koster Van Groos AF, Wyllie PJ (1966) Liquid immiscibility in the system Na_2O-Al_2O_3-SiO_2-CO_2 at pressures to 1 kb. Am J Sci 275:163–185

Koster Van Groos AF, Wyllie P (1969) Melting relations in the system $NaAlSi_3O_8$-$NaCl$-H_2O at 1 kbar pressure. J Geol 77:581–605

Koster Van Groos AF, Wyllie PJ (1973) Liquid immiscibility in the join $NaAlSi_3O_8$-$CaAl_2Si_2O_8$-Na_2CO_3-H_2O. Am J Sci 273:465–487

Kosukhin ON (1978) Low-temperature melt inclusions in the quartz of chambered pegmatites. Sov Geol Geophys 18:56–61

Kwon ST, Tilton GR (1986) Comparative isotopic studies of Cargill and Borden carbonatite complexes from the Kapuskasing gravity high zone, Ontario. Joint Annu Meet GAC, MAC, CGU, Ottawa, 11:92

Lausch J, Möller P, Morteani G (1974) Distribution of rare earth elements between carbonates and gneisses of the Pennine zone in the Zillertal Alps (Tyrol, Austria). N Jahrb Miner Monatsh 1974:490–507

Le Bas MJ (1977): Carbonatite magmas. Mineral Mag 44:133–140

Le Bas MJ, Handley CD (1979) Variation in apatite composition in ijolitic and carbonatitic igneous rocks. Nature (London) 279:54–56

Lloyd FE, Bailey PK (1975) Light element metasomatism of the continental mantle: the evidence and the consequences. Phys Chem Earth 9:389–416

London D (1984) Experimental phase equilibria in the system $LiAlSiO_4$-SiO_2-H_2O: a petrogenetic grid for lithium-rich pegmatites. Am Mineral 69:995–1004

London D (1986a) Magmatic-hydrothermal transition in the Tanco rare-element pegmatite: evidence from fluid inclusions and phase-equilibrium experiments. Am Mineral 71:376–395

London D (1986b) Formation of tourmaline-rich gem pockets in miarolitic pegmatites. Am Mineral 71:396–405

London D (1987) Internal differentiation of rare-element pegmatites: effects on boron, phosphorus, and fluorine. Geochim Cosmochim Acta 51:403–420

London D, Burt DM (1982) Chemical models for lithium aluminosilicate stabilities in pegmatites and granites. Am Mineral 67:494–509

London D, Spooner ETC, Roedder E (1982) Fluid-solid inclusions in spodumene from the Tanco pegmatite, Bernic Lake, Manitoba. Carnegie Inst Washington Yearb 81:334–339

Longstaffe FJ (1982) Stable isotopes in the study of granitic pegmatites and related rocks. In: Černý P (ed) Granitic pegmatites in science and technology. Mineral Assoc Can, Winnipeg, 8:373–404

Manning DAC (1981) The effect of fluorine on liquidus phase relationships in the system Qz-Ab-Or with excess of water at 1 kb. Contrib Mineral Petrol 76:206–215

Manning DAC, Henderson CMB (1984) The behaviour of tungsten in granitic melt vapour systems Contrib Mineral Petrol 86:286–293

Manning DAC, Hamilton DL, Henderson CMB, Dempsey MJ (1980) The probable occurrence of interstitial Al in hydrous, F-bearing and F-free aluminous melts. Contrib Mineral Petrol 75:257–262

Maravic H von (1983) Geochemische und petrographische Untersuchungen zur Genese des niobführenden Karbonatit/Cancrinit-Syenitkomplexes von Lueshe, Kivu/NE-Zaire. Thesis, Tech Univ Berlin D83:330 pp

Middlemost EAK (1974) Petrogenetic model for the origin of carbonatites. Lithos 7:275–278

Mitchell RH, Brunfelt AO (1975) Rare Earth element geochemistry of the Fen alkaline complex, Norway. Contrib Mineral Petrol 52:247–259

Möller P (1983) Lanthanoids as a chemical probe and problems in lanthanoid geochemistry. Distribution and behaviour of lanthanoids in non-magmatic phases. In Sinha SP (ed) Systematics and properties of the lanthanides. Reidel, Dordrecht Lancaster, pp 561–616

Möller P (1988) The dependence of partition coefficients on differences of ionic volumes in crystal-melt systems. Contrib Mineral Petrol 99:62–69

Möller P, Dulski P (1983) Fractionation of Zr and Hf in cassiterite. Chem Geol 40:1–12

Möller P, Muecke GK (1984) Significance of the Eu anomalies in silicate melts and crystal melt equilibria: a re-evaluation. Contrib Mineral Petrol 87:242–250

Möller P, Maus H, Gundlach H (1982) Die Entwicklung von Flußspatmineralistionen im Bereich des Schwarzwaldes. Jahresh Geol Landesamt Baden-Württemberg 24:35–70

Montgomery A (1950) Geochemistry of Ta in the Harding pegmatite, Taos County, New Mexico: Am Mineral 35:853–866

Morey GW (1957) The solubility of solids in gases. Econ Geol 52:225–251

Mysen BO (1977) The solubility of CO_2 and H_2O under predicted magma genesis conditions and some petrological and geophysical implications. Rev Geophys Space Phys 15 (3):351–361

Mysen BO (1986) Structure and petrologically important properties of silicate melts relevant to natural magmatic liquids. In: Short course handbook. Mineral Assoc Canada, Winnipeg, pp 180–209

Mysen BO, Virgo D (1985) Interaction between fluorine and silica in quenched melts on the joints SiO_2-AlF_3 and SiO_2-NaF determined by Raman spectroscopy. Phys Chem Mineral 12:77–85

Nekrasov IYa (1970) Distribution of tantalum and niobium in magmatic and postmagmatic rocks and minerals of the northeastern USSR. Geochem Int 7:378–394

Neuman ER (1980) Petrogenesis of the Oslo Region larvikites and associated rocks. J Petrol 21:499–531

O'Hara MJ, Mathews RE (1981) Geochemical evolution in an advancing, periodically replenished, periodically tapped, continuously fractionated magma chamber. J Geol Soc (London) 138:237–277

Oxburgh ER, Turcotte DL (1968) Mid-ocean ridges and geotherm distribution during mantle convection. J Geophys Res 73:2643–2661

Pichavant M (1981) An experimental study of the effect of boron on a water saturated haplogranite at 1 kbar vapour pressure. Contrib Mineral Petrol 76:430–439

Pichavant M (1983) Melt-fluid interaction deduced from studies of silicate-B_2O_3-H_2O systems at 1 kb. Bull Mineral 106:201–211

Rankama K (1944) On the geochemistry of tantalum. Bull Commun Geol Finl 133:1–78

Rankin AH, Le Bas MJ (1974) Nahcolite ($NaHCO_3$) in inclusions in apatites from some East African ijolites and carbonatites. Mineral Mag 39:564–570
Register ME (1979) Geochemistry and geochronology of the Harding pegmatite, Taos County, New Mexico. Thesis, Dep Geol, Univ New Mexico, 145 pp
Riley GH (1970) Excess ^{87}Sr in pegmatitic phosphates. Geochim Cosmochim Acta 34:727–731
Roedder E (1963) Studies of fluid inclusions II: Freezing data and their interpretation. Econ Geol 58:167–211
Roedder E (1965) Liquid CO_2 inclusions in olivine-bearing nodules and phenocrysts from basalts. Am Mineral 50:1746–1782
Roedder E (1984) Fluid inclusions. Rev Mineral 12:644 pp
Ryabchikov ID, MacKenzie WS (1984) Fluid compositions on the join $NaAlSi_2O_6$-H_2O at 20–30 kbar and 650°C. In: Progress in experimental petrology. Nat Environ Res Council Publ Ser D 25:19–21
Saunders AD, Tarney J, Weaver SD (1980) Transverse geochemical variations across the Antarctic Peninsula: implications for the genesis of calc-alkaline magmas. Earth Planet Sci Lett 46:344–360
Shaw HR, Smith RL, Hildreth W (1976) Thermogravitational mechanism for chemical variations in zoned magma chambers. Geol Soc Am Abstr Progr 8:1102
Shearer CK, Papike JJ, Simon SB, Laul JC (1986) Pegmatite-wall rock interaction, Black Hills, South Dakota: interaction between pegmatite-derived fluids and quartz-mica schist wall rock. Am Mineral 71:518–339
Shearer CK, Papike JJ, Laul JC (1987) Mineral and chemical evolution of a rare-element granite-pegmatite system: Harney Peak Granite, Black Hills, South Dakota. Geochim Cosmochim Acta 51:473–486
Simmons WB, Heinrich EW (1980) Rare earth pegmatites of the South Platte District, Colorado. Col Geol Surv Denver, Resource Ser 11:131 pp
Simmons WB, Lee MT, Brewster RH (1987) Geochemistry and evolution of the South Platte granite-pegmatite system, Jefferson County, Colorado. Geochim Cosmochim Acta 51:455–472
Sörensen H (1979) The alkaline rocks. Wiley & Sons, Chichester, New York, Brisbane, Toronto, 622 pp
Sörensen H (1986) The alkaline rocks – a review. Fortschr Mineral 64 (1):63–86
Stewart DB (1978) Petrogenesis of lithium-rich pegmatites. Am Mineral 63:970–980
Sun SS (1982) Chemical composition and origin of the earth's primitive mantle. Geochim Cosmochim Acta 46:179–192
Sun SS, Hanson GN (1975) Origin of Ross Island basanitoids and limitations upon the heterogeneity of mantle sources for alkali basalts and nephelinites. Contrib Mineral Petrol 52:77–106
Sun SS, Nesbitt RW (1977) Chemical heterogeneity of the Archean mantle, composition of the earth and the mantle evolution. Earth Planet Sci Lett 35:429–448
Tarney J, Wood DA, Saunders AD, Cann JR, Varet J (1980) Nature of mantle heterogeneity in the North Atlantic: evidence from deep sea drilling. Philos Trans R Soc London Ser A 297:179–202
Taylor BE, Foord EE, Friedrichsen H (1979) Stable isotope and fluid-inclusion studies of gem-bearing granitic pegmatite-aplite dikes, San Diego Co., California. Contrib Mineral Petrol 68:187–205
Taylor HP, Epstein S (1962) Relationship between $^{18}O/^{16}O$ ratios in coexisting minerals of igneous and metamorphic rocks. Principles and experimental results. Geol Soc Am Bull 73:461–480
Taylor HP, Frechen J, Degens ET (1967) Oxygen and carbon isotope studies of carbonatites from the Laacher See district, West Germany, and Alnö district, Sweden. Geochim Cosmochim Acta 31:407–430
Taylor RP, Fryer BJ (1983) Rare earth lithogeochemistry of granitoid mineral deposits. CIM Bull 76:74–84
Tingle TN, Fenn PM (1984) Transport and concentration of molybdenum in granite systems: effects of fluorine and sulfur. Geology 12:156–158
Thompson RN, Morrison MA, Hendry GL, Parry SJ (1984) An assessment of the relative roles of crust and mantle in magma genesis: an elemental approach. Philos Trans R Soc London Ser A 310:549–590
Treiman AH (1986) A petrogenetic grid for carbonatites. Geol Assoc Can, Joint Annu Meet Ottawa 1986, Abstr 11:138
Verwoerd WJ (1978) Liquid immiscibility and the carbonate-ijolite relationship: preliminary data on the join $NaFe^{3+}Si_2O_6$-$CaCO_3$ and related compositions. Carnegie Inst Washington Yearb 77:767–774
Vlasov KA (1966) Geochemistry and mineralogy of rare elements and genetic types of their deposits. I. Geochemistry of rare elements. Isr Progr Sci Transl, Jerusalem, 688 pp

Walker RJ, Hanson GN, Papike JJ, O'Neil RJ, Laul JC (1986) Internal evolution of the Tin Mountain pegmatite, Black Hills, South Dakota. Am Mineral 71:440–459

Watkinson DH, Wyllie PJ (1971) Experimental studies of the compositional join $NaAlSiO_4$-$CaCO_3$-H_2O and the genesis of alkali rock-carbonatite complexes. J Petrol 12:357–378

Watson EB, Green TH (1981) Apatite/liquid partition coefficients for the rare earth elements and strontium. Earth Planet Sci Lett 56:405–421

Wedepohl KH (ed) (1969) Tantalum. In: Handbook of geochemistry. Springer, Berlin Heidelberg New York, II: 73 B-G

Wendlandt RF, Harrison WJ (1979) Rare earth partitioning between immiscible carbonate and silicate liquids and CO_2 vapour: results and implications for the formation of light rare earth-enriched rocks. Contrib Mineral Petrol 69:409–419

Winkler HGF (1976) Petrogenesis of metamorphic rocks. Springer, Berlin Heidelberg New York, 334 pp

Wood DA (1979) A variable veined suboceanic upper mantle – genetic significance for mid-oceanic ridge basalts from geochemical evidence. Geology 7:499–503

Woolley AR (1986) The distribution of carbonatites in space and time. Joint Annu Meet GAC, MAC, CGU, Ottawa 11:147

Wyllie PJ (1987) Volcanic rocks: boundaries from experimental petrology. Fortschr Mineral 65:249–284

Wyllie PJ, Huang WL (1975) Peridotite, kimberlite, and carbonatite explained in the system CaO-MgO-SiO_2-CO_2. Geology 3:621–624

Wyllie PJ, Tuttle O (1964) Experimental investigation of silicate systems containing two volatile components. III Am J Sci 262:930–939

Zakharchenko AI (1964) Physicochemical conditions and processes of formation of granitic pegmatites. Geochem Int 1057

Zakharchenko AI (1973) Pegmatite-forming melt-solutions, on the basis of their inclusions in minerals of granitic-chamber pegmatites. COFFI 6:172–183

Znamenskii EB, Konysova VV, Krinberg IA, Popolitov EI, Flerova KV, and Tsykhanskii VD (1962) Distribution of titanium, niobium, and tantalum in granitic rocks containing titanite. Geochemistry 919–925

Geochemistry of Granites Associated with Tantalum and Niobium Mineralization

P.J. POLLARD

Abstract

Tantalum and niobium mineralization is often associated with geochemically specialized granites which are characterized by enrichment in fluorine, and by the development of pervasive, postmagmatic alteration. Three major varieties of granite can be recognized:

1. Alkali granites containing alkali pyroxenes and/or amphiboles which are characterized by high Fe, F, Nb, Zr, Rb, Sn and REE, and by low CaO, Ba, Sr and Ta/Nb. These granites occur principally in anorogenic settings, and are associated mainly with niobium mineralization.

2. Biotite and/or muscovite granites often containing Fe-Li micas which are characterized by high F, Rb and Sn, and by low CaO, Ba, Sr and Eu. These granites occur in both anorogenic and postorogenic settings, and are associated principally with Nb-Ta(-Sn) mineralization.

3. Lepidolite-albite granites often containing topaz which are characterized by high Al_2O_3, F, Li, Rb, Sn, Ta and Ta/Nb, and by low Ba, Sr, Eu, Zr and REE. These granites occur principally in postorogenic settings, but also form marginal facies varieties of biotite and/or muscovite granites in both anorogenic and postorogenic settings. Lepidolite-albite granites are generally associated with Ta(-Nb-Sn) mineralization.

Many granites associated with tantalum and niobium mineralization are zoned vertically, with the finer-grained porphyritic, granophyric, pegmatitic and/or miarolitic rocks which occur in small cupolas grading downwards into medium- and coarse-grained granite. This vertical textural zonation indicates an important concentration of volatile and ore elements in the roof of the plutons from the earliest stages of crystallization.

The mineralogical and geochemical features of rare-metal granites reflect features of their source rock compositions, as well as magmatic fractionation and postmagmatic alteration processes. Trends towards increasing Al_2O_3, Na_2O, F and a variety of rare metals in the most evolved rocks may reflect crystal and/or liquid fractionation during magma evolution and/or postmagmatic alteration by interaction with fluorine-rich hydrothermal fluids.

Detailed petrographic and mineralogical study of a variety of examples is required before geochemical and isotopic modelling of these rare-metal granites can advance beyond the preliminary stage.

Department of Geology James Cook University of North Queensland, Townsville 4811, Australia

Lanthanides, Tantalum and Niobium
P. Möller, P. Černý and F. Saupé (Eds.)

1 Introduction

Tantalum and niobium mineralization is frequently associated with geochemically specialized granites (Tischendorf 1977) which form part of a spectrum of granites related to rare-metal mineralization (Sn, W, Mo, Nb, Ta, Li, Be, Rb, Cs, Th, U, REE). Hypogene tantalum and niobium mineralization is usually associated with the upper, marginal facies of minor intrusive phases which form the late-stage members of strongly fractionated granite suites. These granites are characterized by enrichment in fluorine (±lithium and boron), and by the development of pervasive, postmagmatic alteration. The geochemistry of rare-metal granites thus reflects features of both their magmatic and postmagmatic evolution, and genetic modelling requires consideration of the important contribution of both types of processes in producing their specialized geochemistry.

This contribution provides a brief account of the geology of granites associated with tantalum and niobium mineralization and an overview of their geochemistry, with special emphasis on magmatic and postmagmatic processes. Attention is focussed mainly on mineralogical and geochemical features of the upper, specialized portions of the smaller, late-stage plutons which host the mineralization. The geochemistry of pegmatites, carbonatites and nepheline syenites is treated elsewhere in this volume.

2 Broad-Scale Features

Granites associated with tantalum and niobium mineralization range in age from Precambrian to Mesozoic, and occur within numerous individual metallogenic provinces. Rare-metal granites typically form steep-sided, ovoid to elongate plutons ranging from less than 1 km^2 to several tens of km^2, which generally form the final intrusive phases of multiple-staged granite batholiths. In shallower level environments these plutons form the late intrusive phases of magmatic complexes which are often controlled by ring fractures, and commonly intrude contemporaneous volcanics within shallow level ring complexes. In deeper level environments the rare-metal granites frequently form small satellite phases of granite batholiths, and in some cases form relatively fine-grained, marginal facies varieties of large plutons.

Granite classification on the basis of the molecular proportions of Al_2O_3/($CaO+Na_2O+K_2O$) allows peralkaline ($A < NK$), metaluminous ($CNK > A > NK$) and peraluminous ($A > CNK$) varieties to be recognized. Tantalum and niobium mineralization is associated with all three varieties, with Nb-rich mineralization most prominent in association with peralkaline granites, and Ta-rich mineralization most prominent in association with peraluminous granites.

Granites associated with tantalum and niobium mineralization occur principally in anorogenic and postorogenic tectonic settings. The anorogenic granites occur mainly within stabilized cratons where magmatic activity is not associated with regional deformation, but where magma generation and intrusion may be associated with crustal extension. The localization of individual igneous complexes in these provinces is frequently controlled by regional scale fault systems. The postorogenic

Table 1. Age, rock types and principal metal associations of selected rare-metal provinces

	Rock types	Age	Mineralization
Saudi Arabia	1. Alkali granite	Precambrian	Nb-Zr-REE
	2. Alkali feldspar granite	Precambrian	Sn-Ta-Nb
Nigeria	1. Pegmatite	Precambrian	Sn-Ta-Nb
	2. Alkali granite	Jurassic	Nb-Zr
	3. Biotite granite	Jurassic	Sn-Nb
Mongolia-Transbaikalia	Alkali granite	Permo-Triassic	Nb-Zr
Kazakhstan	Alkali granite		Nb
Azov (Ukraine)	Two-mica granite	Precambrian	Nb-Ta
Brazil (Rondonia)	Biotite granite	Precambrian	Sn-Ta-Nb
South China	Two-mica granite Li-mica granite	Jurassic-Cretaceous	Sn-Ta-Nb-Li
Eastern Transbaikal	Two-mica granite	Jurassic	Sn-Ta-Nb
SW Thailand	1. Two-mica granite	Cretaceous	Sn-Ta
	2. Pegmatite	Cretaceous	Sn-Ta
Western Peninsula Malaysia	Two-mica granite	Triassic	Sn-Ta
Massif Central, France	Topaz-Li-mica granite	Carboniferous	Sn-Ta-Nb-Li
Eastern Siberia	1. Alkali metasomatites	Precambrian	REE-Nb-Ta-Zr
	2. Li-mica granite	Jurassic	Sn-Ta-Nb
	3. Alkali granite	Ordovician?	Nb-Sn

granites generally intrude following the major deformation events in fold belts, and are in many cases also controlled by regional scale faults.

The anorogenic provinces are often characterized by bimodal magmatism comprising basaltic and rhyolitic magmas. The granitic rocks include alkali granite, syenite and alkali feldspar granite. The alkali granites probably evolve from syenitic compositions by plagioclase, clinopyroxene and alkali feldspar fractionation (Bowden 1982; Litvinovskiy et al. 1985). Both peralkaline and peraluminous magmas occur in some provinces (Pavlenko 1974; Bowden and Kinnaird 1984).

In postorogenic provinces granites and alkali feldspar granites are the most common rock types associated with mineralization (Garson and Mitchell 1975; Xu et al. 1984). Magma compositions are usually peraluminous or metaluminous, and peralkaline compositions are relatively uncommon.

The general features of selected rare-metal provinces are summarized in Table 1, and the broad-scale features of several major provinces are outlined below.

2.1 Arabian Shield

Alkali granites associated with Nb-Zr-REE mineralization occur in over 70 igneous complexes within an area of 400,000 km^2 in the Arabian Shield (Drysdall et al. 1984; Jackson et al. 1985; Fig. 1). Mineralization is associated with specialized, peralkaline to metaluminous granites which typically contain alkali amphiboles and/or pyroxenes, with biotite in the more aluminous phases (Jackson et al. 1985). These

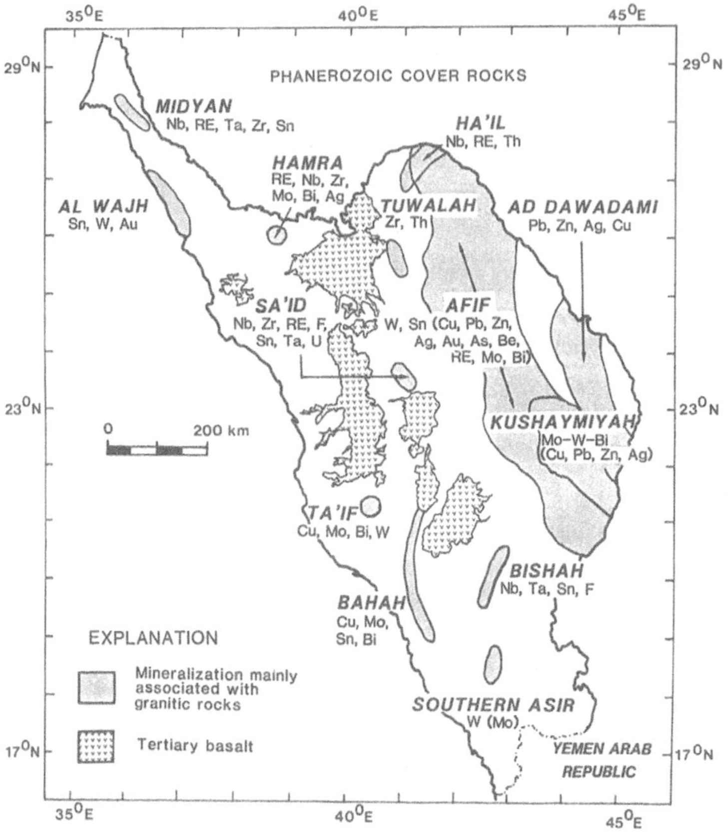

Fig. 1. Map showing the distribution of rare-metal mineralized areas in the Arabian Shield (modified from Jackson 1986)

granites are restricted to basement domains of island-arc character, and were emplaced in a broadly extensional tectonic setting (Jackson et al. 1985). The alkali granites range in age from 686 Ma to 517 Ma, with most in the range 630 to 565 Ma (Jackson et al. 1985).

Rare-metal mineralization, including minor Ta and Nb, is also associated with peraluminous to metaluminous, postorogenic granites within the Arabian Shield (Jackson and Ramsay 1986; Fig. 1). These biotite and two-mica granites occur within all crustal domains.

2.2 Nigeria

The ring complexes of Nigeria (Fig. 2) form one of the most extensively studied anorogenic rare-metal provinces. The mineralized ring complexes form part of a 1250-km-long belt of ring complexes extending across Niger and Nigeria, with the Rb/Sr age decreasing from Ordovician in northern Niger to Late Jurassic in central Nigeria (see compilation of Bowden and Kinnaird 1984; Fig. 2). This well-defined

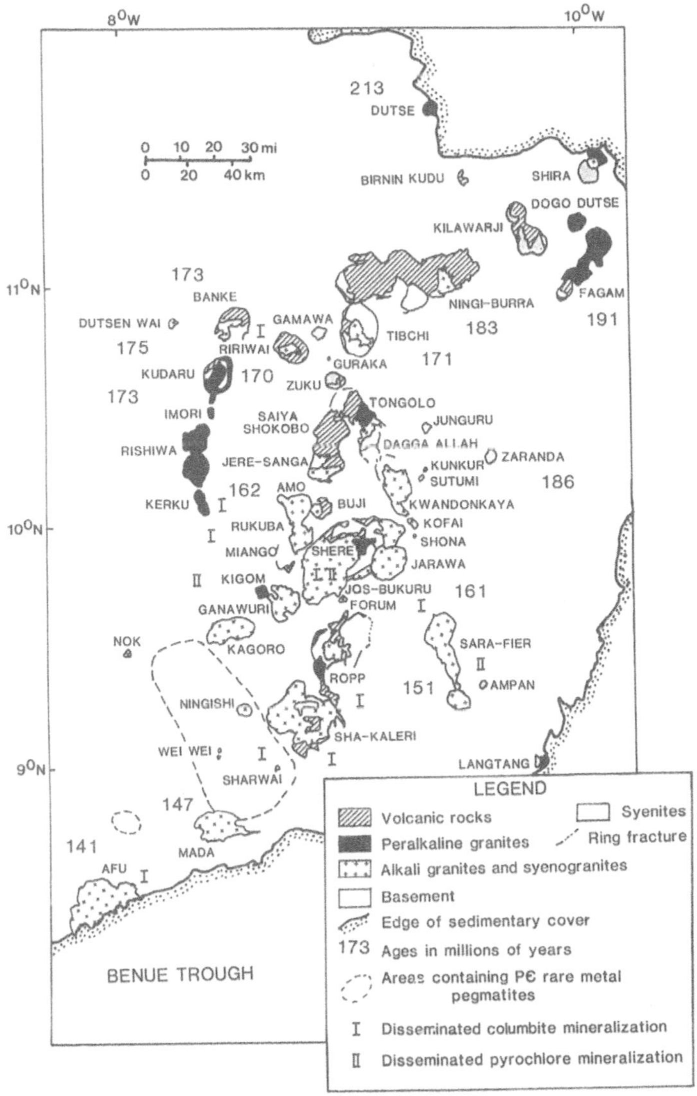

Fig. 2. Map showing the distribution of Mesozoic ring complexes, lithologies and mineralization styles in Nigeria (modified from Kinnaird et al. 1985a)

age trend implies that magmatism results from the relative movement between lithosphere and a sublithospheric hot spot (Bowden and Kinnaird 1984).

Nb-Zr mineralization (pyrochlore type) is associated with late-stage peralkaline intrusions containing alkali amphiboles, while Sn-Nb mineralization (columbite type) is associated with peraluminous biotite granites (Kinnaird et al. 1985a; Fig. 2).

In addition to the Nigerian "Younger Granites", mineralization consisting mainly of Sn-Ta-Nb is associated with pegmatites of the "Older Granite Complex" of Pan African age (600±150 Ma) (Matheis et al. 1982; Fig. 2). Pb isotope studies (Tugarinov et al. 1968) indicate that lead in the Mesozoic granites has been derived from the Precambrian rare-metal pegmatites, or a source similar to that of the pegmatites.

2.3 Mongolia-Transbaikalia

The Mongolia-Transbaikalia alkali granite province (Fig. 3) forms a 1500-km-long belt of intrusions to the east and south of Lake Baikal, which links southwestwards to the Tuva and east Sayan alkaline rocks (Pavlenko 1974; Litvinovskiy et al. 1985). The Mongolia-Transbaikalia province comprises over 350 intrusions with a total area exceeding 10,000 km^2 and is composed of rocks from two major magmatic cycles. The early Permian cycle is predominantly subalkaline in character, while the Permo-Triassic cycle is predominantly alkaline (Litvinovskiy et al. 1984, 1985).

Alkali amphibole- and biotite-bearing granites of the alkaline series are sometimes associated with Nb-Zr-Th mineralization (Severov et al. 1963; Kovalenko et al. 1964). Few details are available concerning the geochemistry of the mineralized granites.

2.4 Eastern Transbaikal

The eastern Transbaikal province of the USSR lies to the east of the Mongolia-Transbaikalia province and extends from the Amur River southwestwards into northern Mongolia (Fig. 3). Although known principally as a tin-tungsten province (Kozlov et al. 1974; Ivankin and Kulagashev 1980), granites associated with Ta-Nb-Li-Be mineralization are also reported (Boguslavskiy et al. 1966; Syritso and Chernik 1967). The rare-metal granites are associated with Mesozoic granites of possible Jurassic age which intrude Palaeozoic metasediments (Syritso and Chernik 1967; Navilkin 1973).

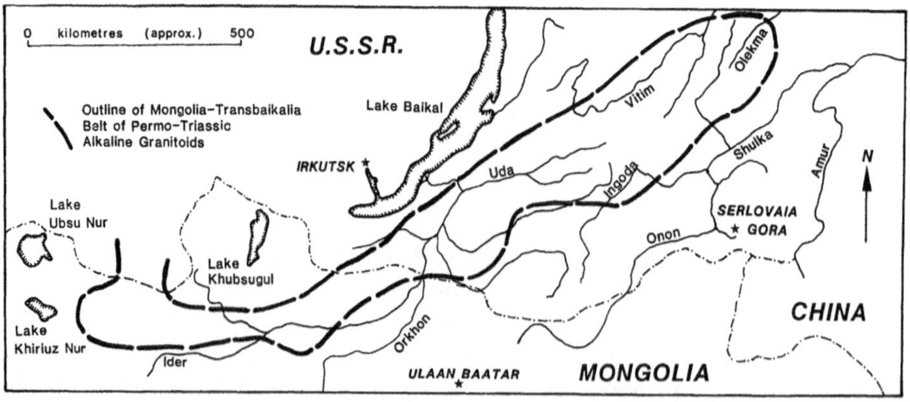

Fig. 3. Sketch map showing location of Permo-Triassic alkaline granitoids in Mongolia-Transbaikalia (modified from Litvinovskiy et al. 1985)

2.5 Southeast Asia

One of the world's major sources of tantalum is the southeast Asian region, where the extensive primary and secondary tin concentrations contain important tantalum values in some areas. Tin mineralization is associated with a series of granite batholiths forming two principal belts (Fig. 4) which extend over 1500 km from northern Thailand to Beilitung in Indonesia. The principal tantalum-rich areas are associated with the western belt granites in Burma, southwestern Thailand and western Peninsula Malaysia (Hutchison 1978; Hosking 1981).

The principal mineralized granites are biotite and two-mica granites of Triassic (Liew and McCulloch 1985) and Cretaceous (Garson and Mitchell 1975) ages.

2.6 South China

The south China region is a major metallogenic province noted for tin, tungsten, tantalum and niobium mineralization. This region consists of a series of northeasterly trending granite belts which range in age from Proterozoic to Cretaceous, and which generally become younger from northwest to southeast (Xu et al. 1980). The rare-metal granites (Xu et al. 1980, 1984) are multiple cyclic and multiple stage

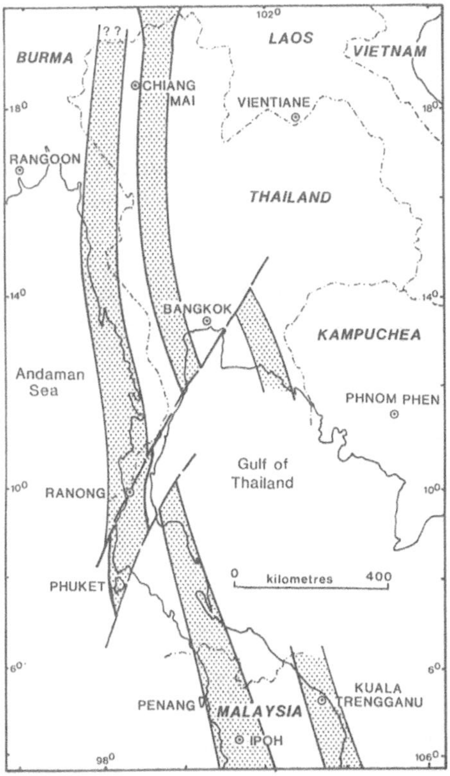

Fig. 4. Map showing part of the two major belts of rare-metal mineralization in southeast Asia (modified from Garson and Mitchell 1975)

bodies which show progressive enrichment in rare metals from Caledonian to Yanshanian, and from early intrusions to late intrusions of the same magmatic cycle. The major rare-metal granites are biotite and two-mica granites of Yanshanian (185–70 Ma) age. Important Ta-Nb concentrations occur mainly in Jiangxi and Hunnan Provinces, and include Yichun (Granite No. 414), Dajishan, Xianghualing (430 mine) and Limu.

3 Rock Types and Facies Variations

Peralkaline rare-metal granites are characterized by the presence of Na-rich pyroxenes and/or amphiboles, principally aegerine and/or arfvedsonite. In addition, a variety of other phases such as riebekite, biotite and astrophyllite may also be present (Pavlenko 1974; Kinnaird et al. 1985a; Jackson et al. 1985). In hypersolvus varieties the single feldspar has generally unmixed to produce vein and patch perthite, while the subsolvus varieties contain microcline perthite and plagioclase with a low An content (albite-oligoclase). Common accessory phases include zircon, xenotime, Ti-oxides, thorite and fluorite.

The peraluminous rare-metal granites contain primary peraluminous minerals such as biotite, muscovite, Li-micas, topaz and tourmaline, but sometimes including garnet and Al-silicates (Pichavant and Manning 1984). The major minerals include quartz, perthitic microcline, plagioclase (albite-oligoclase) and ferromagnesian silicates. Common accessory phases include zircon, monazite, xenotime, ilmenite, Ti-oxides and fluorite.

Granites associated with tantalum and niobium mineralization exhibit a wide variety of facies variations similar to those noted in tin and tungsten granites (Taylor and Pollard 1987). These textural and grain-size variations result from crystallization processes and are generally overprinted by postmagmatic alteration. The general morphology of a typical late-stage granite cupola is shown in Fig. 5.

The upper contact of the pluton is frequently marked by the development of a marginal pegmatitic phase which is usually barren of mineralization. This marginal facies can include the development of downward branching (plumose) alkali

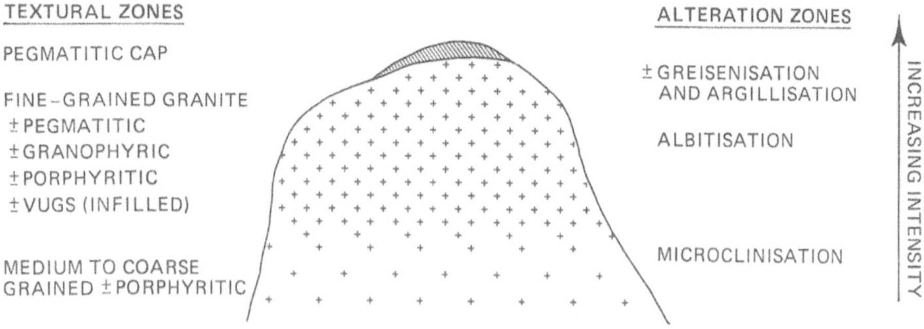

Fig. 5. Typical vertical arrangement of textural and alteration zones in the upper, contact zones of rare-metal granites

feldspar crystals, subhorizontal quartz-feldspar pegmatite interlayered with fine-grained granite, and alternating layers of coarse quartz and quartz-feldspar with individual crystals oriented perpendicular to the contact.

Associated with the marginal pegmatite facies and extending downward for several metres or several tens of metres into the pluton there is usually a zone of relatively fine-grained granite with a range of textural variants. These include porphyritic, granophyric, pegmatitic and miarolitic facies which are often linked by rapid textural gradations. These finer-grained facies varieties usually give way downwards to relatively coarser-grained granite through a gradual coarsening in grain size, or through a rapid, transitional contact. The textural features of these rocks indicate volatile enrichment in the upper portion of the pluton from the earliest stages of crystallization. In many cases, textural evidence indicates that crystallization occurred in the presence of a separate, aqueous-rich fluid phase. These facies variations are of fundamental importance in the interpretation of geochemical data since they suggest that magmatic processes may produce a preconcentration of volatiles and ore elements in the upper part of the magma chamber.

Vertical textural changes are often accompanied by mineralogical changes indicated by changing modal mineralogy (Zalashkova and Sitnin 1967; Deng 1984) and mineral composition (Volkov and Gorbacheva 1980; Hu et al. 1984; Deng 1984). Although these vertical mineralogical zonation features are often complicated by postmagmatic alteration, they suggest that many rare-metal granites crystallized from zoned magma chambers or possibly that crystallization processes themselves produced the vertical zonation. Analogues of the highly fractionated compositions which may occur in the upper levels of these plutons are provided by quenched volcanic and subvolcanic rocks such as ongonites (Kovalenko et al. 1971), topaz rhyolites (Christiansen et al. 1983) and Macusani glasses (Pichavant et al. 1987).

The boron-rich peraluminous granites of southwestern Thailand provide another example of vertical zonation. Disseminated Sn(-Ta) deposits such as Bang Phra and Nok Hook show many of the textural variations outlined above, and grade from essentially biotite to tourmaline-muscovite to muscovite granite in the upper 30–50 m of the plutons (Sirinawin et al. unpublished; Nakapadungrat et al. in press).

Petrographic and geochemical study of the Beauvoir Granite of the Massif Central, France (Cuney et al. 1985) has shown that the granitic mass is made up of three separate intrusions stacked one above the other. Each intrusion has a narrow compositional range, with sharp breaks in some element concentrations occurring at the contacts, suggesting that three magma batches may have been emplaced successively. Further detailed study is required to test whether this situation also occurs in other, poorly exposed rare-metal granites.

4 Postmagmatic Alteration

The intrusion and crystallization of rare-metal granites results in the formation of an aqueous-rich magmatic fluid phase which reacts with the previously crystallized granite to produce pervasive alteration. This alteration frequently occurs throughout the suite of granites related to mineralization as pervasive, incipient alteration, but becomes more intense in the apical zones of the late-stage granites.

Alteration is localized by grain boundaries and microfractures. Macroscopic fractures related to alteration are usually confined to the upper, outer portions of the cupolas.

The major styles of postmagmatic alteration include microclinization, albitization and greisenization (Beus and Zalashkova 1964; Shcherba 1970). Although there are many variations on the theme, alteration is frequently zoned, with microclinization in the deeper levels, albitization in the intermediate levels and greisenization in the upper levels. The thickness of individual zones varies widely, but the total vertical extent may be greater than 300 m. Greisenization is usually only weakly developed in the alkali granite systems.

Postmagmatic alteration is expressed mainly in the replacement of primary feldspars and sometimes quartz by secondary K-feldspar, albite and micas. The most common feldspathic alterations observed are the replacement of plagioclase by microcline and microcline by albite. These are accompanied by replacement of original biotite by muscovite and Li-micas (Syritso and Chernik 1967), or of sodic amphibole by varieties richer in lithium (Kinnaird et al. 1985a). In detail, the alteration patterns developed may be much more complicated than the simplified models suggest. For example, Beus et al. (1962) and Syritso and Chernik (1967) recognize several generations of albitization and microclinization in rare-metal granite cupolas in eastern Siberia. In these tantalum-bearing granites, the green microcline variety amazonite is a common alteration mineral (Beus et al. 1962; Zalashkova and Gerasimovskii 1974).

In cases where massive or fracture-controlled greisen zones are developed, the feldspars are often entirely replaced by phyllosilicates (biotite, muscovite, Li-micas), topaz and quartz. These may in turn be overprinted by later generations of feldspathic alteration. An additional feature of many mineralized granites is the development of intense argillic alteration.

A more detailed discussion of the distribution of rock types and alteration in rare-metal granites is given elsewhere (Pollard this Vol.). However, the details summarized above indicate that both magmatic and postmagmatic processes contribute to the mineralogical and geochemical characteristics of rare-metal granites. In this context, extreme caution should be exercised in the interpretation of geochemical data, particularly where detailed petrographic information is not available.

Tantalum and niobium mineralization is associated mainly with albite-rich zones, although Ta-rich cassiterite is an important component of some greisen bodies. The economically important ore minerals are members of the columbite-tantalite and pyrochlore-microlite groups, together with cassiterite, although a variety of other phases such as fergusonite, struverite and ixiolite may also be present.

5 Geochemical Features of Rare-Metal Granites

The geochemistry of granites related to tantalum and niobium mineralization reflects features of their magmatic and postmagmatic evolution, and a major challenge in the interpretation of geochemical data is the assessment of the relative importance of these processes in producing the specialized geochemistry of the granites. Selected analyses of a variety of granites related to tantalum and niobium mineralization are presented in Tables 2 and 3.

For the purposes of the present discussion the granites are divided into three main groups: alkali granites, biotite and/or muscovite granites and lepidolite-albite granites (Tables 2 and 3). These granites are not restricted to any particular tectonic environment, although in general terms the alkali granites and the biotite and/or muscovite granites are widely developed in anorogenic provinces, while the biotite and/or muscovite granites and the lepidolite-albite granites are widely developed in postorogenic provinces. Lepidolite-albite granites sometimes form marginal facies varieties of the biotite and/or muscovite granites, and some biotite granites can therefore be regarded as precursor granites to the mineralized varieties (e.g. the plusmasitic granites of Saudi Arabia and the biotite granites of eastern Siberia, Table 2).

The alkali granites (Table 2) have low MgO and CaO, and relatively high Fe compared to average low-Ca granites (Turekian and Wedepohl 1961). Their peralkalinity in many cases results from low Al_2O_3 rather than exceptionally high alkalis (Jackson et al. 1985; Kinnaird et al. 1985a). The mineralized Nigerian arfvedsonite-albite granites from the Ririwai Complex, Nigeria contain lower SiO_2 and higher Al_2O_3, Na_2O and F than the unmineralized arfvedsonite granites (Table 2), while the mineralized microgranites from Saudi Arabia contain similar SiO_2 and higher Al_2O_3, Na_2O and F than the average specialized granite (Table 2).

The biotite and/or muscovite granites (Table 2) also have similar or lower TiO_2, MnO, MgO and CaO than low-Ca granite. Na_2O+K_2O is similar to that of the alkali granites, but Al_2O_3 is generally higher, resulting in peraluminous compositions.

The lepidolite-albite granites are characterized by low SiO_2 as well as TiO_2 and MnO, but MgO and CaO are sometimes high relative to the other granite types (Table 2). Al_2O_3 is very high compared to the other granites, and Na_2O is higher than K_2O. The lepidolite-albite granites are usually strongly enriched in F, and the Beauvoir Granite is particularly enriched in P_2O_5 and Li_2O (Table 2). In many respects the lepidolite-albite granites have similar major element compositions to Macusani glasses, but with slightly higher CaO and Na_2O (Table 2). However, it should be noted that the Macusani glasses are strongly enriched in boron, whereas the lepidolite-albite granites generally are not.

The Jian Granite of SE China (Deng 1984) shows the types of major element trends observed in vertically zoned plutons. Moving upwards in the pluton from protolithionite granite to Li-muscovite-albite granite, there is a decrease in SiO_2 and K_2O accompanied by increasing Al_2O_3 and Na_2O, while the upper portion is also rich in fluorine (Table 2). Similar trends are noted in the Ta-rich granites of eastern Siberia (Zalashkova and Sitnin 1967) where the lepidolite-albite granites form apical facies of biotite granites (Table 2, columns 9 and 13.)

Normative compositions for the rare-metal granites in Table 2 are plotted in the Qz-Ab-Or tringle in Fig. 6, together with the minimum compositions for 1000 bar water pressure (Tuttle and Bowen 1958) and the minimum compositions for 1, 2 and 4 wt.% added fluorine at 1000 bar water pressure (Manning 1981). The alkali granites and the biotite and/or muscovite granites plot near the minimum F-free composition for 1000 bar, and extend towards the point with 1 wt.% added fluorine. The lepidolite-albite granites plot near the minimum with 2 wt.% added fluorine, with the

Table 2. Major element geochemistry of granites associated with tantalum and niobium mineralization. (Data sources: *1–5* alkali granites; *6–11* biotite and/or muscovite granites; *12–16* lepidolite-albite granites; *17* Macusani glass; *18–20* vertical sequence from Jian Granite, south China. Note that some element averages are calculated on fewer analyses)

	1	2	3	4	5	6	7	8	9	10
No. of Analyses	6	5		4	11	7		1	29	5
SiO_2	76.95	72.94	74.18	74.15	74.29	76.39	75.41	74.93	73.3	73.15
TiO_2	0.13	0.23	0.32	0.15	0.09	0.07	0.10	-	0.19	0.18
Al_2O_3	11.09	12.43	11.10	12.56	13.05	12.59	12.82	13.72	13.9	13.65
Fe_2O_3	1.63	1.86	2.86	2.16	1.77	0.37	0.77	-	0.68	0.51
FeO	0.83	0.82	1.21	0.63	0.65	1.09	0.59	0.93	1.86	1.30
MnO	0.04	0.07	0.08	0.06	0.03	0.03	0.09	0.02	0.04	0.02
MgO	0.03	0.05	0.09	0.05	0.32	0.01	0.04	0.63	0.30	0.37
CaO	0.18	0.11	0.50	0.50	0.65	0.27	0.62	0.73	0.60	0.41
Na_2O	3.70	5.93	4.25	5.06	4.49	3.75	4.19	4.16	3.50	3.63
K_2O	4.63	4.24	4.61	3.35	4.10	4.36	4.51	3.88	4.70	5.00
P_2O_5	0.01	0.03	0.04	0.03	-	0.01	0.02		0.09	
H_2O^+	0.37	0.50	0.44*		0.45	0.29	0.52*	0.86		
H_2O^-	0.23	0.07			0.18	0.08				
F	0.15	0.70	0.16	0.37	0.48	0.26	0.23		0.13	0.11
Li_2O										

	11	12	13	14	15	16	17	18	19	20
No. of Analyses	22	1	12	2	5	24	1			
SiO_2	74.50	72.83	70.5	69.67	69.74	70.7	72.26	76.41	75.76	71.47
TiO_2	0.06	-	0.01	0.02	0.01	0.01	0.04	-	-	-
Al_2O_3	13.22	15.29	17.0	16.22	16.55	17.4	15.83	13.30	14.26	15.40
Fe_2O_3	0.68	-	0.32	0.30	0.26	0.21	0.04	0.36	0.09	0.66
FeO	1.08	0.29	0.82	1.38	0.49	0.55	0.57	1.30	1.04	0.82
MnO	0.02	0.01	0.13	0.10	0.04	0.08	0.06	0.06	0.12	0.10
MgO	0.14	0.29	0.18	0.07	0.85	0.15	0.02	0.05	0.09	0.40
CaO	0.77	0.54	0.67	0.37	0.82	0.50	0.22	0.46	0.27	0.21
Na_2O	3.84	5.18	5.47	5.72	5.79	6.00	4.14	3.70	3.84	4.92
K_2O	4.72	5.03	3.80	4.00	2.32	3.62	3.66	4.45	4.60	4.05
P_2O_5	0.04				0.53		0.53	0.01	0.01	0.03
H_2O^+	0.37	0.31	0.16*	0.12*	1.08		0.46*			0.82
H_2O^-	0.15	0.29			0.18					
F	0.54	0.13	0.80	0.84	1.25	0.82	1.33			0.74
Li_2O					0.91		0.74			

1 Arfvedsonite granite, Ririwai Complex, Nigeria (Kinnaird et al. 1985a); *2* Arfvedsonite-albite granite, Ririwai Complex, Nigeria (Kinnaird et al. 1985a); *3* Average agpaitic specialized granite, Arabian Shield (Jackson and Ramsay 1986); *4* Alkali microgranites, Arabian Shield (Drysdall et al. 1984); *5* Riebeckite-albite granites, Kazakhstan (Beus et al. 1962); *6* Biotite granites, Ririwai Complex, Nigeria (Kinnaird et al. 1985a); *7* Average plumasitic granite, Arabian Shield (Jackson and Ramsay 1986); *8* Biotite-albite granite, eastern Siberia (Beus et al. 1962); *9* Biotite granites, eastern Siberia (Zalashkova and Sitnin 1967); *10* Biotite (±muscovite) granites, eastern Transbaikal (Syritso and Chernik 1967); *11* Biotite granites, Ukraine (Pyatenko et al. 1967); *12* Fine-grained lepidolite-amazonite-albite apogranite, eastern Siberia (Beus et al. 1962); *13* Lepidolite-amazonite-albite apogranites, eastern Siberia (Zalashkova and Sitnin 1967); *14* Lepidolite-amazonite-albite apogranites, eastern Transbaikal (Syritso and Chernik 1967); *15* Beauvior Granite, France (Aubert 1969; Kosakevitch 1976); *16* Lepidolite-amazonite-albite apogranites (Sitnin 1970); *17* JV1 Macusanite (Pichavant et al. 1987); *18* Protolithionite granite (Deng 1984); *19* Zinnwaldite-albite granite (Deng 1984); *20* Li-muscovite-albite granite (Deng 1984).

*Total H_2O

Table 3. Trace element composition of granites associated with tantalum and niobium mineralization (*1–4* alkali granites; *5–9* biotite and/or muscovite granites; *10–12* lepidolite-albite granites; *13* Macusani glass)

	1	2	3	4	5	6	7	8	9	10	11	12	13
No.[a]	6	5		4	7		29	22	23	12	4	24	1
Li	88	395	51	112	266	184	177	26	132	1558	4355	330	3200
Be	16	50	6	1.5	6	9	5	6	6	8	197	6	41.4
Ba	34		99	28		79							< 10
Cs	1.2	10.5					20		17			32	566
Rb	376	1355	169	528	749	386	441	647	430	1600	> 12000	1500	1177
Sr	5	0	20	18	3	38							1.62
Pb	71	288	14	28	29	32	43		56	40		126	7
Th	90	197	14	43	78	31	22			14			2.3
U			4	49									18.8
Zr	796	3480	795	5849	264	211	134	139	175	15		44	39
Hf	27	144			15								< 2
Nb	265	1335	62	2306	293	56	31	61	61	79	150	111	44
Ta	33	297	1	94		9	7	7	7	149	150	94	26.9
Y	216	315	91	501	220	62							5.2
La	65	71	71	329	89	41							1.92
Ce	155	228	182	784	142	95							4.52
Sc	0.2	0.2	3			4							2.2
Co	1.3	0.4	4	< 5		3							0.71
Cu	120	109	13	16	78	5	27			25			< 10
Zn	388	906	137	537	324	83	36			401			92
Sn	31	56	10	108	46	16	23	26	22	32	1643	154	155

[a] Some element averages based on fewer analyses. Data sources; *1* Arfvedsonite granite, Ririwai Complex, Nigeria (Kinnaird et al. 1985a); *2* Arfvedsonite-albite granite, Ririwai Complex, Nigeria (Kinnaird et al. 1985a); *3* Average agpaitic specialized granite, Arabian Shield (Jackson and Ramsay 1986); *4* Alkali microgranites, Arabian Shield (Drysdall et al. 1984); *5* Biotite granite, Ririwai Complex, Nigeria (Kinnaird et al. 1985a); *6* Average plumasistic granite, Arabian Shield (Jackson and Ramsay 1986); *7* Biotite granites, eastern Siberia (Zalashkova and Sitnin 1967); *8* Biotite granites, Ukraine (Pyatenko et al. 1967); *9* Biotite granites (Sitnin 1970); *10* Lepidolite-amazonite-albite granites, eastern Siberia (Zalashkova and Sitnin 1967); *11* Beauvoir Granite, France (Aubert 1969); *12* Lepidolite-amazonite-albite granites (Sitnin 1970); *13* JV1 Macusanite (Pichavant et al. 1987).

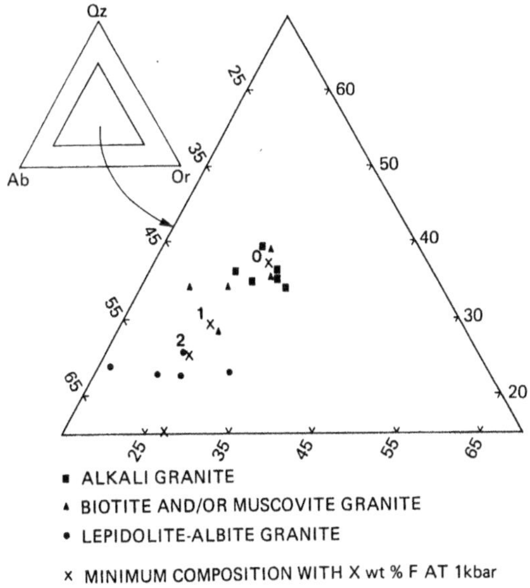

Fig. 6. Normative Qz-Ab-Or diagram for analyses in Table 2

Beauvoir Granite significantly offset towards the Qz-Ab join, possibly reflecting the additional role of lithium (Martin 1983), aluminium and phosphorous in modifying the minimum composition.

The trace element contents of individual rare-metal granites are highly variable (Table 3), reflecting features such as the initial melt composition, degree of fractionation, the intensity of postmagmatic alteration and the presence of mineralization. All the granites are enriched in Rb and depleted in Ba and Sr, and these features are enhanced in the mineralized varieties. Li enrichment is characteristic of lepidolite-albite granites, while enrichment in the other granite types is more erratic. Li typically increases from unmineralized varieties to mineralized varieties (Table 3). Be concentrations vary widely, and Be enrichment is not characteristic of any particular granite type, although it tends to concentrate in the mineralized varieties. Be mineralization is sometimes associated with the exocontact zones of lepidolite-albite granites. Zr and Hf are particularly enriched in the alkali granites, and are typically depleted in the lepidolite-albite granites.

Strong Nb enrichment is characteristic of the alkali granites (Table 3), and Ta may also be strongly concentrated in the late-stage granites. Nb/Ta ratios are usually greater than 4. In the biotite and/or muscovite granites Nb tends to be lower, although Nb/Ta ratios are typically high (> 4). The lepidolite-albite granites have high Ta concentrations and Nb/Ta ratios around 1 or less. The Ta/Nb ratio usually increases in the more albite-rich zones (Fig. 7). Tin is strongly enriched in all granite types, and reaches a maximum in the mineralized granites.

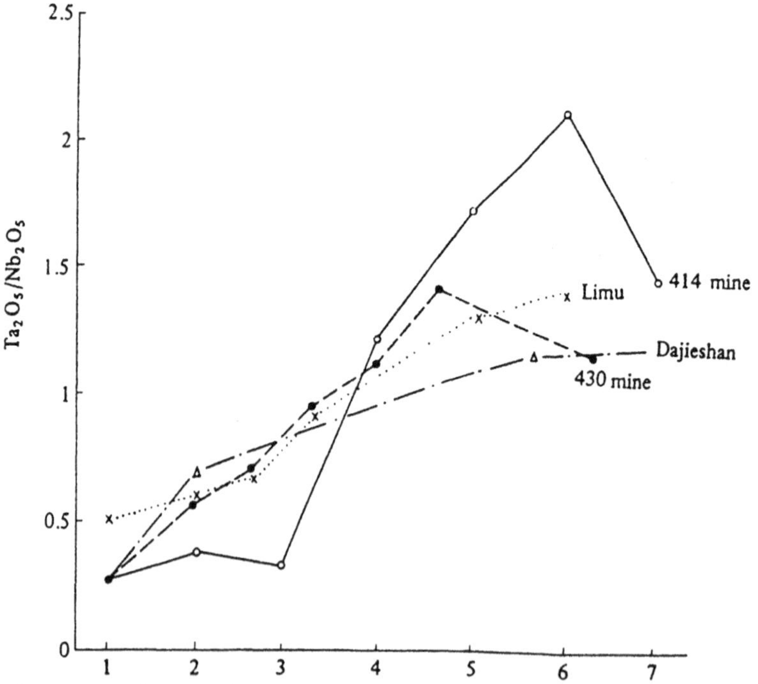

Fig. 7. Ta_2O_5/Nb_2O_5 variation with increasing intensity of albitization (data from Hu et al. 1984). *1* Potash metasomatic zone; *2* Na-K metasomatic zone; *3* weakly albitized zone or Na-K metasomatic zone; *4* moderately albitized zone; *5* intensely albitized zone; *6* intensely albitized and leucomicatized zone; *7* greisenized zone

The data base for many other elements is relatively restricted, and few definite trends emerge. Zn is relatively high in all the alkali granites, but is also enriched in some examples of other granite types. Th is high in the Nigerian granites, while the alkali microgranites from Saudi Arabia have very high U and low Th/U ratios.

REE abundances and distribution patterns are particularly sensitive to the nature of the restite and liquidus phases during magma evolution, to the composition of the postmagmatic fluid phase and to the nature of the minerals formed during postmagmatic alteration. This is reflected in the widely variable REE characteristics of rare-metal granites (e.g. Fig. 8). In general, the alkali granites are enriched in total REE and frequently have negatively sloping chondrite normalized patterns with a marked, negative Eu anomaly. However, some alkali granites have positively sloping or symmetrical patterns which may reflect variable volatile compositions in the fluid phase (Drysdall et al. 1984) and/or the stability of specific LREE-rich minerals such as bastnaesite, synchysite and monazite.

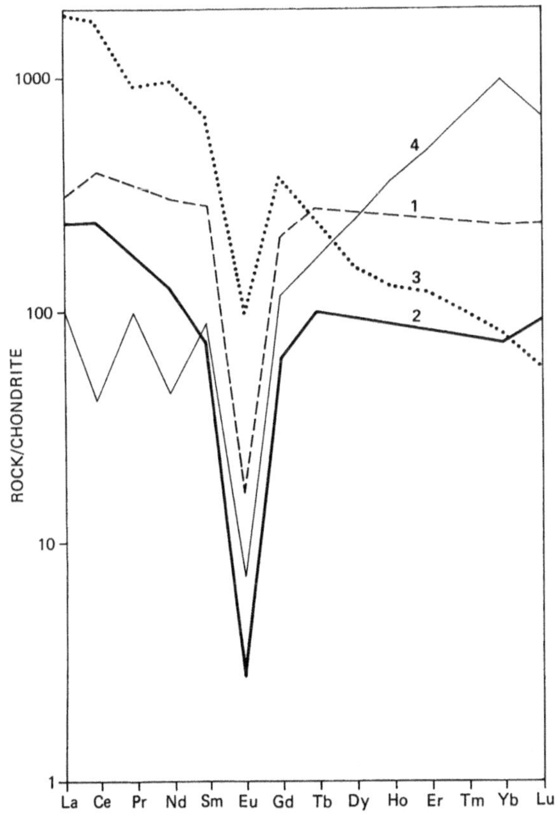

Fig. 8. Chondrite normalized rare-earth element distribution patterns for granites associated with tantalum and niobium mineralization. *1* Arfvedsonite granite, Ririwai Complex, Nigeria (Kinnaird et al. 1985a); average of three analyses. *2* Biotite granite, Ririwai Complex, Nigeria (Kinnaird et al. 1985a); average of two analyses. *3* Microgranite, Umm al Birak, Saudi Arabia (Drysdall et al. 1984); average of two analyses. *4* Microgranite, Ghurayyah, Saudi Arabia (Drysdall et al. 1984); average of two analyses

The peraluminous rare-metal granites often have negatively sloping REE patterns, but at lower total REE abundances than in the alkali granites, and with a more pronounced Eu anomaly (Taylor et al. 1981; Bowden and Kinnaird 1984; Xu et al. 1984 Kinnaird et al. 1985a). REE concentrations in lepidolite-albite granites are low to very low, sometimes less than one times chondrite (Riambault 1985), with negatively sloping, chondrite-normalized patterns.

The Sr-isotope compositions of granites related to Nb and Ta mineralization vary widely, possibly reflecting differences in source compositions and/or the effects of fluid rock interaction. Within postorogenic provinces initial $^{87}Sr/^{86}Sr$ ratios for biotite and two-mica granites are frequently high. For example, in south China the rare-metal granites have initial ratios ranging from 0.709 to 0.741 (Xu et al. 1984), while in southwestern Thailand initial ratios for mineralized granites are higher than 0.728 (Beckinsale 1979). Taken together with geological and geochemical data, these initial ratios are compatible with a dominantly crustal source for these peraluminous granites.

In peralkaline, anorogenic granites the initial $^{87}Sr/^{86}Sr$ ratios also vary widely. In examples with less well-developed hydrothermal systems the relatively low initial ratios (< 0.705) are compatible with a dominantly mantle source for magmas and hydrothermal fluids (Bonin et al. 1978; Taylor et al. 1981). Initial ratios for Saudi Arabian alkali granites range between 0.7017 and 0.7170, with a mean for 12 alkali granite intrusions of 0.7028 (Jackson et al. 1985). Jackson et al. (1985) consider that the alkali granites are formed by partial fusion of calc-alkaline crustal rocks under the influence of a mantle-derived fluid flux. Oxygen isotope data for the alkali granites indicate a major crustal contribution to the original magmas (Sheppard 1986).

In the Nigerian province initial $^{87}Sr/^{86}Sr$ ratios range between 0.705 and 0.752, with those for biotite granites (0.729 ± 0.009) lower than those for arfvedsonite-albite granites (0.752 ± 0.021) in the Ririwai Complex (van Breeman et al. 1975). These high initial ratios are difficult to interpret since $^{18}O/^{16}O$ and D/H data for the arfvedsonite-albite granites do not reflect interaction with meteoric hydrothermal fluids (Sheppard 1986). Although the nature of the processes which gave rise to these high initial $^{87}Sr/^{86}Sr$ ratios is uncertain, it seems likely that the arfvedsonite-albite granites may contain both mantle and crustal components.

5.1 Magmatic Trends

Fractionation trends in rare-metal granite suites are variable, reflecting the initial composition of the parent magmas and the nature of the magmatic liquidus phases. For example, the Nigerian granites have positive correlations for Na_2O and K_2O with silica, and negative correlations for TiO_2, Al_2O_3, total Fe, MnO, MgO, CaO and P_2O_5 against silica (Kinnaird et al. 1985b). Although decreasing TiO_2, MgO and CaO is typical of most rare-metal granite suites, some which are close to minimum melt compositions trend towards increasing Al_2O_3, Na_2O and F with decreasing silica (Pollard et al. 1987), similar patterns are observed in some topaz rhyolites (Christiansen et al. 1983). K_2O is variable in rare-metal granites, sometimes increasing or decreasing slightly with increasing fractionation.

Table 4. Variation in rare metals between successive intrusions in some granites in SE China (Data from Xu et al. 1980)

Granite body	Metallic elements	First stage	Second stage	Third stage	Pre-mineral dykes
Xihuashan,	W	7.0	8.9	70	25
	Be	15.1	17.3	18.9	35.2
Jiangxi	Sn	42.1	32.1	49.2	59.25
	TR_2O_3	0.036		0.041	
province	Nb_2O_5	0.0054		0.005	
	Ta_2O_5	0.0013		0.002	
Denfuxian	W	50.0	69.29	103.61	
Hunan	Nb	25	29	108	
province	Ta	20	13	91	
Mufushan, Jiangxi province	Be	5.4	10		
Qianmutian Zhejiang province	Be	10	7.4	13.1	
Limu	Ta		4.7	11	
Guangxi	Nb		36	73	
province	Sn	6.4	12.1	15.1	
Suzhou	Nb	53	63	84	112
Jiangsu	Ta	2.4	2.5	9.9	1.4
province	Th	50	63	125	368

Ba and Sr decrease markedly towards late-stage granites, while marked increases in Rb, Li, Ta, Nb, Sn, Th, U and Y are often noted (Tables 3 and 4). Zirconium is much higher in peralkaline granites due to the higher solubility of zircon in peralkaline magmas (Watson 1979), and zirconium often increases with fractionation.

Many of the geochemical features of rare-metal granites are compatible with an evolution via separation of plagioclase (decreasing CaO, Sr and Eu) and K-feldspar (decreasing Ba), together with ferromagnesian silicates (TiO_2, MgO, Fe and CaO). This crystal fractionation shifts the bulk composition of the magma towards the ternary Qz-Ab-Or surface. In melts with a high fluorine concentration the Qz field is expanded and the ternary minimum is shifted towards albite-rich compositions (Manning 1981). This can result in magmatic trends towards increasing Al_2O_3, Na_2O and F with decreasing SiO_2. In cases where significant lithium is present in addition to fluorine, the minimum may also be displaced slightly away from the Or apex towards the Qz-Ab join (Martin 1983). This may result in slight decreases in K_2O in the evolving magma. Volatile and incompatible elements are concentrated in residual magmas and may reach concentrations sufficient to promote crystallization of primary topaz, Li-micas, columbite-tantalite and microlite (Kovalenko et al. 1971; Kosakevitch 1976; Cuney et al. 1985).

During the evolution of peraluminous magmas the REE are controlled mainly by the crystallization of zircon, monazite and sometimes apatite. Due to the large partition coefficients of LREE in monazite (Miller and Mittlefehldt 1982), the separation of extremely small amounts of this mineral will result in LREE depletion in the melt. The much smaller partition coefficients for HREE in zircon (Mahood and Hildreth 1983) suggest that HREE enrichment will result unless relatively large amounts of zircon are extracted. In peralkaline magmas the opportunity for HREE enrichment is magnified by the much higher solubility of zircon. MREE enrichment or depletion in these granites may reflect amphibole and/or apatite crystallization (Taylor et al. 1981; Drysdall et al. 1984).

Vertical facies variation in rare-metal granites is often accompanied by vertical geochemical variations (Table 2) which indicate the accumulation of volatile and ore elements towards the roof of the intrusions. Hildreth (1979) has suggested that this roofward accumulation results from gravitational and convective processes in the magma chamber prior to crystallization. Other processes which may be important in this respect include boundary layer diffusion during sidewall crystallization (Huppert and Sparks 1984) and significantly enhanced diffusion rates resulting from the fluorine-rich character of the magmas (Dingwell 1985).

5.2 Postmagmatic Trends

Postmagmatic alteration potentially exerts a major influence on the geochemistry of rare-metal granites. In contrast to porphyry copper systems, where fluid phase evolution occurs relatively early and results in extensive stockworking of the pluton and surrounding rocks (Burnham 1979), many rare-metal granite magmas evolve a fluid phase at a late stage in their crystallization history. This fluid phase is initially confined to grain boundaries, microfractures and vugs (Pollard and Taylor 1986).

The major alteration styles involve ion exchange (Na^+, K^+, H^+) reactions between alkali feldspars and hydrothermal fluids. These reactions are particularly sensitive to the K/Na ratio of the fluid, temperature (Lagache and Weisbrod 1977) and the nature of the anionic species (Pichavant 1983). In F-rich rare-metal granites, microclinization may result from simple cooling of the hydrothermal fluid, while albitization probably results from the precipitation of F-bearing minerals or dilution of F-bearing fluids. The early formed, alkaline hydrothermal fluids are sometimes quartz undersaturated due to temperature-pressure (Kennedy 1950) and compositional (Wyart and Sabatier 1955) effects, resulting in quartz dissolution or replacement accompanying feldspathic alteration of the granite. These fluid-rock interaction processes can result in trends towards increasing Al_2O_3, Na_2O and F with decreasing SiO_2 in progressively more albitized zones towards the roof of the pluton. K_2O frequently decreases during albitization, but may increase substantially in association with the development of amazonite-albite granites in the apical zones of some intrusions (Beus et al. 1962; Syritso and Chernik 1967).

Albitization is usually accompanied by increases in Ta, Nb, Li, Rb, Ta/Nb and Sn. In peralkaline granites there is often an increase in Zr (Table 4, columns 1 and 2), while in peraluminous granites Zr sometimes decreases (Table 3, columns 7 and 10). The behaviour of REE during postmagmatic alteration is to a large extent dependent on the nature of the fluid phase. Preferential enrichment of LREE in Cl-rich fluids and HREE in F- and CO_2-rich fluids gives rise to contrasting REE distribution patterns in alteration systems associated with granites (Taylor and Fryer

1983). Fluids evolved during the crystallization of rare-metal granites are often fluorine-rich, and the observed HREE enrichment and associated Eu depletion in the albitized zones is usually attributed to fluid interaction processes (Aleksiyev 1970; Taylor et al. 1981; Hu et al. 1984).

The development of quartz-mica (± topaz) greisens in rare-metal granites is a postmagmatic process of hydrogen metasomatism accompanied by silicification and topazization. Ta/Nb sometimes decreases (Hu et al. 1984), although Kosals (1969) notes a large increase in Ta/Nb in berylliferous greisens (Ta/Nb 0.13 in albitized granite, increasing to 12.66 in greisens). However, the data of Kosals (1969) shows that the absolute contents of both Ta and Nb decrease during greisenization.

6 Discussion

In considering the genesis of rare-metal granites and their associated mineralization, it is clear that major problems in our understanding exist at a variety of scales. On a broad scale, detailed information concerning the timing, distribution and geochemistry of rare-metal granites is not available for most provinces. It is therefore difficult to discuss the origin of the magmas and their relationship in space and time with specific geotectonic processes. On a smaller scale, there is a major lack of detailed information concerning the timing of textural and mineralogical development within rare-metal plutons, particularly with respect to primary versus secondary features and alteration versus infill. The interpretation of geochemical and isotope data is therefore frequently made on the basis of insufficient petrological study, and the conclusions should be treated with caution. Perhaps the most serious deficiency is the lack of information concerning the development of the tantalum and niobium minerals themselves. Without detailed information on their occurrence, timing and distribution in relation to other magmatic and postmagmatic processes, genetic modelling of these systems cannot extend beyond the preliminary stage.

Two main types of models are generally proposed to account for the geochemical features of rare-metal granites. The model most commonly proposed (Beus and Zalashkova 1964; Syritso and Chernik 1967; Hu et al. 1984) envisages ore formation to result from the upward migration of postmagmatic fluids through the consolidated granite. These fluids produce zoned and overprinting alteration assemblages, leach metals from the lower zones, and deposit them in the upper zones. The major challenges for this type of modelling are to explain the vertical textural and mineralogical variation which provides clear evidence for important volatile enrichment from the earliest stages of crystallization. The geometry of the systems is also difficult to explain within this model, since the major alteration and mineralization is consistently localized in the apical zones of the granite (ridges, cupolas), rather than extending through the roof rocks. It is also difficult to substantiate the leaching model, since the lower, "leached" zones are typically enriched in rare metals, and features such as the upward decrease of Nb and Ta in phyllosilicates (Zalashkova and Sitnin 1967) could reflect the simultaneous crystallization of Ta-Nb minerals in the upper zones which preferentially partition these elements.

The second major model (Kovalenko et al. 1970; Kovalenko and Kovalenko 1984) envisages that the mineralized granites form by simultaneous crystallization of minerals from melt and aqueous-rich fluid under conditions of high fluorine activity. Features such as the "magmatic" nature of contacts between the rare-metal granites and wall rocks, the mineralogical and textural features of these rocks relative to typical fracture-controlled metasomatites, and the discovery of subvolcanic and volcanic analogues of rare-metal granites are cited in support of a predominantly magmatic origin. Within this model the vertical facies variation could reflect a preconcentration of volatiles and ore elements in the apical zones of the intrusion at the magmatic stage, which results in the localization of the most intense alteration and mineralization in these zones at the postmagmatic stage. This model has some difficulty in explaining the extreme enrichments in elements such as Rb via crystallization or emanation differentiation processes (Kovalenko and Kovalenko 1984), although processes such as diffusion in the F-rich magma and the effects of overprinting alteration should also be considered.

Experimental evidence on the nature of fluorine-rich magmas (Manning 1981; Naumov et al. 1971, 1984; Kovalenko and Kovalenko 1984; Dingwell 1985) provides increasing support for the important role of magmatic fractionation and crystallization processes in the formation of rare-metal granites. At the same time, the increasing documentation of fracture-controlled rare-metal metasomatites (Arkhangel'skaya 1985; Prokhorov 1985) indicates the possibility of a substantially postmagmatic origin for some deposits. The major problem to be addressed therefore is the relative roles of these processes in producing the geochemical and mineralogical features of rare-metal granites in specific instances.

In previous sections it was noted that many magmatic and postmagmatic processes in rare-metal granites could give rise to similar patterns of facies development (see also Pollard 1983). The development of albite-rich, quartz-poor rocks with topaz, Li-micas and Ta-Nb minerals could reflect fractionation in a fluorine-rich residual magma, or interaction between granite and a fluorine-rich, alkaline hydrothermal fluid. In both cases the chemical trends towards decreasing SiO_2 with increasing Al_2O_3, Na_2O and F in the most evolved facies may be developed. Similarly, trace element variation may reflect fractionation in the magma or postmagmatic fluid-rock interaction. REE variation is sensitive to these processes, reflecting fractionation of specific mineral phases and/or interaction with fluids of variable composition. Despite this opportunity, there have been few detailed studies of REE in rare-metal granites designed to monitor the effects of these processes.

In most cases, petrological studies of rare-metal granites provide clear evidence for substantial postmagmatic fluid-rock interaction. This indicates important element redistribution in the formation of new mineral phases, and allows for the introduction and removal of various elements. For example, the Ta and Nb concentrations and the Ta/Nb ratio of lepidolite-albite granites are significantly higher than those of Macusani glass (Table 3). Thus, even extremely fractionated magmas such as that which produced the Macusani glasses require significant upgrading via further fractionation in the magma chamber or via postmagmatic processes to approach the Ta and Nb concentrations found in the lepidolite-albite granites.

From the genetic viewpoint, it is vital to distinguish those geochemical features of rare-metal granites which may reflect magmatic preconcentration of volatile and ore elements in the upper portions of the magma chamber, and those which reflect postmagmatic reworking and consequent redistribution. The relative importance of magmatic and postmagmatic processes in the formation of rare-metal deposits varies widely, and detailed field and petrographic study of individual deposits with particular emphasis on transition zones and the timing of textural and mineralogical development is required to provide adequate base data for detailed geochemical interpretation.

Acknowledgments. Some of the rock samples and other materials utilized during the preparation of this chapter were kindly provided by Judith Kinnaird and Hu Shouxi. Reviews by R.G. Taylor and M. Pichavant resulted in considerable improvement of the final manuscript.

References

Aleksiyev EI (1970) Genetic significance of the rare-earth elements in the Younger Granites of northern Nigeria and the Cameroons. Geochem Int 7:127–132

Arkhangel'skaya VV (1985) Distribution of rare-earth elements and yttrium in east Siberian fracture-controlled alkali feldspar metasomatites. Geochem Int 22, 6:61–68

Aubert G (1969) Les coupoles granitiques de Montebras et d'Echassières et la génèse de leurs minéralizations. Mem BRGM 46:345 pp

Beckinsale RD (1979) Granite magmatism in the tin belt of southeast Asia. In: Atherton MP, Tarney J (eds) Origin of granitoid batholiths: geochemical evidence. Shiva, Kent, pp 34–44

Beus AA, Zalashkova N Ye (1964) Postmagmatic high temperature metasomatic processes in granitic rocks. Int Geol Rev 6:668–681

Beus AA, Severov EA, Sitnin AA, Subbotin RD (1962) Albitized and greisenized granites (apogranites). Academic Science Press, Moscow 194 pp (in Russian)

Boguslavskiy IS, Kalenov AD, Egel L Ye (1966) Metasomatic albitite found for the first time in the MPR. Dokl Acad Sci USSR Earth Sci Sec 168:159–162

Bonin B, Grelou-Orsini C, Vialette Y (1978) Age, origin and evolution of the anorogenic complex of Evisa (Corsica): a K-Li-Rb-Sr study. Contrib Mineral Petrol 65:425–432

Bowden P (1982) Magmatic evolution and mineralization in the Nigerian younger granite province. In: Evans AM (ed) Metallization associated with acid magmatism. Wiley & Sons, New York, pp 51–62

Bowden P, Kinnaird JA (1984) The petrology and geochemistry of alkaline granites from Nigeria. Phys Earth Planet Int 35:199–211

Breeman O van, Hutchinson J, Bowden P (1975) Age and origin of Nigerian Mesozoic granites: a Rb-Sr isotopic study. Contrib Mineral Petrol 50:157–172

Burnham CW (1979) Magmas and hydrothermal fluids. In: Barnes HL (ed) Geochemistry of hydrothermal ore deposits, 2nd edn. Wiley & Sons, New York, pp 71–136

Christiansen EH, Burt DM, Sheridan MF, Wilson RT (1983) The petrogenesis of topaz rhyolites from the western United States. Contrib Mineral Petrol 83:16–30

Cuney M, Autran A, Burnol L (1985) Premiers résultats par le sondage GPF de 900m réalisé sur le granite sodo-lithique et fluoré a minéralisation disséminée de Beauvoir. Chron Rech Mineral 481:59–63

Deng S (1984) The genetic series, time and space distributions and related mineralizations of the granites in Hunan Province. In: Geology of granites and their metallogenic relations. Science Press, Beijing, pp 671–688

Dingwell DB (1985) The structure and properties of fluorine-rich silicate melts: implications for granite petrogenesis. In: Taylor RP, Strong DF (eds) Granite-related mineral deposits: geology, petrogenesis and tectonic setting. CIM, Halifax, pp 72–79

Drysdall AR, Jackson NJ, Ramsay CR, Douch CJ, Hackett H (1984) Rare element mineralization related to Precambrian alkali granites in the Arabian Shield. Econ Geol 79:1366–1377

Garson MS, Mitchell AHG (1975) The geology of the tin belt in Peninsular Thailand around Phuket, Phangnga and Takua Pa. Inst Geol Sci Overseas Mem 1:112 p
Hildreth W (1979) The Bishop tuff: evidence for the origin of compositional zonation in silicic magma chambers. Spec Pap Geol Soc Am 180:43–75
Hosking KFG (1981) Tin-bearing minerals of southeast Asia. In:Hassan AH (ed) Complex tin ores and related problems. SE Asia Tin Res Dev Centre Tech Publ 2:17–64
Hu S, Sun M, Yan Z, Xu J, Cao X, Ye Y (1984) An important metallogenetic model for W, Sn and rare granitophile element ore deposits related to metasomatically altered granites. In: Geology of granites and their metallogenetic relations. Science Press, Beijing, pp 519–537
Huppert HE, Sparks RSJ (1984) Double-diffusive convection due to crystallization in magmas. Annu Rev Earth Planet Sci 12:11–37
Hutchison CS (1978) The impurities in southeast Asian tin ore concentrates. Warta Geol 4:39–43
Ivankin PF, Kulagashev AI (1980) Rare-metal ore magmatic systems of Transbaykal. Int Geol Rev 22:339–348
Jackson NJ (1986) Mineralization associated with felsic plutonic rocks in the Arabian Shield. J Afr Earth Sci 4:213–227
Jackson NJ, Ramsay CR (1986) Post-orogenic felsic plutonism, mineralization and chemical specialization in the Arabian Shield. Trans Inst Min Metall Sec B (Appl Earth Sci) 95:B83–93
Jackson NJ, Drysdall AR, Stoeser DB (1985) Alkali granite related Nb-Zr-REE-U-Th mineralization in the Arabian Shield. In: High heat production (HHP) granites, hydrothermal circulation and ore genesis. Inst Mineral Metall, London, pp 459–478
Kennedy GC (1950) A portion of the system silica-water. Econ Geol 45:629–653
Kinnaird JA, Bowden P, Ixer RA, Odling NWA (1985a) Mineralogy, geochemistry and mineralization of the Ririwai complex, northern Nigeria. J Afr Earth Sci 3:185–222
Kinnaird JA, Batchelor RA, Whitely JE, MacKenzie AB (1985b) Geochemistry, mineralization and hydrothermal alteration of the Nigerian high heat producing granites. In: High heat production (HHP) granites, hydrothermal circulation and ore genesis. Ist Mineral Metall, London, pp 169–195
Kosakevitch A (1976) Evolution de la minéralisation en Li, Ta et Nb dans la coupole granitique de Beauvoir (massif d'Echassières, Allier). BRGM, Orleans, 72 pp
Kosals Ya A (1969) Geochemistry of tantalum and columbium in hydrothermal metamorphosis of granites. Int Geol Rev 927–932
Kovalenko VI, Kovalenko NI (1984) Problems of the origin, ore-bearing and evolution of rare-metal granites. Phys Earth Planet Int 35:51–62
Kovalenko VI, Krinberg IA, Mironov VP, Selivanova GI (1964) Behaviour of U, Th, Nb and Ta during albitization of granitoids of the Ognitskii complex (eastern Sayan Mts.). Geochem Int 1:868–874
Kovalenko VI, Kuz'min, Letnikov FA (1970) Magmatic origin of lithium- and fluorine-bearing rare-metal granite. Dokl Acad Sci USSR Earth Sci Sec 190:189–192
Kovalenko VI, Kuz'min MI, Antipin VS, Petrov LL (1971) Topaz-bearing quartz keratophyre (ongonite), a new variety of subvolcanic igneous vein rock. Dokl Acad Sci USSR Earth Sci Sec 199:132–135
Kozlov VD, Sheremet Ye M, Janovskiy VM (1974) Geological characteristics of the Mesozoic plumasite leucocratic granites of the Transbaykalia tin-tungsten belt. Geochem Int 11:997–1008
Lagache M, Weisbrod A (1977) The system: two alkali feldspars-KCl-NaCl-H_2O at moderate to high temperatures and low pressures. Contrib Mineral Petrol 62:77–101
Liew TC, McCulloch MT (1985) Genesis of granitoid batholiths of Peninsular Malaysia and implications for crustal evolution: evidence from a Nd-Sr isotopic and U-Pb zircon study. Geochim Cosmochim Acta 49:587–600
Litvinovskiy BA, Plyusnin GS, Sandimirova GP, Zanvilevich AN, Pakhol'chenko Yu A (1984) New rubidium-strontium dates for Permian alkalic granitoid rocks of Transbaikal. Dokl Acad Sci USSR Earth Sci Sec 277:137–141
Litvinovskiy BA, Kaperskaya Yu N, Zanvilevich AN (1985) Rare-element geochemistry in the rocks of the Mongolia-Transbaykalia alkali granite province. Geochem Int 22:94–112
Mahood G, Hildreth W (1983) Large partition coefficients for trace elements in high-silica rhyolites. Geochem Cosmochim Acta 47:11–30
Manning DAC (1981) The effect of fluorine on liquidus phase relationships in the system Qz-Ab-Or with excess water at 1 kb. Contrib Mineral Petrol 76:206–215
Martin JS (1983) An experimental study on the effect of Li on the granite system. Proc Ussher Soc 5:417–420

Matheis G, Emofurieta WO, Ohiwerei SF (1982) Trace element distribution in tin-bearing pegmatites in southwestern Nigeria. In: Evans AM (ed) Mineralization associated with acid magmatism. Wiley & Sons, New York, pp 205–220
Miller CF, Mittlefehldt DW (1982) Depletion of light rare-earth elements in felsic magmas. Geology 10:129–133
Nakapadungrat S, Chulacharit N, Munthachit Y, Chotigkrai T, Sangsila S (in press) Geology of Sn/W granites in Takua Pa area, southern Thailand. In: Hutchinsan CS (ed) Proc Int Conf Geol Tin Deposits, Nanning, China, Springer, Berlin Heidelberg New York Tokyo
Naumov VB, Kovalenko VI, Kuz'min MI, Vladykin NF, Ivanov GF (1971) Thermometric study of inclusions of melt in topaz from topaz-bearing quartz keratophyre (ongonite). Dokl Acad Sci USSR Earth Sci Sec 199:104–107
Naumov VB, Kovalenko VI, Clocchiatti R, Solovova IP (1984) Crystallization parameters and phase compositions for melt inclusions in ongorhyolite quartz. Geochem Int 21:19–32
Navilkin DV (1973) Geology of USSR. Oliver & Boyd, Edinburgh, 855 pp
Pavlenko AS (1974) The Mongol-Tuva province of alkaline rocks. In: Sorensen H (ed) The alkaline rocks. Wiley & Sons, New York, pp 271–293
Pichavant M (1983) (Na, K) exchange between alkali feldspars and aqueous solutions containing borate and fluoride anions, experimental results at P = 1 kbar. Proc 3rd NATO Adv Stud Inst Feldspars and feldspathoids, Rennes, p 102
Pichavant M, Manning DAC (1984) Petrogenesis of tourmaline and topaz granites: the contribution of experimental data. Phys Earth Planet Int 35:31–50
Pichavant M, Herrera JV, Boulimer S, Briqueu L, Joron JL, Juteau M, Marin L, Michard A, Sheppard SMF, Treiul M, Vernet M (1987) The Macusani glasses, SE Peru: evidence of chemical fractionation in peraluminous magmas. Geochem Soc Spec Publ 1:359–373
Pollard PJ (1983) Magmatic and postmagmatic processes in the development of rocks associated with rare-element deposits. Trans Inst Mining Metal See B (Appl Earth Sci) 92:B1–9
Pollard PJ, Taylor RG (1986) Progressive evolution of alteration and tin mineralization: controls by interstitial permeability and fracture-related tapping of magmatic fluid reservoirs in tin granites. Econ Geol 81:1795–1800
Pollard PJ, Pichavant M, Charoy B (1987) Contrasting evolution of fluorine-rich and boron-rich tin systems. Mineral Deposit 22:315–321
Prokhorov KV (1985) Rare-metal high-temperature sodium metasomatites: conditions of their formation. Int Geol Rev 27:113–125
Pyatenko IK, Sitnin AA, Lavrinenko AF (1967) Geochemical features of metasomatically altered granitic rocks in Azov region. Int Geol Rev 9:1430–1443
Raimbault L (1985) Etude préliminaire des terres rares dans le granite de Beauvoir. Doc BRGM 100:125–128
Severov EA (1963) Concerning the genesis of niobium-containing granites. Int Geol Rev 5:1623–1630
Shcherba GN (1970) Greisens. Int Geol Rev 12:114–150, 239–255
Sheppard SMF (1986) Igneous rocks: III Isotopic case studies of magmatism in Africa, Eurasia and oceanic islands. In: Valley JW, Taylor HP, Jr., O'Neil JR (eds) Stable isotopes in high temperature geological processes. Min Soc Am Rev Mineral 16:319–372
Sitnin AA (1970) On the mobilization of Ta, Nb, Li, Rb, Cs, Be, Sn, W in the process of high-temperature metasomatism of certain granitoids. In: Problems of hydrothermal ore deposition. Int Union Geol Sci A No 2. Schweizerbart, Stuttgart, pp 348–351
Syritso LF, Chernik LN (1967) Evolution in accessory mineral paragenesis during metasomatic alteration of granites in eastern Transbaykalia massifs. Int Geol Rev 9:814–827
Taylor RG, Pollard PJ (1987) Pervasive hydrothermal alteration in tin-bearing granites and implications for the evolution of ore-bearing magmatic fluids. CIM Spec Vol 39
Taylor RP, Fryer BJ (1983) Rare-earth element lithogeochemistry of granitoid mineral deposits. CIM Bull 860:74–86
Taylor RP, Strong DF, Fryer BJ (1981) Volatile control of contrasting trace element distribution in peralkaline granitic and volcanic rocks. Contrib Mineral Petrol 77:267–271
Tischendorf G (1977) Geochemical and petrographic characteristics of silicic magmatic rocks associated with rare-element mineralization. In: Stemprok M, Burnol L, Tischendorf D (eds) Metallization associated with acid magmatism, vol 2. Schweizerbart, Stuttgart, pp 41–96
Tugarinov AI, Pavlenko AS, Kovalenko VI (1968) Geochemical evidence on the origin of apogranites. Geochem Int 5: 1156–1171

Turekian KK, Wedepohl KH (1961) Distribution of elements of some major units of the earth's crust. Geol Soc Am Bull 72:175–192

Tuttle OF, Bowen NL (1958) On the origin of granite in the light of experimental studies in the system $NaAlSi_3O_8$-$KAlSi_3O_8$-SiO_2-H_2O. Geol Soc Am Mem 74, 153 pp

Volkov VN, Gorbacheva SA (1980) Composition of rock-forming biotite and variation in crystallization conditions in a vertically zoned pluton. Geochem Int 17:75–79

Watson EB (1979) Zircon saturation in felsic liquids: experimental results and application to trace element geochemistry. Contrib Mineral Petrol 70:407–419

Wyart J, Sabatier G (1955) Solubilite du quartz dans les solutions des sels alkalins sous pression au-dessus de la temperature crutique. CR Acad Sci 240:2157–2159

Xu K, Guo L, Hu S, Ji S, Shi Y, Wang Z, Sun M, Mou W, Lin C, Ye J (1980) Investigation of the time and spatial distribution of the granitic rocks of southeastern China: their petrographic evolution, petrogenetic types and metallogenetic relations. Dep Geol Nanjing Univ, 56 pp

Xu K, Sun N, Wang D, Hu S, Liu Y, Ji S (1984) Petrogenesis of the granitoids and their metallogenetic relations in south China. In: Geology of granites and their metallogenetic relations. Science Press, Beijing, pp 1–30

Zalashkova N Ye, Gerasimovskii VV (1974) Petrographic and geochemical features of rare-metal amazonite granites. In: Stemprok M (ed) Metallization associated with acid magmatism. Geol Surv (Prague) 1:232–236

Zalashkova N Ye, Sitnin AA (1967) Features of the distribution of rare-elements in tantalum-bearing apogranites and biotite granites. Geochem Int 4:1013–1025

Part III Characterization of Ore Deposits

Rare Earth Mineral Deposits and Their Industrial Importance

P. MÖLLER

Abstract

Rare earth mineral deposits can be subdivided into endogenic, exogenic and metamorphic mineralizations. Today, industrially most important is bastnaesite from the primary deposit of Mountain Pass, California, and Bayan Obo, P.R. China, followed by monazite from the coastal sands of Australia, Brazil, India and Indonesia.

Brief comments on the deposits of Mountain Pass, CA/USA, and Bayan Obo/China and Karonge/Burundi are given.

1 Introduction

The REE are estimated to form about 0.02% of the earth's upper crust by weight. They are much more abundant than Au (4 ppb), Ag (70 ppb) and U (2 ppm) in the crust. The REE are mostly hosted by rock-forming minerals as substitutes for major, lithophile cations. Only at higher concentrations do they form their own minerals, e.g. multiple oxides, hydrous and anhydrous silicates, phosphates and fluorides. The REE are more abundant in alkaline and acidic than in basic and ultrabasic igneous rocks and they are commonly associated with Th, U, Nb, Ta, Ti, Zr, F, Ba, Sr, Pb, Cu, Mo, Sn, W and P. REE compounds are industrially produced from only a few mineral species with high REE contents and good processability. Monazite and bastnaesite are the only minerals of economic importance, while xenotime and apatite are of minor significance, and brannerite, fergusonite, euxenite, gadolinite and REE from pyrochlore group minerals are only of geochemical interest.

REE minerals usually show REE distribution patterns which are distinctively different from that of the earth's crust (Fig. 1). In some REE minerals the light REE are particularly enriched with respect to the heavy REE, while in others it is the reverse. For this reason the evaluation of ores should always quote the total REE content as well as the contents of the individual REE. This means in detail that it is necessary to indicate whether the minerals are rich in Ce-group (light REE) or Y-group elements (heavy REE) and whether the patterns show a negative Eu anomaly or not. This is of major importance, since presently Eu is one of the most valuable of all the REE.

Hahn-Meitner-Institut Berlin GmbH, AG Geochemie, D-1000 Berlin 30

Lanthanides, Tantalum and Niobium
P. Möller, P. Černý and F. Saupé (Eds.)

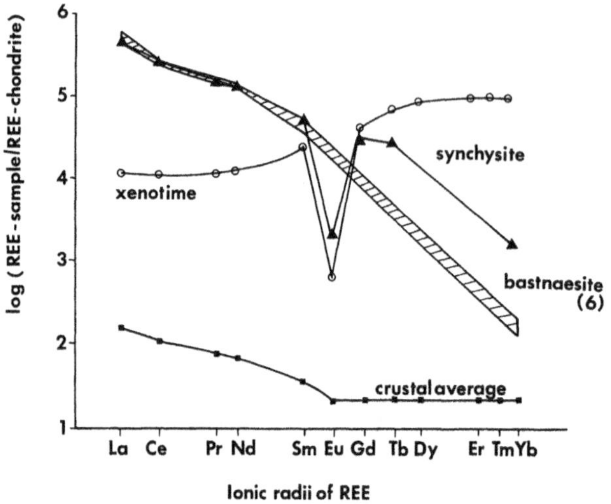

Fig. 1. Typical REE distribution patterns of REE minerals from carbonatites (bastnaesite from Burundi) and pegmatites (synchysite and xenotime). The average REE distribution pattern of the crust is given for comparison

Until 1950, monazite from placers (Brazil, later also Australia and India) and as a by-product from pegmatite mining was the only source of REE. After discovering bastnaesite at Mountain Pass, CA/USA, in 1949, this fluorocarbonate has become the world's main supply of light REE, followed by monazite from placers as a by-product of mineral sands mined for Ti, Zr and Sn minerals in Australia, Brazil, China, India and Malaysia (Robjohns 1984). Since the early 1980s bastnaesite is additionally recovered in increasing quantities as a by-product of iron-mining at Bayan Obo (also reported as Pai-yun-o-po), north of Bao-Tao, P.R. China. The USSR produces most of its rare-earth oxides from carbonatites and alkali rock complexes. Brazil mines deeply weathered carbonatites, as well as monazite placer deposits. Xenotime is produce from sands in Malaysia, Australia, China and Thailand (Hedrick 1985). Heavy REE(Y) have also been produced as by-products of uranium mining in Canada and Australia (Sommerlatte 1969).

From the mid-1970s until the early 1980s roughly 95% of the total world output was produced by only six countries: Australia, Brazil, India, Malaysia, USA and USSR. More than 50% of the world production of concentrates came from one mine: Mountain Pass, CA/USA (Robjohns 1984). In 1984 this mine still produced about 50% of the world production of REE oxides (Robjohns 1986). At the moment, REE concentrates are almost entirely produced from (1) mineralizations related to alkaline rocks and carbonatites and (2) placer deposits.

A selection of localities with REE mineralizations is given in Table 1 and Fig. 2. This list of occurrences and deposits is by no means representative because the economic importance of the various sites is unbalanced. It just demonstrates that REE enrichments are found worldwide. Many more localities are known from which REE concentrates could be mined as by-products, particularly in the USSR and China. There will be no shortage of REE minerals in the foreseeable future.

Table 1. Principal REE-mineral deposits (producing, non-producing and potential) which are shown in Fig. 2

No.	Locality	Host rock	REE-mineral	Remarks	Reference
1	Oka, Ontario, Canada	Carbonatite	Pyrochlore	By-product of Nb-mining	Rose 1979
2	St. Honoré, Ontario, Canada	Carbonatite	Pyrochlore	By-product of Nb-mining	Rose 1979
3	Blind River-Elliot Lake District, Ontario, Canada	Precambrian quartz Pebble conglomerates	Monazite	By-product of U-mining	Rose 1979
4	Blachford Lake Complex, N.W.T., Canada	Peralkaline granites and syenites	Gadolinite, xenotime, yttrofluorite	By-product of Be-, Nb-Ta-, Zr-mining	Černý and Trueman 1985
5	Scrub Oaks, New Jersey, USA	Magnetite-hematite ore-body in metamorphic rocks	Synchysite, xenotime, bastnaesite, apatite	By product of Fe-mining	Klemic et al. 1959
6	Minneville, New York, US	Magnetite-hematite vein in metamorphic rocks	Apatite		Mei at al. 1979
7	Piedmont placer area; VA, NC, SC, AL and GA, USA	Fluvial placers	Monazite		Mertie 1979
8	Atlantic coast, Florida, USA	Littoral sands	Monazite	By-product of rutile	Mertie 1975
9	Lemhi Pass District, Montana, Idaho, USA	Veins in Precambrian quartzites and siltites and younger rocks	Monazite, xenotime	By-product of Th-mining	Staatz et al. 1979
	Little Big Horn River, WY, MT, USA	Conglomerates	Monazite		Ginzburg and Zhuraleva et al. 1959
10	South Platte district, CO, USA	Quartz-monzonite	Fluorite		Simmons et al. 1987
	Powderhorn District, CO, USA	Carbonatite and alkaline rocks	Apatite, Pyrochlore		Armbrustmacher 1980
	Wet Mountains, CO, USA McClure Mountains, CO, USA	Th-veins rel. with alk. rocks and carbonatite dykes	Apatite, synchysite		Staatz 1974
	North-central Colorado, USA	Migmatites	Xenotime, monazite		Heinrich and Wells 1980
11	Gallinas Mountains, New Mexico, USA	Copper sulphide-fluorite vein in Permian sandstones	Bastnaesite	By-product of Cu-mining	Perhac and Heinrich 1964
12	Mountain Pass area, CA, USA	Carbonate-barite body	Bastnaesite, parisite, monazite	REE deposit	Olson et al. 1954
	Music Valley area, USA	Precambrian gneisses	Xenotime, monazite		Evans 1964
13	Llano County, Texas, USA	Pegmatites	Allanite, fergusonite		Ehlman et al. 1964
14	Araxa, Minas Gerais, Brazil	Soil on top of carbonatite	Apatite, pyrochlore	Nb-mining	Guimaraes 1957
15	Catalao, Goyaz, Brazil	Alkaline rock-carbonatite complex	Apatite	Phosphate mining	Beurlen and Cassedanne 1981
16	Atlantic coast, Brazil	Sandy shelf sediments	Monazite	< 20% of heavy minerals	Martins and Coutinho 1981
17	Pocos de Caldas, Minas Gerais, Brazil	Soil on top of alkaline rock	Bastnaesite, cerianite, pyrochlore	Laterites with 4% REO	Ulbrich and Gomes 1981
18	Jos Plateau, Nigeria	Placers derived from alkaline granites	Xenotime	Sn, Ta, Nb mineralization	de Kun 1960

Table 1. *(cont.)*

No.	Locality	Host rock	REE-mineral	Remarks	Reference
19	Mrima Hill and Ruri complex,	Soil on top of carbonatite	Phosphates	Nb mineralization	van Wambeke 1977
	Kenya				
20	Wigu Hill and Mbeya Hill,	Soil on top of	Bastnaesite	Nb reserve (Mbeya)	Harris 1961
	Tanzania	Carbonatites	Monazite	>10% REO in veins (Wigu)	
21	Karonge area, Burundi	Carbonatite	Bastnaesite,		Thoreau et al. 1958
22	Kangankunde Hill, Malawi	Carbonatite	monazite		
			Monazite		Holt 1965
23	Itremo, Malegasy Republic	Pegmatitelike dykes	Bastnaesite, parisite		Chantraine and Radelli
24	SE region of Malegasy Republic	Beach sands	Monazite		1970
25	Be-Li pegmatites, Zimbabwe, Zambia	Pegmatites	Monazite		Murdock 1963
			Samarskite,	Nb-, Ta-mineralization	Gallagher 1968
26	Glenover, Transvaal, RSA	Carbonatite	fergusonite, euxenite		
	Palabora, Transvaal, RSA	Carbonatite	Synchysite		Verwoerd 1967
27	Pilanesberg, Transvaal, RSA	Alkalic rock	Pyrochlore, silicates	Th-mineralization	Coetzee 1976
	Witwatersrand, RSA	Conglomerates	Uraninite, monazite	Au-U-Mining	Backström 1976
28	Steenkampskraal, Cape Prov., RSA	Vein in gneiss	Monazite	Ti-magnetite	Pike 1958
29	Nile delta, Egypt	Black sands	Monazite		Overstreet 1967
30	Sarfartoq, Greenland	Carbonatite	Pyrochlore	Nb, U, P mineralization	Secher and Larsen 1980
31	Ilimaussaq, Greenland	Alkaline igneous complex	Rinkite, monazite	U, Be, Zr, Nb mineralization	Nielsen 1973
			Steenstrupine		
32	Motzfeldt Centre	Alkaline igneous complex	Pyrochlore	Nb reserves	Jones and Larsen 1985
33	Kola Peninsula (Khibina-,	Alkaline igneous complexes	Apatite	Apatite, nepheline	Notholt 1979
	Lovozero-, Kovdor- complex), USSR		Francolite		
34	Kodal, Norway	Jacupirangite dyke	Apatite		Notholt 1979
	Langesundsfjord, Norway	Pegmatite in neph.-syenite	Apatite		Bergstol 1972
35	Synnyr complex, Sibiria, USSR	Alkalic igenous complex	Allanite, apatite		Heinrich and Wells 1980
36	Seligdar, Sibiria, USSR	Veins in metamorphites	Apatite		Faizullin et al. 1977
37	Khan Bogdo, Mongolian P.R.	Arfvedsonite alkali	Polylithionite	Rare-metal-rich	Kovalenko et al. 1977
		granites, pegmatites?	Synchysite, monazite		
38	Mushugai Khuduk, Mongolian P.R.	Carbonatite?	Apatite, bastnaesite		Ontoev et al. 1979
			Fluorite		
39	Bayan Obo, P.R. China	REE-rich iron ores,	Bastnaesite,	Iron mining	Lee 1970; Tu 1981
		alkaline intrusions?	monazite		
40	Central Urals, USSR	Magnetite-bearing skarn	Apatite		Gerasimovskiy and
		orebodies			Mineyev 1981
41	Korea	Beach sands	Monazite		Overstreet 1967
42	Purulia district, W Bengal, India	Placers in bedrock	Monazite	Ilmenite, rutile mining	Overstreet 1967
		Cappings			

43	SW coastal area, India	Black sands	Monazite	Ilmenite, rutile mining	Overstreet 1967
44	W coast area, Sri Lanka	Beach sands	Monazite		Overstreet 1967
45	Malay Peninsula, Indonesia	Beach sands	Monazite	SnO_2-mining	Overstreet 1967
46	Singkep, Billiton, Bangka, Indonesia	Beach Sands	Monazite	SnO_2-mining	Overstreet 1967
47	W Australia	Beach sands	Monazite	rutile, zircon mining	Overstreet 1967
48	E Australia	Beach sands	Monazite	rutile, zircon mining	Overstreet 1967
49	Radium Hill, Australia	Lodes in Precambrian gneisses and schists with intrusives	Navidite	rutile, zircon mining	Parkin and Glasson 1954
50	Mary Kathleen, Queensland Australia	Skarn deposit	Allanite, stillwellite	U-mining	Hawkins 1975
51	Olympic Dam, Australia	Fe-rich sedimentary breccia	Bastnaesite, florencite	Cu-U-Au reserves	Roberts and Hudson 1984
52	Taiwan	Beach sands	Monazite		Overstreet 1967

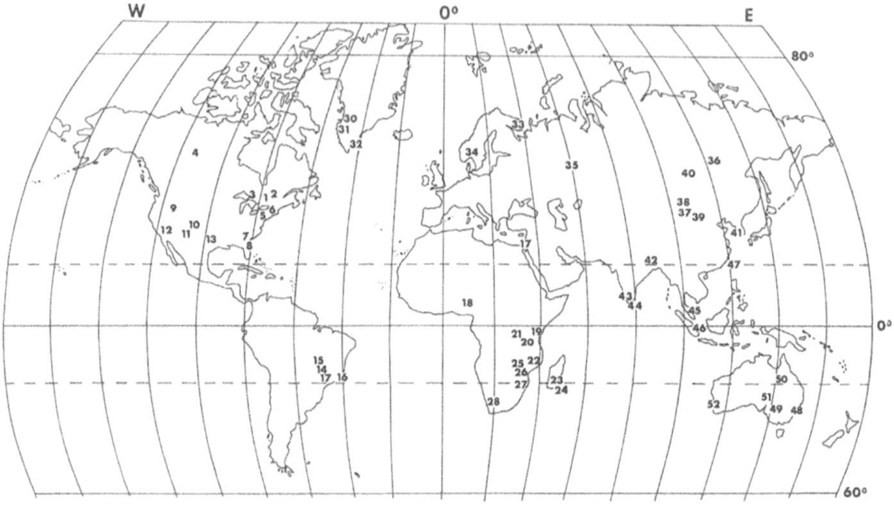

Fig. 2. Location of the REE mineral deposits listed in Table 1. Figures refer to the numbers in Table 1.

2 Classification of REE Deposits

Various schemes to systematize REE mineral deposits have been proposed (Ginzburg and Zhuravleva 1959; Vlasov 1968; Heinrich and Wells 1980; Ovchinnikov and Solodov 1980; Neary and Highley 1984; review in Gmelin 1984). According to them the REE occurrences will be subdivided into endogenic, exogenic and metamorphic mineralizations. The large diversity of REE mineral deposits arise from the fact that most of the mineralizations are polymetallic and the REE represent a minor component only. The locations of most of the mentioned deposits/occurrences are shown in Fig. 2 and their references are quoted in Table 1.

2.1 Endogenic (Igneous and Hydrothermal) Mineralizations

Because of their large ionic radii, the trivalent REE become enriched

1. In partial melts or fluids from the upper mantle, and
2. In melts or late residual fractions of fluids of crustal origin.

The most important sources of REE are deposits associated with carbonatites and alkalic rocks. It is assumed that their melts originate from the upper mantle.

The REE in REE deposits spatially and genetically associated with subalkaline and alkaline granites are thought to originate from the lower crust (Möller, this Vol.). REE-bearing pegmatites are genetically related to granitoids, whereas hydrothermal REE mineralizations can be linked to alkali granites, nepheline syenites, alkali syenites and shonkinites.

In all cases, the REE are enriched in the fluid phases if these fluid phases contain complexing agents such as PO_4^{3-}, CO_3^{2-} and F^-. These are also the main anions that later form REE minerals when P-T conditions of the fluids change adequately.

Endogenic REE deposits are not known to be associated with effusive rocks.

2.1.1 Alkali-Ultrabasic Complexes

Alkali-ultrabasic complexes with loparite-bearing (up to 20% of the rock) malignites (mafic nepheline syenite) and urtites (nephelinite with >70% nepheline, no feldspars), eudialyte-containing (up to 20% of the rock) lujavrites (agpaitic nepheline syenite), apatite-rich (up to 80% of the rock) nepheline, and magnetite rocks are all potential sources of REE (Ginzburg and Zhuravleva 1959). In these rocks the REE form individual minerals which are distributed as accessory components, but occasionally they might be highly enriched within these rocks. Rocks with high contents of knopite (up to 35% of the rock) and titanomagnetite are of special interest because of their enhanced Sm, Gd and Dy contents.

Primary REE-containing accessory minerals (perovskite, loparite, eudialyte) may release their REE during metasomatic processes (nephelinization, carbonatization). As a consequence, pyrochlore, bastnaesite and parisite are crystallized. Extraction of REE from primary ores of this type, however, is difficult. Examples of REE-rich, alkali-ultrabasic rock complexes are:

Khibina massif, Kola Peninsula, USSR;
Lovozero complex, Kola Peninsula, USSR;
Phoscorite and pyroxenite zones of Palabora, Transvaal/RSA;
Kodal, Norway;
Synnyr complex, Eastern Siberia, USSR.

2.1.2 Carbonatites

Carbonatites are of magmatic origin. They are highly enriched in light REE and show REE distribution patterns which are free of Eu anomalies unless altered. Detailed studies by mainly Russian authors (quoted from Gmelin 1984) demonstrate that many of these bodies underwent long-lasting polyphasic alterations.

At the beginning of the crystallization of carbonatites most of the REE are incorporated by apatite and calcite; later on pyrochlore precipitates in association with magnetite, phlogopite, apatite and sulphides. During the latest phases of cooling, the calcite can be partially metasomatically replaced by dolomite or ankerite. During these processes the REE are released from calcite and mainly monazite and fluorocarbonates (bastnaesite, parisite, synchysite) may crystallize from the fluids. For this reason, some of the REE occurrences related to carbonatites are also described as hydrothermal deposits, as for instance the Karonge area, Burundi. Examples of REE-rich carbonatites are:

Mrima Hill and Ruri complex, Kenya;
Glenover, Transvaal, RSA;

Siilinjarvi, Finland;
Mountain Pass, CA/USA;
Oka and St. Honoré, Canada;
Kangankunde Hill, Malawi;
Sarfartoq, Greenland.

2.1.3 Metasomatites

Locally, metasomatites contain a great variety of REE minerals or complex minerals of REE with Nb, Th, Zr and Be. The rare-metal content is extracted from the rock's source environment as soluble complex compounds in alkaline metasomatizing fluids. Subsequent albitization of the intrusive and/or the host rock leads to deposition of REE-containing mineralizations (Ginzburg and Zhuravleva 1959).

Albitites enriched in pyrochlore and zircon are complex ores for Nb, Th, Zr and REE and occur mainly in nepheline syenite massifs. Additional ore minerals in such albitites are: thorite, beckelite, britholite, fersmite, aeschynite and ilmenorutile. An example of this type of occurrence is Miask in Central Urals, USSR.

Fergusonite- and priorite-containing albitites have been described to occur in parts of the cupolas of alkali syenite and alkali granite massifs. The ore minerals are finely disseminated; they form also stringers, veins and stockworks (Ginzburg and Zhuravleva 1959).

Fenitization of country rocks is caused by alkali- and Fe-rich late hydrothermal fluids released from carbonatitic and alkalic intrusions. Typical REE-containing minerals that are formed from these fluids are perovskite, pyrochlore, thorogummite, brockite and REE-miserite (Heinrich and Wells 1980). As an example, the mineralizations of McClure Mt., CO/USA, may be mentioned.

2.1.4 Pegmatites

Some rare-element (granitic) pegmatites contain accessory REE minerals, for example, small amounts of monazite, euxenite, xenotime, allanite and gadolinite in masses of fluorite. The REE-bearing minerals occur in K-feldspar-rich parts of pegmatites where they are associated with albite, muscovite and quartz (Simmons and Heinrich 1980; Simmons et al. 1987). Pegmatites as a whole often show enrichment of heavy REE, which occur as xenotime, fergusonite, euxenite, samarskite and gadolinite. Euxenite and samarskite are in parts intensively altered to Ca-containing minerals of pyrochlore structure during the late stages of pegmatite formation. Examples are:

Pegmatites of Chaffee and Fremont Counties, CO/USA;
South Platte, CO/USA;
Llano County, Texas/USA.

Pegmatitic dykes related to peralkaline rocks such as syenites, nepheline syenites and alkali granites are characterized by pyrochlore-zircon, pyrochlore-ilmenorutile and pyrochlore-aeschynite associations. Only the pyrochlore-zircon pegmatitic dykes with aeschynite are of some economic importance. The REE minerals are mainly concentrated in the albitized parts of the pegmatitic dykes. This group of mineralizations might gain some importance as a future source for Nb and light REE. Examples are:

Motzfeldt Centre, southern Greenland;
Langesundsfjord, Norway;
Mt. Rosa, CO/USA;
Itremo, Madagascar.

Pegmatites related to alkali-ultrabasic complexes often form coarse-grained veins of variable thickness. Rock-forming minerals are microcline, orthoclase, nepheline, arfvedsonite and aegirine. REE are substitutionally present in rock-forming minerals as well as in accessories (mosandrite or lovchorrite). Typical examples are lovchorrite-rinkolite pegmatites of the Khibina massif, Kola Peninsula, USSR. The REE-enriched minerals are largely formed during albitization (Ginzburg and Zhuravleva 1959).

2.1.5 Hydrothermal Mineralization

Hydrothermal deposits with REE-carbonates, -fluocarbonates, -fluorides, -phosphates and -silicates are widespread within foldbelts. Fluorite may also contain considerable amounts of REE and is widespread. The hydrothermal vein mineralizations often show telescoping, i.e. high-temperature mineralizations superimposed on medium- to low-temperature mineralizations. Among all associations the cassiterite-wolframite mineralizations with xenotime and monazite in Bolivia and South Africa are the most prominant example, but these occurrences are industrially insignificant. These veins are usually steeply dipping and bound to calc-alkalic granitoids.

Hydrothermal veins and lenses related to intrusive complexes of shonkinite, alkali syenites and alkali granites can be rich in fluorite and associated with Be-, Nb-, Ta-, REE-, Th- and U minerals. Fe is often dominant in these deposits and is present as siderite, hematite and magnetite. Sometimes the iron ore is profitably mined. The country rocks are often albitized (Ginzburg and Zhuravleva 1959).

The deposits can be carbonate veins, mineralized breccia zones and tabular bodies. Because of the abundance of carbonates and their associations, they are often considered to be related to carbonatites, although "typical carbonatitic bodies" are not known. They differ from "typical carbonatites" by the development of veinlike orebodies, the absence of roundish outline as known for carbonatites, the absence of connections with alkaline ultrabasic rocks, as well as high contents of barite, siderite and fluorite. Examples are:

1. Bastnaesite-barite-carbonate bodies (veins):
   Mountain Pass, CA/USA (see Sect. 3.1);
   Karonge, Burundi (see Sect. 3.3);
   Lugin Gol pluton, Gobi Desert, Mongolian People's Republic;
2. Sulphide-fluorite veins:
   Gallinas Mt., NM/USA;

3. Magnetite-hematite-fluorite bodies:
   Bayan Obo, north of Bao-tou/China (see Sect. 3.2).

Another type of REE-rich vein mineralization is related to alkali granites, alkali-ultrabasic and syenitic rocks with polystage ore deposition. These thorite-rich vein deposits are quite abundant, but their minerals are difficult to recognize because of the intensive staining by large amounts of oxidized iron. They are thought to have formed by volatile, late-stage phases of alkaline magmas. Typical examples are:

Steenkampskraal, Cape Prov./RSA;
Lemhi Pass, ID, MT/USA;
Powderhorn district, CO/USA;
Wet Mountains, CO/USA.

2.2 Exogenetic or Sedimentary Mineralizations

Formation of placers depends on a series of factors such as climate, geomorphology, geotectonics, hydrodynamics of the area and contents of heavy minerals resistant to weathering in source rocks. Liberation and concentration of REE minerals results from the sequence of weathering, erosion, transportation and redeposition.

Weathering processes are substantially controlled by (1) climatic factors and (2) mineral composition of the primary rocks.

Dissolution processes predominate, for instance, in carbonatites, while hydration and hydrolysis are the main weathering processes in silicate rocks. In weathering crusts of alkaline rocks (nepheline syenites, melteigeite, ijolites and carbonatite-ijolite complexes), REE crystallized as supergene carbonates, for instance bastnaesite, tengerite, lanthanite, phosphates (rhabdophane, churchite) or hydrated silicates (vudyavrite). The presence of cerianite indicates oxidizing conditions during alteration. Oxyhydrates of Fe, Mn and others, adsorb REE-hydroxides, but this process led only to "ores" of poor quality. Among this type of residual formations, the most important are laterites (USSR). A unique type of baddeleyite-bauxite deposit with magnetite, U and Th mineralization is given by the weathering crust on top of the nepheline syenite and phonolite massif of the Pocos de Caldas area/Minas Gerais/Brazil. The baddeleyite contains about 1% REE-oxides. Supergene enrichment of Co, Ni, Zn and heavy REE and Y is also known to occur in laterites where heavy REE and Y are hosted by Mn-oxides (Kapustin quoted in Gmelin 1984).

Under tropical conditions the weathering crusts over carbonatites form high-grade deposits of Nb, Zr, P, REE and Fe. The thickness of these layers varies considerably and can reach more than 100 m in tropical climates. The REE content of the residual series (mostly Ce) often amounts to about 10%. The REE are bound to residual pyrochlore and apatite or to secondary minerals such as synchysite. Examples are Araxa (Brazil), Mrima Hill and Ruri complex (Kenya) and Wigu Hill (Tanzania).

Marine or fluviatile placers mostly contain monazite and as minor components, xenotime, fergusonite, euxenite, samarskite, allanite, knopite, pyrochlore and loparite. Monazite-bearing placer deposits are one of the most important sources of REE. Monazite-rich sands have limited extension and monazite is only present as a minor constituent among other heavy mineral phases. Therefore, REE production is only economical if the placers are mined for other heavy minerals such as ilmenite, zircon, rutile, cassiterite or even gold. Average contents of monazite in placers are mostly less than 0.1%, only the Indian ilmenite placers contain 1–2% monazite. The Australian coastal sands may locally reach 1% (Overstreet 1967).

Monazite placers are mined in the southeastern USA, along the southeastern coast of Queensland/Australia and the western coast of Australia. Ilmenite placers are mined along the southern coast of India. Regeneration of river and beach placers has been observed in these areas.

Other "sedimentary" formations also contain considerable amounts of extractable REE, e.g. bituminous shales, phosphorites, asphalt, petroleum, coal etc. For instance, contents of REE in phosphorite are one to seven times, and in ashes and coal two to seven times that of clarkes. Platform phosporites form small bodies with low P_2O_5 ($< 28\%$) but REE contents as high as 0.7%. The marine phosphorites as known in northern Africa are characterized by tremendous horizontal extensions with continuous layers of great thickness and high P_2O_5 contents up to 30%, but REE contents of only 0.015% (Ginzburg and Zhuravleva 1959).

2.3 Metamorphic Mineralizations

Some metamorphic mineralizations are promising deposits because they form huge bodies that contain not only REE, but also U, V, Mo, Ge and Au.

In zones of migmatization, there are mineral associations with interbedded, fine-grained syngenetic allanite and REE-containing titanite. In regions with strong tectonic deformations and metasomatic alteration, these fine-grained minerals recrystallized into cm-sized grains dispersed throughout the host rocks. The REE are thought to have been scavenged by incorporation in Fe-oxyhydrates, or have been adsorbed on gelatinous precipitates in the protolith.

Xenotime-monazite aggregates are found in migmatized gneisses (north-central CO/USA; Music Valley, CA/USA), and monazite mineralization occurs in marble lenses interbedded with gneisses and schists (northern Lemhi Co., ID/USA).

REE together with U, Th and Au occur in metamorphosed placers containing brannerite-thucholite or Au-U-REE-bearing conglomerates such as in the Blind River area, Ont./Canada, or the Witwatersrand, Transvaal/RSA respectively. In any case, REE are only recovered as by-products. Other examples are the REE-bearing conglomerates in the Little Big Horn River area, WY and MT, USA, or the primary fluviatile monazite concentrations in ortho- and paragneisses of the southwestern Atlantic States, USA: the Piedmont monazite belt.

In contract metamorphic aureoles skarns with enhanced REE-mineral contents are found. Examples are:

Hematite skarns with allanite, cerite and fluocerite:
Bastnäs, Sweden;
Magnetite skarns with REE-rich apatite; Minneville, NY/USA;
Tagil-Kushva region, Central Urals;
Uraninite skarns with allanite and stillwellite: Mary Kathleen, Queensland/Australia.

3 Examples of Primary Deposits

A brief discussion of available information on the deposits of Mountain Pass, Karonge, and Bayan Obo is given because they have been referred to specifically during the workshop. Additionally, Morteani and Satir (this Vol.) report on the bastnaesite-fluorite-barite deposit of the Kizilcaören near Eskishir, Turkey.

3.1 Mountain Pass, California/USA

Geological Setting (Olson et al. 1954). The rare-earth deposits in the Mountain Pass district were found in a block of Precambrian metamorphic rocks, 7 km wide and about 40 km long. The complex is bounded to the south and to the east by alluvium and by faults to the east and to the west. The Precambrian rocks consist of garnet-bearing gneisses, biotite granites and migmatites (Fig. 3).

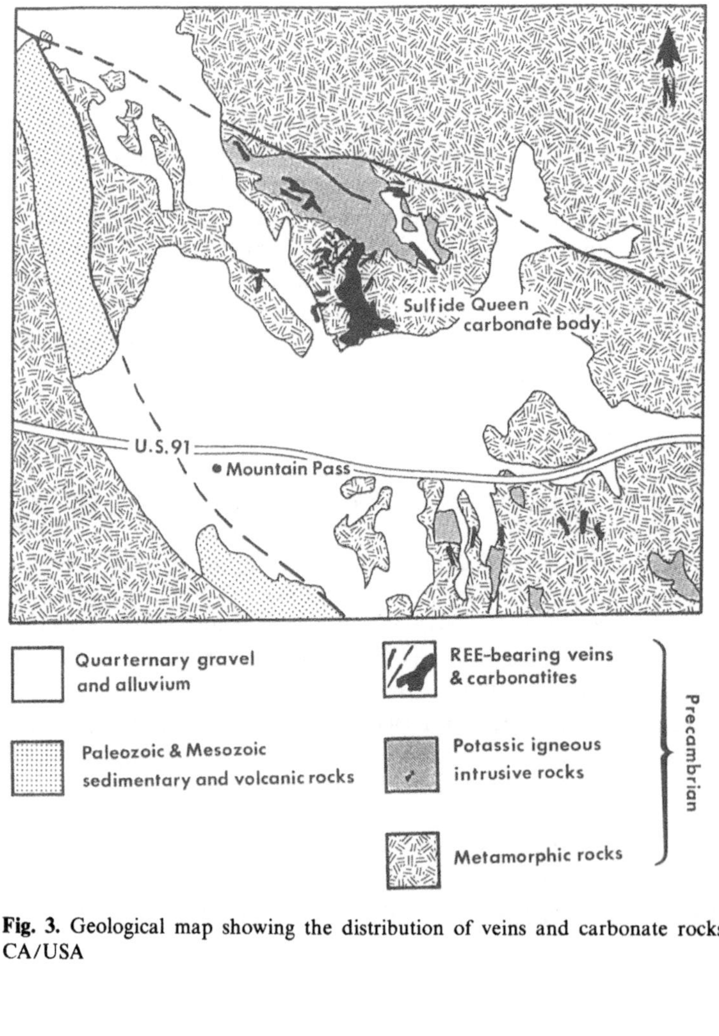

Fig. 3. Geological map showing the distribution of veins and carbonate rocks at Mountain Pass, CA/USA

The metamorphic complex is intruded by a number of stocks and dykes of a K-feldspar-rich igneous suite which includes granite, syenite and shonkinite syenite bodies. They are non-foliated and therefore clearly younger than the metamorphic complex itself.

The REE-bearing carbonates are spatially associated with feldspar-rich, SiO_2-saturated to oversaturated plutons. The most important intrusive carbonatite body is the Sulphide Queen carbonate body. It is related to the largest shonkinite syenite stock. In the vicinity, however, there are several other carbonatite bodies which, like the K-feldspar-rich dykes, cut older rocks discordantly. Because the Sulphide Queen carbonate body contains breccias of all country rock types and shows discordant contacts, it must be the last intrusion of the sequence.

Three types of rocks are distinguished in the carbonatite:

1. Brownish, iron-bearing dolomite;
2. Grey to rose-coloured calcite-barite rock; and
3. Silicified carbonatite which contains up to 60% bastnaesite.

All rock types show unusually high REE, Ba and Sr contents. According to radiometric age determinations, the carbonatites are roughly 1000 Ma younger than the K-feldspar-rich plutons.

Mineralogy (Olson 1954). REE-mineral-bearing rocks are characterized by abundant carbonate minerals, barite and quartz. The carbonate rocks consist of many generations of calcite, ankerite, dolomite, siderite, barite, Sr-barite, Ba-celestite, quartz, bastnaesite, parisite and small quantities of many other minerals, including crocidolite, chlorite, biotite, phlogopite, muscovite, aegirine, sphene, allanite, monazite, magnetite, goethite, hematite, galena, pyrite, chalcopyrite, tetrahedrite, malachite, azurite, cerussite, aragonite, wulfenite, fluorite, strontianite, apatite, thorite, cerite and sahamalite. A rough estimate of the principal constituents in the carbonate rock gives 60% carbonate minerals, 20% barite, 10% REE-bearing fluorocarbonates and 10% quartz and other silicates.

The Mountain Pass carbonatite differs from other carbonatites in the following aspects:

1. Absence of feldspathoids in the associated igneous rocks;
2. Absence of related volcanic rocks;
3. Absence of concentric or ring structures;
4. Absence of Ca-silicate minerals;
5. Extreme concentration of bastnaesite and barite;
6. Scarcity of Ti-Nb minerals;
7. Relatively small amounts of magnetite.

In 1984, roughly 25,300 tons of REE concentrate was produced (Robjohns 1986). This corresponded to roughly one-half of the world's production of REE concentrates. At the same time, the resources for ore containing 12% bastnaesite have been estimated to be 340,000 t. The ore consists of 40–75% calcite, 15–50% barite and 5–15% bastnaesite. Some cross-cutting carbonatite dykes contain up to 90% calcite, 5-30% barite and locally up to 30% bastnaesite. Since bastnaesite occurs only as an accessory mineral in other carbonatites, Mountain Pass has become the most important source for the light REE (La, Ce and Nd).

3.2 Bayan Obo Iron Ore Deposit, P.R. China (Lee 1970; Tu 1981)

The Yin-shan Range is in the west-central part of the Nei Mongol Zizhiqu/northern China (Autonomous Region of Inner Mongolia) which separates the Mongolian Plateau on the north from the North China Plain on the south. The basement consists of Precambrian metamorphic rocks penetrated by granites and pegmatites. Paleozoic rocks with intercalated volcanics unconformably overlie the Precambrian rocks.

The lower upper Precambrian is represented by a miogeosynclinal sequence of the Bayan Obo Group, which consists of quartzite and slate in the lower part; calcareous quartzite, siliceous and clayey limestone and dolomite in the middle part; and slate with interbeds of quartzite in the upper part.

The rare element deposits are found in tabular bodies and irregular bands of iron ore; in pegmatites with microcline, perthite, quartz biotite, muscovite, beryl, and in fluorite, quartz and carbonate veins with aegirine, kataphorite, albite and oligoclase. The REE-rich iron ores are bound to the Precambrian dolomites (Tu 1981) of the Bayan Obo group. The principal iron ore minerals are magnetite, specularite and hematite, which are associated with disseminated REE minerals. Common REE minerals in all of these mineralizations are aeschynite, allanite, apatite, beiyenite (resembles bastnaesite), aborite, parisite, monazite, xenotime and others.

Rare-element deposits in the Yin-shan Range were formed in two periods of mineralization: before the deposition of the Bayan Obo group sediments rare-element-bearing pegmatites were emplaced in the Precambrian rocks. In late Paleozoic times large masses of REE-bearing iron ores were emplaced in the carbonates of the Bayan Obo group and Precambrian schists, probably by high-temperature alkalic solutions (Lee 1970). The alkalic carbonate- and iron-rich solutions from which the iron and rare-element minerals were precipitated by oxidation are genetically related to alkali granites, syenites and gabbroic rocks occurring throughout the mineralized belt. The fluids were rich in La, Ce, Nb, F, Na and P.

Tu (1980) favours a sedimentary origin for the Bayan Obo deposit, because it is strictly strata-bound, lenticular in shape, concordant with country rocks and subjected to folding similar to the country rocks. Lamination, stratification, banding and ooliths are still to be observed. It is suggested that the ore formation was a product of diagenesis in a basin of high alkalinity as indicated by the presence of riebeckite and albite. The ore-bearing dolomite (REE and iron is bound to dolomite only) has a thickness between 200 and 800 m. Facial changes in the dolomite are frequent. Although the Bayan Obo group extends over several hundred kilometres the carbonate horizon consists predominantly of dolomite only in the Bayan Obo district. Thick orebodies are usually accompanied by thick dolomite. The underlying sedimentary formations were shallow open sea sediments, while the ore-bearing dolomite was formed in a closed to semi-closed evaporite basin with a more agitated, oxidizing and alkaline environment. The REE were already present in the sedimentary ore. Hercynian intrusions might have added some Nb and REE to the ore. Despite the striking differences in rare-element contents, Tu (1980) compares the Bayan Obo deposit with the Superior Province type of iron ore deposits.

The eastern orebody is about 1.3 km long, 150 to 210 m wide and at least 500 m deep. Magnetite, specularite and hematite are the principal iron ores which contain a complex REE mineralization of mainly bastnaesite, monazite, fluorite and Nb-rich rutile. The average content is 5.7% REE. The tabular ore bodies and lenses occur in steeply dipping dolomite formations which have an average width of 2–3 km and 18 km lateral extension. With more than 37 million tons of REE-oxide, this area has by far the largest identified reserves of the world (nearly 80% of the total REE reserves after Robjohns 1986).

The magmatic-hydrothermal (Lee 1970) and sedimentary-diagenetic (Tu 1980) concept for the formation of the Bayan Obo REE-iron ore deposit is still debatable (Tu, personal communication 1987)

3.3 Karonge REE Deposit, Burundi (Contribution by L. van Wambeke)

REE mineralization occurs in veins and veinlets forming a dense stockwork in the Karonge river area (van Wambeke 1977). The orebodies are emplaced in a major domal structure. The country rocks are intensely silicified, and Ca- and Mg-carbonates are conspicuously absent from the ore. It can be concluded from the mineral assemblage that this mineralization is related to the last differentiation stage of the carbonatite. The REE mineral sequence is bastnaesite, monazite and rhabdophane which is the youngest mineral. Geochemical data clearly show an REE fractionation with a slight decrease in the Ce/(La + Nd) ratio from bastnaesite to monazite, and a very sharp decrease of this ratio at the end of the deposition corresponding to the formation of rhabdophane (La), which is also enriched in other light REE. The mineral assemblage of REE minerals, barite and goethite is very similar to the highly silicified REE-bearing rock types of Mountain Pass, Ca/USA and Kangankunde, Malawi.

The mine was in production between 1950 and 1975.

4 Conclusion

High quality REE-mineral deposits are rare, but worldwide there are abundant rare metal deposits from which REE can be recovered as by-products. The REE are more abundant in alkaline and acidic than in basic and ultrabasic rocks. REE are commonly associated with other rare elements such as Th, U, Ta, Nb, Ti, Zr, P etc. Monazite and bastnaesite are industrially the most important REE minerals, while xenotime and apatite are of only minor significance. At the moment REE are almost entirely produced from mineralizations related to (1) alkaline rocks and carbonatites and (2) placers. Hydrothermal vein mineralizations have been mined but are economically no longer interesting since REE concentrates became available in huge amounts. The amount of REE concentrates recovered from pegmatites being mined for uranium are also negligible.

REE minerals of alkaline rocks are mainly oxides or silicates of Nb, Ti, Zr, Al, Be and Na; Ca-phosphates and Ca-fluorocarbonates such as bastnaesite, or yttrian members such as synchysite. These minerals are mostly formed as late hydrothermal or hypogenic replacement minerals in alkaline syenitic rocks or in alkalic and subalkalic granitoids and their associated pegmatites. REE deposits spatially and genetically connected with nepheline syenites, alkali syenites, alkali granites and alkali-ultrabasic complexes deserve special attention because of their concentration of both light and heavy REE.

The REE-mineral deposits may be grouped into;

1. Endogenic enrichments of REE associated with alkali-ultrabasic complexes, carbonatites, alkaline pegmatites, metasomatites (albitites, fenites) and hydrothermal vein mineralizations;
2. Exogenic accumulations of REE-rich minerals in fluviatile and coastal sands, and residual soils (e.g. laterites and bauxites) on top of carbonatites;
3. Contact metamorphic rocks, especially skarns, enriched in REE-bearing minerals;
4. Sediments rich in organic material and REE. These rocks become an interesting source in the future because of their availability in large quantities (.e.g. phosphorites, coal ashes etc.).

Acknowledgements. The author gratefully acknowledges comments by W. Chazan, E.J. Martin, G. Roethe, T. Utter and L. van Wambeke which were given during the meeting. The author is indebted to F. Saupé and G. Morteani for improving the manuscript by critical comments.

References

Armbrustmacher TJ (1980) Abundance and distribution of Th in the carbonatite stock at Iron Hill, Powderhorn district, Gunnison County, Colorado. US Geol Surv Prof Pap 1049-B: 1-11

Backström JW (1976) Thorium. S Afr Geol Surv Handb 7:209-212

Bergstol S (1972) Jacupirangite at Kodal, Vestfold, Norway. Potential magnetite, ilmenite, and apatite ore. Mineral Deposit 7:233-46

Beurlen H, Cassedanne JP (1981) The Brazilian mineral resources. Earth Sci Rev 17:177-206

Černý P, Trueman DL (1985) Polylithionite from the rare-metal deposits of the Blachford Lake alkaline complex, N.W.T., Canada. Am Mineral 70:1127-1134

Chantraine J, Radelli L (1970) Tectono-minerogenetic units of the basement of Madagascar. Econ Geol 65:690-699

Coetzee CB (1976) Rare earths. In: Coetzee CB (ed) Mineral resources of the Republic of South Africa. Dep Mines, Geol Surv, Pretoria, pp 199-291

de Kun N (1960) Die Zinn-Niob-Tantal-Zirkon- und Lanthaniden Lagerstätten von Nigerian. N Jahrb Mineral Monatsh 1960:100-107

Ehlman AJ, Walper JL, Williams J (1964) A new, Baringer Hill-type, rare-earth pegmatite from the central mineral region, Texas. Econ Geol 59:1348-1360

Evans JR (1964) Xenotime mineralization in the southern Music Valley area, Riverside County, California. Cal Div Mines Geol Spec Rep 79:1-24

Faizullin RM, Kozlov EN, Kharitonova RS (1977) New data on substantive composition and problems on genesis of apatite-carbonate ores of the Seligdar deposit (Central Aldan district). Geol Geofiz 1977:44-56

Gallagher MJ (1968) Phosphates and other minerals in pegmatites of Rhodesia and Uganda. Mineral Mag 36:50-59

Gerasimovskiy VV, Mineyev DA (1981) Rare earths in apatite from apatite-magnetite ores. Geochem Int 18:67-72

Ginzburg AI, Zhuravleva LN (1959) I. In: Ginzburg AI, Zhuravleva LN, Ivanov IB, Shcherbin VV (eds) Geologiya mestorozhdenii redkikh elementov. Nauka, Moscow 3:59-103

Gmelin Handbook of inorganic chemistry (1984) Sc, La-Lu. Minerals, deposits, mineral index. A8 Vol. Verl. Chemie, Weinheim
Guimaraes D (1957) Relatorio sobre a Jazida de Pirocloro de Barreiro, Araxa, Minas gerais. Bol Dep Nac Prod Mineral 103:1-87
Harris JF (1961) Summary of the geology of Tanganyika; part IV, Economic geology. Tanganyika Geol Surv Mem 1:143 p.
Hawkins HW (1975) Mary Kathleen uranium deposit. In: Knight CL (ed) Economic geology of Australia, and Papua New Guinea, 1. Metals Aust Inst Mining Met, Parkeville/Vict, pp 398-402
Hedrick JB (1985) Rare-earth minerals and metals. Mineral Yearb 1:791-803
Heinrich EW (1966) The geology of carbonatites. Rand McNally, Chicago, 555 pp
Heinrich EW, Wells RG (1980) The diversity of rare-earth mineral deposits and their geological domains. Rare Earth Mod Sci Technol 2:511-516
Holt DN (1965) The Kangankunde Hill rare earth prospect. Malawi Geol Surv Dep Bull 20:130 pp
Jones AP, Larsen LM (1985) Geochemistry and REE minerals of nepheline syenites from the Motzfeldt Centre, South Greenland. Am Mineral 70:1087-1100
Klemic H, Heyl AV, Taylor AR, Stone J (1959) Radioactive rare earth deposit at Scrub Oaks mine, Morris County New York. US Geol Surv Bull 1082-B: 29-58
Kovalenko VI, Vladykin NV, Goreglyad AV (1977) quoted in Gmelin (1984), p 393-395
Lee KY (1970) Some rare element mineral deposits in mainland China. Bull US Geol Surv 1312-N: 34 pp
Martins LR, Coutinho PN (1981) The Brazilian continental margin. Earth Sci Rev 17:87-107
Mei L, Larson RR, Loferski PJ, Klemic H (1979) Analyses and description of concentrate of stillwellite from Minneville, Essex County, N.Y. US Geol Surv Open-File Rep, 8 pp
Mertie JB (1975) Monazite placers in the southeastern Atlantic States. US Geol Surv Bull 1390: 41 pp
Mertie JB (1979) Monazite in the granitic rocks of the southeastern Atlantic States – an example of the use of heavy minerals in geologic exploration. US Geol Surv Prof Pap 1094:1-79
Murdock TG (1963) Mineral resources of the Malegasy Republic. US Bur Mines Inf Circ 8196:1-147
Neary CR, Highley DE (1984) The economic importance of the rare earth elements. In: Henderson P (ed) Rare earth element geochemistry. Elsevier, New York Amsterdam, pp 423-466
Nielsen BL (1973) A survey of the economic geology of Greenland. Geol Surv Greenl Rep 56:45 pp
Notholt AJG (1979) The economic geology and development of igneous phosphate deposits in Europe and the USSR. Econ Geol 74:339-350
Olson JC, Shawe DR, Prey LC, Sharp WN (1954) Rare-earth mineral deposit of the Mountain Pass district San Bernadino County, California. Geol Surv Prof Pap 261:75 pp
Ontoev DO, Luvsandanzan B, Gundsambuu TS (1979) cited in Gmelin (1984), p 395
Ovchinnikov LN, Solodov NA (1980) In: Ovchinnikov LN, Solodov NA (eds) Mestorozhdeniya litifil'nykh redkikh meltallov. Nedra, Moscow, pp 6-28
Overstreet WC (1967) The geological occurrence of monazite. US Geol Surv Prof Pap 539:327 pp
Parkin LW, Glasson KR (1954) The ecology of the Radium Hill uranium mine, South Australia. Econ Geol 49:815-825
Perhac RM, Heinrich EW (1964) Fluorite-bastnaesite deposits of the Gallinas Mountains, New Mexico, and bastnaesite paragenesis. Econ Geol 59:226-239
Pike DR (1958) Thorium and rare earth bearing minerals in the Union of South Africa. Proc 2nd Int Conf Peaceful uses at energy, Geneva VII:91-96
Roberts DE, Hudson GRT (1984) The Olympic Dam copper-uranium-gold deposit, Roxby Down, South Australia. Econ Geol 73:1944-1945
Robjohns N (1984) Rare earths. Min Annu Rev 1984:88-89
Robjohns N (1986) Rare earths. Min Annu Rev 1986:91-92
Rose ER (1979) Rare-earth prospects in Canada. Can Min Met Bull 72:110-116
Secher K, Larsen LM (1980) Geology and mineralogy of the Sarfartoq carbonatite complex, southern West Greenland. Lithos 13:199-212
Simmons WB, Heinrich EW (1980) Rare-earth pegmatites of the South Platte district, Colorado. Col Geol Surv Res Ser 11:131 pp
Simmons WB, Lee MT, Brewster RH (1987) Geochemistry and evolution of the South Platte granite pegmatite system, Jefferson County, Colorado. Geochim Acta 51:455-471
Sommerlatte HWA (1969) Neue Metalle, ihre Lagerstätten und ihre Bedeutung für die Industrie der Bundesrepublik. Erzmetall 22:103-111
Staatz MH (1974) Thorium veins in the United States. Econ Geol 69:494-507

Staatz MH, Sharp BJ, Hetland DL (1979) Geology and mineral resources of the Lemhi Pass thorium district, Idaho and Montana. US Geol Surv Prof Pap 1049-A:90 pp
Thoreau J, Aderca BM, Wambeke L van (1958) Le gisement de terres rares de la Karonge, Urundi. Acad R Sci Outre-Mer, Classe Sci Nat Med 4:648-715
Tu G (1981) On the mineral deposits of Bayan Obo and Shitū. Scientia Sinica 24:228-236
Ulbrich HHGJ, Gomes CB (1981) Alkaline rocks from continental Brazil. Earth Sci Rev 17:135-154
Verwoerd WJ (1967) The carbonatites of South Africa and South West Africa. S Afr Geol Surv Handb 6:452 pp
Vlasov KA (1968) Geochemistry and mineralogy of rare elements and genetic types of their deposits, Vol 3: Genetic types of rare-element deposits. Isr Progr Sci Transl, Jerusalem: 916 pp
Wambeke L van (1977) The Karonge rare earth deposit, Republic of Burundi: new mineralogical-geochemical data and origin of the mineralization. Mineral Deposit 12:373-380

The Bastnaesite-Fluorite-Barite Deposit of the Kizilcaören District, Eskisehir, Turkey

G. MORTEANI and M. SATIR

1 Introduction

There are about 100 minerals known to contain lanthanides in more than trace amounts. But there are only two of commercial importance, monazite, a mixed Th-REE-phosphate and bastnaesite, an REE-fluorocarbonate (M^{3+} Co_3F).

In 1949 a vast deposit of bastnaesite was discovered at Mountain Pass in the Sierra Nevada Mountains in the USA and became by far the most important single source of REE (Olson et al. 1954). Fluorite-bastnaesite mineralization is also known from the Gallinas Mountains, New Mexico (Perhac and Heinrich 1964). The mineralization occurs in epithermal veins and breccia filling in Permian sandstone, but is presumed to be related to shallow Tertiary alkalic intrusives. Around 1977 in the area of Kizilcaören, 80 km east of Eskisehir in northwestern Turkey, a bastnaesite deposit was discovered by the M.T.A. (i.e. Mineral Research and Exploration Institute of Turkey, Ankara) which later proved to be very similar from the geological point of view and which may have economic potential equal to that of Mountain Pass (see below).

In this contribution a short description of the geology and mineralogy of the bastnaesite deposit of Kizilcaören will be given.

2 Geology

In 1959, M.T.A. carried out a regional airborne survey for radioactive minerals which resulted in the discovery of an important radiometric anomaly in the Kizilcaören area, 80 km east of Eskisehir (Ucmak 1970).

A subsequent geological mapping of the anomaly led to the discovery of a bastnaesite-fluorite-barite mineralization by geologists of the M.T.A. (Kaplan 1977; Kulaksiz 1977). A revised preliminary geological map and stratigraphic column, according to Delaloye and Özgenc (1983), Kulaksiz (1977), Stumpfl and Kirikoglu (1986) and our own observations of the Kizilcaören area, is given in Fig. 1.

The basement rocks consist of early Paleozoic anchimetamorphic rocks such as graywackes, sandstones, phyllites, glaucophane schists, calc schists and marbles (Kupfahl 1954; Weingart 1954).

Lehrstuhl für Angewandte Mineralogie und Geochemie, Technische Universität München, Lichtenbergstrasse 4, D-8046 Garching

Lanthanides, Tantalum and Niobium
P. Möller, P. Černý and F. Saupé (Eds.)

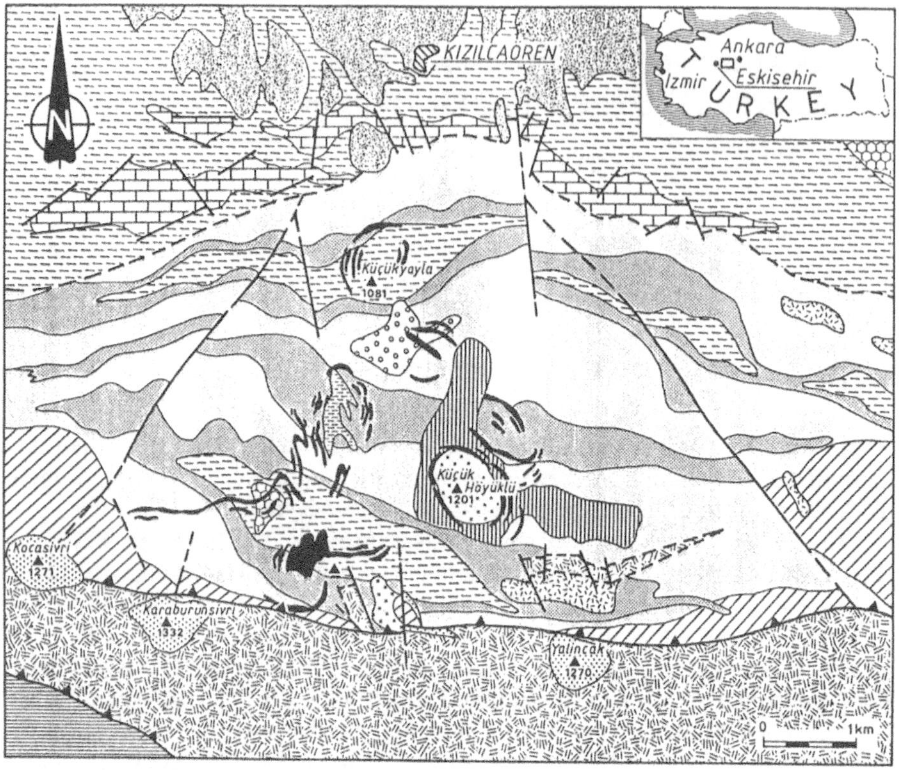
N
KIZILCAOREN
Ankara
Izmir
Eskisehir
TURKEY
Küçükyayla
1081
Küçük
Höyüklü
1201
Kocasivri
1271
Karaburunsivri
1332
Yalıncak
1270
0
1km

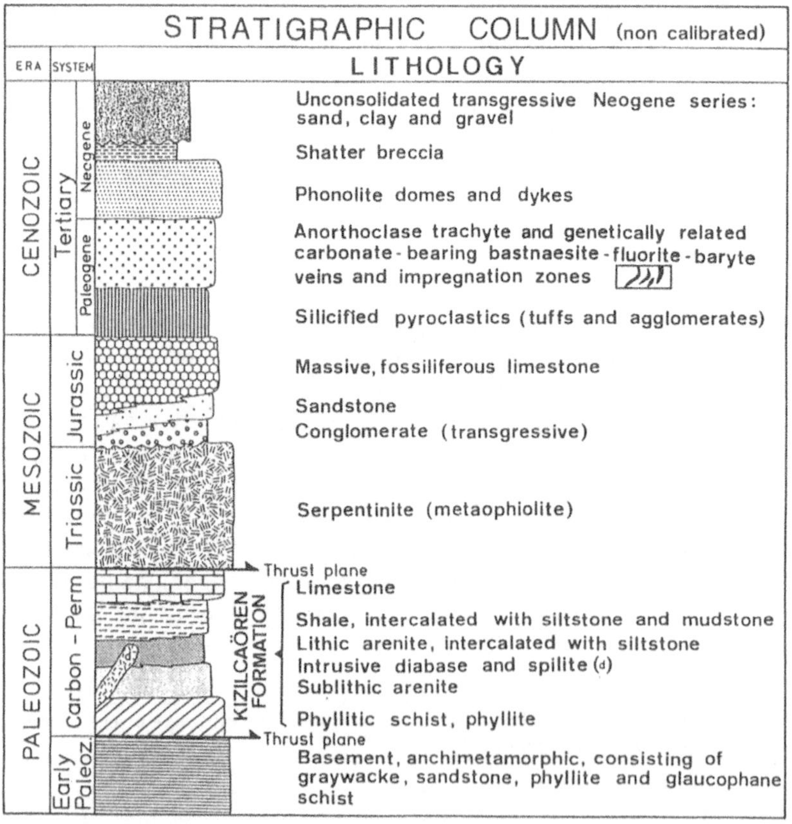
STRATIGRAPHIC COLUMN (non calibrated)
ERA
SYSTEM
LITHOLOGY
CENOZOIC
Tertiary
Neogene
Paleogene
MESOZOIC
Jurassic
Triassic
PALEOZOIC
Carbon - Perm
Early Paleoz.
Unconsolidated transgressive Neogene series: sand, clay and gravel
Shatter breccia
Phonolite domes and dykes
Anorthoclase trachyte and genetically related carbonate-bearing bastnaesite-fluorite-baryte veins and impregnation zones
Silicified pyroclastics (tuffs and agglomerates)
Massive, fossiliferous limestone
Sandstone
Conglomerate (transgressive)
Serpentinite (metaophiolite)
Thrust plane
KIZILCAÖREN FORMATION
Limestone
Shale, intercalated with siltstone and mudstone
Lithic arenite, intercalated with siltstone
Intrusive diabase and spilite (d)
Sublithic arenite
Phyllitic schist, phyllite
Thrust plane
Basement, anchimetamorphic, consisting of graywacke, sandstone, phyllite and glaucophane schist

These are overlain with a tectonic contact by the Permo-Carboniferous Kizilcaören Formation consisting of shales intercalated with sandstones and mudstones, arenites and sublithic arenites intruded by diabases, spilites and picritic basalts.

Triassic metaophiolites were tectonically emplaced between the Kizilcaören Formation and the basement. They are serpentinized peridotites with subordinate pyroxenites. The serpentinization increases from south to north.

Jurassic conglomerates and fossiliferous limestones are transgressive over the older sequences.

During the Tertiary, volcanic activity dominated, with extrusion of phonolites, trachytes, tuffs, agglomerates and volcanic breccias. The phonolites form three volcanic cones in the study area. The trachyte dykes intruded preferentially rocks of the Kizilcaören Formation.

Spatially and probably genetically connected with the trachytic volcanic rocks are carbonatite veins. The carbonatites consist of a primary paragenesis of coarse-grained calcite with minor quantities of diopside, magnetite and apatite. Bastnaesite, fluorite and barite mineralization belong to a later mineralizing event (Kirikoglu 1983; Stumpfl and Kirikoglu 1986). The association of carbonatites has important genetic implications, both for the source of the barite-fluorite-REE ore and for the geotectonic setting of the deposits.

The transgressive Neogene series consists of sands, gravels and conglomerates.

The whole area belongs to the Izmir-Ankara zone (Brinkmann 1966) situated between the Anatolide/Tauride zone in the south and the Pontide zone in the north. The area represents the remnants of a Triassic to early Paleocene ocean which closed along a north dipping subduction zone during the late Paleocene to early Tertiary times (Sengör 1979; Sengör and Yilmaz 1981; Sengör et al. 1984). The late Oligocene alkaline volcanism dated at about 24 Ma by Delaloye and Ozgenc (1983) is related to that event.

On a larger scale the Paleogene magmatism may be related to the development of the Aegean Rift, which began in the early Mesozoic due to the relative movement of the Arabian plate towards the Eurasian plate.

3 Mineralization

The descriptions of the spatial and genetic relations of the bastnaesite-fluorite-barite mineralizations are rather poor in the literature. There is evidence that the mineralized bodies form veinlets and pods along fold zones and within tectonic breccia zones 30–40 m in width. A parallel banding is frequently observed in the veins (Stumpfl and Kirikoglu 1986).

Microscope investigation of some samples showed that the mineralization is polyphase. The main ore minerals are barite, fluorite, pyrite, psilomelane, braunite, bastnaesite, brockite, fluocerite, monazite, limonitized iron ores and various gangue minerals including calcite, quartz and chalcedony.

Fig. 1. Preliminary geological map and stratigraphical column of the area showing bastnaesite, barite and fluorite mineralizations south of the village of Kizilcaören according to Ozkocak (1983), Kulaksiz (1977), Delaloye and Ozgenc (1983), Kirikoglu (1983) and own data

In Figs. 2 and 3 micrograph drawings of typical mineral textures formed by bastnaesite, barite, fluorite and opaques are shown.

The fluorite is frequently followed in time by a barite mineralization. Both minerals show a brittle deformation. The REE minerals, mainly bastnaesite and very minor brockite [Ca, Th, REE $(PO_4)H_2O$] and fluocerite [$(Ce,Le)F_3$] form a very

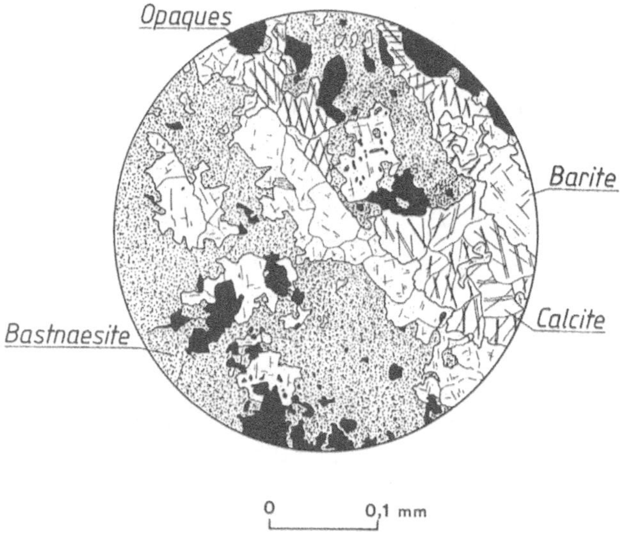

Fig. 2. Micrograph drawing showing the very fine-grained bastnaesite cementing calcite, barite and opaques

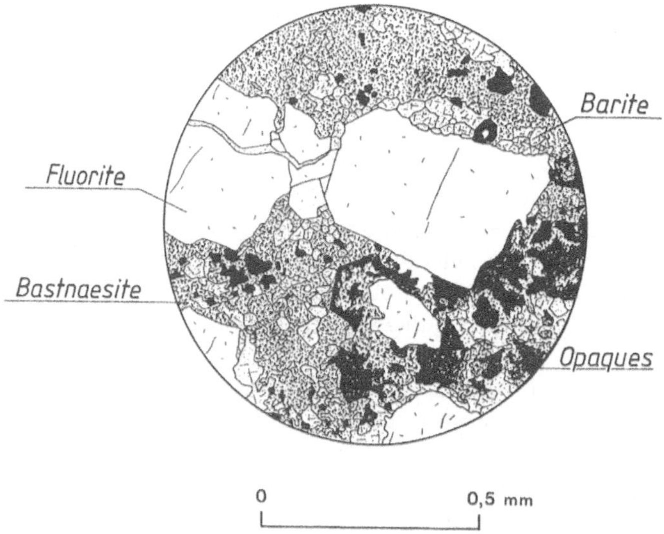

Fig. 3. Micrograph drawing showing subhedral fluorite in a matrix of barite, bastnaesite and probably hydrous iron oxides

fine-grained impregnation. This impregnation locally gives to the rocks an earthy appearance. Minor silification and late carbonatization can also be observed in the mineralized rocks. The microstructures and mineral associations closely resemble those shown by Olson et al. (1954) of the bastnaesite mineralization of the Mountain Pass deposit.

The thickness of the mineralized zones varies from 19 to 31 m. The average ore grade and reserves determined from drill holes, tunnels and trenches by Özkocak (1983) and Ciftci (1984) are given in Table 1. According to Özkocak (1983), the mineralization may have formed by one or a combination of three agents: (1) hydrothermal solutions originating from the nearby granitic to syenitic rocks (not shown in Fig. 1), (2) by the carbonatitic magma and (3) by volcano-sedimentary processes.

Stumpfl and Kirikoglu (1986) suggest a genetic link of fluorite-barite-REE mineralization with the Tertiary to Quarternary alkaline volcanic rocks (trachytes, phonolites, carbonatites) related to the Aegean Rift. They invoke a parental magma enriched in Ba, F and REE to explain the high content of these elements in the volcanics.

Table 1. Oxide content of the average ore from the Kizilcaören bastnaesite, fluorite and barite deposits

CaF_2	:	36–49%
$BaSO_4$	:	4–55%
CeO_2	:	2.5%
La_2O_3	:	1.5%
Nd_2O_3	:	0.5%
ThO_2	:	0.2%
Fe_2O_3	:	8.1%
SiO_2	:	0.35–14.9%
$CaCO_3$	:	2.8–15.3%
MnO	:	2.0%
Others	:	5.9%

From the fact that at Kocayayla Tepe a rich mineralization is found within a carbonatite vein, there should be no doubt that the mineralization is genetically bound to the carbonatite magmatism.

The reserves as calculated by Günalay (1970), Demiröz (1976); Özkocak (1983) and Ciftci and Kumru (1984) are 4 695 000 t of ore with an average grade of 2.78% REE and 384 000 t of ore with an average grade of 0.2% ThO_2 for two of the seven areas in exploration in 1985.

In order to further assess the economic potential and to define the genesis of the Kizilcaören bastnaesite, barite and fluorite deposits a cooperative study of the authors with M.T.A. Ankara is in progress.

References

Brinkmann R (1966) Geotektonische Gliederung von Westanatolien. N Jahrb Geol Paleontol Monatsschr 10:603–618

Ciftci MS (1984) The genesis and benefication studies of Eskisehir – Sivrihisar – Beylikahir fluorite-barite ores by using gravimetric and flotation methods. Thesis Univ Ankara, pp 1–64

Ciftci MS, Kumru C (1984) The concentration of Eskisehir – Sivrihisar – Beylikahir – Devebagirtantepe fluorite ores by using gravimetric and flotation methods. Mineral Res Expl Inst Turkey Dep Technology Rep 5300:1–51

Delaloye M, Özgenc I (1983) Petrography and age determinations of the alkaline volcanic rocks and carbonatite of Kizilcaören district: Beylikahir – Eskisehir, Turkey. Schweiz mineral petrogr. Mitt 63:289–294

Demiröz T (1976) Eskisehir-Sivrihisar-Kizilcaören Köyü güneyi nadir toprak elementleri-Th kompleks cevher yataginin Kocadevebagirtan kesimindeki Th reserv durumu. MTA Enstitüsü Rap 5566:1–48

Günalay ME (1970) Eskisehir ili-Sivrihisar ilcesi-Beylikahir bucagi toryum cevheri isletme projesi. MTA Enstitüsü Rap 4179:1–57

Kaplan H (1977) Rare earth elements and thorium complex deposit of Kizilcaören Village (Sivrihisar – Eskisehir Turkey). Bull Geol Eng 2:69–76

Kirikoglu MS (1983) Fluorit-Baryt-TH-SEE Lagerstätten im Gebiet von Kizilkaören, Provinz Eskisehir, West-Türkei. PhD Thesis, Min Univ Leoben Austria, pp 1–276

Kulaksiz S (1977) The geology of the northwest of Sivrihisar district. Thesis, Hacettepe Univ, Ankara, pp 1–220

Kupfahl HG (1954) Geological report of the Eskisehir and Sivrihisar districts. Mineral Res and Expl Inst Turkey Dep Technol Rep 2247:1–69

Olson IC, Shawe DR, Pray LC, Sharp WN (1954) Rare-Earth mineral deposit of the Mountain Pass District San Bernardino County California. Geol Surv Prof Pap 261:1–75

Özkocak O (1983) Rare earth minerals fluorite, baryte, iron and manganese deposit of Kizilcaören (Eskisehir Province Turkey). Proc Symp New Mineral Raw Materials, pp 94–104

Perhac RM, Heinrich EWM (1964) Fluorite-bastnaesite deposits of the Gallinas Mountains, New Mexico and bastnaesite paragenesis. Econ Geol 59:226–239

Sengör AMC (1979) The North Anatolian Transform Fault: Its age, offset and tectonic significance. Journ Geol Soc London 136:269–282

Sengör AMC, Yilmaz Y (1981) Tethyan evolution of Turkey: a plate tectonic approach. Tectonophysics 75:181–241

Sengör AMC, Satir M, Akkök R (1984) Timing of tectonic events in the Menderes Massif, Western Turkey: Implications for tectonic evolution and evidence for Pan-African basement in Turkey. Tectonics 3:693–707

Stumpfl EF, Kirikoglu MS (1986) Fluorite-Barite-Rare Earths Deposits at Kizilcaören, Turkey. Mitt. österr geol Ges 78:193–200

Ucmak F (1970) Eskisehir ili Sivrihisar ilcesi Beylikahir bucagi toryum cevheri isletme projesi. MTA Enstitüsü Rap 4179:1–52

Weingart W (1954) Report about the revision of the geological map of the Ankara and Sivrihisar. Mineral Res and Expl Inst Turkey Dep Technol Rep 2248:1–35

Characteristics of Pegmatite Deposits of Tantalum

P. ČERNÝ

Abstract

Within the broad spectrum of granitic pegmatites, economic tantalum concentrations are encountered only in those of the orogen-related rare-element class. These are located in Archean greenstone and gneissic belts, Proterozoic terranes skirting cratonic margins and Phanerozoic orogenic belts. Fertile granites generating the rare-element pegmatites are largely late to post-tectonic, post-dating the peak of regional dynamometamorphism. The granite-pegmatite systems are largely confined to deep faults, pre-existing batholithic contacts or lithologic boundaries; they are located in host rocks of the upper-greenschist and lower-amphibolite facies of Abukuma-type terranes.

Tantalum-enriched pegmatites are found in the intermediate to outermost parts of zoned pegmatite aureoles surrounding parental granites. Such pegmatites may belong to the (i) beryl, (ii) complex (spodumene, petalite or amblygonite), (iii) complex (lepidolite), (iv) albite-spodumene or (v) albite types and (subtypes). Type (i) grades into (ii) by increasing complexity of internal zoning, replacement phenomena, mineral paragenesis and geochemical fractionation, the latter including a progressive decrease in Nb/Ta. Tantalum mineralization may be confined to the core-margin assemblage in (i), but it is more commonly enriched in albitic units, greisenlike muscovitic assemblages and lepidolite units, all of them in the internal parts of the pegmatite bodies. Tantalum-bearing minerals are represented mainly by the ferromanganoan oxide species and the members of the broader microlite family, in contrast to the lepidolite subtype (iii) with abundant microlite and only minor manganocolumbite. High grades and variable tonnages of tantalum ore are typical of type (ii) deposits, particularly those containing lepidolite and pollucite.

In contrast to the zoned and/or layered types (i) to (iii), the other two pegmatite types are nearly homogeneous in texture and composition, and they contain the tantalum mineralization dispersed through most of their volumes. Low grades but large tonnages of tantalum reserves may be encountered particularly in type (iv).

Generalized characteristics of the different types of tantalum-bearing pegmatites are complemented by brief descriptions of some better-known specific examples.

Department of Geological Sciences, University of Manitoba, Winnipeg, Manitoba, Canada, R3T 2N2

Lanthanides, Tantalum and Niobium
P. Möller, P. Černý and F. Saupé (Eds.)

1 Introduction

Granitic pegmatites have been the classic source of tantalum since the successful separation of this metal from niobium in 1802. However, new types of tantalum deposits have been identified during the past 4 decades that tend to overshadow the economic significance of pegmatites: granites, granite-related greisens and veins with Ta-bearing cassiterite (reviewed by Taylor 1979), granites with disseminated columbite-tantalite, microlite and other Ta-bearing minerals (e.g. Beus et al. 1962; Drysdall et al. 1986; Pollard this Vol.), and derived placers. Nevertheless, pegmatite deposits of tantalum still play an important role, and will do so in the future. Pegmatites with large volume and high grade of tantalum ore can be economically profitable under average demand conditions, and social or political considerations can occasionally dictate extraction of tantalum-mineral concentrates from minor pegmatite operations that are otherwise not competitive on the world market. For these reasons, this chapter provides a review of not only the preferred pegmatite types that have recently played an important role in worldwide tantalum production, but also of the marginal categories that may have only local significance under exceptional conditions.

2 Tantalum Deposits Within the Pegmatite Spectrum

Among the broad variety of granite-related pegmatites, those associated with typically anorogenic intrusions are characteristically rich in Nb, Y, (REE,Ti) and F. They show only limited Nb-Ta fractionation, and largely insignificant Nb(Ta) concentration. Pegmatites of this kind have been exploited principally for their Y,REE content even during the early times of tantalum recovery (e.g. Barringer Hill, Texas, Landes 1932; South Platte district, Colorado, Simmons and Heinrich 1980; Ytterby and related pegmatites in the Stockholm area, Brotzen 1959a,b).

In the orogen-related association, it is only the rare-element pegmatite class that includes tantalum deposits. The abyssal and muscovite classes as described by Ginsburg et al. (1979) have no significant tantalum mineralization (Fig. 1); the miarolitic class, as originally characterized by the above authors and as indirectly redefined by London (1986a; cf. also Černý 1988) has the potential, but so far no well-defined examples of economic Ta contents are known.

Most pegmatite types of the rare-element class may contain significant tantalum mineralization. In accordance with the classification developed by Beus (1960), Rudenko et al. (1975) and Černý (this Vol.), the pertinent pegmatite types and subtypes can be categorized for the purpose of characterizing their tantalum mineralization as follows (Table 1):

(i) beryl type, with beryl-columbite and beryl-columbite-phosphate subtypes;
(ii) complex type, including spodumene, petalite and amblygonite subtypes;
(iii) complex type, lepidolite subtype;
(iv) albite-spodumene type; and
(v) albite type.

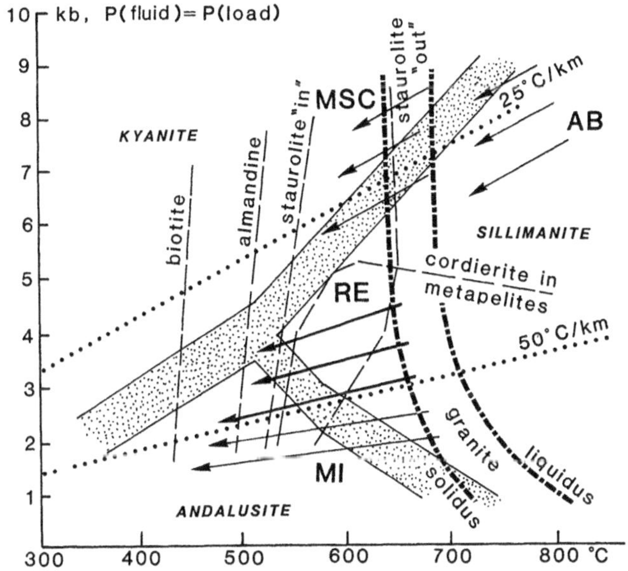

Fig. 1. Occurrences of four orogen-related classes of granitic pegmatites in the P-T metamorphic-igneous grid of their host rocks: *AB* abyssal class; *MSC* muscovite class; *RE* rare-element class; *MI* miarolitic class. *Arrows* indicate direction of progressive fractionation in pegmatite populations and (with the probable exception of *AB*) the approximate range of their primary magmatic crystallization. *Dotted lines* mark the typical Barrovian (25°C/km) and Abukuma (50°C/km) crustal gradients. *Stippled bands* indicate the aluminosilicate stability boundaries after Robie and Hemingway (1984). Granite liquidus and solidus after Jahns (1982)

Table 1. Tantalum-bearing rare-element pegmatite deposits

Pegmatite type (subtype)		%Nb	%Ta	Nb/Ta	Tonnage Nb	Tonnage Ta
(i)	Beryl (beryl-columbite) (beryl-col.-phosphate)	0.01-0.05	0.003-0.025	1.1-0.3	≤ 1,000	≤ 2,000
(ii)	Complex (spodumene) (petalite) (amblygonite)	0.007-0.02	0.013-0.2	0.8-0.12	>4,000	>5,500
(iii)	Complex (lepidolite)	0.002-0.005	0.005-0.015	0.6-0.2	X00	≤ 2,000
(iv)	Albite-spodumene	0.006-0.012	0.004-0.012	1.2-0.5	X,000	≤ 10,000
(v)	Albite	0.01-0.015	0.012-0.025	1.2-0.8	X,000	X,000

Modified from Kuzmenko and Eskova (1968), Kuzmenko ed. (1976), Ovchinnikov and Solodov (1980) and Hatcher and Elliott (1986).

Tantalum-enriched pegmatites commonly occur together in cogenetic groups encompassing two or more of the above types or subtypes; consequently they have several geological features in common. On the other hand, they do differ from each other in most of their internal features. The following sections deal with these two aspects separately.

3 General Features: Common Denominators

All orogen-related pegmatites of the rare-element class with potential economic concentration of tantalum have the same age distribution in geological history. They are found in similar lithologic and geotectonic environments, and in metamorphic terranes of the same grade and facies series. All these pegmatite types are genetically related to granitic intrusions, and they also occupy similar loci within pegmatite aureoles surrounding these parental plutons.

3.1 Age Distribution and Host Lithologies

The discussed pegmatites exhibit a wide range of geological ages, from late Archean to early Miocene. Conspicuous concentrations of rare-element pegmatite fields are related to the Kenoran (~ 2,500 Ma), Karelian-Nullaghinian (~ 2,100 Ma), Hudsonian-Svecofennian (~ 1,700 Ma), late Grenvillian (~ 1,000 Ma), Hercynian (~ 300 Ma) and Cimmerian plus Alpine (~ 200–20 Ma) orogens (cf. Ginsburg et al. 1979 and Černý 1982a, 1988a for details).

In Archean terranes, rare-element pegmatites are found mainly in greenstone belts, synclinoria of metavolcanic-metasedimentary sequences separating granitoid batholiths and gneissic tonalites (Fig. 2). Other Archean occurrences are confined to gneissic metasedimentary troughs locally separating the greenstone-granitoid subprovinces.

Similar environments also host Proterozoic pegmatite fields, but broader belts of eugeosynclinallike suites, bordering margins of Archean cratons or filling rift structures are much more common than true greenstone belts of the Archean type.

The volcanigenic component of the host formations decreases in abundance with decreasing age. Phanerozoic pegmatite-bearing terranes commonly consist of flysch-type sediments more or less devoid of volcanic rocks.

3.2 Metamorphic Environment

Rare-element pegmatite fields are typically hosted by Abukuma-type low-P/high-T metamorphic facies series. The enclosing rocks exhibit metamorphism from the uppermost-greenschist to the lower-amphibolite facies in terranes of relatively steep geothermal gradients (~40–50 °C/km). The A1.2 quartz-andalusite-chlorite subfacies, and particularly the A2.1 andalusite-cordierite-muscovite subfacies of

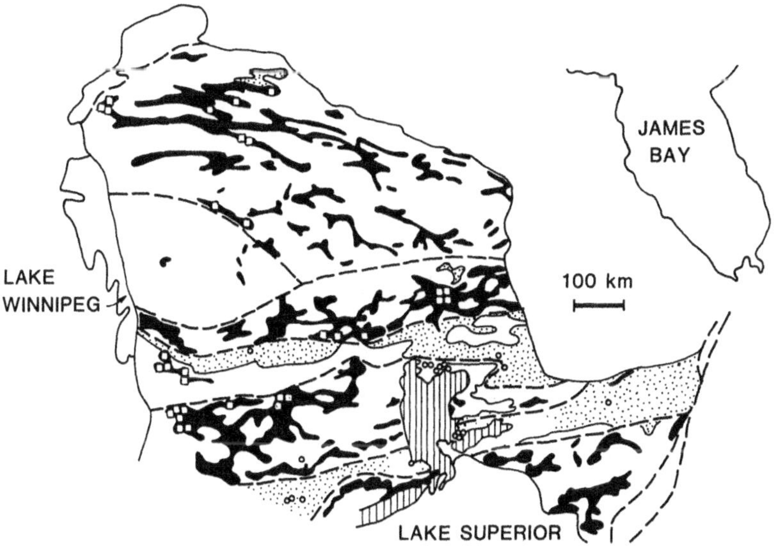

Fig. 2. Distribution of pegmatite districts and fields in the western Superior Province of the Canadian Shield. The number of symbols marking schematically the pegmatite distribution (*open squares* in black greenstone belts, *open circles* in stippled paragneissic belts) is not necessarily proportional to the density of pegmatite populations in the individual areas (Černý 1982b)

Winkler (1967) are characteristic. Fractionation of cogenetic pegmatite groups commonly increases down the metamorphic gradient (Fig. 3), with the parental granites at or near the second sillimanite isograd.

3.3 Parental Granites

Rare-element pegmatites are derived by progressive liquid and crystal/melt fractionation from granitic parents. These fertile plutons are in general leucocratic and commonly heterogeneous, ranging from biotite granite at depth to two-mica and muscovite + garnet facies in their cupolas. The topmost leucogranite is commonly coarse-grained to pegmatitic with abundant graphic feldspar + quartz intergrowths. Sodic aplite and potassic pegmatite are commonly associated with the caps of pegmatitic leucogranite.

The fertile granites are silicic, peraluminous, low in Ca,Mg,Fe,Sn,Ba,Ti and Zr but remarkably enriched in Rb,Be,Ga,Sn,Mn and Y; the contents of Li,Cs,Nb and Ta also are enhanced (Černý et al. 1985). REE abundances are generally low, mostly between 20x and 1x chondritic, with subhorizontal to HREE-depleted patterns; however, diverse fractionation disturbances kinking the patterns are common. Similarly, radiogenic and stable isotope systematics also are disturbed (cf. Černý and Meintzer 1988; and Černý 1988 and this Vol. for specific information and further references).

In contrast to their surrounding haloes of more or less mineralized pegmatites, the fertile granites are typically barren. However, they commonly contain minerals

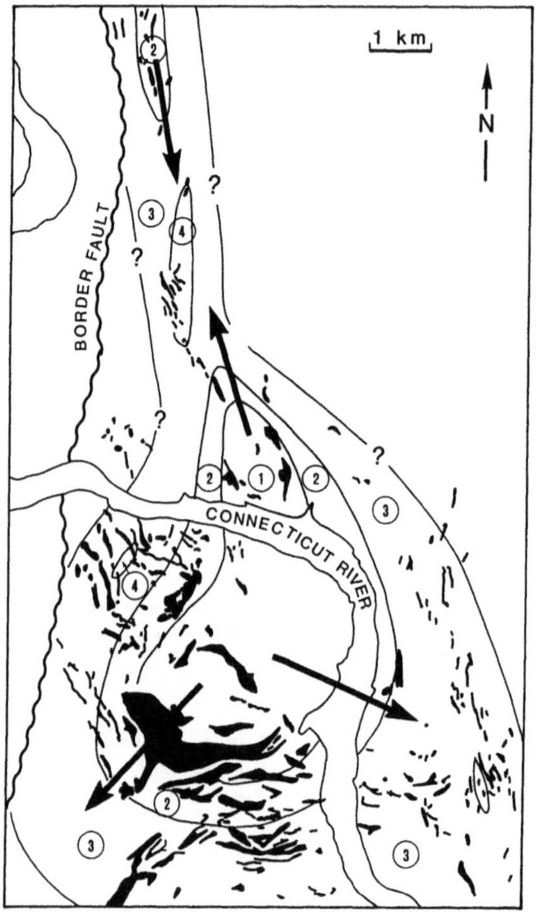

Fig. 3. Distribution and geochemical zoning of pegmatites in the Middletown district, Connecticut (modified from London 1985): *1* unzoned K-feldspar + plagioclase + quartz pegmatitic granite; *2* graphic K-feldspar + quartz pegmatites; *3* zoned K-feldspar + quartz + beryl pegmatites; *4* zoned complex lithium-rich pegmatites. The *arrows* indicate regional decrease in P-T of metamorphic conditions

of Be and Nb-Ta in their potassic pegmatite facies and local pods highly enriched in Li,Rb,Cs,Ta,B and F (e.g. Goad and Černý 1981). These mineralized pegmatite layers and pods bear close resemblance to the highly fractionated dykes populating the surrounding pegmatite aureole.

3.4 Structural Setting

With only a few exceptions, the fertile granites and related pegmatites are late tectonic to post-tectonic within the tectonomagmatic cycle of their host terrane, post-dating the peak of dynamometamorphic evolution and the emplacement of less differentiated granitoid intrusions. The granite-pegmatite systems invade fault systems developed in consolidated metamorphic terranes, and lithological or pre-existing intrusive contacts. For example, in Archean greenstone belts and Proterozoic gneissic troughs, the fertile granites are commonly emplaced along subver-

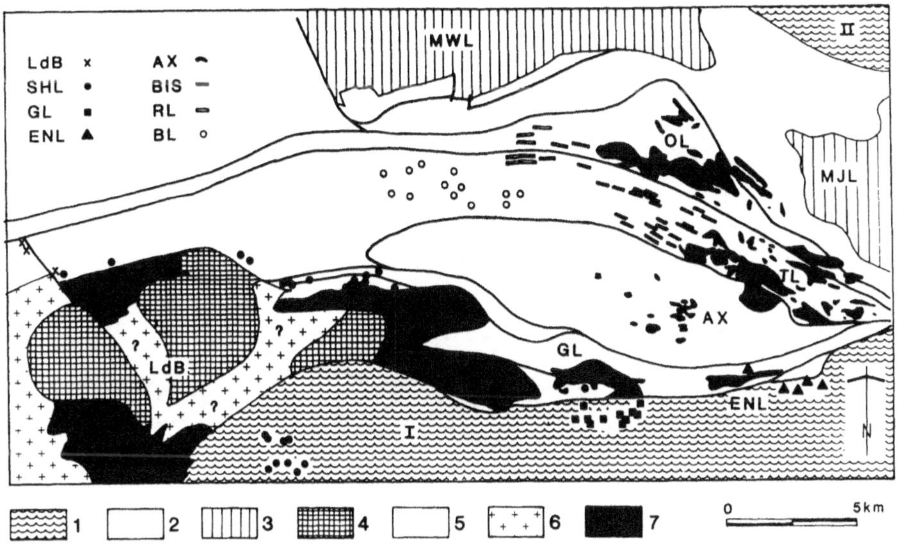

Fig. 4. Distribution of the fertile granites and pegmatites in the Winnipeg River pegmatite district, southeastern Manitoba (modified from Černý et al. 1981): *1* gneissic tonalitic + granitoid basement (?), Winnipeg River batholitic belt (*I*) and paragneiss of the Manigotagan-Ear Falls belt (*II*); *2* Bird River greenstone belt; *3* Maskwa Lake (*MWL*) and Marijane Lake (*MJL*) composite batholiths; Lac du Bonnet (*LdB*) batholith; *4* early hornblende-biotite granodiorite; *5* leucogranite parental to the SHL pegmatite group; *6* biotite granite; *7* pegmatitic granites and their pegmatite aureoles (see *inset* for symbols of different pegmatite groups); *heavy lines*: faults

tical axial faults, in décollements and within contacts of tonalitic or granitic batholiths flanking the belts (Fig. 4). In open basinlike structures, the granites may be related to re-activated faults in the basement, as suggested in the Yellowknife field (Meintzer 1987). In Phanerozoic orogens, the fertile intrusions tend to follow interformational horizons, generating extensive belts of pegmatite fields (Fig. 5).

Deformation, cataclasis and mylonitization do affect mineralized rare-element pegmatites locally. However, in all cases these events can be related to late episodic movements along the host fracture systems (e.g. Eby 1986).

3.5 Position Within Pegmatite Aureoles

Within the limits imposed by local structural conditions, haloes of pegmatites surrounding their granitic parents tend to display regional zoning. From the fertile granite outwards, pegmatites become more fractionated, complex in their internal structure and paragenesis, extensively replaced and mineralized (cf. Trueman and Černý 1982, Černý 1988). Tantalum-bearing pegmatites of type (i) are commonly restricted to the intermediate parts of the regional zoning patterns, whereas the other types are found in the outer reaches and the very margins of such pegmatite groups. Types (i), (ii) and occasionally (iii) are found in cogenetic groups (Heinrich 1953,

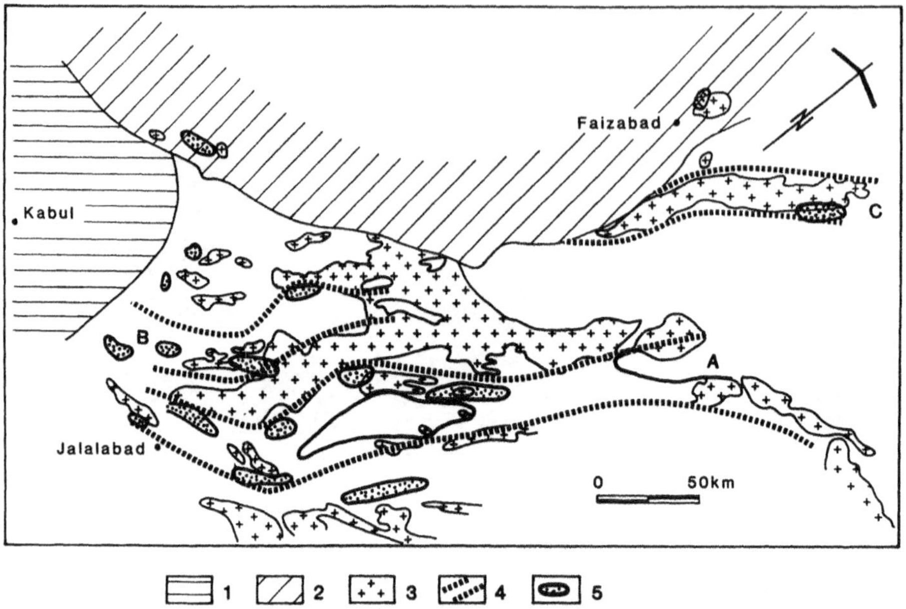

Fig. 5. Rare-element pegmatite belts and fields of eastern Afghanistan (modified by Černý 1982a from Rossovskyi and Chmyrev 1977): *1* Baluchistan-Himalayan fold region with a superimposed Alpine geosyncline; *2* Hercynian fold region of northern Pamir; *3* lower Cretaceous to Paleogene fertile Laghman granites; *4* boundaries of pegmatite belts (*A* Nuristan; *B* Hindukush; *C* Badakhshan belts); *5* pegmatite fields

Solodov 1971; Černý et al. 1981), but (iii), (iv) and particularly (v) tend to be the only highly fractionated and mineralized pegmatite type represented in a given aureole (Foord 1976; Heinrich 1978; Černý 1982b).

4 General Features: Distinguishing Contrasts

Although the principal geological features of tantalum-bearing pegmatites are shared among their different types, the internal characteristics are highly individual and, in some cases at least, not even transitional from type to type. Internal structure of the discussed pegmatites, their level of geochemical evolution, distribution of mineralization, mineralogy of tantalum and exomorphic metasomatism are in many cases specific to the individual types.

4.1 Internal Structure

Concentric zonal structure is characteristic of most pegmatites of the (i) and (ii) categories, and of some in type (iii). A gradual increase in the number of primary units and in the number (and extent) of replacement assemblages is commonly

Table 2. Diversity of pegmatitic tantalum deposits

Pg. type (subtype)	Beryl (i)	Complex (ii) (spd,pet,amb)	Complex (iii) (lcp)	Albite-spodumene (iv)	Albite (v)
Internal structure	→	Increasing complexity		Quasi-homogeneous	
Albitic units	→	Increasing abundance		"Matrix"	Bulk
Metasomatic features	→	Increasing extent		Limited	
Geochemistry	Incr. Li,Rb,Cs,Ga,Hf,Ta,(P),F,B, →	Decr. K/Rb,K/Cs,Nb/Ta,Al/Ga etc.		Variable fractionation	

observed in groups of cogenetic pegmatites from the beryl type (i) to all the complex subtypes (ii) and (iii) (Table 2). Numerous examples of zoned pegmatites, including schematic generalizations, are illustrated by Cameron et al. (1949), Page et al. (1953) and Černý (1989). Some of the most complex pegmatites are described verbally but not illustrated graphically by Heinrich (1976) and Norton (1983).

The albite-spodumene pegmatites (iv) are largely quasi-homogeneous, with only some muscovite and/or phosphate enrichment along the contacts and a slight increase in K-feldspar and spodumene towards the central parts. The latter two minerals are commonly columnar, with preferred orientation subnormal or oblique to the contacts with wall rock (cf. Sundelius 1963; Heinrich 1964; Pye 1965; Luster 1977; Simpson 1977; Černý et al. 1981; Chackowsky 1987 for classic examples).

The albite pegmatite type (v) is also predominantly homogeneous, with only vestigial pods of K-feldspar, quartz or spodumene in the central parts. The bulk of these pegmatites consists of fine-grained saccharoidal albite or cleavelandite with subordinate quartz.

4.2 Geochemical Signature

In general, the zoned pegmatite types show progressive fractionation from (i) to (ii) plus (iii) (Table 2). Within most groups or fields of cogenetic pegmatites, an increase in the Li content is expressed first by Li enrichment in the micas, then by the appearance of ferromanganoan phosphates of Li, and finally by the crystallization of the lithium aluminosilicates. Increasing Rb and Cs contents in potassic minerals accompany this trend, as well as decreasing Fe/Mn,Al/Ga,Zr/Hf and Nb/Ta in silicates, phosphates and oxide minerals. However, the individual fractionation pairs may display considerable disproportions among different groups and fields of rare-element pegmatites (Černý et al. 1985). In contrast, the albite-spodumene pegmatites (iv) may show only limited fractionation ranges within consanguineous groups (Meintzer 1987), but considerable differences between separate genetic units (e.g. Černý et al. 1981; Chackowsky 1987; Černý 1989). In the albite pegmatites (v), geochemical relationships have not been studied in sufficient detail to permit generalizations.

4.3 Tantalum Mineralization Patterns

Columbite is found as a relatively early, coarse-grained to blocky mineral in the core-margin assemblages of only the most primitive pegmatites of the beryl type (i); this columbite commonly shows high Nb/Ta ratios and has hardly any potential as a tantalum ore mineral. In the more differentiated beryl pegmatites, the tantalum-enriched oxide minerals are mainly concentrated in two kinds of units: in albite-rich assemblages (saccharoidal albite or cleavelandite (Fig. 6); and in greisenlike muscovite ± quartz association (Table 3). Both assemblages are located mainly in inner intermediate and core-margin parts of pegmatite bodies. The albitic units extensively corrode adjacent early minerals but are in their bulk primary (London 1986b); the muscovite-rich units may represent primary precipitates but in most cases they are demonstrably secondary, replacing zones rich in K-feldspar (e.g. Wise 1987).

In the complex pegmatite subtypes covered under (ii), the bulk of tantalum mineralization is usually also contained in the above albite- and muscovite-rich units

Fig. 6. A "wave" of layered saccharoidal albite in the Tanco pegmatite, Manitoba: layering is due to variations in the quartz and muscovite contents, and accentuated by local lenses or layers of quartz. The hammer is 40 cm long

Table 3. Styles of tantalum mineralization

Pegmatite type (subtype) beryl (i)	complex (ii) (spd,pet,amb)	complex (iii) (lep)	albite-spodumene (iv)	albite (v)
Core margins				
(Early to) late albite-rich units			"matrix"-albite (muscovite--quartz units)	bulk, dispersed
	Late muscovite-quartz units			
		Late lepidolite units		

(Černý 1982c; Foord and Cook 1988). However, in (Nb), Ta-rich deposits the ore minerals may be found in several intermediate zones rich in albite (e.g. Norton et al. 1964), and lepidolite units are tantalum-rich as well (Jahns and Ewing 1976; Ercit 1986). Tantalum-bearing minerals are in such cases dispersed in non-economic quantities in virtually all zones and replacement units of such pegmatites.

The lepidolite subtype (iii) usually contains the main concentration of tantalum minerals in lepidolite-rich units (e.g. Heinrich 1967). In the quasi-homogeneous albite-spodumene type (iv), tantalum mineralization is disseminated in the fine- to medium-grained albitic matrix and in local muscovite-rich pods. The albite pegmatites (v) also have the tantalum ore minerals dispersed throughout their bodies in the dominant albite-rich mineral assemblages.

4.4 Mineralogy of Tantalum

Assemblages of tantalum-bearing minerals are considerably diversified in the examined spectrum of rare-element pegmatites (Table 4). The number of mineral species and their compositional ranges increase from beryl pegmatites (i) to the complex subtypes designated as (ii). It is mainly the petalite- and spodumene-bearing subtypes that display the broadest variety in tantalum minerals, containing up to 14 different species (cf. the Tanco pegmatite, Sect. 5.2.3). In contrast, the mineralogy of tantalum is rather simple in types (iv) and (v), and the lepidolite subtype (iii) is set apart from all others by its typical dominance of microlite over subordinate manganocolumbite.

The tantalum minerals commonly have a tendency to develop euhedral to subhedral crystals, but anhedral forms filling interstices among albite, micas and quartz are also known (Grice et al. 1972). Simultaneous precipitation of two or three species (including cassiterite) is widespread; however, replacement of early minerals by late secondary phases is also common, particularly in the mineralogically most complex deposits of type (ii) (cf. Černý and Ercit 1985, Table II).

4.5 Exomorphism

Metasomatic phenomena in wall rocks of pegmatite contacts are not a direct function of the tantalum-bearing potential. They depend mainly on the content and

Table 4. Assemblages of tantalum minerals[a]

Pegmatite type (subtype) Beryl (i)	Complex (ii) (spd,pet,amb)	Complex (iii) (lep)	Albite-spodumene (iv)	Albite (v)
Columbite-tantalite Ixiolite (rutile) (tapiolite) (microlite)	Columbite-tantalite Microlite Ixiolite Cassiterite Wodginite (tapiolite) (stibiotantalite) (simpsonite) (cesstibtantite) Secondary minerals of T[b]	Microlite Mangano-columbite (stibiotantalite) (euxenite) (fersmite) (rynersonite)	Columbite-tantalite Cassiterite (microlite)	Columbite-tantalite (rutile) (cassiterite) (microlite)

[a] See Čenrý and Ercit (this Vol.) for a systematic review of niobium and tantalum minerals; see Černý and Ercit (1985) for examples of mineral assemblages in specific pegmatite localities.
[b] See Černý and Ercit (this Vol.) for detailed information.

behaviour of volatile components, and on the host-rock lithology. Silicate rocks of contrasting bulk composition such as amphibolites, gabbros or serpentinites are the most reactive (e.g. Černý and Burt 1984), and pegmatites enriched in F and B affect the wall rock most extensively (e.g. Shearer et al. 1986). A hydrous B-, F-bearing fluid may be expelled from a complex pegmatite at several stages of its consolidation, transporting Li, Rb, Cs, K, As, Be, U and other minor elements and generating several assemblages of exomorphic minerals such as muscovite, biotite, bityite, tourmaline, holmquistite, chlorite, carbonates and others (Zagorskyi et al. 1974; Zagorskyi and Khaltuyeva 1976; London 1986c; Morgan and London 1987).

Other conditions being equal (i.e. mainly the same type of wall rock), the most extensive exomorphic aureoles can be expected around the complex pegmatites (ii) and (iii), whereas holmquistite should occur with types (ii), (iii) and (iv). Beryl-type pegmatites (i) are usually the least reactive; the albite pegmatites (v) have not been examined enough for a meaningful generalization.

As shown by Beus (1968) and more recently by Gaupp et al. (1984), alkali metals and the volatile components generate the most extensive dispersion haloes and exomorphic mineral assemblages. Other elements prove to be much less mobile, and only traces of Nb and Ta are found in the most immediate vicinity of pegmatite contacts. Except in unspecified cases mentioned by Nikitin (1957), tantalum mineralization is confined to within the pegmatite bodies.

5 Specific Characteristics of Ta-Bearing Pegmatite Types

A brief review is presented in this section of the principal features of the individual tantalum-bearing pegmatite types. Besides general characteristics, specific examples are given of pegmatite localities that have produced tantalum in the past, are exploited at present or have a potential in this respect for the future. Information available on the different deposits is quite variable, dependent on the extent of past research and on the degree of confidentiality imposed by industrial interests on the economically important parameters. However, the following subsections should provide a meaningful review of the pegmatites examined here.

5.1 Beryl Type (i)

This pegmatite type represents a modest level in the structural and geochemical evolution of rare-element pegmatites. It is usually located just outside the margins of parent granite intrusions, and in any case in intermediate portions of zoned pegmatite groups. Individual pegmatite bodies are tabular (fracture-filling) or, in less competent-behaving host rocks, lenticular to bulbous in form.

Internal zoning is usually restricted to an aplitic, coarse-grained granitic or muscovite-rich border zone, a wall zone of graphic K-feldspar + quartz integrowth in albite + quartz + muscovite matrix, a core-margin unit of blocky K-feldspar and a quartz core. In pegmatites of more advanced textural evolution, the graphic granite may be absent, with only megacrystic K-feldspar in the wall zone. Pods and layers of saccharoidal albite or cleavelandite are common in the core margin; they usually carry beryl, garnet or tourmaline (schorl) and coarse muscovite. Ferromanganoan phosphates ($\pm$Ca,Mg,Li), if present, are found in the core margin or associated with the albitic units. The border zone may carry minor biotite but muscovite is the dominant mica. In phosphate-poor pegmatites, pods of muscovite associated with the quartz core may show increased Fe, Mn and Li contents.

A simple assemblage of Nb, Ta-bearing oxide minerals (Table 4) is usually associated with the albite-rich units. Nb, Ta-mineralization is less common in muscovite pods but if present, its fractionation level is perceptibly more advanced than in the albite association. Nevertheless, the overall Nb/Ta(wt.) ratio in the beryl-type pegmatites is rather high, ore grades are low and the volume of individual pegmatite bodies is small; all of which make the type (i) pegmatites unattractive as a tantalum source.

5.1.1 Greer Lake Group, Southeastern Manitoba

The Greer Lake pegmatitic granite and the related pegmatite group are located 135 km east-northeast of Winnipeg. They are a part of the Cat Lake-Winnipeg River pegmatite field situated in the westernmost English River subprovince of the Archean Superior Province. Most of the fertile granites and derived pegmatite groups are located in schists of the Bird River Greenstone Belt, and were generated after the peak of regional metamorphism, ca. 2500 Ma ago. A detailed geological description is given by Černý et al. (1981, 1986).

The Greer Lake pegmatitic granite intrudes metabasalts but the pegmatite group is distributed in adjacent tonalitic and granitoid gneisses. The pegmatites are dominantly concordant intrusions with shallow to intermediate dip, pinching and swelling within the foliation of the gneisses. Typical dimensions of the mineralized pegmatites are well under 100 × 10 m in outcrop; the largest (but barren) intrusion reaches 400 × 15 m.

A typical member of this group has a thin border zone of medium-grained oligoclase, microcline-perthite and quartz. Large crystals of graphic microcline-perthite + quartz intergrowths, embedded in coarse-grained albite + quartz ± biotite ±muscovite, constitute an extensive wall zone. The core consists of blocky microcline-perthite and quartz, locally segregated into a blocky core margin of microcline-perthite surrounding a lenticular quartz core.

Layers and pods of fine- to medium-grained albite with quartz, muscovite and garnet are located at the outer margins of the blocky core, or within and around the segregated microcline-perthite core margin. Cleavelandite and book muscovite are commonly found in this latter location.

Most of the rare-element-bearing and other accessory minerals are found in these albite-rich assemblages. Garnet, beryl and cordierite are associated with subordinate but widespread Nb, Ta, Ti, Sn-bearing oxide minerals, monazite, gahnite, zircon, xenotime, and extremely rare apatite, pyrite and chalcopyrite.

The tantalum-bearing minerals are represented by columbite-tantalite, ixiolite, microlite, cassiterite, niobian and tantalian rutile, and ilmenite. The composition of most of these phases is extremely varied, in correlation with the overall fractionation level of the parent pegmatite (Fig. 7). For example, ferrocolumbite is common in dykes with an average K/Rb of blocky K-feldspar > 40, but manganotantalite occurs in Li-enriched pegmatites with the same ratio at < 25. In the first case, Ta/(Ta + Nb) (wt.) of microlite coexisting with columbite is 0.6–0.7; however, microlite replacing manganotantalite in the highly evolved dykes has Ta/(Ta + Nb) 0.90. In the more primitive pegmatites, titanium is enriched in the ferrocolumbite and particularly in ixiolite, to the degree of exsolving as niobian rutile.

The Greer Lake pegmatite group was extensively explored in the early 1980s, at which time some of its pegmatites were found to be marginally interesting as tantalum resources.

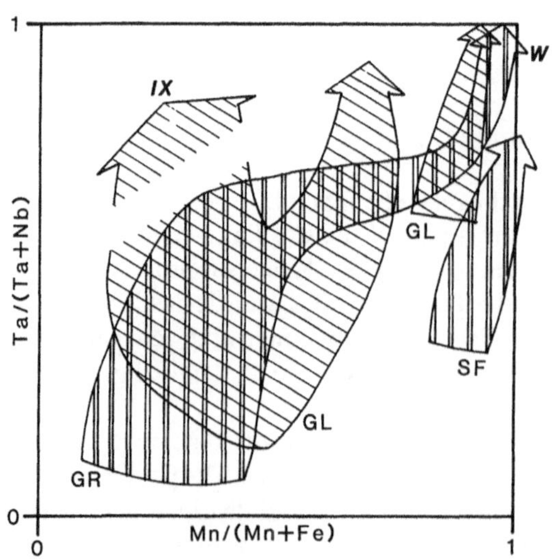

Fig. 7. Fractionation trends in columbite-tantalite, ixiolite (*IX*) and wodginite (*W*) in the Greer Lake pegmatitic granite (*GR*) and its highly fractionated pods and offshoots (*SF*), and in the surrounding pegmatite group (*GL*), expressed in the columbite quadrilateral (atomic ratios) (after Černý et al. 1986)

5.1.2 PEG Group, Upper Ross Lake, Northwest Territories

The PEG group of over 500 exposed pegmatites is located between the Upper Ross and Redout Lakes, about 75 km east-northeast of Yellowknife, N.W.T. It spreads over a distance of 4 km westward from the Redout pluton, which is a two-mica granite considered parental to the group (Fig. 8). This classic case of regional zoning, located in the southern part of the Archean Slave Province of the Canadian Shield, was described by Hutchinson (1955) and recently further detailed by Meintzer (1987) and Wise (1987) (for published reports see also Ayres and Černý 1982; Meintzer et al. 1984; Wise et al. 1986, 1988).

Beryl-type (i) pegmatites comprise two zones in the regional sequence, the western and broader one being populated mainly by the beryl-columbite subtype grading locally into the beryl-columbite-phosphate subtype (zone #4 in Fig. 8). Pegmatites of this zone are generally lenticular, with intermediate angles of dip, from 0.5 × 15 m to 15 × 300 m in outcrop dimensions. They are discordant to the foliation of the enclosing granodiorite gneiss or, locally, amphibolite. The pegmatites are poorly to well zoned. In the texturally best differentiated intrusions, a fine-grained quartz + muscovite + plagioclase wall zone is followed by a coarse-grained quartz + cleavelandite + K-feldspar + muscovite outer intermediate zone, an inner intermediate zone of the same composition but coarser in grain size and beryl-bearing, and a core of quartz + K-feldspar + beryl. The core may be segregated into a K-feldspar + beryl-bearing core-margin and monomineralic quartz core.

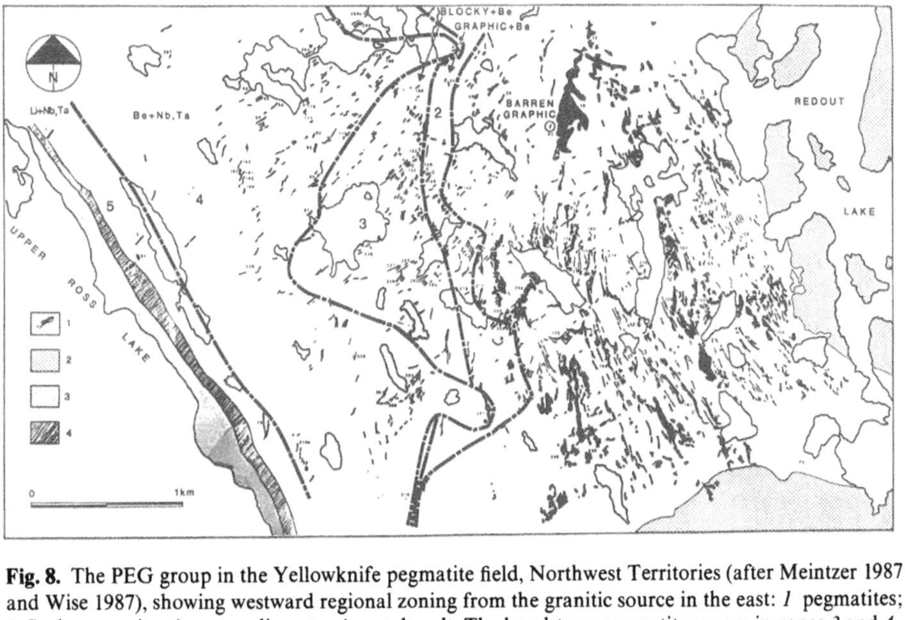

Fig. 8. The PEG group in the Yellowknife pegmatite field, Northwest Territories (after Meintzer 1987 and Wise 1987), showing westward regional zoning from the granitic source in the east: *1* pegmatites; *2* Redout granite; *3* metasediments; *4* metabasalt. The beryl-type pegmatites occur in zones *3* and *4*

Tantalum-bearing minerals are found dispersed in the inner intermediate zone, and particularly in local accumulations of cleavelandite or fine-grained muscovite. Ferrocolumbite, ferrotantalite, ferrotapiolite and rare manganocolumbite plus microlite are commonly associated with beryl and triphylite, occasionally also cassiterite, garnet, gahnite or triphylite + beusite intergrowths.

Figure 9 shows that all ferromanganoan oxide minerals of Ta are poor in Mn, despite the broad range of the Ta/(Ta + Nb) ratios. A moderate correlation is observed between increasing Ta in the ore minerals and decreasing K/Rb in the blocky K-feldspars of the different pegmatites. This latter ratio varies through zone 4 from 90 to 11, close to the total range exhibited by the previously described Greer Lake group (Černý et al. 1986). Other geochemical parameters of the two beryl-type populations are also closely similar, except the general lack of phosphates at Greer Lake.

The PEG-group pegmatites have been intermittently explored since their discovery in the early 1940s till 1982, and several pegmatites were mined in the late 1940s. Mining operations and ore-dressing tests yielded heavy concentrates from 0.03 to 0.21 wt.% of the bulk samples, in our case with the Nb_2O_5/Ta_2O_5 (wt.) ratio

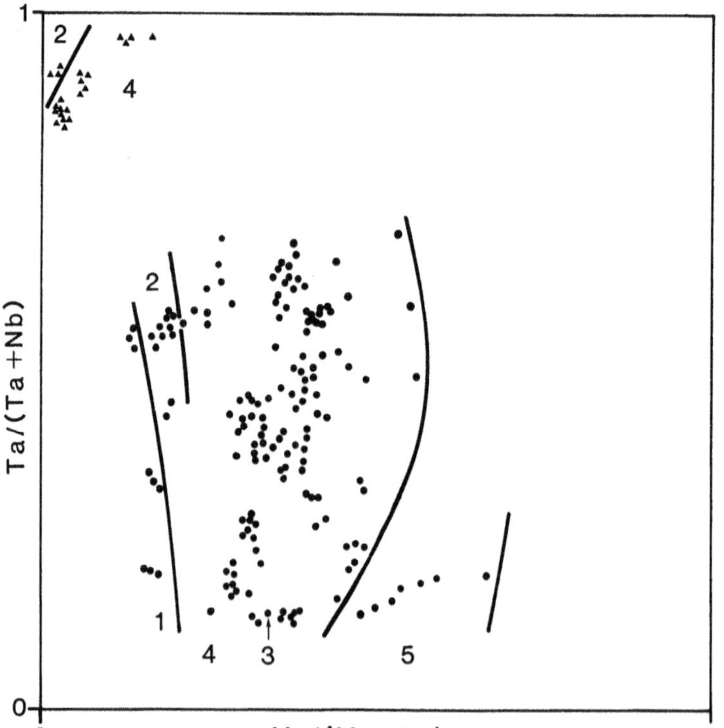

Fig. 9. Compositional evolution of columbite-tantalite and ferrotapiolite of the PEG group, Northwest Territories in the columbite quadrilateral (atomic ratios). *Numbers* correspond to the numbers of regional zones in Fig. 8. The beryl-type pegmatites occur in zones 3 and 4 (modified from Wise 1987)

of 0.15. However, the small size of the pegmatites and their remote location prevented expansion of the mining activities.

5.1.3 Other Examples

The beryl-type (i) deposits are the most widespread type of rare-element pegmatites. They have been exploited mainly for beryl, as their (Nb)Ta potential is usually subordinate. Nevertheless, numerous localities are described in the literature in sufficient detail to provide thorough information on their characteristics in different geological terranes. The reader is referred to the reports on the Meyers Ranch pegmatites, Colorado (Hanley et al. 1950), Donkerhoek area in Namibia (Schneiderhöhn 1961), Crystal Mountains field, Colorado (Thurston 1955), the Don Patch, Bull Moose, Big Chief and similar pegmatites of the southern Black Hills, South Dakota (Hanley et al. 1950; Page et al. 1953; Norton et al. 1964), and the pegmatites of Connecticut (Cameron and Shainin 1947).

5.2 Complex Type–Spodumene, Petalite and Amblygonite Subtypes (ii)

Complex rare-element pegmatites are defined as the zoned lithium-rich types containing aluminosilicates and/or aluminum phosphates of Li. Subtypes are distinguished according to the dominant lithium-bearing phase, which is indicative of PTX conditions of the pegmatite consolidation or bulk chemistry: spodumene (high P, low T), petalite (low P, high T; London 1984), amblygonite (high $\mu PFO_2 \pm NaLi_{-1}$; Burt and London 1982), and lepidolite (high μLiF and μKF at relatively low acidity, with low μKF/μHF stabilizing also topaz; London 1982). From the viewpoint of tantalum mineralization, the first three subtypes can be treated together; however, the lepidolite subtype is distinctly different and will be dealt with in section 5.3.

The three subtypes discussed here represent the granitic pegmatites most fractionated in terms of geochemistry and internal structure, with the most diversified mineralogy in general, and mineralogy of tantalum in particular. These subtypes are invariably found at the outer margins of regionally zoned groups of cogenetic pegmatites. Individual pegmatite intrusions are usually bulbous, lenticular or tabular in shape, and they may reach gigantic dimensions.

Internal zoning is usually complex, consisting of 6 to 12 units; zones of primary crystallization are dominant over secondary replacement bodies. The zoning is largely concentric but commonly asymmetric, and locally complicated by layered units that disrupt the pattern of concentric envelopes (e.g. Norton 1983). Outer zones of granitic and graphic texture are absent in most cases, but extensive units of saccharoidal or platy albite, and of secondary muscovite or lithium micas are found in the interior. The diversity of zoning is also enhanced by multiple intermediate zones with gigantic crystal size and different proportions of quartz, K-feldspar and lithium aluminosilicates. In some cases, a conspicuous concentration of K-feldspar is found in central parts of the quartz core.

Three types of internal units host the main concentrations of tantalum minerals: albitic bodies (cleavelandite or saccharoidal), secondary muscovite units replacing zones rich in K-feldspar and units consisting of lithian micas of either secondary or primary nature.

The mineralogy of tantalum may be extremely diversified (Table 4): 6 to 8 species are commonly present, and in some deposits their number reaches 12 to 14. This is due to steep fractionation gradients between the outer and inner units of a given pegmatite in terms of Nb/Ta, Fe/Mn, Ti/Ta and Sn/Ta, and in the role of F in the different zones and replacement units. Columbite-tantalite and related ferromanganoan oxide minerals are typical of F-poor environments, whereas microlite-group phases are dominant in assemblages rich in lithian micas. Abundance of Sn may generate members of the wodginite and ixiolite groups; Ti has the same effect, particularly during the early stages of crystallization. Extreme fractionation not only drives the overall mineral chemistry from (Fe,Nb)-dominant to (Mn,Ta)-dominant compositions, but on occasion leads to a severe depletion of (Fe,Mn) and precipitation of simpsonite. Further complexity, particularly in the broader microlite and stibiotantalite mineral families, is caused by increased activities of Cs, Ba, Sb, Bi, U, Sn, P and B. Last but not least, secondary tantalum-bearing minerals replacing the primary phases are particularly typical of the complex pegmatites (see Černý and Ercit this Vol., Sect. 4 for details).

The tantalum-bearing minerals are usually associated with cassiterite (which may be Ta-bearing as well), hafnian zircon, beryl, apatite and other miscellaneous phosphates, locally also with minor uraninite, monazite, rutile, gahnite or ilmenite.

The Nb/Ta ratio of the ore-forming mineral assemblages is commonly low, the Ta values significantly high, and the total resources reach respectable tonnages in giant members of the discussed pegmatite subtypes. This is the most desirable kind of pegmatitic tantalum deposits.

5.2.1 Hugo Pegmatite, South Dakota

The Hugo pegmatite is one of several mineralized pegmatites in the northeastern margin of the pegmatite field of the southern Black Hills. It is closely adjacent to the main intrusion of the parental Harney Peak pegmatitic granite. The granite-pegmatite system is of Hudsonian age (~1,700 Ma), emplaced into late Archean to early Proterozoic metasediments and metavolcanic rocks. The most comprehensive reference to the Hugo deposit is that by Norton et al. (1962), the petrology of the granite and its pegmatite aureole is discussed by Shearer et al. (1987).

The Hugo pegmatite is an approximately turnip-shaped intrusion about 175 × 135 m in plan, emplaced into micaschists that are recrystallized and tourmalinized along contacts. Seven zones, two replacement units and two types of fracture filling were recognized by Norton et al. (1962), but it is possible that the Li-mica units were generated by a single metasomatic event (Fig. 10). The successive zones of the pegmatite change progressively in composition and texture from the contacts to the core; transitions between all units are gradational in both respects. The border and wall zones (1,2) consist of albite + quartz + muscovite, with columnar tourmaline in

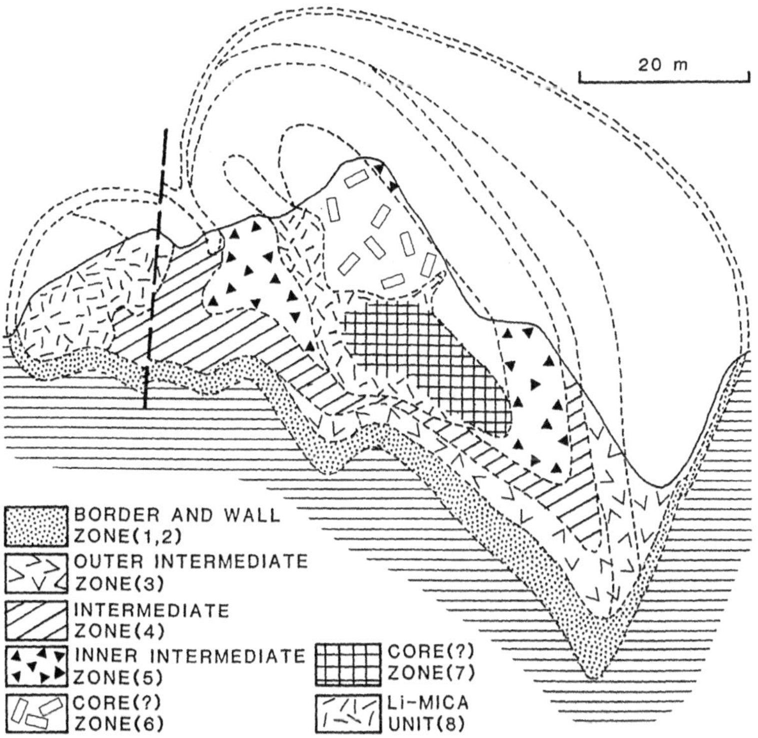

Fig. 10. Vertical section through the Hugo pegmatite near Keystone, Black Hills of South Dakota (modified by Černý 1989 from Norton et al. 1962)

the border unit (1). K-feldspar occurs first in zone (3), whereas the muscovite content decreases. Zone (3) is subdivided according to the plagioclase/K-feldspar ratio that decreases upwards (Fig. 10). Early amblygonite and spodumene appear in zone (4) and reach a maximum accumulation in zone (5). Units (5,6) are dominated by quartz, but K-feldspar is the principal primary mineral in zone (7). The metasomatic assemblage of cleavelandite + lithian mica extends through zone (7) and also penetrates zones (4,5,6), with minor incursions into (2,3).

Columbite-tantalite, tantalian cassiterite and minor microlite are the only tantalum minerals identified to date. They are dispersed in the metasomatic assemblage of unit (8) but concentrated in mineable quantities in zones (3) and (4), in association with albite, muscovite, beryl, apatite and minor triphylite, tourmaline and sulphides. Some of the columbite-tantalite in zone (4) forms widely scattered, huge crystal aggregates reaching over 100 kg in weight; in contrast, the mineral is medium-grained in unit (3) and particularly in unit (8). Pre-1960 analyses generally show intermediate Nb/Ta ratios for columbite-tantalite but no systematic study has been conducted with more recently developed techniques. The K/Rb range of K-feldspar from the interior zones, 25 to 9, and other indicators of rare alkali fractionation, such as micas and beryl, document an advanced evolution of the parent pegmatite melt.

No reliable records have been kept on tantalum concentrate production during most of the mining of this deposit, which commenced in 1909 with extraction of amblygonite. The total appears not to have been higher than a few thousand kilograms. The Hugo pegmatite may serve as an example of a complex, spodumene-subtype pegmatite which can yield tantalum concentrate only as a by-product.

5.2.2 Harding Pegmatite, New Mexico

The Harding pegmatite is located in the western part of the Picuris Range, southwest of Taos, New Mexico. It is hosted by a thick sequence of metasediments and metavolcanic rocks of Precambrian (Aphebian) age, penetrated by several generations of granitoid intrusions that span a 300-Ma range. The pegmatite deposit is close to 1,400 Ma in age. Most of the studies conducted on the Harding pegmatite have been summarized in Jahns and Ewing (1976), Brookins et al. (1979) and Lumpkin et al. (1986).

The Harding pegmatite is a flat, subhorizontal intrusion emplaced parallel to foliation along a contact between amphibolite and quartz-mica schist. Its known dimensions of 330×500×25 m are considered minimal at least in the longest direction, as a substantial part of the pegmatite has undoubtedly been eroded (Fig. 11). The pegmatite consists of six primary zones and two replacement units, of

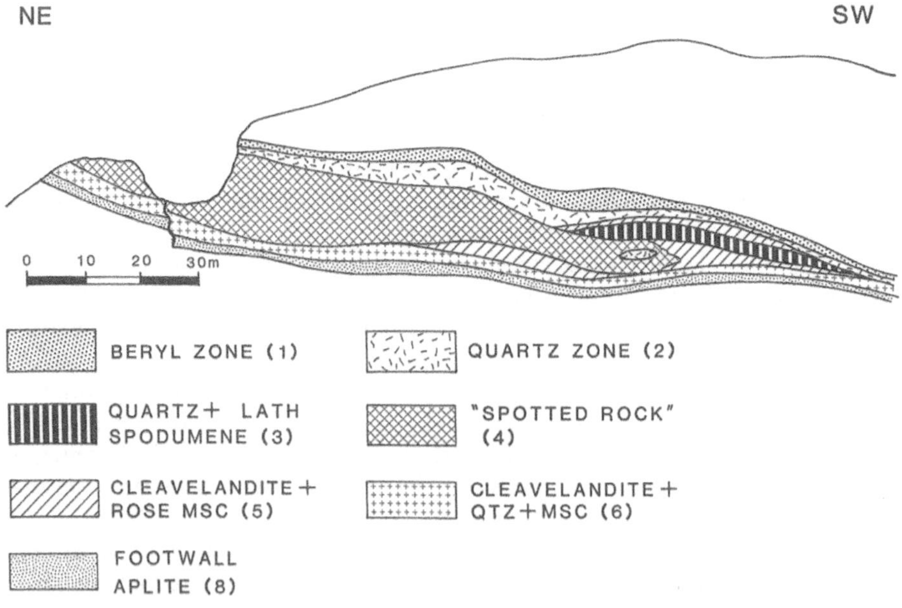

Fig. 11. Vertical section through the main part of the Harding pegmatite, New Mexico (modified from Jahns and Ewing 1976 and Brookins et al. 1979)

generally subhorizontal attitudes parallel to the pegmatite contacts, and with compositions deviating from a typical zoning sequence (Fig. 11).

Three primary zones constitute the hanging-wall portion of the pegmatite. A thin, fine-grained border unit plus a coarse-grained wall unit constitute the beryl zone (1); quartz, albite and muscovite are associated here with accessory beryl, K-feldspar, apatite and columbite-tantalite. A quartz zone (2) also contains accessory feldspars and muscovite; it is underlain by the quartz-lath spodumene zone (3) with accessory microlite, beryl, apatite and lepidolite which are mainly distributed along its base.

The core of the pegmatite consists of the microcline-spodumene zone (4), so-called spotted rock, which also contains substantial quartz, minor apatite, albite, microlite and columbite-tantalite. Lithian muscovite and lepidolite of unit (5) commonly replace this zone, locally forming nearly pure micaceous masses. The core is underlain by the discontinuous but extensive zone of cleavelandite and quartz (6), and by local patches of perthite zone (7) with minor quartz and albite. The footwall aplite zone (8) consists of albite, quartz and accessory beryl, apatite and columbite-tantalite.

The cleavelandite metasomatic unit of albite, quartz and minor muscovite occurs primarily in the core (4) but is also found in the beryl zone (1). The rose muscovite-cleavelandite unit mainly replaces parts of the quartz-lath spodumene zone (3). Both metasomatic units carry microlite, and locally columbite-tantalite.

Microlite and columbite-tantalite, both predominantly fine-grained, are the ore minerals of tantalum. A thorough study of microlite by Lumpkin et al. (1986) shows this mineral to be relatively uniform in most compositional parameters when compared to some other pegmatite deposits. The contents of W,Sn,Sb,Bi,Y and Ce are generally very low, but Ti and U are considerably enriched in samples from the metasomatic units (up to 10 wt.% UO_3 and 3.2 wt.% TiO_2). Zoning of individual crystals is widespread, as is alteration and replacement along grain margins and fractures. The $Ta_2O_5/(Ta_2O_5+Nb_2O_5)$ (wt.) ratio ranges from 0.85 to 0.97. The limited data published for columbite-tantalite show the same ratio ranging from 0.45 to 0.94, with those from the "spotted rock" (4) averaging close to 0.8.

In the 1942–1947 period, the Harding pegmatite produced about 10 metric tonnes of microlite concentrate averaging 68 wt.% Ta_2O_5 and 7 wt.% Nb_2O_5; the nearby pegmatite-derived placers yielded some 200 kg of tantalite-columbite concentrate with 43 wt.% Ta_2O_5. The tantalum reserves are far from exhausted, but the pegmatite is now the property of the University of New Mexico and is not available for industrial exploitation.

5.2.3 Tanco Pegmatite, Manitoba

This deposit represents an extremely fractionated pollucite-bearing variety of the complex, petalite subtype pegmatite with substantial volumes of lithian micas and amblygonite-montebrasite. The pegmatite is located about 180 km east-northeast of Winnipeg, in the Bird River greenstone belt of the Archean Superior Province of the Canadian Shield. U-Pb and Rb-Sr dating of nearby granitoid intrusion and pegmatites, and limited work on Tanco itself, place these rocks into the 2,650–2,550 Ma interval of the Kenoran orogeny. The most recent reviews of the overall pegmatite geology are those of Crouse et al. (1979) and Černý (1982c). London (1986b) provides an analysis of the pegmatite's internal evolution, and Ercit (1986) gives a thorough review of its tantalum mineralization.

The pegmatite was emplaced into members of two subhorizontal fracture sets segmenting the subvertically foliated metagabbroic host rock and adjacent volcaniclastic units of upper greenschist facies of metamorphism. The pegmatite constitutes a bi-lobate, shallowly north-dipping and east- and west-plunging lenticular body which fingers out into swarms of subparallel dykes along most of its margins. The maximum dimensions of the intrusion are 1,440 × 820 × ~100 m.

Nine units have been distinguished in the pegmatite (Fig. 12). The fine-grained albite + quartz border zone (1) with accessory apatite and tourmaline, and the texturally complex K-feldspar + albite + quartz wall zone (2) with accessory beryl, columbite, muscovite and lithian muscovite form concentric envelopes that are commonly much thicker along footwall contacts. Lower and upper intermediate zones (4) and (5) share a gradational boundary, and can also be considered as a single shell with only a gradual upward increase in the K-feldspar/(albite + mica) ratio: the

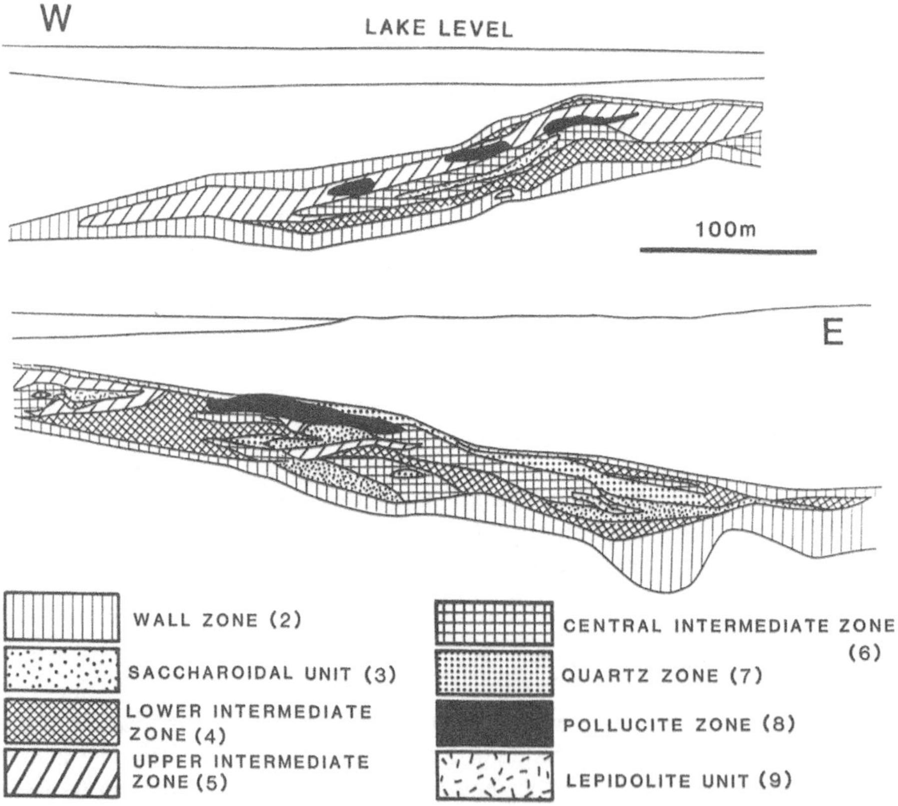

Fig. 12. Longitudinal section through the Tanco pegmatite, southeastern Manitoba (modified from Černý 1982c)

upper zone consists mainly of spodumene + quartz intergrowth after petalite and quartz, with K-feldspar and amblygonite, all of giant crystal size.

In contrast, the remaining units occur as more or less discontinuous layers. The central intermediate zone (6), composed of primary K-feldspar + quartz but with extensive muscovitization associated with minerals of Be,Sn and Ta, is centrally located. It is in part penetrated and extensively underlain by the saccharoidal albite zone (3) which also contains high concentrations of Be,Sn,Ta,Zr and Ti minerals (Fig. 6). The quartz zone (7) consists of numerous lenses scattered near the core and in the upper parts of the pegmatite, whereas the pollucite zone (8) forms nearly monomineralic lenses that structurally belong to the upper intermediate zone. The lepidolite unit (9) also constitutes several lenticular bodies in the upper central parts of the pegmatite, which were developed mainly at the expense of the K-feldspathic central intermediate zone; beryl, cassiterite and Ta-oxide minerals are typical accessory phases.

As shown in Fig. 12, most of the central crest of the pegmatite is occupied by the petalite-rich units (4) and (5). In both flanks, these two units are separated by the central intermediate zone (6) and overlain by quartz and pollucite bodies of units (7) and (9).

Units (2) and (4) to (8) can be ascribed to a primary zonal sequence. Units (1) and (3) are also primary, although extensively reactive along their contacts with other units. Unit (9) is largely metasomatic. The primary zones (2), (4) and (5) also contain internal metasomatic assemblages carrying mainly albite and micas.

Tantalum minerals are found in most units of the deposit but attain economic significance mainly in the saccharoidal albite (3), muscovitized parts of the central intermediate zone (6), and to a lesser degree in the lepidolite unit (9). The mineralogy of Ta is extremely diversified in Tanco: 14 different species, transitional through intermediate compositions in related series or groups, have been identified (ferrocolumbite, manganocolumbite, manganotantalite, ferrotapiolite, wodginite, titanowodginite, tantalowodginite, ixiolite, tantalian rutile, tantalian cassiterite, microlite, cesstibtantite, simpsonite, rankamaite and possibly stibiotantalite (Ercit 1986 and unpublished data). Ixiolite is represented by its stannian and titanian types; microlite also encompasses stannian, uranian, antimonian and plumbian varieties.

The Ta-ore minerals are generally fine-grained, ranging from subhedral and skeletal or poikilitic forms to interstitial anhedral fillings. Consequently, liberation and extraction of the tantalum oxide minerals has been one of the main difficulties in processing the Tanco ore.

Compositions of the Ta-bearing minerals show a generally decreasing Nb/Ta and increasing degree of structural order from the outer pegmatite units inwards; Ti and Sc contents decrease in this direction but the Fe/Mn ratio is locally erratic. Enrichment in Fe is noticeable in unit (3) and in the late alteration products replacing simpsonite (Fig. 13.) Subordinate Sn, Pb, U and Sb in microlite do not follow any simple fractionation trends but show definite preferences for specific pegmatite units. From zone (2) through (3), (4), (5,8) to (6), the mean Ta/(Ta + Nb) ratio (wt.) increases from 0.89 to 0.95 in microlite, from 0.85 to 0.90 in wodginite, and from 0.45 to 0.80 in columbite-tantalite.

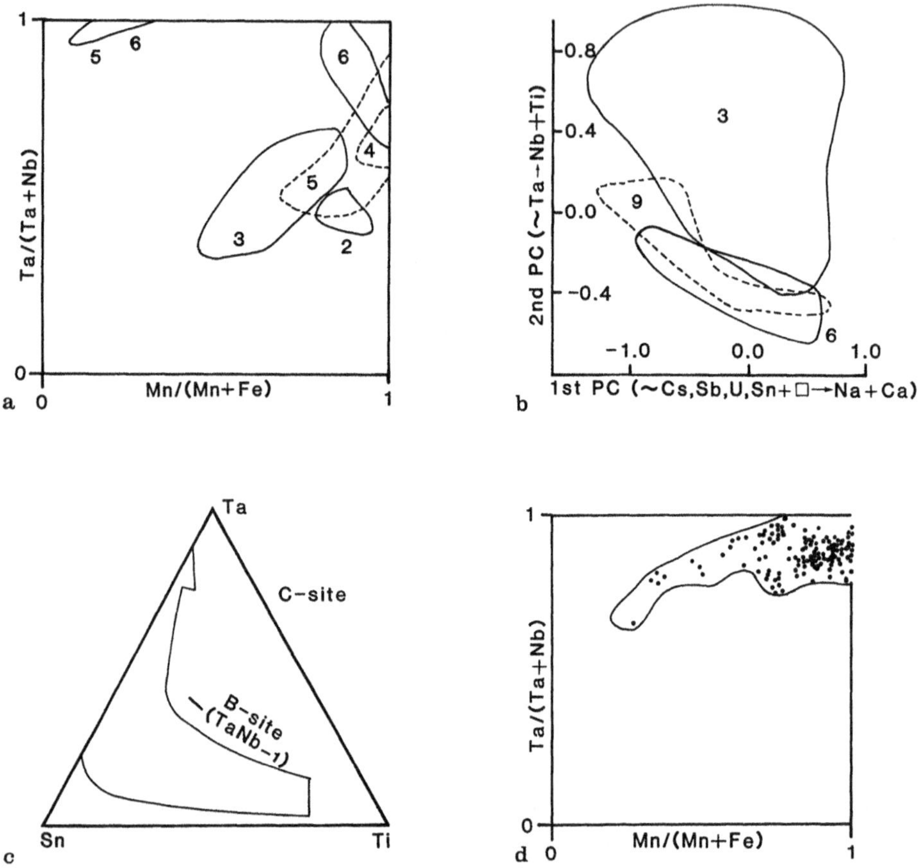

Fig. 13. Fractionation of Nb,Ta-oxide minerals in the Tanco pegmatite (atomic proportions): **a** columbite-tantalite; **b** microlite; **c** C-site of wodginite with concurrent B-site substitution; **d** Tanco wodginite population (*data points*) represented in the columbite quadrilateral (based on total Fe) in comparison with the general compositional field of the wodginite group (*outline*) (modified from Ercit 1986)

Tantalum concentrates were produced from the Tanco deposit since the late 1960s until 1983, and the production was resumed in 1988. The preproduction reserves amounted to 2,071,358 t of ore containing 0.216 wt.% Ta_2O_5.

5.2.4 Greenbushes Pegmatite, Western Australia

The Greenbushes pegmatite is located in the extreme southwest of western Australia, approximately 200 km south of Perth. It is situated in the Archean Yilgarn block of the Western Australian Shield, in amphibolite-facies, staurolite- and kyanite-bearing gneissic terrane of the Balingup metamorphic belt, within a north-trending shear zone. No obvious granitic parent outcrops in the close vicinity but pegmatitic granites are mapped within 6 to 8 km to the southwest. The pegmatite was dated at about 2,650 and 2,530 Ma by different methods. Geochemical and petrological investigation of the deposit is only in its initial stages; to date, the only comprehensive reports on the pegmatite in the accessible literature are those by Blockley (1980), Hatcher and Bolitho (1982) and Hatcher and Elliott (1986).

The north-northwest-striking pegmatite dips approximately 40° west-southwest, with considerable deviations in its extremities and in closely associated satellitic bodies (Fig. 14). The main intrusion is 3,300 m long and reaches up to 230 m in width; it extends at least 700 m downdip, and conceivably much further.

The pegmatite consists essentially of four major units (Fig. 15). The border unit (1), particularly well developed along the footwall contact, consists of albite + quartz + tourmaline and accessory garnet, apatite and muscovite. The megacrystalline K-feldspar unit (2) has K-feldspar and quartz as the main constituents, and tourmaline + spodumene in accessory amounts. The medium- to coarse-grained albitic unit (3) consists of albite, quartz and tourmaline, with accessory spodumene, apatite, beryl and the minerals of Sn, Ta and Nb; it represents the main Ta-bearing unit in the deposit. The medium-grained spodumene unit (4) is comprised mainly of spodumene + quartz, with feldspars and apatite as accessory constituents. The concentration of spodumene reaches 50 wt.% over substantial segments of this unit.

In general, the pegmatite lacks any classic zoning pattern but is crudely layered, and mostly sheared, banded and recrystallized into gneissic texture, particularly in the spodumene unit. Contact relationships with the host rocks and relict igneous textures within the deposit suggest that it was emplaced after the main regional dynamometamorphism. Deformation and recrystallization of the pegmatite followed after its consolidation, probably during a re-activation of the host shear zone.

The ore mineralogy seems to be rather simple: tantalite is the main phase, and stibiotantalite is abundant in a particular segment of the pegmatite. Cassiterite is tantalum-enriched as well. Wodginite, tapiolite, microlite and holtite were also identified but they probably constitute only a minor proportion of the concentrates. The high level of tantalum fractionation, indicated by the Nb_2O_5/Ta_2O_5 (wt.) ratios of tantalite (1.25 to 0.36) and stibiotantalite (0.20), is in agreement with the advanced fractionation of alkali metals (whole rock K/Rb from 15 to 4).

The Greenbushes pegmatite has been mined for tin since the 1890s but only from the surface alluvial deposits; intense weathering of the pegmatite to an average depth of 60 m permitted to apply the existing technology also to this "soft-rock pegmatite". Exploration of the unweathered pegmatite began in 1977 and led to the establishment of huge reserves in the northern third of the deposit: 13,500,000 t with 0.059 wt.% Ta_2O_5, 0.044 wt.% Nb_2O_5 and 0.15 wt.% SnO_2. In the spodumene units spatially separated from the tantalum mineralization, 5,900,000 t grading 4.0% Li_2O were defined within a 28,300,000 t volume of 2.8% Li_2O ore.

5.2.5 Other Examples

Complex-type pegmatites constitute only 2% of rare-element pegmatite fields (Ginsburg et al. 1979), and only a small proportion of these are enriched in tantalum to an economically interesting degree. Nevertheless, a fair number of representatives

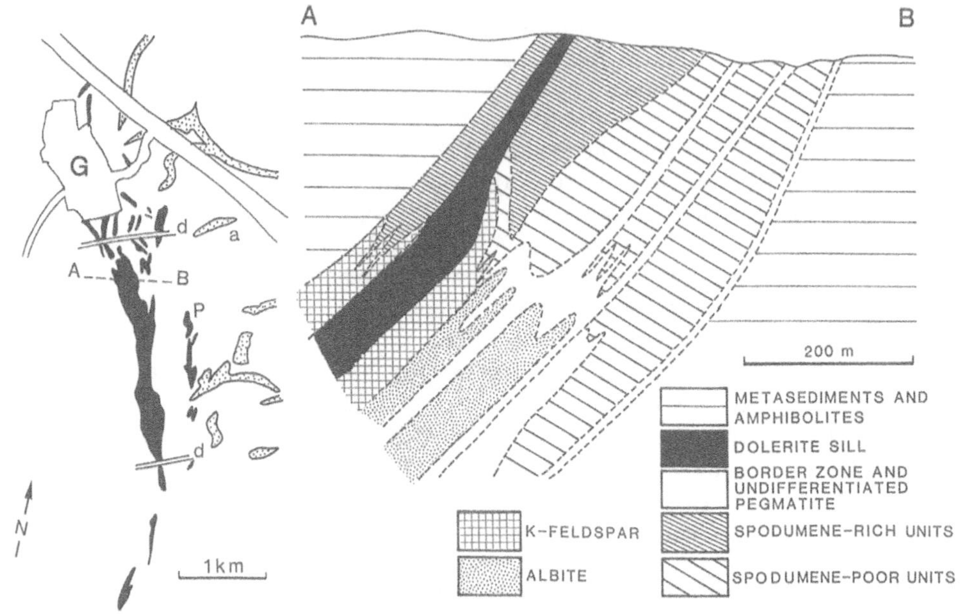

Fig. 14. *Left*: The Greenbushes pegmatite, its satellites to the east and the nearby alluvial tin + tantalum placers (*dotted*). *P* pegmatite; *G* Greenbushes townsite; *d* dolerite dykes. *Right*: Section through the Greenbushes pegmatite (*A-B line* on the map) (after Hatcher and Bolitho 1982 and Hatcher and Elliot 1986)

of the spodumene, petalite and amblygonite subtype are thoroughly documented in the literature and provide additional illustration of the discussed kind of deposits with tantalum potential. The Varuträsk pegmatite in Sweden (Quensel 1956) and the Bikita deposit in Zimbabwe (Cooper 1964) are in many respects similar to the previously described Tanco. Viitaniemi in Finland (Lahti 1981) and Peerless in South Dakota (Sheridan et al. 1957) are Ta-poor but well-characterized pegmatites of the amblygonite subtype. The Etta pegmatite of South Dakota (Norton et al. 1964) is a close relative to the Hugo deposit, whereas the Mongolian Altai #3 pegmatite (Beus 1960; Solodov 1962, 1971; Wang et al. 1981) and the McAllister deposit of Alabama (Foord and Cook 1989) are distinguished by specific features of their internal structure. The present level of knowledge of the gigantic Manono pegmatite system in Zaire is so far rather sketchy (Thoreau 1950).

5.3 Complex Type – Lepidolite Subtype (iii)

As mentioned in Section 5.2, the lepidolite subtype is characterized by high activity of F (but low P) that destabilizes the lithium aluminosilicates. Topaz, which is not uncommonly associated with lepidolite, is generated at increased levels of acidity (London 1982).

The lepidolite pegmatites are generally located at the outskirts of zoned pegmatite groups, and commonly at the extreme margins of the pegmatite population, as the F- (and routinely also B-)enriched melts from which they crystallize have low viscosity and are relatively highly mobile.

High fluidity may also be a reason for the relatively small average size of lepidolite pegmatites, as their melt readily penetrates a multiplicity of available structural sites, thus becoming dispersed into swarms of low-volume intrusions. Thin lenticular and tabular bodies are the norm, although somewhat bulbous shapes also are encountered.

Internal structure is variable, and to a degree dependent on the overall abundance of Li. In the presence of relatively restricted quantities of Li, lepidolite-rich and highly evolved mineralized pods develop only in the central parts of pegmatite dykes, flanked by extremely steep fractionation gradients in the surrounding zones of considerable modal simplicity (Stanek 1965; Foord 1976; Goad and Černý 1981). In subvertically dipping pegmatites, the lepidolite units may be central in horizontal sections but shifted upwards in the vertical direction (Beus 1948; Jahns 1953; Jolliff et al. 1986). In lithium-rich pegmatites, the internal structure may be layered and rather irregular (cf. Sect. 5.3.1). Also, the lithium- (and boron-)rich pegmatites of the lepidolite subtype are considerably sodic in their bulk composition, with albite vastly predominant over K-felsdpar (in accordance with experimental data summarized by Pichavant et al. 1987). In contrast, the modes of lithium-poor dykes are close to granitic.

The tantalum mineralization is usually associated with fine-grained lepidolite ± albite units, and it is predominantly represented by extremely tantalum-rich microlite. However, cassiterite may be the only primary tantalum-bearing mineral in some districts. Of the ferromanganoan minerals, subordinate to minor manganocolumbite is typical, preferentially incorporating Nb. Stibiotantalite or bismutotantalite are also characteristic but usually in only minor to trace amounts.

The overall fractionation level of lepidolite-subtype pegmatites is usually high to extreme, in terms of K/Rb, K/Cs, Fe/Mn, Zr/Hf and other indicators. The ore grades are highly variable, although locally reaching respectable values, and the ore mineral concentrates have a very low Nb/Ta ratio. These features are, however, offset by the commonly small size of the deposits, and by the difficulties encountered in extracting the highly friable microlite which tends to slime.

5.3.1 Brown Derby No. 1 Pegmatite, Colorado

This deposit is located 10 km northeast of Parlin, Gunnison County, Colorado, about 200 km southwest of Denver. It belongs to the most complex, fractionated and mineralized pegmatites of the Quartz Creek district, the age of which is close to 1,300 Ma. The deposit is situated in the outermost fringe of the district which is in turn centred on a plug-shaped intrusion of the Black Wonder pegmatitic granite. The pegmatite has been studied extensively (Hanley et al. 1950; Staatz and Trites 1955; Heinrich 1967; Rosenberg 1972).

The No. 1 pegmatite is the uppermost and easternmost of a group of three subparallel Brown Derby pegmatites that strike northeast and dip 20°–35° southeast. The three Brown Derby dykes branch both along the strike and down dip, and they may be interconnected. They penetrate a sequence of interlayered hornblende and biotite schists. The No. 1 pegmatite has a strike length of about 320 m and ranges from 1 to 12 m in thickness. Local changes in thickness are common and abrupt, usually associated with convolute changes in strike and dip.

The internal structure of No. 1 is strongly asymmetric. Gradual changes also appear in individual units along the strike. The pegmatite seems to be layered rather than concentrically zoned (Fig. 15). Heinrich (1967) distinguishes two principal types of internal units: (1) those consisting mainly of quartz, K-feldspar, non-cleavelanditic albite, biotite and muscovite, with accessory schorl or dark green tourmaline, blue beryl and columbite, that are considered to be predominantly of primary origin, and (2) those composed chiefly of quartz, cleavelandite, several types of lithium micas and multicoloured tourmalines – elbaites, with accessory topaz, microlite, stibiotantalite, rose muscovite, fluorite and pink beryl; these are considered to be largely of replacement origin. The cleavelandite + lepidolite units are crudely conformable in strike and dip with the overall attitude of this pegmatite but detailed examination reveals cross-cutting features. The above relationships are further complicated by the local presence of an interior quartz-rich unit with blocky topaz, K-feldspar (Rosenberg 1972) and pollucite (Černý 1989). Textural relationships indicate primary crystallization of all the above components which resemble a blocky core typical of zoned pegmatites. However, sections such as that shown in Fig. 15 are difficult to interpret and await re-investigation in the light of recent experimental results on Li-,B- and F-bearing pegmatite systems.

As indicated by the presence of pollucite, the Brown Derby No. 1 pegmatite is highly fractionated. The K/Rb ratio of its blocky K-feldspars ranges from 30 to 7, and decreases to 14–4 in its micas. The composition of subordinate mangan-

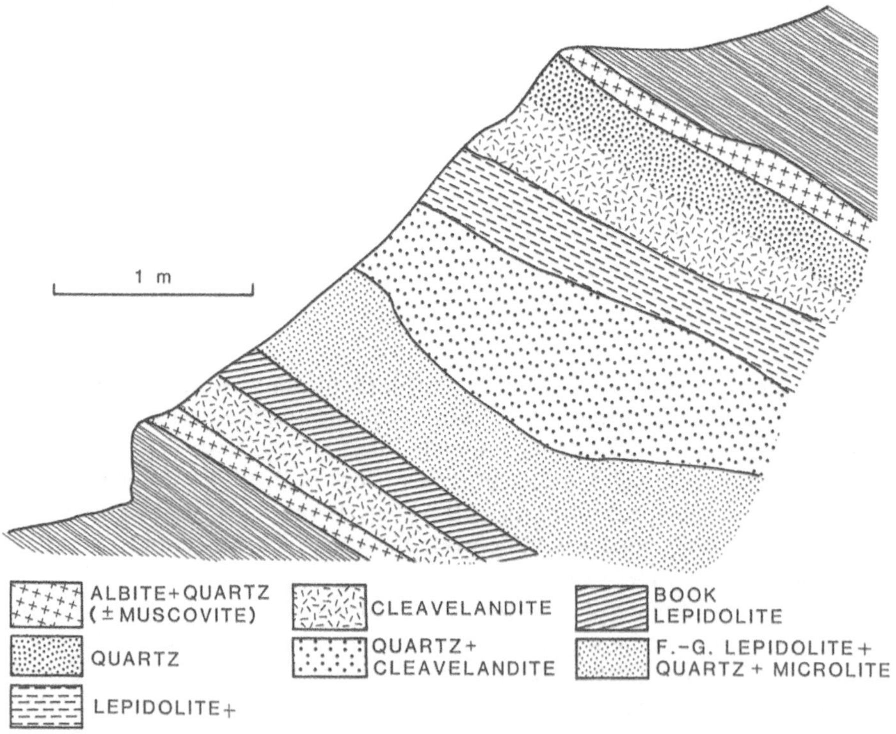

Fig. 15. Section through the Brown Derby No. 1 pegmatite, Colorado (modified from Heinrich 1967)

ocolumbite (Nb/Ta = 3.3), the presence of stibiotantalite, traces of samarskite and the overall predominance of Ta-rich (and in part U-bearing) microlite all indicate a high level of Nb-Ta fractionation, to an overall value of much less than 1.0.

Microlite, beryl and lepidolite were produced from the Brown Derby No. 1 pegmatite in the early 1940s. No exact production figures are available, but the concentrates did not substantially exceed 300 t of lepidolite and several tonnes each of beryl and microlite.

5.3.2 Other Examples

The lepidolite subtype (iii) of rare-element pegmatites is not generally widespread, but it tends to be fairly abundant in specific pegmatite fields. Lepidolite pegmatites with economically significant tantalum mineralization are scarce but several localities can be quoted that would cover a broader spectrum of structural and paragenetic varieties. The Phangnga pegmatite field in Thailand (e.g. Garson et al. 1969) and the Wodgina pegmatites of Australia (Blockley 1980) are among the tin-rich, lepidolite-subtype deposits of tantalum. The Mesa Grande and Pala districts of southern California (Foord 1976; Stern et al. 1986; Shigley et al. 1986) and the lepidolite-rich pegmatites of Borshtchovochnyi Kryazh (Tatarinov 1974) are transitional into miarolitic dykes famous for their gem stock rather than rare-element mineralization. The Pidlite dyke in New Mexico (Jahns 1953) and the Bob Ingersoll No. 1 of the southern Black Hills, South Dakota (Jolliff et al. 1986, 1987) represent examples of steeply dipping bodies with fairly concentric internal structure but subordinate tantalum mineralization. The Khukh-Del-Ula field of topaz-bearing lepidolite pegmatites in Mongolia may be one of the largest in the world, and some of its aspects are reported in excellent detail; however, some of its geological features and the extent of tantalum mineralization are difficult to fathom (Vladykin et al. 1974).

5.4 Albite-Spodumene Type (iv)

The albite-spodumene type of rare-element pegmatites differs considerably from pegmatites of the preceding categories. Individual bodies are invariably tabular, up to 3 km long, with extreme length/width ratios (50–100) and remarkable down-dip extent (up to 1 km). They constitute pegmatite populations of their own. If associated with complex pegmatites, they are in a very pronounced minority and usually farther away from the parent granite than any other type. They commonly occur in subparallel swarms considerably elongate parallel to the pegmatite attitude. The swarms tend to approximate the course of intrusive contacts of parent granitic plutons or follow prominent regional joint or fault systems.

Internal zoning of individual pegmatites is simple, and in many cases very inconspicuous. Well-developed zoning shows a thin and fine-grained quartz + albite wall zone, followed by the dominant zone of medium-grained albite + quartz (± mica) aggregate with subparallel spodumene and (subordinate) K-feldspar megacrysts; a central zone of blocky K-feldspar (± quartz) is subordinate or may be completely missing. Asymmetric zoning may also be developed, consisting of units similar to those quoted above (Solodov 1962, 1971).

Besides the largest lithium concentrations known in any pegmatite type, the albite-spodumene pegmatites may carry Be,Sn and/or Nb-Ta mineralization. Whereas beryl is usually associated with the central parts of dykes which are enriched in K-feldspar, cassiterite and columbite-tantalite (rarely microlite or fergusonite) are dispersed in the albite + quartz (± mica) matrix of the dominant spodumene-rich zone. In pegmatites with symmetric internal structure, units of saccharoidal albite and muscovite-rich assemblages that in part replace K-feldspar-bearing zones may constitute the principal Sn- and Nb-Ta-bearing ores. The overall degree of fractionation is considerably variable; K/Rb in blocky K-feldspar shows different ranges in unrelated pegmatite populations (e.g. 35–15 at Kings Mountain, North Carolina, Černý 1989; 9–5.5 in the Gods River pegmatite, Chackowsky 1987 and 160–44 in the Violet-Thompson dyke, Manitoba, Černý et al. 1981; 59 in East Siberia, Kuzmenko 1976; 47–26 in the Georgia Lake field of Ontario, Zayachkivsky 1985).

As shown in Table 1, ore grades of the albite-spodumene deposits are the lowest of all tantalum-bearing pegmatites. Also, the Nb/Ta ratio of the concentrates is not particularly attractive. However, the relatively uniform distribution of tantalum-bearing minerals and the considerable volume of some pegmatites make this type an attractive exploration target.

5.4.1 Kings Mountain Pegmatites, North Carolina

These deposits are located in the vicinity of Kings Mountain, Cleveland County, North Carolina. They belong to the extensive "tin-spodumene belt" which is confined to the Hercynian Kings Mountain metamorphic belt of the Piedmont, within the gigantic pegmatite province of the Appalachians. The Kings Mountain segment represents one of the narrowest portions of the tin-spodumene belt, but it contains the majority of its resources. The pegmatites, best exposed in an open-pit operation SSW of Kings Mountain, have been subject to extensive geological, mineralogical and resource studies (Kesler 1961, 1976; Simpson 1977; Luster 1977; Kunasz 1982). However, geochemical and petrological research has been so far limited.

The Cherryville quartz monzonite adjacent to the northwest side of the pegmatite belt is considered to be parental to the pegmatites because lithium-free to lithium-poor pegmatites grade from its interior and margins towards spodumene-rich deposits. Rb-Sr ages of the quartz monzonite and the pegmatites overlap at about 345 ± 15 Ma.

The pegmatites are located predominantly in amphibolites, within a sequence of micaschists and layered amphibolites of the intermediate amphibolite facies of metamorphism. They range from small stringers to tabular pinching and swelling bodies as much as 90 m thick and almost 1 km long (Fig. 16), and they extend at least 200 m downdip without a perceptible change in their overall characteristics. The pegmatites are oriented mainly parallel to the northeast-trending strike and steep northwesterly dip of the host-rock foliation. However, pegmatite segments of highly

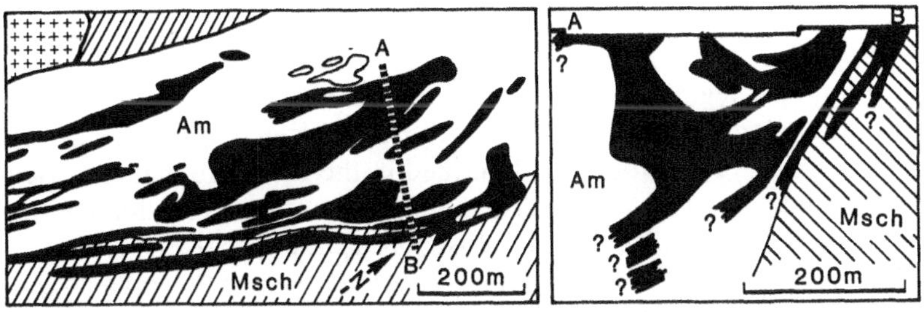

Fig. 16. Schematic map and vertical section of the lithium pegmatite deposits at Kings Mountain, North Carolina: *Am* amphibolite; *Msch* micaschists; *crosses* Cherryville monzogranite; *open outline* barren pegmatites; *black* albite-spodumene pegmatites (modified from Kesler 1961)

variable dip are typical of the amphibolite host with irregular foliation and fracturing.

The internal structure and primary mineralogy of the Kings Mountain pegmatites are very simple. Zoning is moderate, poorly expressed and restricted mainly to the pegmatite margins. The thin border zone of albite + quartz usually contains minor Fe,Mn-phosphates. Spodumene and K-feldspar are present in the layered intermediate zone which consists of a layered sodic aplitic unit and intercalated layers and pods of potassic pegmatite. The core zone accounts for more than 90% of the major pegmatites: it consists of columnar phenocrysts of K-feldspar and green spodumene (up to 100 cm in length but averaging at 30 cm) imbedded with subordinate muscovite in a fine- to medium-grained matrix of albite and quartz. Both spodumene and K-feldspar are corroded and veined by the matrix.

The above primary mineral assemblages show faint to extensive deformation. Structural interpretations are far from uniform but all authors agree that the pegmatites were intruded before the conclusion of four episodes of regional folding. The pegmatites cut across the metamorphic fabric of their hosts. However, the ubiquitous subparallel orientation of spodumene and K-feldspar phenocrysts, parallel to wall-rock foliation and to axial planes of major folds, indicates pegmatite crystallization (or recrystallization) in a stress field related in orientation to that responsible for the regional fabric.

The limited data on the concentrations of trace elements in rock-forming minerals indicate an advanced level of fractionation. K/Rb is variable from 35 to 15 in the phenocrystic K-feldspar, and the columbite-tantalite has Nb/Ta (wt.) between 8 and 0.6.

Despite the relative enrichment of lithium in the core zone and the increased Fe, Mn and P contents along the margins, the entire width of the pegmatites is mined and processed as a lithium ore. Current reserve estimates on the Foote Minerals Co. property alone (mined since the 1940s) amount to 26,000,000 metric tonnes grading 0.7% Li. No data are available for the contents and reserves of cassiterite and columbite-tantalite, but they may become economically exploitable in the future (Kunasz 1982).

5.4.2 Other Examples

The Kings Mountain pegmatites are somewhat exceptional because of their structural setting and deformation. Many other (albeit much less explored) localities of the albite-spodumene pegmatites are described in the literature that are essentially post-tectonic intrusions. However, all of them display oriented spodumene and K-feldspar suggestive of crystallization in a tensional regime. The reader is referred to Sundelius (1963) on the Peg Claims in Maine, Heinrich (1964) on the Volta Grande deposits in Brazil, Mulligan (1965) on the Preissac-Lacorne district in Quebec and Pye (1965) on the Georgia Lake field in northwestern Ontario. Individual pegmatites characterized in the literature are the Violet-Thompson dyke (Černý et al. 1981) and the Gods River pegmatite (Chackowsky 1987) in Manitoba, and several localities in eastern Siberia, central Asia and Transbaikalia (Kuzmenko 1976). The cassiterite- and spodumene-bearing pegmatites of the Kamativi district, Zimbabwe also seem to belong to this category, although they show a more complex internal structure (Rijks and v.d. Veen 1972; A.J. Anderson, pers. commun. 1987).

5.5 **Albite Type** (v)

Occurrences of the albite pegmatites (v) are relatively uncommon; this is clearly the least widespread of all pegmatite types of the rare-element class. Albite pegmatites are found as singular dykes within populations of the complex type (ii), or rarely constituting groups of their own. In the first case, they tend to be located farther away from granitic sources than the complex pegmatites.

Most albite-type pegmatites form small intrusions a few hundred metres long and less than 5 m thick. Similar to the albite-spodumene type (iv), the length/width ratio is high, spanning an even broader range of 30 to 120. The attitude is widely variable from subvertical to subhorizontal. Internal zoning varies from well-developed to non-existent; the latter cases are typical mainly of albitic bodies sensu stricto, in the virtual absence of K-feldspar and spodumene.

Zoned dykes of albite pegmatites commonly have fine-grained albite + quartz border units, followed by a more or less monomineralic albite wall zone. Cleavelandite is typical of the hanging-wall segment and saccharoidal albite of the footwall portion. Subordinate quartz lenses and blocky K-feldspar are centrally located, occasionally with layers of fine- to medium-grained muscovite. Pegmatites with minor to subordinate spodumene, classified by Solodov (1971) as a separate subtype, have this mineral also restricted to the interior units. Dominantly albitic pegmatites may contain only scattered pods of K-feldspar + quartz, muscovite and rarely spodumene + albite + quartz in their central parts.

The mineralogy of tantalum is relatively simple: columbite-tantalite, niobian to tantalian rutile, rarely microlite or euxenite. Tantalum-bearing cassiterite may locally attain considerable significance. Tantalum mineralization is dispersed throughout the albitic zones and in the muscovite unit; in zoned intrusions, Nb/Ta decreases from the wall-rock contacts inwards. Besides some phosphates and sulphides, the main accessory mineral in the albite pegmatites is beryl which may be considerably enriched in this pegmatite type (Beus 1960; Solodov 1971). However, the beryl tends to occur in K-feldspar-enriched units, spatially separated from the tantalum mineralization.

The relatively high ore grades (Table 1) are offset by mediocre Nb/Ta ratios that may be considerably variable within a single pegmatite (2–0.2). The thin sheetlike shape of the pegmatites is not amenable to easy underground mining. Secondary placer deposits generated by weathering of albite-type pegmatites are therefore much more attractive for industrial development.

5.5.1 Hengshan Pegmatite Field, Guangdong, P.R.C.

This pegmatite field is located about 100 km northwest of the city of Guangzhou (Canton), just north of the district town centre of Zhaoquing. The host terrane consists of a thick metasedimentary sequence of schists, phyllites and quartzites of Simian age that underwent Caledonian regional metamorphism (andalusite grade) and suffered thermal effects from plutonic granodiorite and quartz-monzonite intrusions (~100 Ma). Several periods of folding and subsequent faulting generated a complex regional structure prior to the emplacement of pegmatites, which are themselves locally offset by late faults. No data seem to have been published to date; this section is based on the information and data provided by Mr. Xie Yanbao (Geological Survey of Guangdong Province) in 1983, and on the field observations by the present author.

Four pegmatite types are encountered in the field: barren biotite-muscovite-K-feldspar pegmatites mainly in migmatitic gneiss south of the andalusite-grade terrane, muscovite-K-feldspar and muscovite-K-feldspar-albite pegmatites with poor Nb, Ta mineralization (San Shen Keng district with pegmatites marginal to a granodiorite pluton), and (muscovite-) albite pegmatites emplaced into the andalusite-bearing metasediments at moderate to considerable distances from any outcropping granitoid intrusions (Bi Long, Chun Xing and Da Wei Keng districts). So far, genetic links of the pegmatites to potential granitic parents are obscure.

The last pegmatite type mentioned above forms predominantly shallow-dipping, locally highly contorted and sheetlike bodies ranging to 1120 × 450 m in extent but largely much less than 10 m thick (Fig. 17). The pegmatites are relatively homogeneous, consisting predominantly of saccharoidal albite + quartz, with subordinate to minor K-feldspar, and minor muscovite in local schlieren and pods. Spodumene and amblygonite are present only in accessory quantities, as are beryl, zircon, xenotime, monazite and uraninite. Minerals of niobium and tantalum are represented by Fe,Nb-rich manganotantalite, niobian rutile and microlite. They are relatively uniformly distributed throughout the saccharoidal albite and muscovitic units, and the entire thickness of the pegmatite dykes is commonly mined as an orebody.

The pegmatites have been explored with increasing intensity since their discovery in 1972, and have been exploited during the past 10 years. Despite their marginal value on the national scale of tantalum production, they are an important component of the local economy. The difficult and labor-intensive mining methods are balanced by the high relative ore grades that range from 0.030 to 0.055 wt.% Ta_2O_5, and from 0.013 to 0.019 wt.% Nb_2O_5.

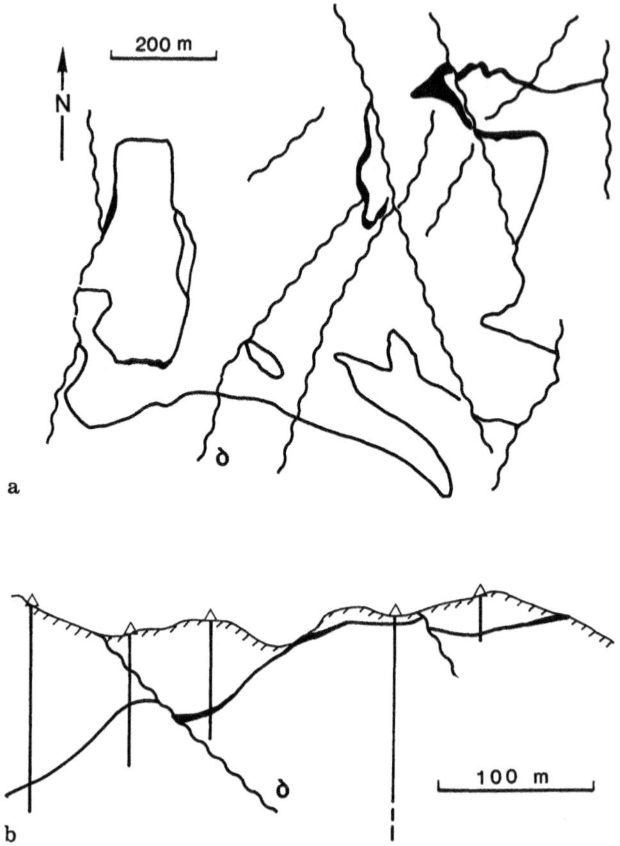

Fig. 17. Schematic map (**a**) and vertical section (**b**) of the albite pegmatites of the Hengshan field, P.R.C.: contourlike outcrops of three subhorizontal pegmatites in a hilly terrain (after Xie Yanbao, pers. commun., 1983)

5.5.2 Other Examples

The number of thoroughly examined albite pegmatites (v) in the literature is rather low, in part because of the relative scarcity of this type. Solodov (1962, 1971) published maps and sections of several localities and general characteristics of the type. Černý et al. (1981) and Chackowsky (1987) described two singular occurrences of cassiterite-rich albite pegmatites from two pegmatite fields of Manitoba. Kuzmenko (1976) gave brief accounts of this pegmatite type from Kazakhstan, Kola Peninsula, Ural Mts., Transbaikalia and from a pegmatite field in the "Far East of the USSR", populated only by albite- and lepidolite-type pegmatites.

6 Concluding Remarks

This final section concerns some general problems of mineralized rare-element pegmatites related to their tantalum potential. Petrogenetic aspects are significant in this connection, as is an assessment of the role of granitic pegmatites in future tantalum production.

6.1 Petrogenesis of Rare-Element Pegmatites

In principle, three lines of genetic interpretation can be distinguished in the study of rare-element granitic pegmatites: igneous differentiation from granitic parents, diverse variants of direct metamorphic-anatectic origin and derivation from mantle sources.

The anatectic hypotheses are based on largely arbitrary assumptions, unrealistic geochemical processes and arguments which are not compatible with many fundamental features of the regional distribution, intrusive style and composition of pegmatites and their parent melts (Zasedatelev 1974, 1977; Norton 1973, 1981; Stewart 1978; Gaupp et al. 1984). Rejection of the theory of igneous differentiation is commonly based on misconceptions of the composition of the original granite magma in relation to the chemistry of consolidated granitic rocks, and on the underestimation of the efficacy of diverse fractionation mechanisms operating in granitic melts (cf. Černý 1982b, 1989 for discussion of some of the above papers).

Derivation of pegmatites from subcrustal sources applies to some peralkaline types and carbonatite-related rocks (Ginsburg et al. 1978; Zou et al. 1985). A mantle component may be involved in the rare-element assemblage of some pegmatites related to bimodal anorogenic intrusive suites. However, the close relationship between orogen-related fertile leucogranites and their aureoles of LILE-enriched rare-element pegmatites excludes any possibility of material contribution from the mantle, proposed by, e.g. Shmakin (1983).

In contrast to the above, the classic concept of generating rare-element granitic pegmatites by differentiation of granitic magmas is supported by field relationships, bulk petrochemical links and diverse lines of geochemical evidence:

1. Examples of rare-element pegmatites forming pods within parent granites; transecting margins of parent pluton into its metamorphic envelope; constituting cupolas of plutonic intrusions; and forming zoned aureoles of granitic plutons, progressively differentiated from the interior through the marginal into the extreme exterior dykes, are too numerous to be quoted here (for a selection see Černý 1982b, Černý and Meintzer 1988).

2. Rare-element pegmatites have essentially granitic compositions similar to those of the associated leucogranites. The bulk compositions closely correspond to low-temperature minima of the experimental systems $Ab+An+Or+Qz+H_2O$ (Orville 1960; James and Hamilton 1969), $Ab+Or+Qz+F+H_2O$ (Manning and Henderson 1981), $Ab+Qz+Ecr$(eucryptite)$+H_2O$ (Stewart 1978) and $Ab+Or+Qz+B_2O_3+H_2O$ (Manning and Pichavant 1988). Progressive fractionation of leucogranitic melt into compositions close to highly evolved rare-element pegmatites was recently documented on the example of the Macusani glass and enclosing tuff (Pichavant et al. 1987).

3. Progressive evolution of fertile granites into rare-element pegmatites is documented by numerous studies of trace-element fractionation (e.g. Goad and Černý 1981; Černý et al. 1985; Shearer et al. 1987). Geothermometry, geobarometry, isotope and fluid inclusion studies also show pegmatite crystallization as an extension of plutonic granite consolidation to lower temperatures and pressures, in the presence of increased concentration of fluxing components and volatiles (e.g. London 1986b; Shearer et al. 1987). These agents, such as Li,(Rb,Cs),Be,B,P,F and H_2O, depress the liquidus and solidus of the pegmatite melt and reduce its viscosity (cf. London 1987 and Pichavant et al. 1987 for reviews). These two aspects are the principal reasons for regional zoning of cogenetic pegmatite groups: the most fluid and thermally stable melts can migrate farthest away from plutonic sources.

Some of the above facets of pegmatite genesis are discussed in more detail in Černý (1982a,b, 1989). Černý et al. (1985) and Černý and Meintzer (1988).

6.2 Origin of Tantalum-Bearing Internal Pegmatite Units

After more than a century of research, the nature of the parent medium which generates tantalum mineralization, its provenance and timing in the pegmatite evolution are still much debated and, to a degree, controversial.

6.2.1 Internal vs External Derivation

Essentially two different interpretations have evolved for the albitic and micaceous assemblages hosting the bulk of tantalum mineralization in rare-element pegmatites. They are considered as either inherent components of the pegmatite consolidation in a more or less closed system (e.g. Solodov 1971; Ginsburg et al. 1979; Černý 1982a) or products of a late "*Nachschub*" of fluids from an external source, granitic or metamorphic (Nikitin 1957; Beus 1960; Norton 1981; Gaupp et al. 1984; Möller this Vol.).

Three principal arguments can be quoted in favour of the internal generation of albitic and micaceous units:

1. Pegmatite bulk compositions correspond to the minima in experimental granitic melts, including shifts in the Ab/Or ratio and in the quartz content caused by the presence of Li,F and/or B (cf. Pichavant et al. 1987 for summary of experimental data); they show no compositional deviations caused by albitization or greisenization from external sources, subsequent to the primary pegmatite consolidation.

2. Furthermore, fractionation of trace elements typically carried by the albite- and mica-rich units (K/Rb,Ti/Sn,Fe/Mn,Nb/Ta,Zr/Hf,Al/Ga) is commensurate with fractionation levels attained by the primary assemblages (K/Rb,K/Cs,Rb/Tl, Al/Ga), indicating evolution of both from a common parent melt; discrepancies should be encountered in case of superimposed, and relatively to totally heterogeneous processes.

3. Finally, albitizing or greisen-generating fluids migrating into the pegmatites from outside would react with any rock of granitic mineralogy and chemistry,

depositing albite, muscovite and related rare-element minerals outside the pegmatites. Such phenomena are not known in metamorphic country rocks hosting exterior rare-element pegmatites; the few small-scale occurrences noted by Nikitin (1957) are the result of pegmatite exomorphism rather than a reverse process.

The reader is referred to papers by London (1986b, 1987) for recent experimental results on the magmatic to supercritical transition in a closed (to exomorphic) pegmatite system.

6.2.2 Primary Crystallization vs Metasomatism

Concerning the nature of the albitic and micaceous units in relation to other pegmatite zones, two different concepts are again available. Some investigators advocate a metasomatic origin, based mainly on textural evidence (e.g. Schaller 1925; Landes 1933; Zalashkova 1957; Beus 1960; Solodov 1962, 1971; Stewart 1978; Crouse et al. 1979; Černý 1982a,c; Foord and Cook 1988). Other authors consider late but primary crystallization a more viable proposition (Melentyev and Delitsyn 1969; Norton 1983).

There is no doubt that margins of albite-rich units (and those of saccharoidal albite in particular) show vigorous reaction with pre-existing mineral assemblages, especially with K-feldspar; metasomatic origin of beryl-bearing albite aggregates after K-feldspar in fluorine-bearing fluids was experimentally documented by Beus et al. (1963). However, albitic units also show primary growth features indicative of crystallization from a solid substrate into fluid-filled space (e.g. Norton et al. 1964; Uebel 1977; Lenton 1979; Černý et al. 1981; Thomas and Spooner 1988). Fluid-inclusion studies and experimental work of London (1986b, 1987) show that in the presence of lithium and boron, an $Li_2B_4O_7$-fluxed silicate liquid rich in albite can evolve during magmatic crystallization of K-enriched assemblages, and generate albite-rich units upon loss of boron (crystallization of tourmaline) and concomitant release of aqueous fluid. If such a role of boron could also be extended to Li-poor pegmatites, a mechanism universally applicable to the generation of diverse albitic units in mineralized rare-element pegmatites would become available (cf. London 1986b, 1987 and Černý 1989 for further comments).

As for the mica-rich units, their origin is so far judged mainly on textural evidence because experimental data are ambiguous. In hydrous K, Na-aluminosilicate systems, muscovite ± albite would form instead of K-feldspar at reduced K-salinity, disregarding the presence or absence of a pre-existing K-feldspar phase (Hemley and Jones 1964; Burt 1981). Similarly, lepidolite ± albite may form instead of K-feldspar because of moderate changes in $\mu LiF/\mu KF$ in an otherwise identical system (London 1982). Under univariant conditions of assemblage boundaries, K-feldspar + muscovite or K-feldspar + lepidolite may form in equilibrium. Furthermore, lepidolite is found to be stable in silicate melts (Glyuk et al. 1980) as well as under subsolidus conditions (Munoz 1971), similar to muscovite relationships.

Thus, it is not surprising that some authors advocate a replacement origin for muscovite and lepidolite units (Norton et al. 1962; Stewart 1978; Černý 1982c; Thomas and Spooner 1988) but others propose primary crystallization from immiscible liquid (Melentyev and Delitsyn 1969), or from residual "magmalike phase", although acknowledging the existence of metasomatic lepidolite-bearing assemblages (Norton 1983).

Coprecipitation of incompatible lithophile elements such as Nb and Ta with albite, muscovite or lepidolite in the pegmatite units discussed above is in agreement with experimental data. Boron enhances the solubility of these elements in Na-aluminosilicate melts, but crystallization of tourmaline would reduce it (London 1986b). Acidic F-bearing fluids are capable of transporting complexed Nb and Ta but these elements are relatively insoluble in the saline KF-rich solutions required to crystallize lithian micas (London 1982).

6.3 Geochemical Features and Evolution of the Tantalum Ore Minerals

As shown in Table 4 and throughout the text, rare-element pegmatites commonly display diverse assemblages of tantalum-bearing minerals that may be specific for individual types or subtypes. Moreover, members of compositionally variable mineral groups may evolve within pegmatite groups, individual pegmatite bodies or single crystals; these trends tend to follow fractionation paths which also seem to be specific for at least some pegmatite types and subtypes.

Such fractionation trends are at present best known for columbite-tantalite, complemented by minerals of the tapiolite series and the ixiolite and wodginite groups, in terms of Mn/(Mn + Fe) and Ta/(Ta + Nb) of the columbite quadrilateral. Figure 18 shows a preliminary attempt at a generalized summary of our present

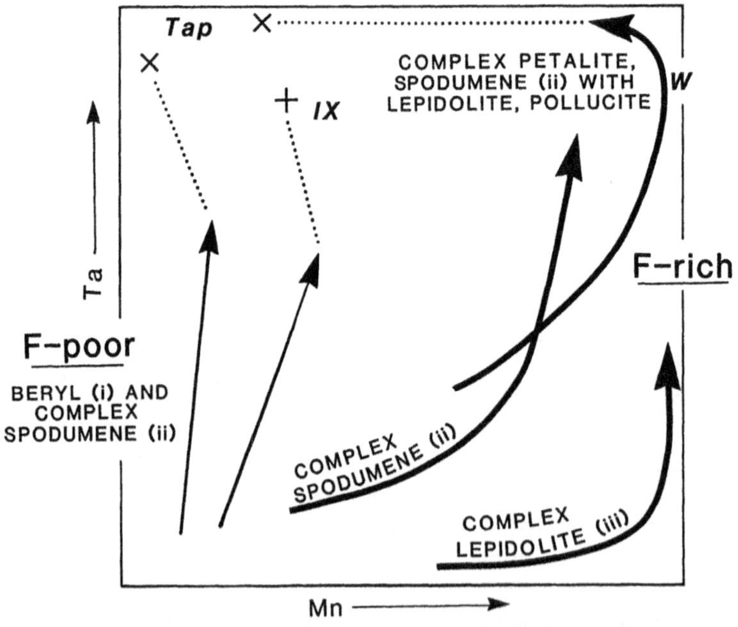

Fig. 18. General fractionation trends of columbite-tantalite and related phases in the columbite quadrilateral (atomic ratios). *Tap* tapilite; *IX* ixiolite; *W* wodginite. Mineral assemblages and compositions are distinctly influenced by the activity of fluorine (based on data of Černý and Ercit 1985 and unpublished)

knowledge, based on the data collected by Černý and Ercit (1985) and Černý et al. (1985, 1986; see these three references for additional sources of data).

As illustrated in Fig. 18, activity of fluorine is definitely an important factor in Nb-Ta fractionation, although not necessarily the only one. For example, the complex (ii) spodumene-subtype pegmatites of the Yellowknife field evolved at extremely low fugacities of F_2 (86% of primary montebrasite with < 3 wt.% F_2, scarcity of microlite and total absence of lepidolite). Nevertheless, Nb-Ta fractionation progressed in them from ferrocolumbite to niobian ferrotantalite, ixiolite and ferrotapiolite (Wise et al. 1986, 1988). In contrast, increased activity of F_2 seems to be a conditio sine qua non for effective Fe-Mn fractionation, demonstrated particularly by the lepidolite subtype (iii) pegmatites. Fugacity of H_2 should not be of particular importance here, as it is in non-pegmatitic systems (Boctor 1985). The values of Fe^{2+}/Fe^{3+} in columbite-tantalite and associated minerals (except for wodginites and ixiolites) are extremely high, and precipitation of iron due to its oxidation to the ferric state is not a generally significant factor in consolidation of granitic pegmatites.

Figure 18 incorporates some data on ixiolite and wodginite but only as a complement to columbite-tantalite and tapiolite fractionation. To date, the data base is too narrow for meaningful generalizations of evolution trends in these mineral groups that also involve variations in the Sn, Ti, W, Sc and Fe^{3+} contents. Figure 13 shows an example of wodginite behaviour at a locality where it is abundant and widespread (Ercit 1986). Similarly, microlite has been examined in sufficient detail at only two deposits (Ercit 1986; Lumpkin et al. 1986), and generalizations may be difficult even when a broader spectrum of data becomes available: many cations such as Sb, Bi, Cs or Pb may prove to be locality-dependent rather than widespread and persistent participants of general fractionation trends (Fig. 13).

6.4 Economic Considerations

From the viewpoint of international tantalum markets, only those pegmatite deposits can play a significant role that may produce ore concentrates competitively with the economics of tantalum-bearing cassiterite deposits and tantalum-mineralized granitic intrusions. Two categories of rare-element granitic pegmatites have this potential: the complex pegmatites (ii) of the spodumene and petalite subtypes that occasionally host respectable volumes of high-grade tantalum mineralization, and the albite-spodumene type (iv) which commonly constitutes large (and occasionally gigantic) intrusions with a low but persistent grade of tantalum mineralization throughout.

The above statement does not mean that the other pegmatite types should be ignored. However, the probability of locating tantalum mineralization of worldwide significance in the other types of deposits is poor. Nevertheless, the beryl (i), lepidolite (iii) and albite (v) pegmatite types may play important roles at a more restricted territorial scale, where the conditions of international market are not of primary economic importance.

Acknowledgements. The author is indebted to Drs. M. Cuney, G. Matheis and other participants of the REE,Nb,Ta workshop for their discussions that contributed to improving this review, and to Dr. D.L. Trueman for critical comments on the manuscript. Besides literature sources, this review is based on the author's research supported over the past 20 years by the National Science and Engineering Research Council of Canada Grants, Canada Department of Energy, Mines and Resources Research Agreements, Canada-Manitoba Mineral Development Agreements, Department of Indian and Northern Affairs (Yellowknife) Research Contracts and Research Grants from the Tantalum Mining Corporation of Canada, Ltd. The enormous contribution of graduate students, initiated by J.D. Grice and later expanded by A.J. Anderson, L.E. Chackowsky, T.S. Ercit, K. Ferreira, B.E. Goad, P.G. Lenton, R.E. Meintzer, B.J. Paul, M.A. Wise and E.K. Ucakuwun are deeply appreciated. The author also thanks his colleagues who shared with him their understanding of tantalum-bearing granitic pegmatites, particularly A.A. Beus, D.M. Burt, R.B. Ferguson, E.E. Foord, R. Gaupp, E. Wm. Heinrich, D. London, R.F. Martin, J.J. Norton, Wm. B. Simmons, D.B. Stewart, D.L. Trueman and A.C. Turnock.

References

Anderson AJ (1984) Geochemistry, mineralogy and petrology of the Cross Lake pegmatite field, Manitoba. M Sc Thesis, Univ Manitoba, Winnipeg

Ayres LD, Černý P (1982) Metallogeny of granitic rocks in the Canadian Shield. Can Min 20:439–536

Beus AA (1948) Vertical zoning of pegmatites on the example of the pegmatite field Aksu Pushtiru (Turkestan Range). Dokl Acad Sci USSR 60:1235–1238

Beus AA (1960) Geochemistry of beryllium and genetic types of beryllium deposits. Acad Sci USSR Moscow Engl Transl Freeman, 1966, 401 pp

Beus AA (1968) Geochemical exploration for endogenic deposits of rare elements on the example of tantalum. Nedra Moscow, Engl Transl GSC Libr, Ottawa

Beus AA, Severov EA, Sitnin AA, Subbotin KD (1962) Albitized and greisenized granites (apogranites). Acad Sci USSR Moscow, 191 pp

Beus AA, Sobolev BP, Dikov JP (1963) On the geochemistry of beryllium in high-temperature post-magmatic mineralization events. Geokhimya 1963:297–304 (in Russian)

Blockley JG (1980) The tin deposits of western Australia, with special reference to the associated granites. Geol Sur W Austr Min Res Bull 12:184

Boctor NZ (1985) Rhodonite solubility and thermodynamic properties of aqueous $MnCl_2$ in the system MnO-SiO_2-HCl-H_2O. Geochim Cosmochim Acta 49:565–575

Brookins DG, Chakoumakos B, Cook CW, Ewing RC, Landis GP, Register ME (1979) The Harding pegmatite: summary of recent research. N Mex Geol Soc Guideb 30th Field Conf, Santa Fe County, pp 127–133

Brotzen O (1959a) Outline of mineralization in zoned granitic pegmatites. A qualitative and comparative study. Geol Fören Förhandl 81:1–99

Brotzen O (1959b) Mineral association in granitic pegmatites: a statistical study. Geol Fören Förhandl 81:231–296

Burnol L (1974) Géochimie du béryllium et types de concentration dans les leucogranites du Massif Central français. Bur Rech Geol Min Mem 85, 168 pp

Burt DM (1981) Acidity-salinity diagrams – application to greisen and porphyry deposits. Econ Geol 76:832–843

Burt DM, London D (1982) Subsolidus equilibria. In: Černý P (ed) Granitic pegmatites in science and industry. Min Assoc Can Short Course Handb 8:329–346

Cameron EN, Shainin VE (1947) The beryl resources of Connecticut. Econ. Geol. 42:353–367

Cameron EN, Jahns RH, McNair A, Page LR (1949) Internal structure of granitic pegmatites. Econ Geol Monogr 2, 115 pp

Černý P (1982a) Anatomy and classification of granitic pegmatites. In: Černý P (ed) Granitic pegmatites in science and industry. Min Assoc Can Short Course Handb 8:1–39

Černý P (1982b) Petrogenesis of granitic pegmatites. In: Černý P (ed) Granitic pegmatites in science and industry. Min Assoc Can Short Course Handb 8:405–461

Černý P (1982c) The Tanco pegmatite at Bernic Lake, southeastern Manitoba. In: Černý P (ed) Granitic pegmatites in science and industry. Min Assoc Can Short Course Handb 8:527–543
Černý P (1988) Rare-element granitic pegmatites. In: Brown AC (ed) Monograph series on geology of ore deposits (in press)
Černý P, Burt DM (1984) Paragenesis, crystallochemical characteristics, and chemical evolution of micas in granitic pegmatites. In: Bailey SW (ed) Micas, Reviews in Mineralogy 13, pp 257–297
Černý P, Ercit TS (1985) Some recent advances in the mineralogy and geochemistry of Nb and Ta in rare-element granitic pegmatites. Bull Mineral 108:499–532
Černý P, Meintzer RE (1988) Fertile granites in the Archean and Proterozoic fields of rare-element pegmatites: crustal environment, geochemistry and petrogenetic relationships. In: Taylor RP, Strong DF (eds) Recent advances in the geology of granite-related mineral deposits. CIM Special Volume 39:170–207
Černý P, Trueman DL, Ziehlke DV, Goad BE, Paul BJ (1981) The Cat Lake-Winnipeg River and the Wekusko Lake pegmatite fields, Manitoba. Man Dep Energ Mines Mineral Res Div Econ Geol Rep ER80–1
Černý P, Meintzer RE, Anderson AJ (1985) Extreme fractionation in rare-element granitic pegmatites: selected examples of data and mechanisms. Can Min 23:381–421
Černý P, Goad BE, Hawthorne FC, Chapman R (1986) Fractionation trends of the Nb- and Ta-bearing oxide minerals in the Greer Lake pegmatitic granite and its pegmatite aureole, southeastern Manitoba. Am Mineral 71:501–517
Chackowsky LE (1987) Mineralogy, geochemistry and petrology of pegmatitic granites and pegmatites at Red Sucker Lake and Gods Lake, northeastern Manitoba, M Sc Thesis, Univ Manitoba, Winnipeg
Cooper DG (1964) The geology of the Bikita pegmatites. In: Geology of some ore deposits in Southern Africa 2:441-462
Crouse RA, Černý P, Trueman DL, Burt RO (1979) The Tanco pegmatite, southeastern Manitoba; Can Inst Min Metall Bull 1979, 2:1–10
Drysdall AR, Ramsay CR, Stoeser DB (eds) (1986) Felsic plutonic rocks and associated mineralization of the Kingdom of Saudi Arabia. J Afr Earth Sci 4:1–290
Eby R (1986) A structural and petrofabric study of sheared pegmatite dikes, Red Cross lake, Manitoba. B Sc Thesis, Univ Manitoba, Winnipeg
Ercit TS (1986) The simpsonite paragenesis; the crystal chemistry and geochemistry of extreme Ta fractionation. Unpubl PhD Thesis, Univ Manitoba, Winnipeg
Foord EE (1976) Mineralogy and petrogenesis of layered pegmatite-aplite dykes in the Mesa Grande district, San Diego County, California PhD Thesis, Stanford Univ
Foord EE, Cook RB (1989) Mineralogy and paragenesis of the McAllister Sn-Ta pegmatite deposit, Coosa County, Alabama. Can Min 27 (in press)
Garson MS, Bradshaw N, Ratawong S (1969) Lepidolite pegmatites in the Phangnga area of peninsular Thailand. In: Fox W (ed) 2nd Tech Conf Tin. Int Tin Counc Dep Mines Res, Thailand
Gaupp R, Möller P, Morteani G (1984) Tantal-Pegmatite: geologische, petrologische und geochemische Untersuchungen. Monogr Ser Min Dep 23. Borntraeger, Berlin Stuttgart
Ginsburg AI, Zdorik TB, Feldman LG (1978) Rare-metal carbonatites and rare-metal pegmatites. Int Geol Rev 21:291–298
Ginsburg AI, Timofeyev IN, Feldman LG (1979) Principles of geology of the granitic pegmatites. Nedra, Moscow (in Russian)
Glyuk DS, Trufanova LG, Bazarova SB (1980) Phase relations in the granite-H_2O-LiF system at 1,000 kg/cm^2. Geochem Int 17:35–48
Goad BE, Černý P (1981) Peraluminous pegmatitic granites and their pegmatite aureoles in the Winnipeg River district, southeastern Manitoba. Can Min 19:177–194
Grice JD, Černý P, Ferguson RB (1972) The Tanco pegmatite at Bernic Lake, Manitoba. II. Wodginite, tantalite, pseudo-ixiolite and related minerals. Can Min 11:609–642
Hanley JB, Heinrich EWm, Page RL (1950) Pegmatite investigations in Colorado, Wyoming, and Utah, 1942–1944. US Geol Surv Prof Pap 227
Hatcher MI, Bolitho BC (1982) The Greenbushes pegmatite, southwest Western Australia. In: Černý P (ed) Granitic pegmatites in science and industry. Min Assoc Can Short Course Handb 8:513–525
Hatcher MI, Elliott A (1986) Greenbushes – a new world source of lithium. 7th Int Cong Ind Min Proc, pp 217–232

Heinrich EWm (1953) Zoning in pegmatite districts. Am Mineral 38:68–87
Heinrich EWm (1964) Tin-tantalum-lithium pegmatites of the São João del Rei district, Minas Gerais, Brazil. Econ Geol 59:982–1002
Heinrich EWm (1967) Micas of the Brown Derby pegmatites, Gunnison County, Colorado. Am Mineral 52:1110–1121
Heinrich EWm (1976) A comparison of three major lithium pegmatites: Varuträsk, Bikita, and Bernic Lake. US Geol Surv Prof Pap 1005:50–54
Heinrich EWm (1978) Mineralogy and structure of lithium pegmatites. J Min Recife 7:59–65
Hemley JJ, Jones WR (1964) Chemical aspects of hydrothermal alteration with emphasis on hydrogen metasomatism. Econ Geol 59:538–569
Hutchinson RW (1955) Regional zonation of pegmatites near Ross Lake, District of MacKenzie, Northwest Territories. Geol Surv Can Bull 34:50 pp
Jahns RH (1953) The genesis of pegmatites II: quantitative analysis of lithium-bearing pegmatite, Mora County, New Mexico. Am Mineral 38:1078–1112
Jahns RH (1953) The study of pegmatites. Econ Geol 50th Anniv Vol 1955:1025–1130
Jahns RH (1982) Internal evolution of granitic pegmatites. In: Černý P (ed) Granitic pegmatites in science and industry. Min Assoc Can Short Course Handb 8:293–346
Jahns RH, Ewing RC (1976) The Harding mine, Taos County, New Mexico. N Mex Geol Soc Guideb, 27th Field Conf, Vermejo Park, pp 263–276
James RS, Hamilton DL (1969) Phase relations in the system $NaAlSi_3O_8$-$KAlSi_3O_8$-$CaAl_2Si_2O_8$-SiO_2 at 1 kilobar water vapour pressure. Contrib Mineral Petrol 21:111–141
Jolliff BL, Papike JJ, Shearer CK (1986) Tourmaline as a recorder of pegmatite evolution: Bob Ingersoll pegmatite, Black Hills, South Dakota. Am Mineral 71:472–500
Jolliff BL, Papike JJ, Shearer CK (1987) Fractionation trends in mica and tourmaline as indicators of pegmatite internal evolution: Bob Ingersoll pegmatite, Black Hills, South Dakota. Geochim Cosmochim Acta 51:519–534
Kesler TL (1961) Exploration of the Kings Mountain pegmatites. Min Eng 13:1062–1068
Kesler TL (1976) Occurrence, development and long-range outlook of lithium-pegmatite ore in the Carolinas. US Geol Surv Prof Pap 1005:45–50
Kunasz I (1982) Foote Mineral Company – Kings Mountain operation. In: Černý P (ed) Granitic pegmatites in science and industry. Min Assoc Can Short Course Handb 8:505–511
Kuzmenko MV (ed) (1976) Rare-element granitic pegmatite fields (geochemical specialization and distribution). Nauka Moscow, 332 pp (in Russian)
Kuzmenko MV (1978) Geochemistry of tantalum and genesis of endogenous tantalum deposits. Nauka, Moscow 212 pp (in Russian)
Kuzmenko MV, Eskova EM (1968) Tantalum and niobium; genetic types of deposits and geochemistry. Nauka Moscow 338 pp
Lahti S (1981) On the granitic pegmatites of the Eräjärvi area in Orivesi, southern Finland. Bull Geol Surv Finl 314:82 pp
Landes KK (1932) The Barringer Hill, Texas, pegmatite. Am Mineral 17:381–390
Landes KK (1933) Origin and classification of pegmatites of central Maine. Am Mineral 10:33–55, 95–103
Lenton PG (1979) Mineralogy and petrology of the Buck Claim lithium pegmatite, Bernic Lake, southeastern Manitoba. Unpubl M Sc Thesis, Univ Manitoba, Winnipeg
London D (1982) Stability of spodumene in acidic and saline fluorine-rich environments. Carnegie Inst Geophys Lab Annu Rep 81:331–334
London D (1984) Experimental phase equilibria in the system $LiAlSiO_4$-SiO_2-H_2O: a petrogenetic grid for lithium-rich pegmatites. Am Mineral 69:995–1004
London D (1985) Pegmatites of the Middletown district, Connecticut. Field trip guide
London D (1986a) Formation of tourmaline-rich gem pockets in miarolitic pegmatites. Am Mineral 71:396–405
London D (1986b) Magmatic-hydrothermal transition in the Tanco rare-element pegmatite: evidence from fluid inclusions and phase-equilibrium experiments. Am Mineral 71:376–395
London D (1986c) Holmquistite as a guide to pegmatitic rare metal deposits. Econ Geol 81:704–712
London D (1987) Internal differentiation of rare-element pegmatites: effects of boron, phosphorus, and fluorine. Geochim Cosmochim Acta 51:403–420
Lumpkin GR, Chakoumakos BC, Ewing RC (1986) Mineralogy and radiation effects of microlite from the Harding pegmatite, Taos County, New Mexico. Am Mineral 71:569–588

Luster GR (1977) Lithologic variability of the Kings Mountain pegmatite, North Carolina. Unpubl M Sc Thesis, Pennsylvania State Univ
Manning DAC, Henderson CMB (1981) The effect of the addition of fluorine on liquidus phase relationships in the system Qz-Ab-Or with excess water at 1 kb. Prog Exp Petrol 1978-80, 5th Rep NERC Publ Ser D 18:16-32
Manning DAC, Pichavant M (1988) Volatiles and their bearing on the behaviour of metals in granitic systems. In: Taylor RP, Strong DF (eds) Recent advances in the geology of Granite-related mineral deposits. CIM Special Volume 39:13-24
Meintzer RE (1987) The mineralogy and geochemistry of the granitoid rocks and related pegmatites of the Yellowknife pegmatite field, Northwest Territories. PhD Thesis, Univ Manitoba, Winnipeg
Meintzer RE, Wise MA, Černý P (1984) Distribution and structural setting of fertile granites and related pegmatites in the Yellowknife pegmatite field, District of MacKenzie. Curr Res Pt A, Geol Surv Can Pap 84-1A:373-381
Melentyev GB, Delitsyn LM (1969) Problem of liquation in magma. Dokl Acad Sci USSR AGI New York Earth Sci Ser 186:215-217
Morgan GB VI, London D (1987) Alteration of amphibolitic wall rocks around the Tanco rare-element pegmatite, Bernic Lake, Manitoba, Am Mineral 72:1097-1121
Mulligan R (1965) Geology of Canadian lithium deposits. Geol Surv Can Econ Geol Rep 21:131 pp
Munoz JL (1971) Hydrothermal stability relations of synthetic lepidolite. Am Mineral 56:2069-2087
Nikitin VD (1957) Characteristics of rare-metal mineralization in pegmatite veins. Zapiski Vses Min Obshtch 86:18-29 (in Russian)
Norton JJ (1973) Lithium, cesium, and rubidium – the rare alkali metals; In: Brobst DA, Pratt WP (eds) United States mineral resources. US Geol Surv Prof Pap 820:365-378
Norton JJ (1981) Origin of lithium-rich pegmatitic magmas, southern Black Hills, South Dakota. Geol Soc Am Rocky Mt Sec 34 Annu Mect, Rapid City Abstr Progr, p 221
Norton JJ (1983) Sequence of mineral assemblages in differentiated granitic pegmatites. Econ Geol 78:854-874
Norton JJ, Page RL, Brobst DA (1962) Geology of the Hugo pegmatite, Keystone, South Dakota. US Geol Surv Prof Pap 297-B:49-128
Norton JJ et al. (1964) Geology and mineral deposits of some pegmatites in the southern Black Hills, South Dakota. US Geol Surv Prof Pap 297-E:293-341
Orville PM (1960) Petrology of several pegmatites in the Keystone district, Black Hills, South Dakota. Geol Soc Am Bull 71:1467-1490
Ovchinnikov LN, Solodov NA (eds) (1980) Deposits of lithophile rare metals. Nedra, Moscow (in Russian)
Page RL et al. (1953) Pegmatite investigations, 1942-1945, Black Hills, South Dakota. US Geol Surv Prof Pap 247:1-228
Pichavant M (1981) An experimental study of the effect of boron on a water-saturated haplogranite at 1 kb vapour pressure. Contrib Mineral Petrol 76:430-439
Pichavant M, Herrera JV, Boulmier S, Briqueu L, Joron J-L, Juteau M, Marin L, Michard A, Sheppard SMF, Treuil M, Vernet M (1987) The Macusani glasses, SE Peru: evidence of chemical fractionation in peraluminous magmas. In: Magmatic processes: physicochemical principles. Geochem Soc Spec Publ 1:359-373
Pye EG (1965) Georgia Lake area. Ontario Dep Mines Rep 31:1-113
Quensel P (1956) The paragenesis of the Varuträsk pegmatite. Ark Min Geol 2:9-125
Rijks HRP, Veen AH v.d. (1972) The geology of the tin-bearing pegmatites in the eastern part of the Kamativi district, Rhodesia. Mineral Deposit 7:383-395
Robie RA, Hemingway BS (1984) Entropies of kyanite, andalusite and sillimanite: additional constraints on the pressure and temperature of the Al_2SiO_5 triple point. Am Mineral 69:298-306
Rosenberg PE (1972) Paragenesis of the topaz-bearing portion of the Brown Derby No. 1 pegmatite, Gunnison County, Colorado, Am Mineral 57:571-583
Rossovskyi LN, Chmyrev VM (1977) Distribution patterns of rare-metal pegmatites in the Hindu Kush (Afghanistan). Int Geol Rev 19:511-520
Rudenko SA, Romanov VA, Morakhovskyi VN, Tarasov EB, Galkin GA, Dorokhin VK (1975) Conditions of formation and controls of distribution of muscovite objects of the North-Baikal muscovite province, and some general problems of pegmatite consolidation. In: Gordiyenko VV (ed) Muscovite pegmatites of the USSR. Nauka, Leningrad, pp 174-182 (in Russian)

Schaller WT (1925) The genesis of lithium pegmatites. Am J Sci 5th Ser 10:269–279
Schneiderhöhn H (1961) Die Erzlagerstätten der Erde, vol 2. Die Pegmatite. Fischer, Stuttgart
Shearer CK, Papike JJ, Simon SB, Laul JC (1986) Pegmatite-wall rock interactions, Black Hills, South Dakota: Interaction between pegmatite-derived fluids and quartz-mica schist wall rock. Am Mineral 71:518–539
Shearer CK, Papike JJ, Laul JC (1987) Mineralogical and chemical evolution of a rare-element granite-pegmatite system: Harney Peak granite, Black Hills, South Dakota. Geochim Cosmochim Acta 51:473–486
Sheridan DM, Stephens HG, Staatz MH, Norton JJ (1957) Geology and beryl deposits of the Peerless pegmatite, Pennington County, South Dakota. US Geol Surv Prof Pap 297-A:1–47
Shigley JE, Kampf AR, Foord EE, London D (1986) Field trip guide book to gem pegmatites of southern California. 14th Gen Meet Int Mineral Assoc 1986, Stanford Univ, pp 1–39
Shmakin BM (1983) Geochemistry and origin of granitic pegmatites. Geochem Int 20:1–8
Simmons WmB, Heinrich EWm (1980) Rare-earth pegmatites of the South Platte district, Colorado. Col Geol Surv Res Ser 11:1–131
Simpson MG (1977) Structural analysis of the Kings Mountain belt, Kings Mountain, North Carolina. Unpubl M Sc Thesis, Univ S Car
Solodov NA (1962) Internal structure and geochemistry of rare-element granitic pegmatites. Acad Sci USSR Moscow (in Russian)
Solodov NA (1971) Scientific principles of perspective evaluation of rare-element pegmatites. Nauka, Moscow (in Russian)
Staatz MH, Trites AF (1955) Geology of the Quartz Creek pegmatite district, Gunnison County, Colorado. US Geol Surv Prof Pap 265:1–111
Staněk J (1965) Contributions to the mineralogy of the South-Moravian County (Pegmatite from Dobrá Voda, near Velké Meziříčí). Folia Fac Sci Nat Univ Purkynianae Brunensis 6, Geol Op 8, 1 (in Czech)
Stern LA, Brown GE, Jr., Bird DK, Jahns RH, Foord EE, Shigley JE, Spaulding LB Jr (1986) Mineralogy California. Am Mineral 71:406–427
Stewart DB (1978) Petrogenesis of lithium-rich pegmatites. Am Mineral 63:970–980
Sundelius HW (1963) The Peg claims spodumene pegmatites, Maine. Econ Geol 58:84–106
Tatarinov AV (1974) Types of miarolitic pegmatites of the Borshtchovoshnyi Kryazh. Zap Vses Min Obsch 103:52–67
Taylor RG (1979) Geology of tin deposits. Dev Econ Geol 11. Elsevier, Amsterdam
Thomas AV, Spooner ETC (1988) Occurrence, petrology and fluid inclusion characteristics of tantalum mineralization in the Tanco granitic pegmatite, s.e. Manitoba. In: Taylor RP, Strong DF (eds) Recent advances in the geology of Granite-related mineral deposits. CIM Spec Vol 39:208–222
Thoreau J (1950) La pegmatite stannifère de Manono, Katanga. CR Trav Congr Sci, Elisabethville 41:1–33
Thurston WR (1955) Pegmatites of the Crystal Mountains District, Larimer County, Colorado. US Geol Surv Bull 1011:1–185
Trueman DL, Černý P (1982) Exploration for rare-element granitic pegmatites. In: Černý P (ed) Granitic pegmatites in science and industry. Min Assoc Can Short Course Handb 8:463–494
Uebel P-J (1977) Internal structure of pegmatites, its origin and nomenclature. N Jahrb Mineral Abn 131:83–113
Vladykin NV, Dorfman MD, Kovalenko VI (1974) Mineralogy, geochemistry and genesis of rare-element topaz-lepidolite-albite pegmatites of the Mongolian People's Republic. Trud Min Mus Acad Sci USSR 23:6–49 (in Russian)
Wang X-J, Zou T-R, Xu J-G, Yu X-Y, Qiu Y-Z (1981) Study of pegmatite minerals from the Altai region. Sci Publ House, Beijing, pp 1–140 (in Chinese)
Winkler HGF (1967) Petrogenesis of metamorphic rocks; 2nd edn. Springer, Berlin Heidelberg New York, 334 pp
Wise MA (1987) Geochemistry and crystal chemistry of Nb, Ta and Sn minerals from the Yellowknife pegmatite field, N.W.T. Ph D Thesis, Univ Manitoba, Winnipeg
Wise MA, Meintzer RE, Černý P (1986) The Yellowknife pegmatite field: mineralogy and geochemistry of Nb-Ta-Sn oxide minerals. Contrib Geol NWT 2, Geol Div INAC Yellowknife NAP EGS 1985–86:15–25

Wise MA, Meintzer RE, Černý P (1988) Phosphate mineralogy of the Yellowknife pegmatite field. Contrib Geol NWT 3, Geol Div INAC Yellowknife NAP EGS 1986-87:123-129

Zagorskyi VY, Khaltuyeva K (1976) Chemistry of metasomatic alteration of amphibolites in exocontacts of tantalum-containing pegmatites. Chem Abstr 90:1-87

Zagorskyi VY, Makrygin AI, Matveeva LN (1974) On the rubidium-rich Li-Fe-Mg-micas from exocontacts of rare-element pegmatites. 1973 Annu Rep Geochem Inst Sib Br Acad Sci USSR:143-147 (in Russian)

Zalashkova NE (1957) Albitization stages in granitic pegmatites on the example of an Altai pegmatite field. Trudy IMGRE 1957:155-164 (in Russian)

Zasedatelev AM (1974) Possible accumulation of lithium in host rocks of lithium pegmatite veins during old sedimentation processes. Dokl Acad Sci USSR 218:196-198

Zasedatelev AM (1977) Quantitative model of metamorphic generation of rare-metal pegmatite with lithium mineralization. Dokl Acad Sci USSR 236:219-221

Zayachkivsky B (1985) Granitoids and rare-element pegmatites of the Georgia Lake area, northwest Ontario. M Sc Thesis, Lakehead Univ, Thunder Bay, Ontario

Zou, Tianren, Yang, Yueqing, Guo, Yongquan, Ni, Yunxiang (1985) China's crust- and mantle-source pegmatites and their discriminating criteria. Geochem Beijing 4:1-17

Geologic Characteristics and Genetic Problems Associated with the Development of Granite-Hosted Deposits of Tantalum and Niobium

P.J. POLLARD

Abstract

Granite-hosted deposits of tantalum and niobium form massive or sheetlike zones of disseminated mineralization in the upper portions of some geochemically specialized granite plutons. Such mineralization usually reaches a maximum in association with albite and fluorite (±lithium)-rich zones in small, late-stage intrusions. These zones are formed in some cases by crystallization of fluorine-rich magmas, and in other cases may result from postmagmatic alteration.

The detailed textural and timing information required to assess the influence of magmatic and postmagmatic processes in the development of these deposits is lacking in many cases. Two major genetic models for mineralization envisage metal accumulation to result either from enrichment within the upper portion of a zoned magma chamber at the magmatic stage, or by leaching from lower levels of the granite with precipitation in the upper levels at the postmagmatic stage. Further detailed mineralogical, textural, geochemical and isotopic studies are required before the relative importance of these alternatives can be fully assessed.

1 Introduction

Granite-hosted deposits of tantalum and niobium, together with associated alluvial concentrations are of major economic significance as a source for these metals. Tantalum and niobium also form important by-products of many tin mining operations and these currently form the chief commercial source of tantalum. Niobium and tantalum deposits typically form massive or sheetlike zones of disseminated mineralization in the upper portions of geochemically specialized granite plutons. These are in some cases associated with zones of fracture-controlled rare-metal mineralization which extends into the surrounding country rocks.

Much of the research concerning the geology and geochemistry of granite-hosted deposits of tantalum and niobium has been carried out in the USSR and China, although in recent years there have also been several studies carried out in other parts of the world. This contribution outlines the major features of granite-hosted deposits of tantalum and niobium, and discusses the models which have been

Department of Geology, James Cook University of North Queensland, Townsville, 4811, Australia

Lanthanides, Tantalum and Niobium
P. Möller, P. Černý and F. Saupé (Eds.)

proposed to account for their formation. Deposits associated with pegmatites, nepheline syenites and carbonatites are treated elsewhere in this volume.

Details of the geologic setting and geochemical characteristics of rare-metal granites are discussed by Pollard (this Vol.), and only the essential features are summarized here. Granites associated with tantalum and niobium mineralization can be divided into three principal types on the basis of their mineralogical and chemical composition:

1. Alkali granites containing alkali pyroxenes and/or amphiboles (aegirine, riebeckite, arfvedsonite etc.) which are characterized by high $Na_2O + K_2O$, Fe, F, Nb, Zr, Rb, Sn and REE, and by low CaO, Ba, Sr and Ta/Nb. These granites occur principally in anorogenic settings and are associated mainly with niobium mineralization in the form of pyrochlore.
2. Biotite and/or muscovite granites which frequently contain Fe-Li micas and which are characterized by high F, Rb and Sn, and by low CaO, Ba, Sr and Eu. These granites occur in both anorogenic and postorogenic settings and are associated principally with Nb-Ta(-Sn) mineralization in the form of columbite-tantalite and cassiterite.
3. Lepidolite-albite granites containing topaz which are characterized by high Al_2O_3, F, Li, Rb, Sn, Ta and Ta/Nb, and by low Ba, Sr, Eu, Zr and REE. These granites occur principally in postorogenic settings but also form marginal facies varieties of biotite and/or muscovite granites in both anorogenic and postorogenic settings. Lepidolite-albite granites are generally associated with Ta(-Sn) mineralization in the form of columbite-tantalite and Ta-rich cassiterite.

2 Morphology of Rare-Metal Plutons

Despite differences in the mineralogy and geochemistry of the granites associated with niobium and tantalum mineralization, the styles of associated mineralization are in all cases very similar. Mineralization is typically associated with intrusive complexes which become increasingly geochemically specialized towards the later, finer-grained phases. The granites are characterized by the widespread development of postmagmatic alteration which occurs as both pervasive and fracture-related styles. Disseminated mineralization is localized within the upper contact zones of small, steep-sided plutons, or within irregularities on the upper surface of larger plutons. These irregularities take the form of dome-shaped protruberences (cupolas) or elongate ridge structures which form local high points on the upper surface of the pluton. Within these areas the granites exhibit important textural and mineralogical variations which can usually be rationalized into a vertical zonation scheme (Fig. 1).

The upper contact of the pluton is frequently marked by the development of a marginal pegmatite facies (stockscheider), which takes a variety of forms in detail. The most common occurrence is a subhorizontal quartz-feldspar pegmatite which may be interlayered with fine-grained granite. The pegmatite component sometimes consists of alternating layers of coarse quartz and quartz-feldspar, with individual crystals oriented perpendicular to the contact. In other cases, this marginal facies

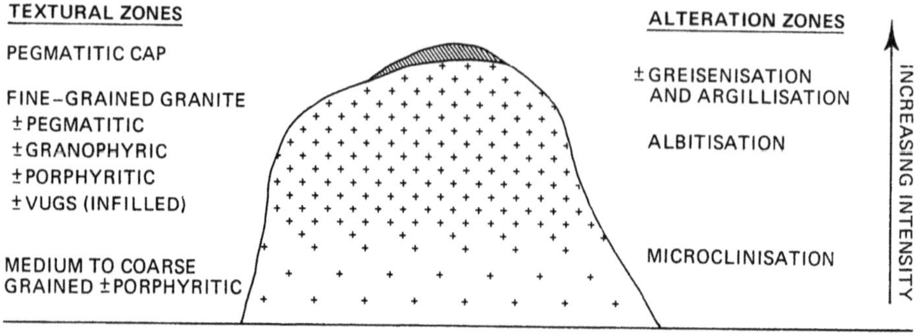

Fig. 1. Generalized scheme of the vertical distribution of rock types and alteration in rare-metal granites

includes the development of downward branching (plumose) alkali feldspar in a matrix of fine-grained granite, and more rarely the development of "brain rock" composed of narrow (< 1 cm) alternating layers of microgranite and oriented quartz. Another variation includes the presence of layering parallel to the upper contact of arfvedsonite-albite granite (Bowden and Kinnaird 1984), with individual layers composed of quartz-rich and amphibole-rich zones within a network matrix of K-feldspar. A less common feature reported by Hu et al. (1984) is the presence of a quartz "crust" containing fluorite, wolframite, triplite and sulphides immediately above the pegmatite zone in the 414 Li-Nb-Ta deposit.

The thickness of marginal pegmatite zones varies from a few centimetres to greater than 5 m, and becomes narrower away from the apex of the intrusion. They appear to have crystallized at a very early stage, and seldom contain any tantalum or niobium mineralization. Similar marginal features commonly occur in tin and tungsten granites (Taylor and Pollard 1987), and in Climax-type porphyry molybdenum deposits (Shannon et al. 1982).

The pegmatite zone usually gives way downwards to a zone of relatively fine-grained granite which may have a number of textural styles, and which may contain segregations up to a metre in diameter of coarse-grained pegmatite. In some cases the fine-grained granites are composed of domains of granophyric quartz-feldspar intergrowths within a ramifying network of coarser-grained "pegmatitic" quartz, feldspar, mica, topaz etc. Miarolitic cavities are another common feature, and these are frequently infilled by hydrothermal minerals including cassiterite.

The fine-grained granite zone is typically 10–100 m thick in larger plutons, but may be the dominant rock type in small, isolated plutons. It is this zone which contains the bulk of the disseminated niobium and tantalum mineralization. The fine-grained granites usually pass downwards into coarser-grained granite through a gradual coarsening in grain size, or through a more abrupt, gradational contact. Deep drilling in the Li-Ta-Sn-rich Beauvoir Granite (Cuney et al. 1985) has shown that this granite is composed of at least three separate intrusions of slightly different composition stacked one above the other. Although important to genetic models concerning magma evolution, this type of relationship has not been demonstrated elsewhere despite excellent vertical exposure in some cases.

The vertical textural variation in rare-metal granites is very important because it indicates an enrichment in volatiles in the upper part of the magma chamber from the earliest stages of crystallization. The development of granophyric, pegmatitic and miarolitic textures is compatible with crystallization in the presence of an aqueous-rich fluid phase, and suggests that volatiles and ore elements may be preconcentrated towards the top of the magma chamber prior to crystallization.

3 Major Rock Types and Distribution of Alternation

Rare-metal granites are frequently referred to as apogranites, and these are defined by Beus (1982) as strongly albitized granites composed of 35–70% albite in which the abundance of primary microcline perthite and plagioclase is less than 20%. This genetic terminology should be used with caution since experimental (Manning 1981) and petrological (Kovalenko et al. 1971; Kosakevitch 1976; Cuney et al. 1985) studies indicate that fluorine-rich magmas can crystallize below the crest of the alkali feldspar solvus, and give rise to primary magmatic rocks similar in composition to apogranites. It is therefore vital that careful petrographic studies are undertaken in an attempt to distinguish primary from secondary albite and thus allow meaningful interpretation of these rocks.

Rock types within rare-metal plutons are often classified according to the presence of typical minerals and/or according to the principal alteration styles. In practice the terminology is sometimes confused by the manner in which phyllosilicates are classified and by the application of genetic terminology without adequate documentation. Beus (1982) recognizes eight major types of rare-metal granites associated with tantalum and niobium mineralization (Table 1) utilizing typical mineral assemblages and mica composition. Individual plutons may contain several of these facies due to internal, vertical zonation of the granite. An indication of the extent of this vertical zonation of rock types is given in Tables 2 and 3, and in Fig. 2.

Despite differences in magma compositions the distribution of rock types in many rare-metal granites is very similar. The most common occurrence is a microcline-rich zone at the deepest levels which is succeeded upwards by an albite-rich zone, a zone of greisenization, and/or by late, fracture-controlled zones of albitization and/or microclinization (Beus and Zalashkova 1964; Shcherba 1970). In some cases, ore-bearing pegmatites of several generations occur around the upper endo- and exocontact zones.

3.1 Microcline-Rich Zones

Microclinization is usually expressed by the replacement of plagioclase by microcline, which typically occurs by replacement inwards from the margin or by the development of microcline patches in the turbid, more calcic core zones of plagioclase. In many examples from the literature it is difficult to assess the intensity of

Table 1. Classification of rare-metal deposits (Beus 1982)

Paragenetic type	Rock-forming minerals	Typomorphic minerals	Typical rare-metal minerals	Contents of some rare-metals (ppm)	Locations
1. Muscovite-albite	Albite microcline quartz	Muscovite	Beryl, bertrandite cassiterite or wolframite sometimes columbite-tantalite	BeO up to 600 SnO_2 up to 800 Ta_2O_5 up to 100 Ta:Nb = 1:2 to 5:1	USSR (Eastern Siberia), ARE (Eastern Desert), PRC (Hunan (Province) North-East of the USSR, France (Central Massif)
2. Lepidolite-amblygonite-albite	Albite microcline quartz	Lepidolite zinnwaldite topaz apatite amblygonite	Columbite cassiterite, monazite, zircon herderite spodumene	Li 40 to 900 Ta_2O_5 up to 100 Nb_2O_5 up to 100 Ta:Nb = 1:1–1:2 SnO_2 500–1500 BeO up to 450	
3. Lepidolite-albite	Albite microcline quartz	Lepidolite, Li-muscovite topaz	Tantalite cassiterite thorite	Li up to 5000 Ta_2O_5 up to 400 Nb_2O_5 up to 200 Ta:Nb = 1:1–1:2 SnO_2 up to 500	People's Republic of China Thailand?
4. Lepidolite-amaxonite-albite	Albite amazonite quartz	Lepidolite Li-muscovite topaz	Cassiterite tantalite, microlite, Ta-hatchettolite	Li up to 1500 Ta_2O_5 up to 400 Nb_2O_5 up to 200 Ta:Nb = 1:2–2:1 SnO_2 up to 500	USSR (Eastern Siberia)
5. Zinnwaldite (cryophyllite) albite	Albite microcline (amazonite) quartz	Zinnwaldite cryophyllite	Columbite cassiterite zircon, thorite	Li up to 1500 Nb_2O_5 up to 200 Ta_2O_5 up to 80 Ta:Nb = 1:2 to 1:5	S.E. Uganda, USSR (Eastern Siberia)
6. Biotite (protolithionite) amazonite-albite	Same	Biotite (protolithionite)	Columbite zircon (malacon) thorite	Zr up to 250 Nb_2O_5 up to 500 Ta_2O_5 to 50 Ta:Nb = 1:5 to 1:10	USSR (Kazakh-stan), ARE (Eastern Desert)
7. Biotite (protolithionite) albite)	Albite microcline quartz	Biotite (protolithionite) zinnwaldite	Columbite pyrochlore zircon (malacon) thorite	Zr up to 1000 Nb_2O_5 up to 3000 Ta_2O_5 up to 300 Ta:Nb = 1:4 to 1:10 and less	USSR (Kazakh-stan, Ukraine), Northern Nigeria Madagascar
8. Riebeckite albite	Same	Riebeckite	Pyrochlore columbite, zircon thorite, gagarinite	Zr up to 2000 Approximately the same Ta:Nb = 1:10 and less	USSR (Kazakh-stan, Eastern Siberia), Northern Nigeria

Table 2. Vertical zonation in a tantalum-bearing granite (Ontoev 1978)

Rock type	Depth (m)	Ta (g/t)
Amazonite-albitic apogranite with lepidolite	0–50	147
Amazonite-albitic apogranite with lepidolite and muscovite	50–100	65
Amazonite-albitic apogranite with muscovite	100–150	49
Porphyroblastic microcline-albitic apogranite	150–200	20
Two-mica granite	200–250	20
Biotitic granite of the major intrusion	> 250	4.1

Table 3. Vertical zonation in rare-metal granites (Data from Syritso and Chernik 1967 and Beus 1982)

Rock type	Depth (m)
Eastern Transbaikal (1)	
Quartz-muscovite greisen	0
Quartz-albitites	100
Medium porphyritic muscovite granite with microclinization	300
Eastern Transbaikal (2)	
Lithionite-amazonite apogranite	
Lithionite-amazonite-albite apogranite	
Muscovite-lithionite-amazonite-albite apogranite	
Muscovite-microcline-albite-apogranite	
Kazakhstan	
Quartz-microcline pegmatoid zone	
Biotite-albite apogranite with columbite and malacon	
Albitized biotite granite	
Porphyroblastic biotite granite	
Eastern Siberia (1)	
Quartz-amazonite zone along contact	1
Lepidolite-amazonite-albite apogranite with topaz, mangatantalite, microlite, cassiterite and columbite-tantalite	100–250
Muscovite-albite apogranite with columbite-tantalite	200–250
Muscovite granite	
Two-mica granite with perthite porphyroblasts	
Eastern Siberia (2)	
Quartz-amazonite pegmatoid	0.5–1
Lepidolite-amazonite-albite apogranite with microlite and cassiterite	100–200
Zinnwaldite-amazonite-albite apogranite with columbite and thorite	200–300
Biotite-amazonite-albite apogranite with low amount of columbite, thorite and cyrtolite	200
Biotite granite with pink perthite porphyroblasts	

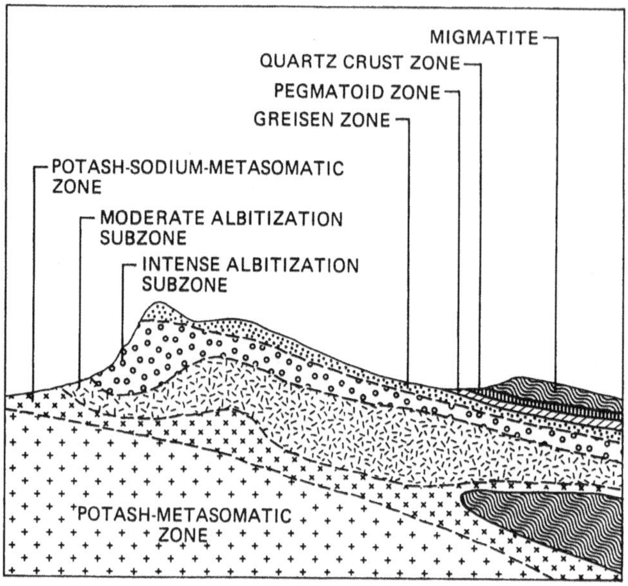

Fig. 2. Cross-section of the 414 Nb-Ta-Li deposit, south China showing vertical distribution of rock types (Hu et al. 1984)

microclinization because there is no clear distinction made between the proportion of microcline replacing plagioclase, and that of microcline developed by ordering of preexisting orthoclase. Since the latter does not necessarily imply a change in bulk composition, it need not reflect exchange between minerals and hydrothermal fluids.

The development of microcline "porphyroblasts" is often reported as the earliest alteration feature in the deepest levels of the pluton. Syritso and Chernik (1967) describe microcline crystals 0.8–3.6 cm in diameter which enclose relicts of plagioclase and a fine-grained granite matrix. Although there has been considerable debate concerning the origin of such crystals (Stone and Austin 1961; Booth 1968), with many regarded as original phenocryst phases, their localization adjacent to incipient fractures (Beus and Zalashkova 1964) in some cases favours a postmagmatic origin.

Early microclinization is often accompanied by a change in phyllosilicate composition from biotite to muscovite (Syritso and Chernik 1967) or Li-mica. In examples where depths are reported, microclinization often occurs 200–300 m below the apex of the pluton (Tables 2 and 3), although in some cases (Kinnaird et al. 1985) early microclinization is absent.

At higher levels in the system microclinization is sometimes expressed by the development of amazonite in the albite-rich zone, and by the development of fracture-controlled microclinites (e.g. Kinnaird et al. 1985).

3.2 Albite-Rich Zones

The development of albite-rich zones in rare-metal granites is extremely important since it is these zones which host the bulk of the niobium and tantalum mineralization. The interpretation of mineral textures in the albite-rich zones is therefore vital to an understanding of the mechanisms of ore formation.

In many cases there is clear evidence for the formation of albite-rich zones via the replacement of preexisting microcline. This alteration usually develops via the coarsening of vein and patch perthite, and/or by the nucleation and growth of new albite at the expense of microcline. In the Ririwai biotite granite there is a gradual increase in the abundance of perthite albite in alkali feldspar over a vertical distance of nearly 300 m (Kinnaird et al. 1985), suggesting an increasing intensity of albitization towards the roof of the pluton. In some examples this process can result in the development of albitites (>70% albite).

Albitization is accompanied by changes in mica compositions, usually expressed by the development of muscovite after biotite, or Li-micas (zinnwaldite, cryophyllite, lepidolite) after muscovite (Koval et al. 1971, 1972; Syritso et al. 1971). In alkali granites, Kinnaird et al. (1985) note that arfvedsonite becomes increasingly Li-rich in the albitized zone. Albitization is also often accompanied by the replacement of quartz by albite, and/or by the dissolution of quartz and subsequent infilling of the cavities by albite and other hydrothermal minerals.

Albite-rich zones in rare-metal granites can also form via the crystallization of fluorine-rich residual magmas. One effect of the addition of fluorine to granite magmas is to shift the minimum towards albite-rich compositions, while the minimum temperature may be lowered sufficiently to enable the crystallization of microcline and albite (Manning 1981). This effect is demonstrated in the occurrence of ongonites which contain quartz, topaz, microcline, albite, columbite-tantalite and cassiterite in a glassy matrix (Kovalenko et al. 1971), and in the evolution in composition of some rare-metal granites (Kovalenko and Kovalenko 1984; Cuney et al. 1985) and topaz rhyolites (Christiansen et al. 1983).

The importance of textural relationships in the interpretation of the origin of some albite-rich zones can be illustrated by the textural features of some Li-mica-albite granites. Bowden and Kinnaird (1984) describe a "snowball" texture in zinnwaldite-amazonite-albite granite from the Afu Complex, Nigeria, which consists of lath-shaped albite surrounding and enclosed by larger, anhedral crystals of untwinned microcline and unstrained quartz. Beus et al. (1962) describe a variation of this texture where albite laths are arranged regularly along the growth zones of quartz "porphyroblasts", and similar textures have been noted in Proterozoic rare-metal granites from Uganda (Beus 1982). At the Granite 414 Li-Nb-Ta deposit in south China, Hu et al. (1984) note a "whirlpool" or "snowball" texture where lath-shaped albite is included in quartz, microcline and topaz phenocrysts (Fig. 3).

These unusual textures are interpreted by Beus et al. (1962) and Beus and Zalashkova (1964) to result from the porphyroblastic growth of quartz during metasomatism of the granite. This may be a viable mechanism where the albite inclusions are randomly oriented since porphyroblasts can include preexisting matrix material during growth. However, the arrangement of albite laths along growth zones in quartz, microcline and topaz at the Granite 414 deposit (Fig. 3) is

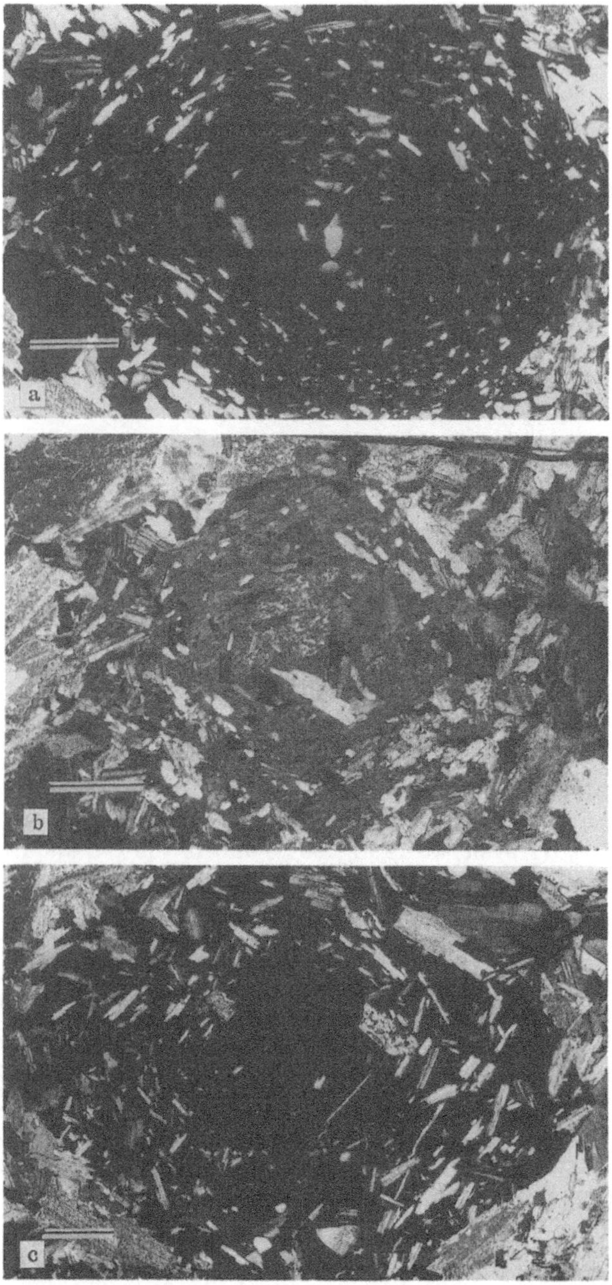

Fig. 3. Photomicrographs of (**a**) quartz, (**b**) alkali feldspar and (**c**) topaz from the upper, albite-rich zone of the 414 Nb-Ta-Li deposit, south China showing the zonal arrangement of albite laths in phenocryst minerals. *Bar* = 0.5 mm

much more likely to have formed during the simultaneous crystallization of these minerals from the magma. A thorough investigation of these textural relationships is vital to an understanding of the development of individual albite-rich zones and the associated mineralization.

3.3 Greisenization and Argillization

In comparison with tin and tungsten systems, the development of greisenization in granites associated with tantalum and niobium mineralization is usually much more restricted, particularly in alkali granites. Greisenization usually occurs in the upper, endo- and exocontact zones of the pluton as both pervasive and fracture-controlled alteration. During greisenization the granite feldspars are usually replaced by quartz, muscovite and/or Li-micas and topaz. Greisen bodies sometimes contain minor amounts of columbite-tantalite, but in most cases other ore minerals such as (Ta-rich) cassiterite, wolframite and beryl are more important components. Base-metal sulphides are also important constituents of many greisens.

Argillic alteration is a crucial component of many rare-metal granites since it is the softness of the argillic-altered material which enables these low-grade deposits to be mined via cheap, alluvial style techniques. Argillic alteration is intensely developed in the roof zone of many rare-metal plutons, consisting of replacement of feldspars by clay minerals (kaolinite, montmorillonite, smectite). The turbid nature of the feldspars in many granites results from the incipient stage of this alteration.

There is considerable uncertainty concerning the genesis of argillic alteration in many granite-related mineral deposits. In some cases it appears to result largely from weathering in tropical climates, while in other cases a hydrothermal component is also suspected. For example, the granite-hosted Sn(-Ta) deposits of southwestern Thailand have intense argillic alteration throughout the upper portions of the mineralizing plutons. However, in some cases they also contain clay-rich stockwork veins carrying cassiterite (Sirinawin et al. unpublished; Fig. 4) which may represent a late stage of hydrothermal alteration and mineralization.

4 Controls on Mineralization

Granite-hosted tantalum and niobium mineralization occurs mainly in disseminated style deposits, and only occasionally as fracture-controlled mineralization. Disseminated mineralization is usually confined to the upper portion of the pluton, while fracture-controlled mineralization may extend through the granite, into the surrounding host rocks. Unfortunately, many descriptions of mineralization in rare-metal deposits make no distinction between original, magmatic minerals and alteration and infill minerals. It is therefore difficult to assess the origin of the ore minerals in most cases.

In alkali granites the chief ore mineral is pyrochlore, although columbite-tantalite and cassiterite are important in some deposits (e.g. Drysdall et al. 1984). In biotite granites such as those of the Nigerian province (Kinnaird 1985) columbite

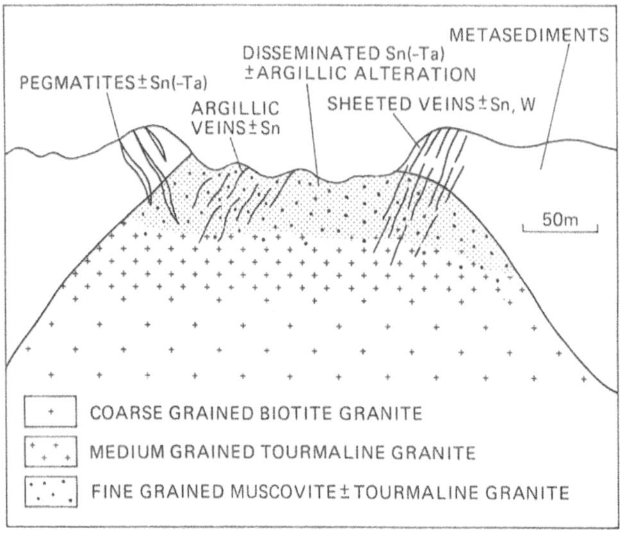

Fig. 4. Schematic cross-section of granite-hosted Sn(-Ta) deposits in southwestern Thailand (after Sirinawin et al. unpublished)

and cassiterite are the chief ore minerals, accompanied by thorite, zircon and xenotime. These deposits usually have low Ta/Nb ratios, and are not presently of economic significance for tantalum.

The various types of muscovite and Li-mica granites (Table 1) are characterized by the development of minerals of the columbite-tantalite series, together with cassiterite which is itself often Ta-rich. Minerals of more restricted occurrence include pyrochlore-microlite, thorite and a variety of lithium, beryllium and phosphate minerals including spodumene, herderite, beryl, bertrandite and amblygonite. Deposits of this type have relatively high Ta/Nb ratios (Table 1) and are exploited principally for tantalum and tin.

4.1 Disseminated Mineralization

Disseminated tantalum and niobium mineralization occurs principally in the albite-rich zone near the top of rare-metal plutons. The highest grade mineralization usually occurs in the top 50 m, and the grade decreases gradually, laterally and vertically away from the apex of the intrusion. Columbite-tantalite and/or microlite mineralization frequently extends 200-300 m below the upper contact (Tables 2 and 3). The definition of ore grade in these styles of deposits varies widely, reflecting a variety of economic and other factors. In many cases deposits are only economic where argillic alteration allows low-cost mining techniques (sluicing, gravel pumping) to be employed, or where other valuable commodities such as tin, lithium, clay or feldspar are also recovered.

In the Beauvoir Granite (Kosakevitch 1976; Cuney et al. 1985) early crystallizing albite, lepidolite and topaz are closely associated with columbite-tantalite and cassiterite, forming an important Ta-Sn-Li resource 200–300 m thick. In this case, as with the ongonites, the ore minerals are thought to have crystallized from a specialized, residual magma, and the distribution of mineralization is therefore largely controlled by magmatic processes. Some replacement of columbite-tantalite by pyrochlore-microlite is also evident (Kosakevitch 1976).

In most cases the ore minerals are dispersed through the albite-rich zone as interstitial phases, or as inclusions in feldspars, micas, topaz and cassiterite. Although they are frequently regarded as secondary, hydrothermal minerals (Beus et al. 1962), the detailed textural and timing information required to assess the replacement or infill origin of columbite-tantalite and pyrochlore is often lacking. Clearly, there is a need for detailed study to place the timing of development of the ore minerals in the context of magma crystallization and hydrothermal alteration processes.

4.2 Structurally Controlled Mineralization

Structurally-controlled mineralization is sometimes present in the upper, endo- and exocontact zones of rare-metal plutons, and a variety of deposit styles can occur in response to the timing and mechanism of fracture development. In general, there is a sharp decrease in the intensity of tantalum and niobium mineralization associated with fracturing. The major exception is where fracturing occurs at a very early (magmatic) stage, and highly specialized melt is channelled into the country rocks to crystallize as combinations of microgranites and pegmatites containing disseminated mineralization (e.g. southwestern Thailand, Garson and Mitchell 1975; Walwa, Australia, Taylor 1979; Saudi Arabia, Drysdall et al. 1984).

The vein styles include a variety of albite-, microcline-, and/or mica-rich bodies associated with wall-rock alteration. Although some veins may contain concentrations of tantalum and niobium (Ginzberg and Fel'dman 1977), many are mineralized principally with cassiterite, wolframite, beryl etc. In the Ririwai complex of Nigeria (Kinnaird et al. 1985) the major vein system in columbite-bearing granite is associated with early microclinization and later greisenization. This vein system contains only minor niobium and tantalum, and is principally a zinc-rich tin deposit (Bowden and Kinnaird 1978).

5 Genetic Modelling

Two major types of models are generally proposed to account for the formation of granite-hosted deposits of niobium and tantalum (Fig. 5), and each stresses the importance of different features of rare-metal granites. The model proposed by Beus and co-workers (Beus 1962, 1982; Beus et al. 1962; Beus and Zalashkova 1964; Zalashkova and Sitnin 1967) considers that ore formation results essentially from the upward movement of postmagmatic solutions which leach metals from the lower zones of the pluton and concentrate them in the upper zones. Thus, an essentially homogeneous distribution of rare metals in the consolidated granite is modified during postmagmatic alteration, and significantly upgraded in the roof of the pluton.

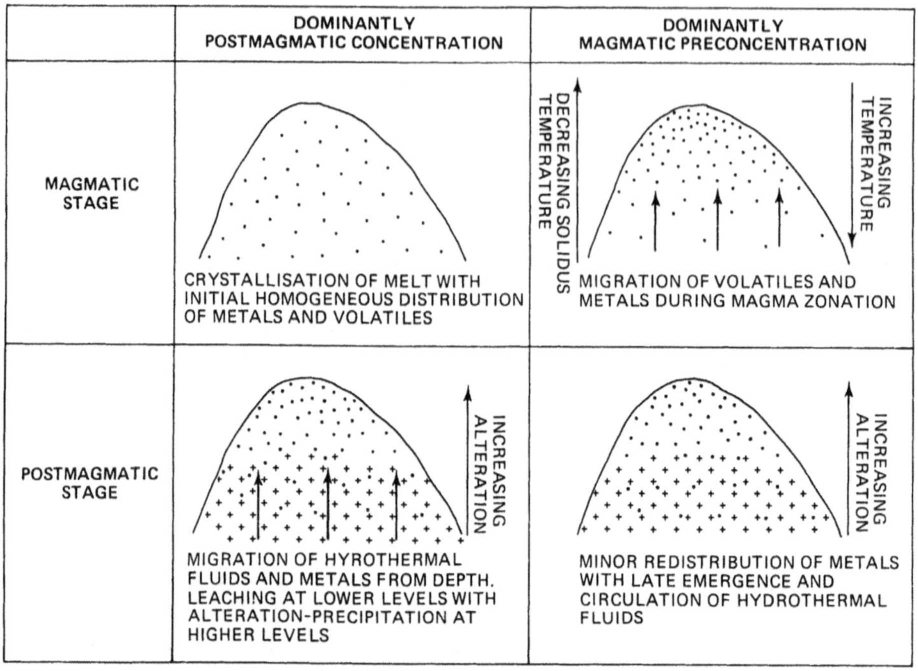

Fig. 5. Major hypotheses concerning the mechanisms of metal concentration in rare-metal plutons

The other major model (Kovalenko et al. 1970; Kovalenko 1978; Kovalenko and Kovalenko 1984) considers that the concentration of rare metals in the upper zones of the pluton occurs during crystallization in the presence of an aqueous-rich fluid phase, and that postmagmatic leaching does not play a significant role in upgrading metal values.

5.1 Magmatic Processes

Experimental studies on melt-fluid and mineral-fluid systems provide a framework for the interpretation of many features of rare-metal granites. In particular, the role of fluorine is vital to an understanding of tantalum and niobium mineralization in these fluorine-rich granites.

Kovalenko et al. (1970) note that the mineralogical and textural features of rare-metal granites are quite different to those of fracture-controlled, rare-metal metasomatites, where there is a tendency for the development of monomineralic compositions. They also point out the clustering of bulk compositions of the albite-rich rocks near the ternary Qz-Ab-Or minimum with high water pressures. This trend towards albite-rich compositions is compatible with experimental work on the addition of fluorine to the haplogranite system at 1 kbar water pressure (Manning 1981).

As noted previously, the textural features of many rare-metal granites and the distribution of rock types are consistent with crystallization in the presence of an aqueous-rich fluid phase, and suggest a concentration of volatiles in the upper portion of the magma chamber from a very early stage. In addition to lowering the liquidus and solidus temperatures, fluorine causes a reorganization of melt structure (e.g. Manning 1981), promoting depolymerization and the creation of new complex species in the melt. In this process, the number of structurally favourable sites for the incorporation of lithophile elements increases, resulting in enhanced transport of these elements. Fluorine also promotes an increase in melt diffusivities (Dingwell 1985), and experiments in Soret diffusion indicate a concentration of sodium and fluorine in the cooler end of the cell. These data provide support for a model whereby chemical gradients are established in the pluton at an early stage, partly as a result of diffusion-controlled processes, with the accumulation of fluorine, sodium, and ore-forming elements towards the roof of the pluton.

The upward increase in fluorine causes a decrease in liquidus and solidus temperatures, which results in a protracted crystallization history, allowing increased opportunity for the operation of magmatic fractionation processes. Fluorine partitions towards the melt ($K_D = 5$ to 10; Hards 1976) and will be released to the aqueous fluid phase only during the final stages of crystallization, unless bound in crystalline phases (topaz, micas). The evolution of a hydrous, fluorine-rich fluid phase potentially can enhance geochemical gradients in the magma chamber and cause metal values to be upgraded in the upper portions.

5.2 Postmagmatic Processes

In the presence of fluorine, the silicate solute content of the fluid phase in equilibrium with the magma is significantly increased in comparison with the pure H_2O system, and the fluid phase is enriched in Na, K, Si and particularly Al (Dingwell and Scarfe 1983). In the presence of chlorine anions, the K/Na ratio of the fluid phase buffered by two alkali feldspars is temperature-dependent (Lagache and Weisbrod 1977), and migration of the hydrothermal fluid down a temperature gradient will cause K for Na exchange (microclinization) of the wall rocks.

In the presence of fluorine anions the fluid phase in equilibrium with two alkali feldspars is strongly enriched in sodium by comparison with the chlorine system (Pichavant 1983). However, the limited solubility of NaF in high-temperature, aqueous solutions suggests that most fluorine will be present as species other than NaF. Any process which causes a reduction in the fluorine concentration of the fluid, such as precipitation of fluorine-rich minerals (topaz, fluorite, micas), or dilution by other fluids, can potentially result in albitization of the wall rocks. In addition, the potential transport of rare-metal elements as complexes with sodium and fluorine suggests that metal precipitation may be accompanied by albitization in association with the development of fluorine-rich minerals.

The solubility of quartz in hydrothermal fluids is sensitive to temperature, pressure and fluid composition. Kennedy (1950) has shown that quartz solubility increases significantly with decreasing temperature between 450° and 360°C at low pressures. In addition, the presence of a variety of fluorine and carbonate complexes in hydrothermal fluids can substantially increase quartz solubility (Wyart and Sabatier 1955). These processes have the potential to dissolve large amounts of quartz during hydrothermal alteration in rare-metal granites, and may help to explain the quartz-depleted character of many albite- and microline-rich rocks in rare-metal deposits. Although these types of alteration are commonly observed in other types of rare-metal deposits (e.g. tin and tungsten deposits), their importance in producing niobium and tantalum mineralization remains uncertain.

Perhaps the major problem for the postmagmatic concentration model concerns the behaviour of the ore elements. There are no studies which have demonstrated that leaching of niobium and tantalum occurs in the lower zones, and no textural evidence that columbite-tantalite in the upper, albite-rich zone is primarily of secondary origin.

6 Discussion

The broad-scale features of granite-hosted deposits of tantalum and niobium are relatively well established, and there is a reasonably consistent distribution of facies and rock types in these systems. The importance of magmatic and postmagmatic processes varies within different deposits due to differences in the environments of emplacement (depth, host rocks), the degree of fractionation, the timing of fluid-phase saturation and associated fracturing, and the potential influence of externally derived fluids. In detail, many of the features of these rock types can be interpreted in terms of magmatic and/or postmagmatic processes, and in many cases these can produce complementary patterns of mineralogical and geochemical variation.

It seems clear from the experimental studies summarized earlier that the fluorine-rich character of the residual magmas plays a vital role in extending the crystallization history to relatively low temperatures, and in producing extremely specialized magma compositions through reorganization of melt structure and the enhancement of diffusion-controlled processes. Many fluorine-rich magmas are relatively anhydrous, and crystallization does not generate the large amounts of mechanical energy typical of porphyry copper systems. The water-rich fluid phase which emerges late in the crystallization history is therefore initially confined to grain boundaries, microfractures and vugs (Pollard and Taylor 1986), rather than being channelled out of the granite by a system of stockwork fractures. The fluids can therefore react with the granite during decreasing temperature to produce pervasive alteration of the pluton.

Data concerning controls on the transport, partitioning and deposition of tantalum and niobium are inadequate for detailed modelling, as are the isotopic data required to assess the sources of the fluids and the fluid-rock ratios. However, it appears that in company with topaz and Li-micas, the deposition of some tantalum and niobium minerals spans the magmatic and postmagmatic temperature ranges. The general lack of columbite-tantalite as a component of true hydrothermal veins suggests a dominantly magmatic origin for members of this series, while Ta-rich cassiterite may precipitate dominantly at the postmagmatic stage.

Rare-metal granite magmas may crystallize directly as topaz Li-mica granite with disseminated columbite-tantalite, while subsequent reequilibration in the presence of a fluid phase may promote feldspar ordering and mica alteration without substantial redistribution of metal values. Alternatively, at higher water contents or lower pressures, the primary mineralogy may be overprinted by a variety of alteration styles depending on factors such as temperature, pH, fO_2 and fluid-flow pathways. This overprinting may redistribute metal values and promote the development of secondary alteration and infill ore minerals, although the scale of this redistribution has yet to be established.

The relative importance of these processes in individual cases is not clear, particularly with respect to the distribution and redistribution of rare metals. A resolution to these problems can only be gained via careful petrological study of a variety of deposits with special emphasis on mineral textures, to draw a distinction between magma crystallization and the development of alteration and infill minerals, particularly with respect to the ore minerals. Carefully controlled rare-earth element and isotope studies are also required in order to assess the role of fluid-rock interaction in modifying initial compositions.

Acknowledgements. J.A. Kinnaird and Hu Shouxi kindly provided rock samples and other materials which were of great assistance in the preparation of this chapter. R.G. Taylor, P. Černý and D.L. Trueman made many valuable comments which improved the final presentation.

References

Beus AA (1962) Wall rock alterations of hydrothermal-pneumatolytic deposits of rare elements. Int Geol Rev 4:1144–1153

Beus AA (1970) Metasomatic zoning in deposits of rare elements of the albitite formation. In: Pouba Z Stemprok M (eds) Problems of hydrothermal ore deposition. Int Union Geol Sci Ser A, 2. Schweizerbart, Stuttgart, pp 325–329

Beus AA (1982) Metallogeny of Precambrian rare-metal granitoids. Rev Bras Geosci 12:410–413

Beus AA, Zalashkova N Ye (1964) Postmagmatic high temperature metasomatic processes in granitic rocks. Int Geol Rev 6:668–681

Beus AA, Severov EA, Sitnin AA, Subbotin RD (1962) Albitized and greisenized granites (apogranites). Academic Science Press, Moscow, 194 pp (in Russian)

Booth B (1968) Petrogenetic significance of alkali feldspar megacrysts and their inclusions in Cornubian granites. Nature (London) 217:1036–1038

Bowden P, Kinnaird JA (1978) Younger granites of Nigeria – a zinc-rich tin province. Trans IMM Sec B (Appl Earth Sci) 78:B66–69

Bowden P, Kinnaird JA (1984) Geology and mineralization of the Nigerian anorogenic ring complexes. Geol Jahrb R B 56:68 p

Christiansen EH, Burt DM, Sheridan MF, Wilson RT (1983) The petrogenesis of topaz rhyolites from the western United States. Contrib Mineral Petrol 83:16–30

Cuney M, Autran A, Burnol L (1985) Premiers resultats apportés par le sondage GPF de 900m realisé sur le granite sodolithique et fluoré à minéralization disseminée de Beauvoir. Chron Rech Mineral 481:59–63

Dingwell DB (1985) The structure and properties of fluorine-rich silicate melts: implications for granite petrogenesis. In: Taylor RP, Strong DF (eds) Granite-related mineral deposits: geology, petrogenesis and tectonic setting (ext abs). CIM, Halifax, pp 72–79

Dingwell DB, Scarfe CM (1983) Major element partitioning in the system haplogranite-HF-H_2O: implications for leucogranites and high-silica rhyolites. EOS 64:312

Drysdall AR, Jackson NJ, Ramsay CR, Douch CJ, Hackett H (1984) Rare element mineralization related to Precambrian alkali granites in the Arabian Shield. Econ Geol 79:1366–1377

Garson MS, Mitchell AHG (1975) The geology of the tin belt in Peninsular Thailand around Phuket, Phangnga and Takua Pa. Inst Geol Sci Overseas Mem 1:112 pp

Ginzberg AI, Fel'dman LG (1977) Deposits of tantalum and niobium. In Smirnov VI (ed) Ore deposits of the USSR, Vol. 3. Pitman, London, pp 372–424

Hards NJ (1976) Distribution of elements between the fluid phase and the silicate melt phase of granites and nepheline syenites. NERC Rep Progr Exp Petrol 3:88–90
Hu S, Cao X, Yan Z, Ouyan Z (1984) The relations between H-Na-K metasomatic alterations and Li-Nb-Ta metallizations of the granite No 414 in south China. 27th Int Geol Congr (preprint)
Kennedy GC (1950) A portion of the system silica-water. Econ Geol 45:629–653
Kinnaird JA (1985) Hydrothermal alteration and mineralization of the alkaline anorogenic ring complexes of Nigeria. J Afr Earth Sci 3:229–251
Kinnaird JA, Bowden P, Ixer RA, Odling NWA (1985) Mineralogy, geochemistry and mineralization of the Ririwai complex, northern Nigeria. J Afr Earth Sci 3:185–222
Kosakevitch A (1976) Évolution de la minéralization en Li, Ta et Nb dans la coupole granitique de Beauvoir (massif d'Echassiéres, Allier). BRGM, Orleans, 72 pp
Koval PV, Yurchenko SA, Nicolayeva DKh (1971) Paragenetic types of muscovite from apogranites of Transbaykal. Int Geol Rev 13:1233–1241
Koval PV, Kovalenko VI, Kuz'min MI, Pisarskaya VA, Yurchenko SA (1972) Mineral parageneses, composition and nomenclature of micas in rare metal and albite-bearing granitoids. Dokl Acad Sci USSR Earth Sci Sec 202:163–166
Kovalenko VI (1978) The genesis of rare metal granitoids and related ore deposits. In: Stemprok M, Burnol L, Tischendorf G (eds) Metallization associated with acid magmatism. Geol Surv (Prague) 3:235:258
Kovalenko VI, Kovalenko NI (1984) Problems of the origin, ore-bearing and evolution of rare-metal granitoids. Phys Earth Planet Int 35:51–62
Kovalenko VI, Kuz'min MI, Letnikov FA (1970) Magmatic origin of lithium- and fluorine-bearing rare-metal granite. Dokl Acad Sci USSR Earth Sci Sec 190:189–192
Kovalenko VI, Kuz'min MI, Antipin VS, Petrov LL (1971) Topaz-bearing quartz keratophyre (ongonite), a new variety of subvolcanic igneous vein rock. Dokl Acad Sci USSR Earth Sci Sec 199:132–135
Lagache M, Weisbrod A (1977) The system: two alkali feldspars-KCl-NaCl-H_2O at moderate to high temperatures and low pressures. Contrib Mineral Petrol 62:77–101
Manning DAC (1981) The effect of fluorine on liquidus phase relationships in the system Qz-Ab-Or with excess water at 1 kb. Contrib Mineral Petrol 76:206–215
Ontoev DO (1978) Relation of multi-stage deposits of tungsten, molybdenum and tin to the history of granitoid formation. In: Stemprok M, Burnol L, Tischendorf G (eds) Metallization associated with acid magmatism. Geol Serv (Prague) 3:97–108
Pichavant M (1983) (Na,K) exchange between alkali feldspars and aqueous solutions containing borate and fluoride anions, experimental results at P = 1kbar. 3rd NATO Adv Stud Inst Feldspars and feldspathoids, Rennes, p 102
Pollard PJ, Taylor RG (1986) Progressive evolution of alteration and tin mineralization: controls by interstitial permeability and fracture-related tapping of magmatic fluid reservoirs in tin granites. Econ Geol 81:1795–1800
Shannon JR, Walker BM, Carten RB, Geraghty EP (1982) Unidirectional solidification textures and their significance in determining relative ages of intrusions at the Henderson Mine, Colorado. Geology 10:293–297
Shcherba GN (1970) Greisens. Int Geol Rev 12:114–150, 239–255
Stone M, Austin WGC (1961) The metasomatic origin of the potash feldspar megacrysts in the granites of southwestern England. J Geol 69:464–472
Syritso LF, Chernik LN (1967) Evolution in accessory mineral paragenesis during metasomatic alteration of granites in eastern Transbaikalia massifs. Int Geol Rev 9:814–827
Syritso LF, Zalashkova, NYe, Zorina ML, Sokolova YeP (1971) Micas in metasomatically altered granites of the acid series. Int Geol Rev 13:717–729
Taylor RG (1979) Geology of tin deposits. Elsevier, Amsterdam, 543 pp
Taylor RG, Pollard PJ (1987) Pervasive hydrothermal alteration in tin-bearing granites and implications for the evolution of ore-bearing magmatic fluids. In: Taylor RP, Strong DF (eds) Recent advances in the geology of granite-related mineral deposits. CIM Spec Vol 39
Wyart J, Sabatier G (1955) Solubilité du quartz dans les solutions de sels alcalins sous pression au-dessus de la température critique. CR Acad Sci 240:2157–2159
Zalashkova NYe, Sitnin AA (1967) Features of the distribution of rare elements in tantalum-bearing apogranites and biotite granites. Geochem Int 4:1013–1025

Geological Setting of Tantalum-Bearing Mineralization in the Caledonides of SE Ireland

P. McARDLE[1] and P.S. KENNAN[2]

There is only one known example of tantalum-bearing mineralization in Ireland. It occurs in a suite of lithium pegmatites close to the margin of the Leinster Granite in SE Ireland (Steiger and von Knorring 1974; Luecke 1981; Kennan et al. 1986). The geology of SE Ireland (Fig. 1) is essentially composed of rocks which were deformed during the Caledonian orogeny in late Silurian and early Devonian times. These older rocks are unconformably overlain by, or in places faulted against, Upper Palaeozoic rocks. The major Leinster Granite (Brindley 1973), to the margin of which the tantalum-bearing pegmatites are spatially related, was emplaced into a sequence of Cambrian to Silurian sedimentary and volcanic rocks (Brück et al. 1979).

A succession of three formations belonging to the Ribband Group (Cambro-Ordovician) has been described from the district containing the mineralized pegmatites (McArdle 1981). The Ballybeg Pelitic Formation (2000–2500 m thick), forming the base of the Ribband Group, consists of thinly bedded turbiditic siltstones. Within the metamorphic aureole of the Leinster Granite these siltstones are represented by staurolite-biotite-andalusite schists. The uppermost part of the formation is characterized by significant amounts of tourmalinite, dacitic tuff and spectacular andalusite-quartz segregations. The lithium pegmatites are spatially related to this stratigraphic level. The overlying Kilcarry Volcanic Formation (500 m) is composed of andesitic lavas with minor dacitic tuff bands and, occurring within the Leinster Granite aureole, is represented by hornblende schists. The uppermost Monaughrim Semi-Pelite Formation consists of more than 250 m of impure shelf sandstones with some distinctive calcareous horizons. Also confined to the Leinster Granite aureole, these latter rocks now consist of garnet-biotite semi-pelitic schists and hornfelses.

The wall rocks of the mineralized pegmatites have suffered a polyphase tectonic evolution during the Caledonian orogeny (McArdle 1984). The overall structural pattern as now displayed was determined during the initial two phases (D1 and D2). During the first (D1) phase the rocks developed a slaty cleavage and associated minor structures under low-grade metamorphic conditions. The cleavage dips to the SE and is parallel to lithostratigraphic contacts and the margin of the Leinster Granite. An important shear zone, the East Carlow Deformation Zone (McArdle and Kennedy 1985), developed during the second phase (D2). It occurs along the

[1]Geological Survey of Ireland, Beggars Bush, Haddington Road, Dublin 4, Ireland
[2]Dept. of Geology, University College, Stillorgan Road, Dublin 4, Ireland

Lanthanides, Tantalum and Niobium
P. Möller, P. Černý and F. Saupé (Eds.)

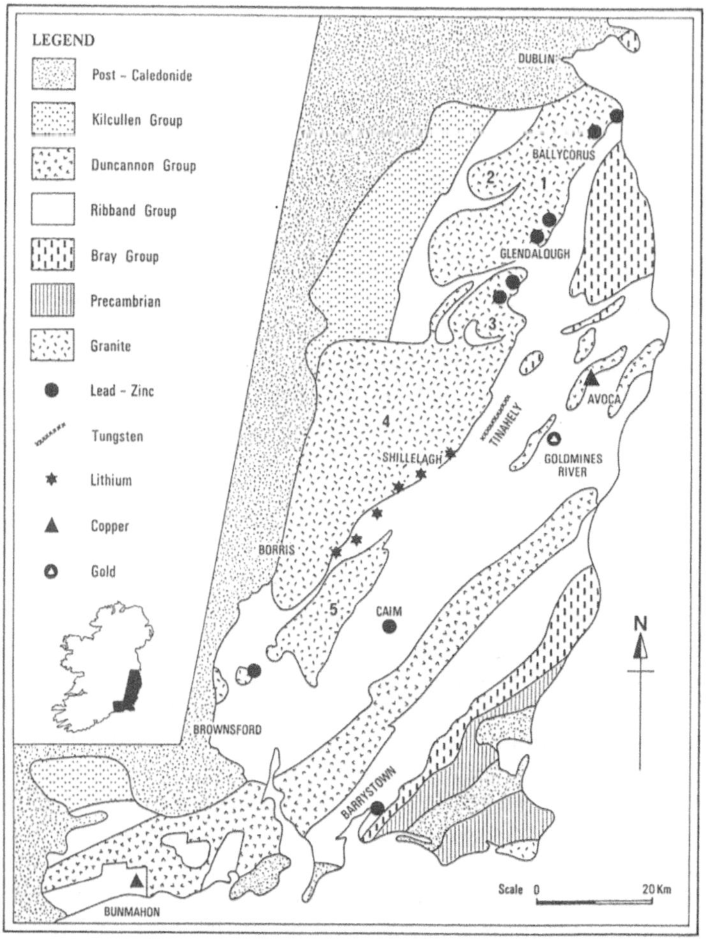

Fig. 1. Lithium and other metal deposits in the Caledonides of SE Ireland (McArdle et al. in press)

margin of Unit 4 (Fig. 1) of the Leinster Granite. It is a dip-slip ductile zone which formed during emplacement of the Granite, under metamorphic conditions ranging from amphibolite to greenschist facies. A suite of granitic, appinitic and lamprophyric intrusions were emplaced into this zone. It forms the locus for the lithium pegmatites described here as well as tungsten mineralization further NE at Tinahely (McArdle and Kennan, 1988 and Fig. 1).

The Leinster Granite (Fig. 1) is composed of five dome-shaped units synchronously intruded during D2 deformation in late Silurian – early Devonian times (Brindley 1973; Brück and O'Connor 1977). Though the granite has a uniform composition over large areas, various distinctive granite types can be recognized. Type 1 granite is a fine-grained quartz diorite which is locally gradational in grain size and composition to Type 2 granite. Type 2 granite forms the main bulk of the Leinster Granite and is a coarse-grained, two-mica granodiorite-granite (adamellite). Type 3 and 4 granites are less important muscovite-rich granite varieties (Brück and O'Connor 1977; see also Brindley 1954).

Six significant if sub-economic lithium pegmatites have been confirmed to date, largely by diamond drilling (Steiger and von Knorring 1974; McArdle 1984). These pegmatites (Fig. 1), as well as seven other reported occurrences, are confined to the vicinity of the Kilcarry Volcanic Formation. The wall rocks contain tourmalinite and disseminated sulphides (pyrrhotite, chalcopyrite, arsenopyrite and pyrite) which are stratabound at the top of the Ballybeg Pelitic Formation (McArdle 1984). Individual pegmatites occur in granitic sheets conformably lying in the aureole schists or else are present on the margin of the Leinster Granite; none are known within the Leinster Granite further than 1 km from its margin. The pegmatites are not internally zoned, but they are composite. They form simple tabular sheets which are generally steeply dipping. The white spodumene contains 0.2% ferric oxide and forms crystals up to 0.3 m long. A bulk sample was found to contain 24% spodumene, equivalent to about 1.5% lithia, 35% quartz, 25% albite, 6% microcline, 9% muscovite and 1% accessories (Steiger and von Knorring 1974). Tin and niobium-tantalum minerals are constant accessories in these pegmatites in the form of cassiterite and columbite-tantalite. Additional accessories include manganapatite, spessartine, beryl, bertrandite, lithium mica, lithiophilite, tourmaline and pyrite. One pegmatite (Stranakelly) is greisened in part and contains lithia mica, cleavelandite and quartz. Mica-rich areas contain 0.2% by weight of tantalum minerals – uraniferous microlite and manganotantalite (Steiger and von Knorring 1974). A general model for the evolution of these pegmatites, which also encompasses the origin of other metal deposits in SE Ireland (Fig. 1), has been discussed by McArdle and Kennan (1988) and Kennan et al. (1986).

Acknowledgement. The contribution by P. McArdle is published with the permission of the Director, Geological Survey of Ireland.

References

Brindley JC (1954) The geology of the northern end of the Leinster granite: Part I – internal structural features. Proc R. Ir Acad 56B:159–190

Brindley JC (1973) The structural setting of the Leinster granite, Ireland – a review. Sci Proc R Dubl in Soc 5A:27–36

Brück PM, O'Connor PJ (1977) The Leinster batholith: geology and geochemistry of the northern units. Geol Surv Ir Bull 2:107–141

Brück PM, Colthurst JRJ, Feely M, Gardiner PRR, Penny SR, Reeves TJ, Shannon PM, Smith DG, Vanguestaine M (1979) Southeast Ireland: Lower Palaeozoic stratigraphy and depositional history. In: Harris AL, Holland CH, Leake BE (eds) The caledonides of the British Isles – reviewed. Geol Soc London Spec Publ 8:533–544

Kennan PS, McArdle P, Williams FM, Doyle EM (1986) A review of metal deposits associated with the Leinster granite and a model for their genesis. In: Andrew CJ, Crowe RWA, Finlay S, Pennell WM, Pyne JF (eds) Geology and genesis of mineral deposits in Ireland. Ir Assoc Econ Geol, pp 201–210

Luecke W (1981) Lithium pegmatites in the Leinster granite (southeast Ireland). Chem Geol 34:195–233

McArdle P (1981) The country rocks flanking the Leinster granite between Aughrim and Ballymurphy. Geol Surv Ir Bull 3:85–95

McArdle P (1984) A study of the geological setting of mineralization associated with the margin of the Tullow Lowlands Unit of the Leinster granite. Unpubl PhD Thesis, Nat Univ Ireland, 330 pp

McArdle P, Kennan PS (1988) Controls on the occurrence and distribution of tungsten and lithium deposits on the southeast margin of the Leinster granite, Ireland. In: Boissonnas J, Omenetto P (eds) Studies in economic geology and geochemistry (EC research programme 1982–1985). SGA Spec Publ 5, Springer, Berlin Heidelberg New York Tokyo 199–209

McArdle P, Kennedy MJ (1985) The East Carlow Deformation Zone and its regional implications. Geol Surv Ir Bull 3:237–255

McArdle P, Williams FM, Doyle EM, Kennan PS, Moore-Lewy G, Gallagher V, O'Connor PJ (in press). The metallogenic relationships of mineralization along the southeast margin of the Leinster granite. Geol Surv Ir Rep Ser RS 86/2:57 pp

Steiger R, Knorring O von (1974) A lithium pegmatite belt in Ireland. J Earth Sci (Leeds) 8:433–443

Part IV Prospecting

Prospecting for Rare-Earth Element Deposits

P. MÖLLER

Abstract

Prospection for REE and Y is largely connected to that of other metals, together with which REE are found in nature. However, a few dominantly REE deposits are also known. One of the major difficulties is the recognition of REE minerals in the field, since many REE(Y) ores are composed of rare exotic minerals not always known to prospectors.

Concepts for prospection and characteristics of rare-metal deposits are reviewed from literature.

The wide range of geological environments in which primary and secondary REE deposits occur requires different adapted concepts for prospecting. Prospection for primary REE-bearing mineralizations has to encompass various physical and chemical aspects. REE incorporation in rock-forming minerals leads to only rare formations of REE deposits. Since all kinds of deposits, from which REE minerals are recovered as by-products, are industrially interesting, prospection on REE is subdued to that of other metals. However, occasionally even the REE may be the goal of prospection. REE minerals are not familiar to most of the prospectors. Therefore it needs relatively simple, qualitative tests for detecting REE in minerals and rocks. Otherwise the presence of REE-bearing minerals might be overlooked in the geological and exploration work (Rose 1976, 1979).

Heinrich (1958) made general suggestions to prospectors how to find REE(Y)-rich mineralizations:

1. Most REE(Y) ores are somewhat radioactive, largely owing to the co-presence of Th, to a lesser extent to U. Some mineralizations, however, are weakly or non-radioactive.

2. Although a few of the REE(Y) minerals themselves fluoresce, associated species, such as fluorite, feldspar and zircon may fluoresce particularly in shades of red or green owing to their REE(Y) contents. Xenotime may show orange-yellow fluorescence, whereas monazite appears emerald-green under the unfiltered light of a mercury-vapour lamp.

3. REE(Y), particularly those of light REE, tend to be associated with minerals containing Ba, Sr, Nb, Ti, Zr, F, PO_4^{3-} and CO_3^{2-}.

Hahn-Meitner-Institut Berlin GmbH, AG Geochemie, Postfach 390128, D-1000 Berlin 30

Lanthanides, Tantalum and Niobium
P. Möller, P. Černý and F. Saupé (Eds.)

4. REE(Y) deposits are found in any type of mineral deposits and in any geological environment. Light REE deposits, however, are preferentially associated with igneous alkalic rocks.

5. The leaves of hickory trees concentrate REE to an extraordinary degree and thus serve as a biogeochemical indicator.

The application of radioactivity and fluorescence in the field in order to find indications of REE minerals are also discussed by Adams and Staatz (1973).

REE minerals are often recovered as by-products of the processing of ores of other rare metals such as Ta, Nb, Zr etc. For rare-metal provinces in general, Ginzburg (quoted in Gmelin 1984) gives the following geochemical characteristics:

1. Usually different genetic types of rare-metal deposits occur together in one province.

2. The same rare-metal elements that are enriched in deposits of magmatic origin mostly occur also in increased concentrations in sedimentary or volcanoclastic depositions found nearby.

3. Rare-metal provinces show increased contents of rare metals in many rock-forming minerals (high "regional Clarke values"). Most of the REE and the rare metals are bound to accessories.

Isomorphic substitution hinders the formation of individual minerals. In magmatic rocks coefficients of relative enrichment $L > 2-3$ (L = element content in a given rock divided by the Clarke value) in general indicate that there is a potential ore-bearing capacity. In case of the alkali granites, agpaitic nepheline syenites or carbonatites L has to exceed 10 for REE(Y) in order to be indicative for unknown deposits (Borodin and Ovchinnikov, quoted in Gmelin 1984). Early-formed spessartine is suggested to serve as a geochemical indicator for yttrium and heavy REE in pegmatites (Ginzburg 1955).

Alluvial black sands consist of heavy dark minerals such as rare-earth minerals together with varying amounts of ilmenite, rutile, zircon and magnetite. They leave visible seams or streaks of black sand on stream beaches or shore. Since many black-sand minerals are magnetic and others are radioactive, magnetometric and radiometric techniques are used in large-area reconnaissance work. This is followed by panning of sediments and examination of the concentrate with a hand microscope (Parker and Baroch 1971).

When sampling stream sediments in areas where the basement shows high levels of indicative trace elements (REE, Nb, Sr, Ba etc.), it may become impossible to detect geochemically the local anomalies due to the rare-element-enriched environment. In case of carbonatites intruding alkaline plutons, usually having rather high background contents of trace elements, it is difficult to localize the potential mineralizations associated with carbonatites.

Considering the high price for Eu today, emphasis is laid on finding Eu-rich minerals or deposits rich in heavy REE. The presence of cenosite seems to be a mineral indicator reflecting high heavy REE(Y) contents (Staatz et al. 1974) whereas, as a general rule, Eu is always deficient with respect to neighbouring REE elements in rare-metal minerals from pegmatites.

References

Adams JW, Staatz H (1973) Rare earth elements. In: Brobst DA, Pratt WP (eds) United States mineral resources (Washington). Geol Surv Prof Pap 820:547–556
Ginzburg AI (1955) About minerals – geochemical indicators and their significance in prospecting for ores of rare metals in pegmatites. Econ Geol 50:748–750
Gmelin (1984) Handbook of inorganic chemistry. Rare Earth, 39, A8:314–316
Heinrich EW (1958) Economic geology of the rare earth elements. Mineral Mag 98:265–273
Parker JG, Baroch CT (1971) The rare earth elements, yttrium, and thorium; a materials survey. Bur Min Inf Circ 8476:92 pp
Rose ER (1976) A field test for rare-earth elements. Geol Surv Can Pap: 75–16
Rose ER (1979) Rare-earth prospects in Canada. Can Mineral Metall Bull 72:110–116
Staatz MH, Shaw VE, Wahlberg JS (1974) Distribution and occurrence of rare earths in the thorium veins on Hall Mountain, Idaho. J Res US Geol Surv 2:677–683

Exploration for Granite-Hosted Tantalum Deposits: An Approach via District Analysis

P.J. Pollard

Abstract

District analysis is a rapid and inexpensive technique for generating exploration targets in the search for granite-hosted tantalum deposits. This technique involves the collection, collation and synthesis of data concerning specific characteristics of the deposits, and allows exploration attention to be focussed quickly on areas where there is the maximum chance for successful discovery. The major features of granite-hosted tantalum deposits which can be utilized in this manner include their association with specific granite types (ilmenite-series, low magnetic susceptibility, high F, Th and U), their localization within domes and cupolas on the upper surface of the batholith and their association with specific alteration styles and peripheral mineralization types.

1 Introduction

In common with exploration for other styles of deposits, exploration for granite-hosted tantalum deposits involves several basic facets which are designed to focus attention on areas in which there is the maximum possibility for successful discovery. Given the management decision to explore for a particular commodity, it becomes the task of the exploration group to design and implement a program which utilizes the specific characteristics of the particular deposit style(s) being sought to optimize the chances of success.

This contribution examines some of the features of granite-hosted tantalum deposits and indicates some of the approaches which should be considered in exploration. In particular, attention is focussed on a rapid and inexpensive technique known as district analysis (Taylor 1979; Taylor and Pollard 1986), which is particularly successful in generating exploration targets for these and other styles of deposits.

Department of Geology, James Cook University of North Queensland, Townsville, 4811, Australia

Lanthanides, Tantalum and Niobium
P. Möller, P. Černý and F. Saupé (Eds.)

2 Definition of Exploration Objectives

Any effective exploration program must be preceded by a clear definition of objectives as to the nature of the commodities sought, economic constraints (cash flow, mine life, return on investment etc.) and geographic limitations. Most commodities occur within a number of different types of deposits, and a comparison of tonnage-grade information for known deposits will usually indicate the deposit type(s) which may fulfill the exploration objectives.

Unfortunately, tonnage-grade information for granite-hosted tantalum deposits is extremely difficult to obtain due to restrictions on the release of such information, or because this type of information is not collected systematically. In addition, the character of the deposits does not readily allow the estimation of grades without substantial drilling information, because grades usually decline gradually with depth. The definition of ore grade is also dependent on a variety of factors (taxation, infrastructure, mining method, associated elements etc.), and thus varies widely between different countries and between individual deposits.

Despite this difficulty of size and grade estimate, it seems clear that granite-hosted tantalum deposits offer large, but low-grade exploration targets. Tonnages can frequently be in the 10–100 million tonnes range, with grades usually less than 0.02% Ta. They therefore represent marginal deposits which usually require a specific combination of special features to make them viable. The most common of these are the presence of additional valuable elements such as tin, lithium or clay, and the presence of intense argillic alteration which allows low-cost, alluvial style mining and processing techniques to be employed.

These factors must all be considered in the early stages to enable the exploration objectives to be clearly defined. It is assumed here that the management decision to explore for granite-hosted tantalum deposits has been made, and the task is therefore to define those special features of the deposit type which can be employed in the exploration program.

3 Modelling and Area Selection

The characteristics of rare-metal provinces and granite-hosted rare-metal deposits are described in detail elsewhere in this volume (Pollard this Vol.), and only those characteristics which are applicable to exploration will be dealt with here. The objective is to build up an exploration model which will enable the characteristics of the deposits themselves to guide the selection of relevant exploration techniques.

Two generalized models of granite-hosted tantalum deposits are given in Fig. 1. The deposits are usually associated with the late-stage members of geochemically specialized, peraluminous granite suites which intrude following the major phases of deformation within orogenic belts. Nb-Ta mineralization is also associated with peralkaline to peraluminous granites in anorogenic provinces, although these seldom give rise to commercial concentrations of tantalum and will not be dealt with here. Ta-bearing granites (as opposed to Nb-Ta granites) are generally geologically young, ranging in age from Devonian to Mesozoic. The mineralized granites

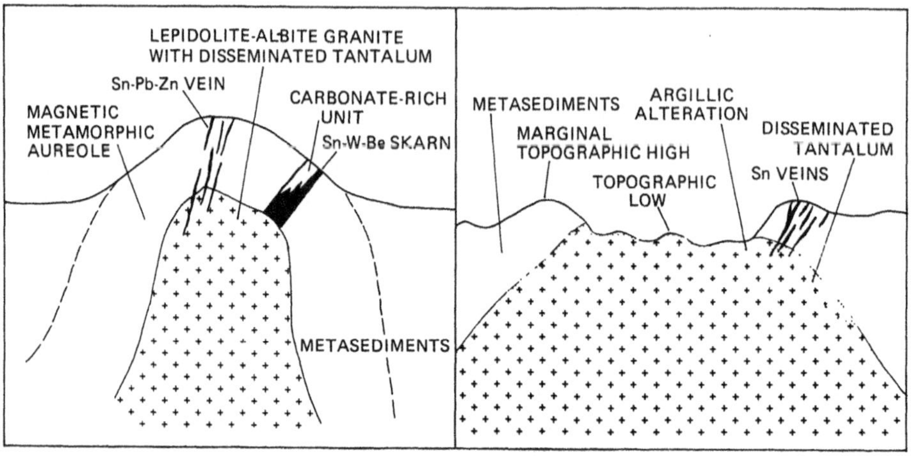

Fig. 1. Examples of exploration models for concealed lepidolite-albite granite, and exposed tantalum-bearing granite associated with intense argillic alteration

typically form steep-sided plutons of a few square kilometres which are often localized by fractures and fracture intersections. These plutons frequently rise above the level of the main batholith to form domes and cupolas which host the mineralization.

Vertical zonation of the tantalum-bearing granite is usually indicated by an upward increase in the content of albite and topaz, and by the increasingly Li-rich character of the micas (biotite, zinnwaldite, cryophyllite, lepidolite). The Sn(-Ta) granites of southwestern Thailand are an unusual, boron-rich variety and contain significant tourmaline in the upper, albite-rich zone.

The principal ore minerals in tantalum-bearing granites include members of the columbite-tantalite and pyrochlore-microlite series, often with associated Ta-rich cassiterite and struverite. Accessory minerals include ilmenite, monazite, thorite and zircon in the deeper levels, with minerals such as apatite, beryl, cyrtolite, fergusonite, amblygonite, herderite and fluorite in the upper levels. Hypogene tantalum mineralization is usually confined within the pluton, although the surrounding rocks may contain minor Ta-bearing pegmatites and/or Sn-W-Be greisens and skarns, with outer Sn-Pb-Zn veins.

3.1 Regional Scale

The selection or rejection of the region in which to explore is often made on the basis of geographic location and economic factors. The most common approach is to select known metallogenic provinces noted for granite-hosted tantalum mineralization, or provinces which have favourable geological characteristics but no known mineralization.

Individual provinces can be broken down into a series of districts based on the distribution of known deposits, deposit types, production etc., and it may be decided to exclude a number of these districts from the exploration program.

If sufficient information is available, the granites in the region can be divided into prospective and non-prospective types on the basis of mineralogical, geochemical and/or geophysical data. The granites associated with tantalum mineralization are geochemically specialized, with high contents of elements such as Rb, Li, Cs, F, Sn, Ta and sometimes B, P, Be and W. They are often depleted in Ba, Sr, Ti and REE. Ca, Mn and Mg are variable, reflecting the content of Li-micas, apatite, herderite, wolframite and other accessories. K/Na is also variable, but generally low, except where amazonite-albite assemblages occur in the upper portion of the granites.

The mineralized granite suites can be classified into the ilmenite-series of Ishihara (1977) since they contain ilmenite and monazite as characteristic accessory minerals. Their mineralogy and geochemistry is reflected in their low magnetic signature, and their low potassium and relatively high thorium and uranium response to radiometric techniques. Despite their low magnetic signatures, the granites often have magnetic aureoles due to the remnant magnetic effect caused by contact metamorphism (Webster 1984). Large areas of granite can be rejected on the basis of these criteria, and exploration should be concentrated in areas where the prospective type granites occur, or are suspected to occur.

3.2 Local Scale

Granite-hosted tantalum deposits typically occur within the upper portion of small, late-stage plutons, or in local high points (ridges, domes) on the upper surface of larger plutons. In order to optimize the chances of success, the exploration strategy should be oriented towards locating these features exposed at the surface, or concealed by a thin veneer of cover. They usually occur within a few kilometres of outcrops of earlier intrusive phases of the batholith, and may be expressed in a number of ways. The most common expressions include the presence of hills which are resistant to erosion because of metamorphic and/or hydrothermal alteration (small magnetic highs?), the presence of porphyry dykes, base-metal and tin occurrences (leakage) and characteristic circular or box-shaped fracture patterns resulting from the intrusion of the underlying pluton. Another common feature is the presence of alluvial deposits in an area with no known primary occurrences. Care should be exercised in some areas where argillic alteration (hydrothermal and/or supergene) is important, since deposits may have a subdued or negative relief, even forming swampy areas covered by their own elluvial concentrations.

The use of the above features of the deposits in the initial stages of exploration is a rapid and inexpensive technique of rejecting large areas of less favourable ground. The compilation and use of data for this purpose is sometimes called district analysis, and is outlined in more detail in the next section.

In areas where there is little information available it may be appropriate to initiate stream sediment sampling, and soil and rock-chip sampling as targets are defined. The most useful elements apart from tantalum for these types of surveys include tin, lithium and fluorine. Another approach which should be considered in areas where intense argillic alteration is expected is the use of airborne or satellite-based multi-spectral scanners for the discrimination of specific clay minerals or clay mineral mixtures. Within the region to be explored, these techniques will require detailed orientation studies.

4 District Analysis

District analysis is a technique which is extremely useful in exploration as a means of compiling and analyzing data which are important to an understanding of mineralization potential and for the application of exploration models. District analysis becomes particularly important in regions where there has been an accumulation of geological and exploration data, or where important new exploration concepts have been developed for a particular deposit type.

Numerous approaches are available, and the selection and application of techniques is controlled by the aims of the investigation. The techniques which are particularly useful in the search for granite-hosted tantalum deposits are discussed below.

4.1 Data Collection

The objective of this initial phase is to gather data for every known mineral occurrence in the region, together with relevant geological information. The sources of data include published information (journals, geological survey reports, maps, monographs etc.) and unpublished information (open file reports, company reports, theses etc.) which forms the data base for further study. It is sometimes useful at this stage to compile information sheets concerning all known deposits in the region. These sheets should contain as much information as possible concerning rock types, mineralogy, alteration, production, grade, mining method, reserves etc., as they form an important data base during the collation stage. The data collection stage of the investigation may be supplemented by reconnaissance field trips to establish specific details, with visits to the major, known deposits.

4.2 Data Collation

The second major phase is the collation of data into a series of maps which are designed to bring out specific points of interest. Maps for consideration for the present purpose include:

1. Distribution of all known mineral occurrences (pits, prospects, mines);
2. Distribution of all economic occurrences;
3. Intensity of mineral occurrences per unit area;
4. Distribution of known production;
5. Major structural trends of granites and mineralization;

6. Distribution of individual elements or groups of elements (Ta, Sn, Li, Pb, Zn, etc.);

7. Distribution of individual minerals (ore and gangue minerals), or groups of minerals to represent alteration styles or deposit types;

8. Distribution of deposit types; this may include structural, mineralogical or genetic classifications;

9. Distribution of primary versus secondary deposits;

10. Distribution of selected geological features relevant to the deposit model: granites, shape of the upper surface of batholiths (ridges, domes etc.; Taylor and Steveson 1972), porphyry dykes, topographic anomalies (hills near the edges of batholiths), etc.;

11. Specific geochemical, geophysical or petrological data of special interest: ilmenite-series granites, non-magnetic granites, radioactive granites (U, Th), argillic alteration;

12. Tonnage-grade diagrams for the region and subcentres within the region;

13. Distribution of past and present exploration activity to highlight areas of intense interest and areas which may have been given only cursory attention.

Even at this early stage a number of interesting areas will emerge which warrent further attention via field inspection. The alluvial versus primary deposit map seldom fails to reveal the presence of alluvial concentrations without known primary deposits, or the presence of primary deposits without known secondary accumulations. These situations always merit early attention for locating primary and secondary targets.

An extremely important aspect is the production of tonnage-grade diagrams, although given the almost total lack of tonnage-grade data for granite-hosted tantalum deposits, this may require considerable research and consultation with local miners. Geologists familiar with the area can often quickly compute rough estimates from the size of excavations and a general knowledge of mine grades. All deposits of high grade and moderate to high tonnages should be evaluated to ensure that they have been correctly identified (e.g. numerous small workings on high-grade areas within a much larger, lower-grade deposit).

Granite-hosted tantalum mineralization is closely associated with domes and ridges or small intrusions which form local high points on the surface of the batholith. A great deal of attention is therefore applied to locating these structures and predicting where similar, buried structures may occur. The outcrop pattern of the granites, together with the topography can be used to predict where ridge and dome structures are likely to occur beneath cover. The presence of these structures is also often revealed by fracture patterns (sometimes occupied by a vein system), porphyry dykes, contact metamorphism and hydrothermal alteration (resistant hills), and by the presence of secondary concentrations. All these features should be plotted on maps of appropriate scale, paying particular attention to those areas within several kilometres of outcrops of granites of the prospective type.

4.3 Data Synthesis

The aim of this phase is to evaluate the resource potential, and to isolate prospective target regions and even specific drilling targets. A set of model criteria are selected which highlight geological, geophysical and physiographic characteristics of the deposits, and these are applied to the collated data in order to isolate areas which contain the highest number of favourable characteristics. A set of model criteria such as that shown in Table 1 is useful for granite-hosted tantalum deposits because it contains features which may relate to exposed deposits (e.g. occurrence of lepidolite-albite granite, tantalum occurrence), and features which may relate to unexposed deposits (e.g. ridge extension, metamorphic hill, vein- or skarn-type mineralization). This list is not exhaustive, and should be adapted to the needs of the particular region. For example, in south China the Ta-bearing granites often take on a peculiar spotted or blotchy appearance in the weathering zone due to staining by manganese oxides. This feature may have been described from granites in the region being explored, and if found to relate to Ta-bearing granites should be included in the exploration model.

Table 1. Selection of possible model criteria for granite-hosted tantalum deposits. These criteria should be refined as the specific characteristics of deposits in a region are established

1.	Tantalum-rich province
2.	Prospective granite type within 5-km radius
3.	Occurrence of lepidolite-albite granite
4.	Disseminated tantalum deposit in district
5.	Tantalum occurrence
6.	Alluvial deposit downstream (Sn-Ta)
7.	Ridge extension under country rocks
8.	Metamorphic hill (possible magnetic anomaly)
9.	Radiometric anomaly (U, Th)
10.	Topographic low surrounded by hills (argillic alteration?)
11.	Minor Sn-Ta pegmatite
12.	Sn-W-Be veins or skarn
13.	Sn-Pb-Zn veins
14.	Circular or box-shaped fracture pattern
15.	Porphyry dykes

In many cases a scoring system can be adopted which gives some indication of the relative merits of individual areas with respect to the model criteria. This can be a simple addition of favourable features, or can apply a weighting to each feature. For example, a lead-zinc vein may be assigned a lower weighting than an occurrence of lepidolite-albite granite. A further set of maps which contour or otherwise highlight the most prospective areas can be constructed, and exploration attention is then focussed on the most prospective zones. These are followed up by field inspection, which may include mapping, stream sediment, soil and rock-chip sampling, and in some cases by drilling.

5 Discussion

The district analysis technique outlined here is a rapid and inexpensive technique which is capable of generating exploration targets through an analysis of available information, together with a component of well-directed fieldwork. It is particularly suitable for a variety of granite-related mineral deposits and can be supplemented by geophysical and geochemical sampling techniques. Throughout the program, a careful reevaluation of the exploration criteria should be maintained so that the exploration model(s) can be refined and expanded as particular characteristics of deposits within the region emerge. The most important aspects of this technique are the specific goal-oriented nature of each of the phases, and the suitability for application at any scale.

Acknowledgements. The district analysis approach summarized here has been applied and refined over a number of years by the Tin-Tungsten Research Group at James Cook University of North Queensland. Discussions with R.G. Taylor and N.M. Tate considerably clarified the exploration approach. D.I. Trueman and P. Černý made many helpful suggestions for improvement of the final manuscript.

References

Ishihara S (1977) The magnetite-series and ilmenite-series granitic rocks. Min Geol 27:293–305

Taylor RG (1979) Geology of tin deposits. Elsevier, Amsterdam, 543 pp

Taylor RG, Pollard PJ (1986) Recent advances in exploration modelling for tin deposits and their application to the southeast Asian environment. Geol Soc Malaysia Bull 19:327–347

Taylor RG, Steveson BG (1972) An analysis of metal distribution and zoning in the Herberton tinfield, north Queensland. Econ Geol 67:1234–1240

Webster SS (1984) A magnetic signature for tin deposits in south-east Australia. Exp Geophys 15:15–31

Exploration Strategy and Methods for Pegmatite Deposits of Tantalum

P. ČERNÝ

Abstract

Economic concentrations of tantalum in granitic pegmatites are restricted to the rare-element class of orogenic pegmatite associations. These pegmatites are particularly abundant in the Kenoran greenstone belts and parametasedimentary gneissic troughs, Hudsonian cratonic margins and rifts, and Variscan fold belts. Abukuma-type metamorphic facies series is the typical metamorphic environment, and a close spatial relationship to major structural breaks plus interformational boundaries is noted. Within the provinces, individual pegmatite fields comprise largely posttectonic intrusions of fertile granites and their pegmatite aureoles, related to fault systems or lithologic boundaries within andalusite-, cordierite- and staurolite-bearing schists of mainly lower-amphibolite facies. Within the fields, individual granite + pegmatite groups may contain tantalum mineralization within the pegmatitic-granite cupolas of the parental granites themselves, but it is usually found only in pegmatites populating intermediate to outer parts of the aureoles. Directions of decreasing metamorphic grade of the host rocks; increasing textural and paragenetic complexity of the pegmatites; progressive changes in their typomorphic minerals; and increasing fractionation point to areas with better tantalum potential. Within individual pegmatitic granites and pegmatite bodies, a variety of geochemical indicators is available to assess the general fractionation level conducive to Ta enrichment relative to Nb, and to absolute concentrations of Ta in the bulk composition. Geochemical and geophysical methods are applicable to searching for, and outlining, pegmatite deposits. Special techniques are available for exploration in deeply weathered terranes and in glaciated regions.

1 Introduction

The purpose of this chapter is to provide a strategy blueprint for a search for pegmatitic tantalum deposits, aimed at virgin terranes of different magnitudes: pegmatite provinces, fields, groups and individual intrusions. Different exploration techniques applicable at different regional levels are also discussed. Both aspects of

Department of Geological Sciences, University of Manitoba, Winnipeg, Manitoba, Canada, R3T 2N2

Lanthanides, Tantalum and Niobium
P. Möller, P. Černý and F. Saupé (Eds.)

the present review, and in particular the prospecting and exploration methods, are focussed on regions of moderate climatic conditions that permit easy access to unaltered bedrock lithologies, with significant segments of the pegmatite populations outcropping on the surface. Consequently, the final sections of this study are devoted to brief reviews of methods employed in the search for hidden deposits, and in terranes affected by special weathering and/or erosional conditions.

The fundamental premise constraining the present exploration concept is the igneous derivation of tantalum-bearing pegmatites from specialized fertile granites, and their virtual restriction to orogenic environments (Beus et al. 1968; Kuzmenko 1976; Ginsburg et al. 1979; Černý 1982b, 1989 and this Vol.). Within the spectrum of orogen-related pegmatites, tantalum deposits are further restricted to only the rare-element class, characterized by specific features of paragenesis, geochemistry and geotectonic plus metamorphic environment (cf. references above, and the review in Černý this Vol.). Thus, the present review deals exclusively with the rare-element pegmatite class, as shown in Table 1. Individual pegmatite types and

Table 1. Classification of granitic pegmatites (modified from Černý 1989)

	Class	Type	Subtype	
Anorogenic[a]	Deep-level			
	Rare-element			
	Miarolitic			
Orogenic[b]	Abyssal[c]			
	Muscovite[d]	Ceramic		
		Muscovite		
		Complex		
	Rare-element[e]	Rare-earth	Allanite-monazite	
			(gadolinite?)[g]	
		(i) Beryl	Beryl-columbite	
			Beryl-columbite-phosphate	
		Complex	Spodumene	(ii)
			Petalite	(ii)
			Amblygonite	(ii)
			Lepidolite	(iii)
		(iv) Albite-spodumene		
		(v) Albite		
	Miarolitic[f]	Lepidolite-topaz-tourmaline		
		Beryl-topaz		

[a] Associated with anorogenic granite intrusions; individual classes not fully characterized at present.
[b] Associated mainly, but not exclusively, with orogenic, late to post-tectonic granites.
[c] Mainly migmatitic in upper amphibolite to granulite facies.
[d] Geochemically dominantly primitive pegmatites in Barrovian metamorphic facies series.
[e] Geochemically largely highly evolved pegmatites in Abukuma-type metamorphic facies series.
[f] Traditionally considered shallow-seated, but probably geochemically conditioned; transitional from the rare-element class.
[g] Questionable; gadolinite subtype with prominent Nb, Y, F signature probably belongs to the rare-element class of the anorogenic family.
Roman numerals correspond to the classification of tantalum-bearing pegmatites, as described and used in the text.

subtypes that may host economic tantalum concentrations are termed and numbered here as in the companion paper on characteristics of pegmatite tantalum deposits (Černý this Vol.).

2 Orogen-Scale Considerations: Geotectonic Distribution of Pegmatite Provinces and Belts

Pegmatite provinces comprise the total pegmatite population within a single metallogenic province, a large-scale geological unit with uniform fundamental features of geological evolution and style of mineralization. Within a province, the pegmatites may (and usually do) belong to different classes that have formed at different stages of the province's evolution, but regional systems belonging to the same class display repetitive properties throughout the province's territory. The best-defined pegmatite provinces with a dominant rare-element class population are those of Archean cratons, such as the Superior and Slave Provinces in Canada (Figs. 9 and 28 in Ayres and Černý 1982), Pilbara and Yilgarn cratons in western Australia (Fig. 2 in Blockley 1980; Fig. 65 in Kuzmenko 1976), and the Zimbabwian and Kaapvaal cratons in Africa (Anhaeusser 1976). The Proterozoic Svecofennian province in southern Finland and central Sweden (Fig. 8 in Nurmi and Haapala 1986), Moldanubian province of Hercynian age in central Europe (Fig. 51 in Černý 1989) and the Pamir-Hindukush province of Cimmerian-Alpine age in Afghanistan and the USSR (Fig. 5 in Černý this Vol.), are typical representatives from younger orogens.

Regional populations – pegmatite fields – are locally concentrated into structurally controlled provincial subunits termed pegmatite belts: they are related to large-scale structures such as deep fault zones, mobilized cratonic margins or mobilized troughs transecting consolidated shields. Within the belts, individual pegmatite fields largely belong to a single class formed during the same tectonomagmatic cycle. The pegmatite systems of the Archean English River and Quetico belts in the Superior Province of the Canadian Shield (Fig. 28 in Ayres and Černý 1982), the Hercynian "tin-spodumene belt" in the Appalachian province (Jahns et al. 1952; Kesler 1961, 1976) and the Hercynian pegmatites of Mongolian Atlai, P.R.C. (Kuzmenko 1976; Fig. 11 in Černý 1982b) are among the best examples of rare-element pegmatite belts.

However, rare-element pegmatites are not ubiquitous components of all metallogenic provinces. The compound effect of petrogenetic conditions and erosion levels restrict the distribution and surface exposure of the rare-element pegmatite systems.

2.1 Age Distribution

Rare-element pegmatites are particularly abundant in the Archean shields affected by the Kenoran (2,750–2,550 Ma) orogenic events, as exemplified by the examples quoted in the preceding introduction. The other conspicuous concentrations populate the Hudsonian (1,800–1,600 Ma) orogens (such as the Churchill province in Canada, Svecofennian province of Scandinavia, South Siberian belt in the USSR and Inner Mongolian province, P.R.C.), the Hercynian (330–260 Ma) systems (such as the Appalachian province of eastern North America; Moldanubian province of central Europe; Pyrenean and Massif Central provinces of western Europe; and the Urals, central and eastern Kazakhstan, Turkestan and Altai provinces of the USSR) and the Cimmerian (190–100 Ma) to Alpine (85–20 Ma) orogenic belts (eastern Transbaikalia in the USSR, and the aforementioned Pamir-Hindukush province).

Rare-element pegmatite populations of other orogens, such as the Nullaghinian-Karelian (2,200–1,900 Ma), Grenvillian (largely 1,200–900 Ma), Riphean (600–500 Ma) and Caledonian (480–350 Ma) systems, are rather modest, in part because of their much more restricted global extent. Differences in erosional effects are among the main reasons for limited exposure of geologically young granite-pegmatite associations. The specific conditions of crustal evolution in the Kenoran event (intense heat flow and degassing, massive evolution of late and post-tectonic granites) coupled with the areally extensive deep erosion of the Archean cratonic nuclei contribute to the abundance of rare-element pegmatites in this age bracket.

2.2 Lithologic and Tectonic Environments

In Archean terranes, greenstone belts (Fig. 1), metasedimentary gneissic troughs and metasedimentary-metavolcanic basins are the dominant units hosting rare-element pegmatites. The greenstone belts have been interpreted as back-arc basins closed by collision of volcanic island arcs and continental margins, or compressed volcanic- and sediment-filled rifts (as reviewed by Condie 1981, Černý 1989). The second hypothesis was recently modified and strengthened by Ayres and Thurston (1985). The thick turbidite sequences of metasedimentary troughs separating granite-greenstone terranes are interpreted as linear basins adjacent to volcanic island chains (Ayres 1978), or as aulacogenlike rifts (Ginsburg et al. 1979). Broad basins formed in initially extensional regimes rarely host extensive pegmatite fields; the Yellowknife terrane (Henderson 1981) is rather exceptional in this respect (Meintzer 1987).

Proterozoic pegmatite provinces have only a limited number of Archean-type greenstone belts. The rare-element pegmatites are hosted mainly in metavolcanic-metasedimentary successions of basins and belts (as in the Churchill province of the Canadian Shield and the Damara province in Africa). These are considered as rifted, ocean-invaded and subsequently closed ensialic troughs, or formations filling collision boundaries between Archean continents (Gibb and Walcott 1971; Dimroth 1981; Baragar and Scoates 1981).

Phanerozoic rare-element pegmatite provinces are located in essentially two environments. Eugeosynclinal-type sediments containing only minor volcanic components host the pegmatites in intra-arc and foredeep sequences of active continental margins, whereas flysch sediments are the typical hosts along originally trailing continental edges (Fig. 2). Continent-arc or continent-continent collisions are responsible for the basin closure, leading to deformation, metamorphism and granite plus pegmatite generation. The anatectic event may also involve remobilization of the basement.

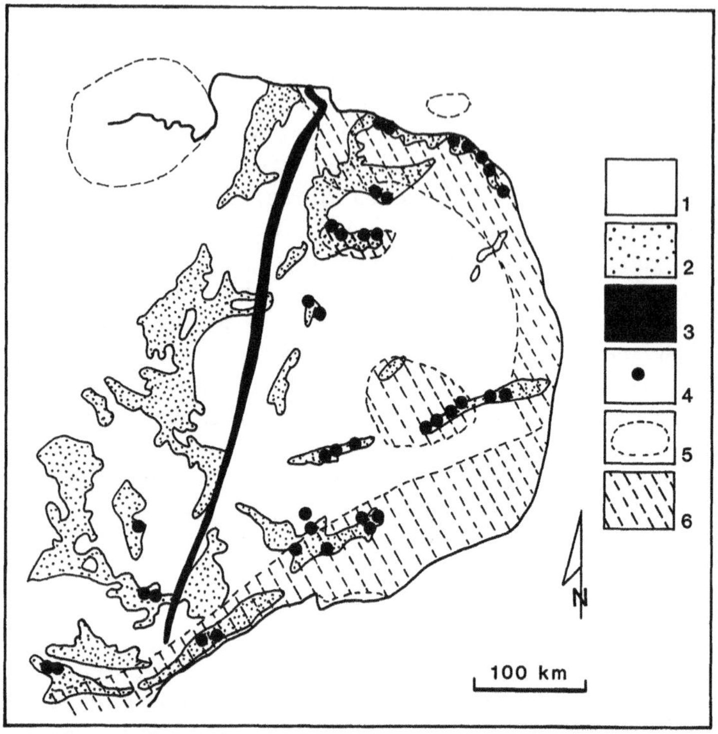

Fig. 1. Distribution of pegmatite fields in the Archean Zimbabwe craton (after Anhaeusser 1976): *1* tonalitic and granitoid terrane; *2* greenstone belts; *3* the Great Dyke; *4* pegmatite groups or fields of the rare-element class; *5* pegmatite fields of the muscovite class; *6* areas of upper-greenschist to lower-amphibolite facies (after Saggerson and Turner 1976). Modified from Černý and Meintzer (1988)

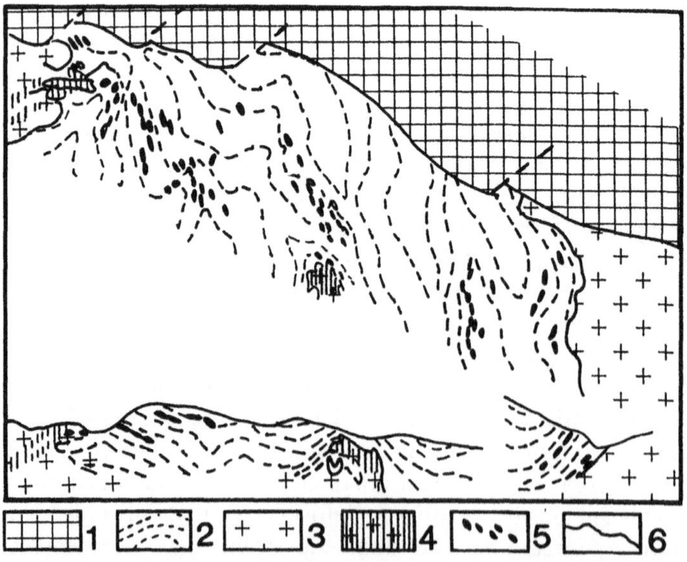

Fig. 2. A Li, Sn, Ta-bearing rare-element pegmatite field of Cimmerian age in Central Asia in plan (*top*) and segmented vertical section (*bottom*; Černý 1982b as modified from Kuzmenko 1976): *1* uplifted Precambrian basement block; *2* Upper Triassic phyllites and micaschists; *3* biotite, two-mica and muscovite granites; *4* pegmatitic granites; *5* pegmatites; *6* erosional surface

2.3 Metamorphic Environment

Rare-element pegmatite populations are confined to regions of relatively steep geothermal gradients (ca. 40° to 50°C/km). Abukuma-type facies series is therefore the typical metamorphic environment, with the andalusite-sillimanite succession in a prograde metamorphism of pelitic sediments. Kyanite-bearing schists host rare-element pegmatites only in exceptional cases, as in the Mtoko region of Zimbabwe (Hornung and von Knorring 1962) and in the Greenbushes area (Hatcher and Elliott 1986). Geochronology of metamorphic and igneous events of these areas has not yet been worked out; the kyanite-generating metamorphism could have preceded the pegmatite-producing tectonomagmatic event, or, if contemporaneous, it could represent extreme high-P conditions within the uncertainty of the andalusite-kyanite or sillimanite-kyanite field boundaries (at about 4 kb; see Fig. 1 in Černý this Vol.).

3 Regional-Scale Approaches: Constitution of Pegmatite Fields

Rare-element *pegmatite fields* are territorially defined pegmatite populations related to the same type of granitic parents, which are usually generated during a single tectonomagmatic stage of the regional geological evolution. These granite plus pegmatite systems are consequently of about the same age, and are commonly related to the same structural framework. However, their degree of fractionation may be extensively diversified, depending on the level of emplacement and/or depth of erosion of the individual intrusions.

Tectonic environment, metamorphic grade of host rocks and intrusive style of the parental fertile granites are mainly responsible for the distribution of rare-element pegmatite fields in orogenic terranes of different ages, as discussed below.

3.1 Structural Affiliation

In late Archean greenstone belts, pegmatite fields are located mainly along deep axial fault systems and associated dislocations; lithological boundaries, particularly batholithic contacts bordering the greenstone belts, also are among preferred loci (Fig. 3; also Fig. 4 in Černý this Vol.; Figs. 2 and 4 in Černý and Meintzer 1988). In the Archean metasedimentary gneissic troughs, rare-element pegmatite fields are commonly found along the metamorphic and/or tectonic boundaries of these troughs, or in their central parts. In open metasedimentary-metavolcanic basins, the granite-pegmatite distribution may be controlled by the tectonic features of the basement.

Hudsonian rare-element pegmatite fields in early Proterozoic provinces are found in mobilized eugeosynclinal like basins and troughs disposed along margins of Archean cratons. Fault systems, lithological boundaries and fault-related domes are among typical host structures (e.g. Fig.5 in Černý and Meintzer 1988). However, quasi-diapiric emplacement of fertile granites with no obvious relationship to

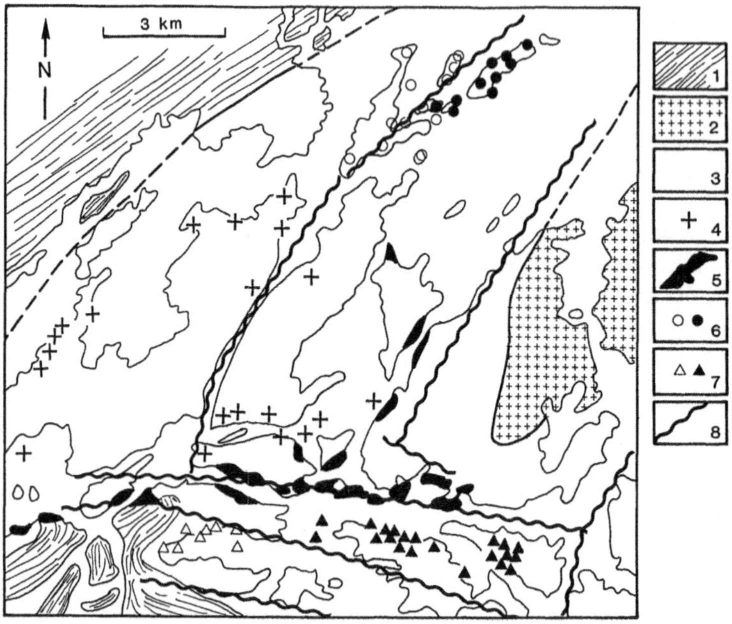

Fig. 3. Archean rare-element pegmatite field at Cross Lake, central Manitoba (modified from Anderson 1984): *1* upper amphibolite to lower-granulite facies gneiss and migmatite; *2* batholithic granitoid intrusions; *3* greenstone belt (undivided); *4* Nelson River pegmatite series; *5* Minago River pegmatitic granite series; *6* barren (○) and beryl-columbite subtype (●) pegmatites of the Northern series; *7* biotite-peristerite (△) to spodumene-subtype (▲) pegmatites of the Southern series; *8* fault boundaries of major tectonic blocks (after Lenton, pers. commun. 1986)

observable pre-existing structural breaks is also encountered in broad metasedimentary basins.

Phanerozoic fields of rare-element pegmatites are confined to granitoid belts within linear orogens of the Appalachian or Alpine types (Fig. 4). These belts are hosted by sedimentary sequences subjected to progressive metamorphism (as in the Afghan Hindukush; Fig. 5 Černý this Vol.), or by remobilized segments of older crystalline terranes (e.g. in the Moldanubian province of central Europe; Fig. 51 in Černý 1989).

3.2 Distribution of the Fertile Granites

Granites parental to rare-element pegmatites are largely late-tectonic to post-tectonic intrusions that post-date the peak conditions of regional metamorphism. Their distribution is controlled by shear and fault systems and lithologic contacts, typically within the transition from ductile to brittle regimes of host-rock behaviour. Numerous examples of elongate fertile granites emplaced along faults, interformational and intrusive contacts, domes and plugs at structurally weakened loci, and of plutons within hinge traces related to basement activation are shown by Černý and

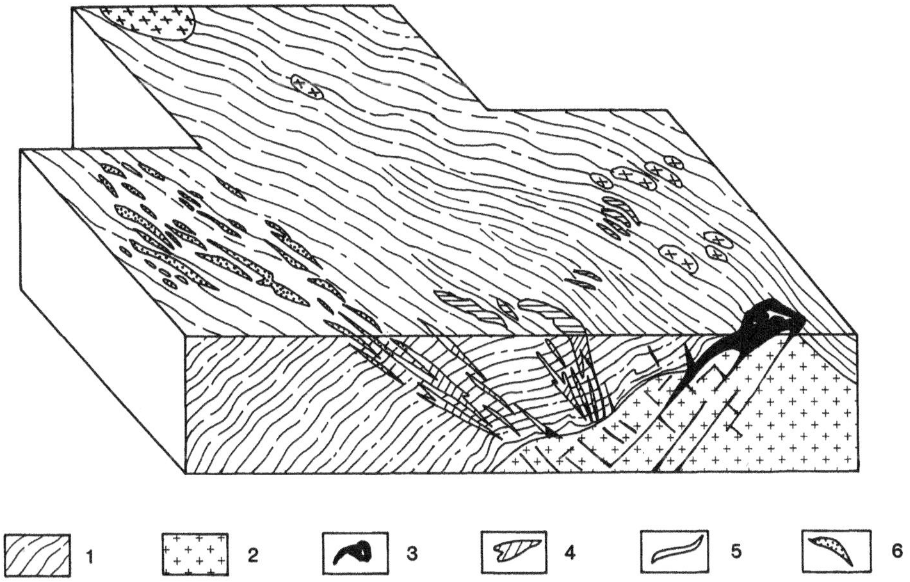

Fig. 4. Segment of a zoned pegmatite field in Transbaikalia (modified by Trueman and Černý 1982 from Beus 1960): *1* Jurassic metaconglomerates and schists; *2* two-mica granite; *3* graphic biotite + microcline pegmatitic granite and pegmatite; *4* blocky microcline + biotite pegmatites, in part albitized; *5* albite + muscovite pegmatites with extensive replacement; *6* spodumene-subtype pegmatites with extensive replacement

Meintzer 1988 (Figs. 1 to 5) Meintzer (1987) and Černý (1989, Figs. 24,35,37,45,49, 52; this Vol., Figs. 3 to 5; 1982b, Figs. 3, 5, 7, 11).

Pegmatite aureoles surrounding the fertile granites penetrate either structures related to those channelling the granite themselves, or fracture systems generated (or opened) by the forcible intrusion of the granites in their roofs. However, detailed structural studies of the granite-pegmatite relationship are practically non-existent.

3.3 Metamorphic Environment

Pegmatites of the rare-element class, confined in general to the Abukuma-type metamorphic facies series, are restricted to the A 2.1 andalusite + cordierite + muscovite subfacies of Winkler's (1967) cordierite-amphibolite facies, rarely extending into the chlorite-bearing A 1.2 schists below the cordierite and staurolite isograds (Fig. 1). Some rare-element pegmatite fields are located in terranes of lower geothermal gradients, marked by the staurolite stability extending above the temperatures of the first sillimanite boundary, but still located within the andalusite-sillimanite succession. As noted above in Section 2.3, kyanite-bearing host rocks are extremely rare.

In regions with simple, tectonically undisturbed metamorphic zoning, fractionation trends of pegmatite populations invariably progress with decreasing metamorphic grade. The primitive members (including parent granites) occur at or above the first sillimanite isograd, and the most fractionated and complex pegmatites are located in the transition into the upper greenschist facies (e.g. Fig. 49 in Černý 1989). However, regional faulting post-dating pegmatite emplacement commonly develops mosaic patterns of vertically displaced blocks that complicate the fractionation trends and occasionally even the distribution of pegmatite types and subtypes on the erosion surface (Fig. 3).

4 Local-Scale Analysis: Zoning in the Pegmatite Groups

Pegmatite groups comprise populations genetically related to a specific intrusion of fertile granite, whereas the term *pegmatite series* is preferred for geochemically defined and spatially isolated pegmatite swarms lacking outcrops of granitic parent. Pegmatite groups commonly number from tens to hundreds of individual bodies that range from a few to about 100 per km^2, and may maintain such densities of occurrence over areas of tens of km^2. Rational exploration for tantalum deposits within a pegmatite group may therefore require identification of parental granite (or estimating its approximate location, if hidden) and understanding the principal features of regional zoning within its pegmatite aureole.

4.1 Characteristics of the Fertile Granites

Fertile granites generating swarms of rare-element pegmatites are typically subsolvus-type, leucocratic rocks of variable texture and composition. Biotite-bearing, fine-grained phases commonly grade into two-mica or muscovite granites coarsening to pegmatitic grain size. Sodic aplites and local zoned segregations of potassic pegmatite commonly occur in the pegmatitic leucogranite facies, which is concentrated in the uppermost parts of individual intrusions. Vertical or concentric layering, upwards and outwards from the most primitive phases to the pegmatitic types, is widespread.

Accessory garnet, tourmaline, cordierite, gahnite and/or andalusite are typical of most leucogranite phases. They are locally accompanied by beryl, columbite-tantalite, monazite, lepidolite, arsenopyrite or löllingite, and/or triphylite, particularly in the potassic pegmatite facies.

The chemical composition of the fertile granites is restricted to a truly granitic range. Average bulk compositions projected on the Ab-Or-Q triangle are concentrated on the potassic side of the 0.5 to 2 kb minima. In fine-grained, two-mica and muscovite leucogranites, $K_2O + Na_2O$ is about 7.5–8.5 wt.%, SiO_2 is typically 74–76 wt.% and K_2O is slightly to distinctly predominant over Na_2O. The pegmatitic leucogranite facies is usually alkali-rich, with up to 9.0 wt.% $K_2O + Na_2O$, K_2O/Na_2O commonly ≤ 1.0, and SiO_2 73–74 wt.%. CaO and MgO are very low in all types. All granite phases are peraluminous, with CIPW normative corundum reaching up to 6% in pegmatitic leucogranites.

The compositional variations among different mineralogical and textural facies of the fertile granites are very restricted for the major components, even within a single intrusion. However, the increase in some "pegmatophile" elements (Li 1–3,540, Rb 32–5,775, Cs 3–51, Be 1–604, Ga 19–90, Sn < 1–112, Y 3–102 ppm) and the depletion of others (Sr 445–< 1, Ba 900–6, Ti 4,300–< 100, Zr 77–< 1 ppm) is rather dramatic. Extreme fractionation is indicated by the ratios of geochemically coherent as well as antipathetic elements (K/Rb 270–42, K/Cs 15,400–1,600, K/Ba 48–18,200, Rb/Sr 1.6–185, Al/Ga 3,100–1,180, Zr/Hf 64–14, Mg/Li 50–1.7).

REE abundances are low, usually less than 40x chondritic, and mostly between 20x and 1x. Subhorizontal and HREE-depleted patterns seem to be most abundant, but diverse fractionation disturbances kinking the patterns are common. Most types of patterns have distinct to prominent negative Eu anomalies. Individual fractionation series within a single intrusion commonly show an overall decrease in REE abundances from the fine-grained leucogranites to the potassic pegmatites.

Rb-Sr isotope systematics show the initial $^{87}Sr/^{86}Sr$ ratios consistently higher than mantle values for given absolute ages. Disturbances of isotope ratios are common, particularly in the pegmatitic facies. Sm-Nd and U-Pb data are scarce, and the former are also subject to disturbances. Oxygen isotope relationships are highly variable, reflecting different degrees of exchange with host rocks and subsolidus re-equilibration. Whole-rock values of $\delta^{18}O$ vary from +8.1 to +14.1‰.

Beus et al. (1968) evaluated major and trace components of different granite types from six pegmatite-bearing regions of the USSR to establish geochemical signatures for granites capable of fractionating tantalum-rich pegmatites. Only those granites were selected for the study whose genetic links to the different classes of deposits were clearly indicated by the field relationships. The elements and ratios found to be best for this purpose are summarized in Table 2 (after Beus et al. 1968). The Rb, Li (< 25 ppm) and Mg/Li (> 100) values are, according to the authors, the most contrasting and thus reliable indicators; they compare well with the data from other terranes (Černý and Meintzer 1988). It is generally recommended to analyze at least 20 samples per rock type to obtain a reasonable data base for meaningful statistical evaluation (Beus and Sitnin 1968).

Individual examples of fertile granites parental to tantalum-enriched pegmatites are described in Blockley (1980), Černý et al. (1981), Goad and Černý (1981),

Table 2. Geochemical indicators of fertile granites (after Beus and Sitnin 1968)

Element, ratio	Li		Rb	Sn	V	Mg/Li		Zr/Sn	
ppm, range	< 25	> 100	< 150	> 20	> 4	< 30	> 100	< 30	> 100
Parental to Ta-bearing apogranites	< 1%[a]	64%	< 1%	28%	1%	100%	< 0.1%	83%	< 0.1%
Parental to Ta-bearing pegmatites	< 1	41	5	26	1	39	0.4	93	14
Parental to Ta-poor pegmatites	26	27	24	16	1	27	30	87	5
Parental to Ta-free greisens	26	23	30	21	14	44	12	46	70
Barren granites	54	2	43	1	30	0.1	98	7	48

[a] Percentages of data, based on analyses of at least 20 samples, within the indicated range.

Černý and Brisbin (1982). Isotope data are provided in particular by Clark and Cheung (1980), Kistler et al. (1981), Longstaffe et al. (1981), Clark (1982), Longstaffe (1982), Walker (1984), Meintzer (1987) and Shearer et al. (1987). Geological, geochemical and genetic aspects of these granites are reviewed by Černý (1982b) and Černý and Meintzer (1988); the above generalizations are extracted from these two publications.

4.2 Zoning in the Pegmatite Aureoles

Given suitable channelways, pegmatite melts tend to migrate from the fertile-granite intrusions upwards because of their lower density, higher fluidity and better thermal stability relative to the granite magma. Consequently, pegmatite aureoles are typically disposed above the top and flanks of their plutonic parents. The viscosity of pegmatite melts and their liquidus decrease with enrichment in H_2O, F, B, P, Li (Rb,Cs) and Be. Thus, the most fractionated batches of pegmatite melt are the most mobile, and usually migrate much farther away from the granitic parent than the more primitive melts. This results in zonal distribution of pegmatite types and subtypes in the roofs of fertile granites, as diagramatically shown in Fig. 5.

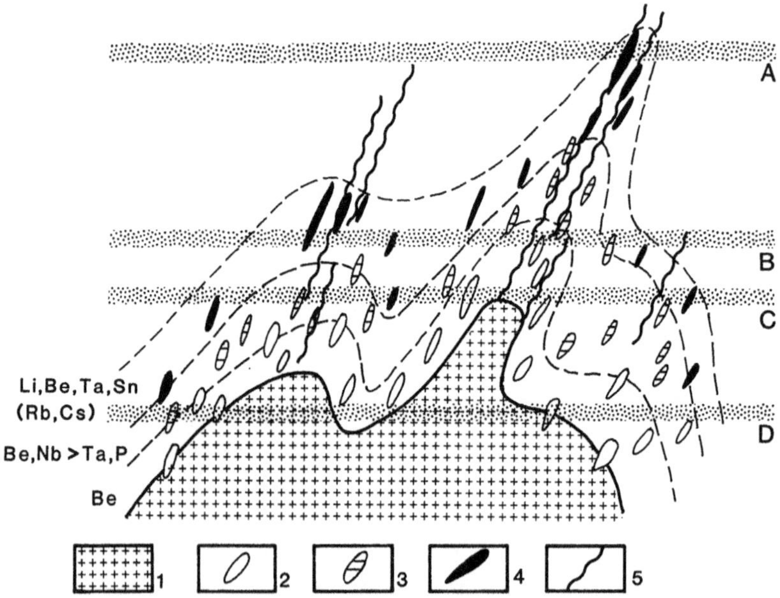

Fig. 5. Schematic vertical section through a zoned fertile granite-pegmatite system: *1* fertile granite, commonly with leucocratic and pegmatitic cupolas; *2* barren to beryl-bearing pegmatites; *3* beryl-type, columbite- to phosphate-bearing pegmatites; *4* complex spodumene- (or petalite-)bearing pegmatites with tin, tantalum and locally cesium mineralization; *5* faults. **A** Shallow erosion level exposing only some of the most fractionated distal pegmatites (e.g. Tanco, Greenbushes and Bikita regions); **B** deeper erosion exposing pegmatites of three types in a concentric but asymmetric pattern; **C** a still deeper erosion level exposing the topmost part of the parent granite, with an apparent lack of zoning in the pegmatite population on its left-hand flank, and only a few small tantalum-bearing dykes; **D** deep erosion extensively exposing the fertile granite but only a few primitive proximal pegmatites along its margins

The geometry and degree of symmetry in the pegmatite distribution are of course strongly dependent on the local structural environment into which a pegmatite group is emplaced. Structural patterns must be considered in defining pegmatite groups in terms of their areal extent and fractionation trends. Elements of asymmetry and apparent random distribution of pegmatite types may be introduced by the morphology of the plutonic parent and by the vagaries of erosion, even if the three-dimensional zoning is quite regularly developed relative to the plutonic contacts (Fig. 5).

4.2.1. Textural Evolution

The textural complexity of granitic pegmatites generally increases with progressive fractionation and enrichment in rare elements. Thus, even if accessory mineralization is scarce or is too fine-grained for effective field determination, changes in the number of internal primary zones and in the number, variety and extent of metasomatic assemblages may be indicative of zoning disposition.

From the granite margin outwards, structural characteristics of pegmatites evolve from homogeneous granitic, through primitively zoned with granitic and graphic textures, to fully differentiated with multiple zones and core. The number and intensity of metasomatic processes increase simultaneously with this evolution of zoning; it is mainly the late and in part metasomatic assemblages (albitic units, muscovite + quartz complexes) that carry significant tantalum mineralization.

This trend of increasing textural complexity, elegantly portrayed by Vlasov (1952; Fig. 11 in Černý 1982a), applies to the rare-earth to complex (ii) pegmatites of Table 1. Textural features of the albite-spodumene (iv) and albite (v) pegmatites are specific for these types and show no transition from the types (i) to (iii) (see Černý this Vol. for details).

4.2.2 Paragenesis

Paragenetic and geochemical features also become more complex outward from the granite. From the observations recorded mainly by Heinrich (1953), Beus (1960), Nedumov (1964), Solodov (1971) and Varlamoff (1972), the following generalized spatial sequence of pegmatites can be outlined for zoned groups:

1. Barren pegmatites of granitic texture and composition, characterized by magnetite and biotite;
2. Microcline + plagioclase pegmatites: partly graphic, simply zoned, slightly albitized and characterized by biotite, ilmenite and Y, REE, Be, Nb-Ta minerals (rare-earth type);
3. Microcline pegmatites: partly graphic, simply zoned, slightly albitized and characterized by muscovite and beryl, schorl [beryl type (i)];
4. Microcline + albite pegmatites: well-zoned, partly albitized and characterized by muscovite, beryl and Nb-Ta minerals, schorl, cassiterite and Fe, Mn (±Li) phosphates [beryl-columbite-phosphate subtype of (i)];

5. Microcline + albite pegmatites: well-zoned including extensive replacement units, and characterized by Li, Rb, Cs, Be, Sn, Ta, B, P and F mineralization [complex type, spodumene, petalite and amblygonite subtypes (ii)];

6. Albite + microcline pegmatites: well-zoned but extensively replaced, and characterized by lepidolite and minerals of Be, Sn, Ta ±Cs [lepidolite subtype (iii)];

7. Albite + spodumene pegmatites: relatively homogeneous, and characterized by low-level Be, Ta, Sn mineralization [albite-spodumene type (iv)];

8. Albite pegmatites: relatively homogeneous, largely fine-grained and characterized by Sn, Ta and Be mineralization [albite types (v)]; and

9. Chiefly quartz veins with minor feldspar(s) and one or more of beryl, cassiterite and wolframite, and with gold and sulphides.

This sequence has not been observed in its entirety in a single group, but has been assembled from segments documented in studies of many occurrences. The number and types of members which compose any given group is normally controlled by a common geochemical factor.

Figure 6 shows an example of the highly asymmetric but well-zoned Rush Lake pegmatite group, which flanks the pronouncedly zoned Osis Lake fertile granite in the Superior Province of southeastern Manitoba (Goad and Černý 1981; Černý and Brisbin 1982). From the pegmatitic leucogranite outwards, barren pegmatites with tourmaline and rare triphylite grade into the beryl-type dykes (i) and complex spodumene- and petalite-subtype pegmatites with tantalum and cesium mineralization.

Additional details on the regional paragenetic zoning in pegmatite groups are provided by Solodov (1962, 1971), Ginsburg et al. (1979), Černý (1982a,b), Trueman and Černý (1982) and Černý (1989).

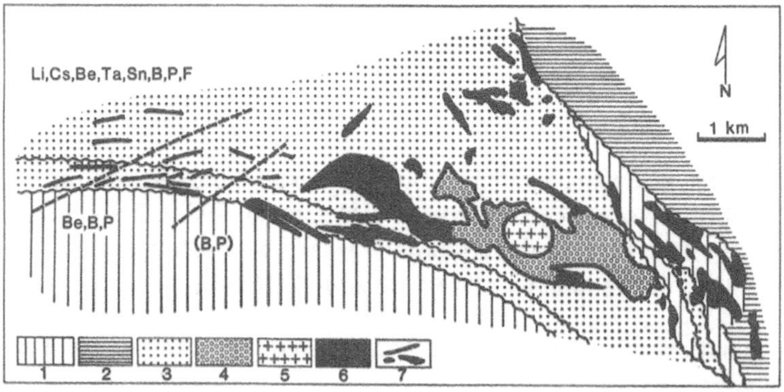

Fig. 6. The Osis Lake granite, southeastern Manitoba and its asymmetrically zoned pegmatite aureole (modified after Černý et al. 1981 and Černý and Brisbin 1982): *1* metaconglomerates of the Bernic Lake formation; *2* metamorphosed lithic arenites and metaconglomerates of the Flanders Lake formation; *3* metaturbidites of the Booster Lake formation; *4* coarse-grained garnet + muscovite leucogranite; *5* biotite granite; *6* pegmatitic garnet + tourmaline + muscovite leucogranite; *7* pegmatite dykes. See Fig. 4 of Černý (this Vol.) for the regional setting

4.2.3 Mineral Typomorphism

In many cases, the occurrence of a number of accessory minerals in granitic pegmatites is restricted to only a specific stage in the pegmatite evolution, and their presence is typical of certain pegmatite types. On the other hand, many minerals may occur in quite a variety of pegmatite types but their properties are variable with the changing geochemistry and mineral paragenesis of their parents. Both types of minerals are useful in establishing the zoning pattern in pegmatite districts.

Plagioclase becomes more sodic with increasing pegmatite fractionation (unless affected by contamination from wall rocks) in both primary and metasomatic varieties. The number, variety and volumes of albitization also increase.

Rose-quartz in quartz cores is restricted to barren, tourmaline-bearing and Be, Nb, Ta-mineralized pegmatites, including those with Li, Fe, Mn-phosphates. It is not known from spodumene-rich or Li, F-bearing pegmatites.

Brownish and dirty-green muscovite is usually indicative of poorly mineralized primitive pegmatite types, according to Gordiyenko (1970). Coarse-flaked, yellowish-green and silvery muscovite is typical of Be, (Nb,Ta)-bearing (i) and complex (ii) pegmatites. The numbers of mica generations and their compositional diversity (muscovite - lithian muscovite - mixed types - lepidolite) increase with progressive fractionation into the complex (ii) and (iii) pegmatites.

Black tourmaline is restricted to the barren and the relatively simple beryl (i) pegmatites, and to the border and wall zones of more evolved types. It grades into blue and green in albitized pegmatites with Sn, Nb and Ta mineralization, and the green, pink and colourless elbaites indicate increased Li, Rb and Cs potentials of complex pegmatites (ii).

Beryl appears as a primary, coarse columnar mineral of greenish, yellow or brownish colour in simple, blocky pegmatites with negligible albitization. With increasing fractionation and enrichment of its parent pegmatite in rare alkalis, late generations of beryl become associated with albitic units and lepidolite replacement assemblages and turn pale-coloured to white. White to pink, stubby or tabular beryl is typical of complex pegmatites (ii) and (iii) with Ta, Li, Rb, Cs mineralization.

Columbite-tantalite is the dominant or exclusive Nb-Ta mineral of moderately fractionated pegmatites and may be accompanied or replaced by microlite. Tapiolite and niobian-tantalian rutile may also be associated. However, it is only in the complex pegmatites (ii) and (iii) where the association of Ta(Nb) minerals becomes conspicuously diversified. Wodginite, several species of the pyrochlore group in addition to microlite (uranmicrolite, plumbomicrolite), stibiotantalite, bismutotantalite, rynersonite, simpsonite, cesstibtantite and their alteration products are typical for the extremely fractionated pegmatite types (cf. Černý and Ercit this Vol. and Černý this Vol.).

Green spodumene in small columnar crystals uniformly distributed throughout pegmatite bodies (phenocrystic type of Heinrich 1975) is typical of albite-spodumene pegmatites (iv). White columnar spodumene in zones adjacent to quartz cores (Heinrich's zonal type) is characteristic of complex pegmatites (ii). Tabular pseudomorphs of fibrous white spodumene + quartz after petalite are most typical of the rare-element pegmatites consolidated in a low-pressure crystallization regime (Heinrich's pseudomorphic spodumene type).

Gordiyenko (1970) notes blue apatite as being indicative of at least Be, Nb-Ta mineralization, and an increase in intensity of blue colour with the extent of rare-element mineralization.

Further details, including comments on exomorphic minerals and empirical significance of grain size are available in Beus et al. (1968), Solodov (1971) and Trueman and Černý (1982).

4.2.4 Geochemical Trends

In pegmatite areas lacking distinctive textural or mineralogical trends, systematic sampling and the analysis of suitable mineral species can be used to determine cryptic zoning. In this respect common rock-forming minerals containing fractionation-sensitive pairs of elements can be utilized. This approach complements, of course, field observations in paragenetically defined zoning patterns and it leads to quantitative identification of pegmatite types.

Examples include K/Rb ratios in K-feldspar and K/Rb, K/Cs or Mg/Li in muscovite; these data can be compared graphically as in Fig. 7 to show progressive fractionation. Extensive studies utilizing Li, Rb and Cs were published by Utchakin et al. (1962) and Shmakin (e.g. 1979). As respective zones within a given pegmatite display differing levels of fractionation, care should be exercised in a regional program in that as much as possible, only similar zones are sampled in individual members of a pegmatite field.

Figure 8 shows an example of regional zoning in a pegmatite group, established on combined paragenetic, textural and geochemical data (Meintzer 1987; Wise 1987). The Blaisdell Lake group consists of three swarms that are virtually separated

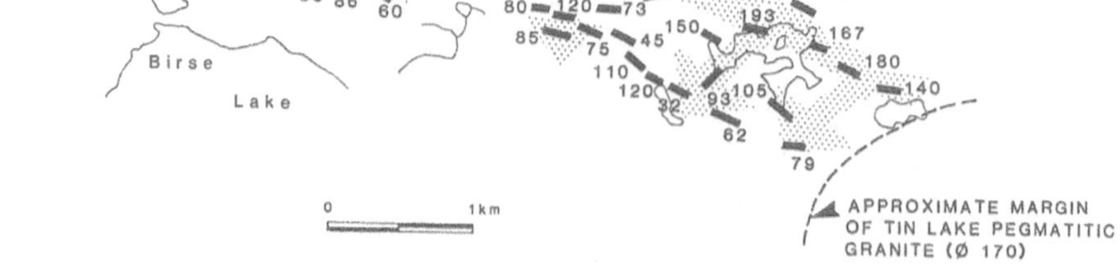

Fig. 7. Geochemical evolution of the Birse Lake pegmatite group flanking, and transitional from, the Tin Lake leucogranite in the Winnipeg River district, southeastern Manitoba, as modified by Trueman and Černý (1982) from Černý et al. (1981) and unpublished data. *Numbers* indicate average K/Rb ratio of blocky K-feldspar (3–4 samples per pegmatite) in individual dykes (*black bars*) of the group. Stippled arrows indicate general fractionation trends

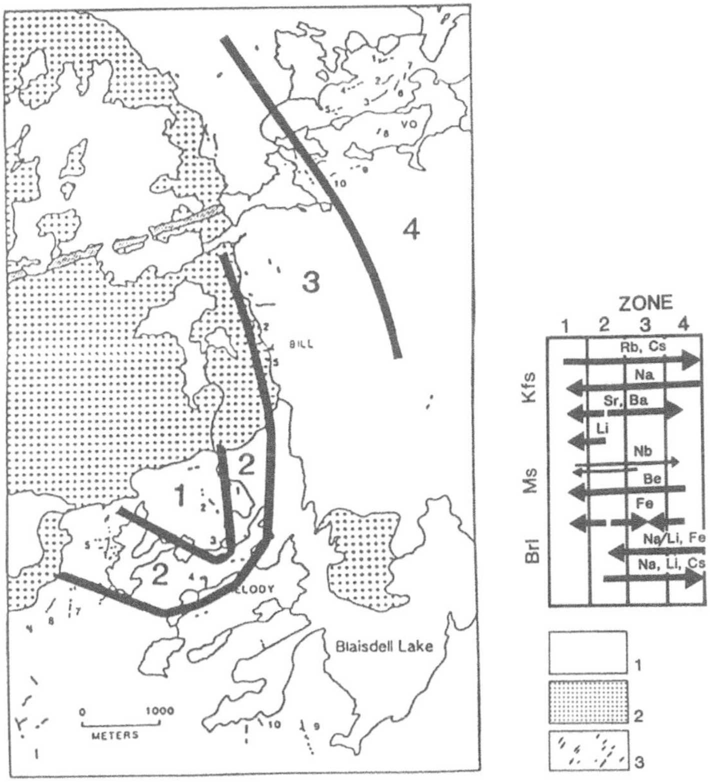

Fig. 8. Zoning in the Blaisdell Lake group (Melody, Bill and VO swarms) of rare-element pegmatites, flanking the Wedge granite in the Archean Yellowknife pegmatite field, Northwest Territories (modified from Meintzer 1987 and Wise 1987): *1* nodular biotite schist of the Burwash formation; *2* two-mica garnetiferous leucogranite; *3* pegmatites. The diagram illustrates the direction of regional increase in fractionation-sensitive elements and fractionation ratios in K-feldspar, muscovite and beryl in pegmatites subdivided into four zones: *1* barren dykes; *2* beryl-columbite subtype; *3* beryl-columbite and beryl-columbite-phosphate subtypes; *4* complex, spodumene subtype. Minerals of the columbite-tantalite group show increasing Mn and Ta contents and structural disorder from *zone 1* to *zone 4*

in areal extent; nevertheless, an overall zoning pattern can be established using all applicable criteria.

Several quantitative approaches were attempted during the past 2 decades to establish ranges of geochemical parameters for individual pegmatite types of the rare-element class. All criteria used in these studies can be used with advantage for defining regional zoning patterns.

Gordiyenko (1971) collected data on the Li, Rb and Cs contents of K-feldspar and muscovite from a diversity of pegmatite fields and provinces and arrived at a characterization of four different rare-element pegmatite types in them. Figure 9 provides also a comparison of the barren rare-element pegmatites with their counterparts from the muscovite-class populations, revealing a definite separation of their geochemical features.

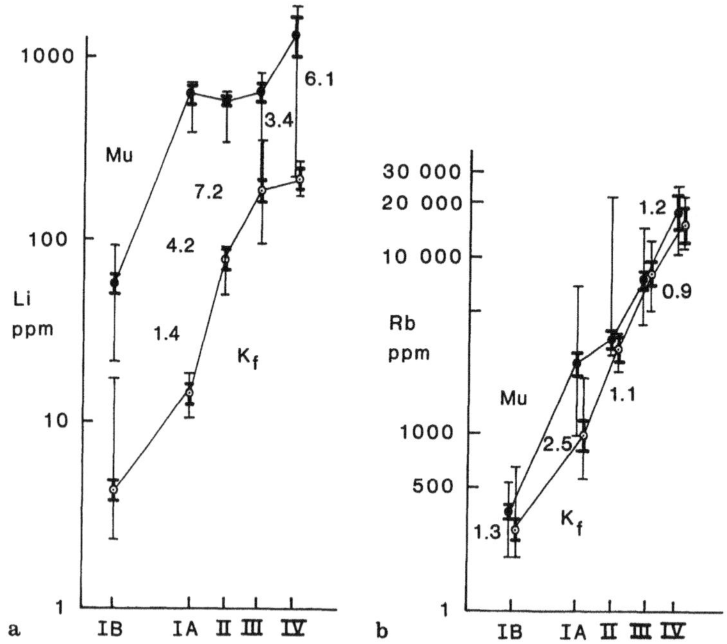

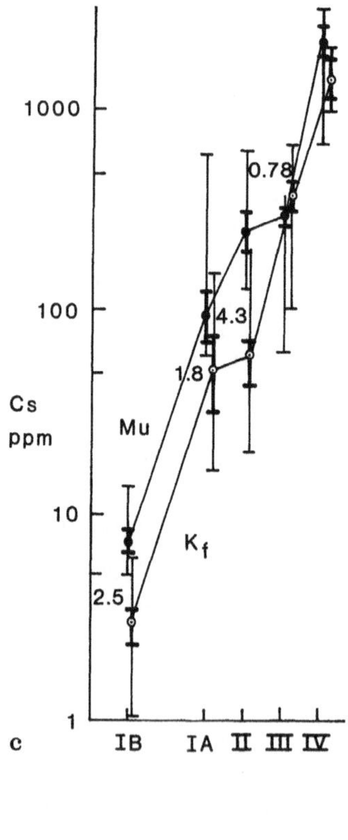

Fig. 9. Li, Rb and Cs contents of muscovite (*Mu*) and K-feldspar (*Kf*) typical of different pegmatite categories, as modified by Trueman and Cerný (1982) from Gordiyenko (1971): *IB* barren pegmatites of the muscovite-class provinces; *IA* barren pegmatites of the rare-element-class provinces; *II* rare-element pegmatites of the beryl type (i) (beryl-columbite subtype in particular); *III* complex rare-element pegmatites of the spodumene subtype (ii); *IV* complex rare-element pegmatites of the spodumene (ii) and lepidolite (iii) subtypes, enriched in tantalum and cesium. *Solid* and *open dots* average values of the mean θ; *heavy vertical bar* the 95% interval of θ; *thin vertical bar* range of arithmetic means with 95% confidence level. *Numeric values* indicate partitioning coefficient Mu/Kf

Gordiyenko's data base contains muscovite and K-feldspar of different generations which are, particularly in the more complex pegmatites, far from uniform in trace element contents. This makes the graph of Fig. 9 quite universal and reliable in comparing data from several generations of these minerals. On the other hand, selection of a specific generation of each mineral for establishing the distinguishing parameters would considerably decrease the individual data spreads.

A more recent method proposed by Gordiyenko (1976) is based on the Na_2O content (wt.%) and the K/Cs ratio of K-feldspars alone. It covers pegmatites of non-mineralized ceramic and muscovite class and those of the rare-element-class provinces with a more detailed subdivision than the earlier work. Two separate diagrams are presented for Precambrian and Phanerozoic pegmatite provinces (Fig. 10a and b respectively).

Utilization of the fractionation ratio K/Cs instead of the absolute Cs content helps to eliminate the influence of possible sample contamination by quartz and non-perthitic albite, but reliable determinations of the high K and trace contents of Cs are required. However, a certain cushion is available here because the overall range of the K/Cs ratio and its rate of change among individual pegmatite types are extensive.

The difference between the geochemical signatures of the Precambrian and Phanerozoic pegmatites is most probably the result of statistical differences rather than a reflection on the true characteristics of these two pegmatite populations. Gordiyenko's data base for Phanerozoic localities amounts to only 13% of his

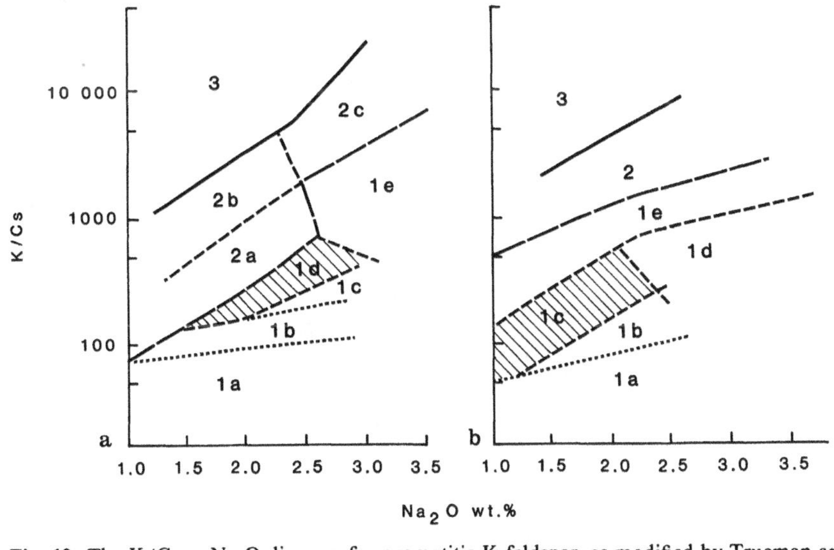

Fig. 10. The K/Cs vs Na_2O diagram for pegmatitic K-feldspar, as modified by Trueman and Černý (1982) from Gordiyenko (1976). *Heavy solid line* boundary between the muscovite class (*3*) and rare-element class (*1,2*) pegmatite provinces; *heavy long-dashed line* boundary between geochemical series within the rare-element class (*1* cesium-enriched; *2* poor in cesium). *Short-dashed* and *dotted lines* separate pegmatite varieties: *1a, b, c* Li, Rb, Cs, Be, Ta with pollucite (types ii, iii of this work); *1d* Li, Be, Ta without pollucite (ii, iii); *1e* barren; *2a* spodumene-bearing (ii); *2b* beryl-bearing (i); *2c* barren. *Ruled areas* overlap of *1c* and *1d*. **a** Precambrian; **b** Phanerozoic pegmatites

Precambrian file; the Precambrian diagram is the better of the two, and is probably uniformly applicable to pegmatites of any age bracket.

Černý et al. (1981) employed, as a characterization tool, the K/Rb vs Cs relationship in blocky K-feldspar from the inner intermediate zones adjacent to quartz cores (or from the blocky central parts if quartz cores were not developed). As shown in Fig. 11, the compositional fields characterizing many pegmatite groups can overlap considerably. This is not surprising in a pegmatite field with closely related mineralization types; nor is it surprising in a graph employing Rb data which are not particularly discriminatory (Gordiyenko 1976). Despite this drawback, critical K/Rb + Cs values can be indicated in Fig. 11 which appear to be minimal for the potentially economic mineralization in almost any of the pegmatite groups of the district.

The well-known variation in the alkali contents of beryl can also be used for estimating the fractionation level of granitic pegmatites. Černý (1975) quantified the alkali contents of beryl and suggested a correlation of the Na/Li vs Cs values of the latest, most fractionated beryl generations with several types of rare-element pegmatites, a modified graphic representation of this relationship is shown in Fig. 12. The practical value of this graph in exploration is, however, questionable because of the extreme variability of beryl composition within a single pegmatite (even within a single crystal) and because of the considerable overlap of multigeneration sequences from different pegmatite types. Well-defined late-generation beryl samples must be selected here to obtain reliable results.

Garnet compositions are similarly plagued by extensive variations in the Fe/Mn ratio which is their most sensitive characteristic. This ratio decreases with progressive pegmatite fractionation, and it may show dramatic changes even within a single pegmatite (Baldwin and von Knorring 1983).

Further details on the above and related geochemical methods are available in Beus et al. (1968), Solodov (1971) and Trueman and Černý (1982).

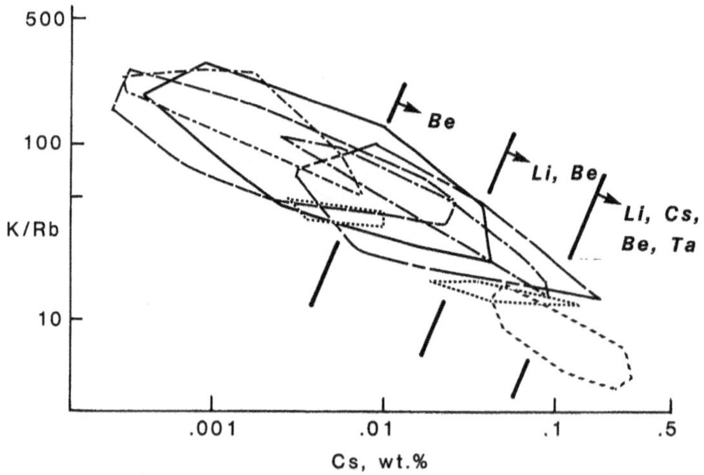

Fig. 11. K/Rb vs Cs of the blocky K-feldspar in pegmatites of the Winnipeg River district, southeastern Manitoba, summarized from Černý et al. (1981). Individual fields mark compositions of feldspar samples from different pegmatite groups, based on 6 to 69 analyses each. *Heavy lines* indicate approximate appearance of significant mineralization

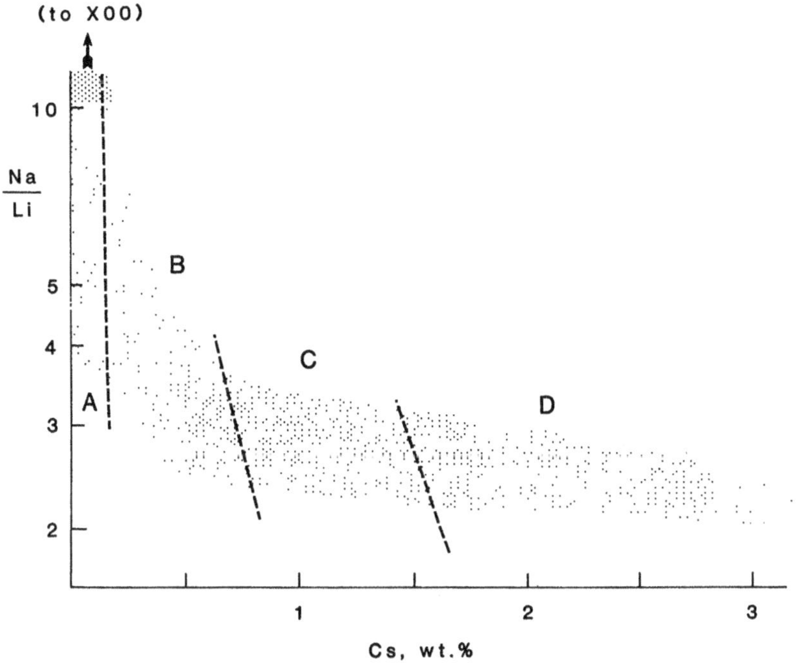

Fig. 12. Na/Li vs Cs in late generations of beryl (modified from Černý 1975). The decrease in Na/Li is fairly proportional to the increase in the absolute Na and Li contents. *A* Barren and beryl-type (i) pegmatites (± columbite-tantalite, phosphates) poor in rare alkalis; *B* beryl-columbite and beryl-columbite-phosphate subtypes (i), up to spodumene-bearing pegmatites, enriched in rare alkalis (ii); *C* spodumene and petalite subtypes (ii); *D* spodumene, petalite, amblygonite and lepidolite subtypes (ii, iii) with pollucite and Nb< Ta mineralization

5 Deposit-Scale Identification: Pegmatitic Granites and Pegmatites

Once a regional disposition of different pegmatite types has been established, the identification of prospective individual deposits can be initiated within the zones of interest. Rare as they are, however, tantalum-enriched deposits located within fertile granites must be mentioned here, in addition to pegmatites proper.

5.1 Mineralization in Pegmatitic Granites

The outer pegmatitic envelopes of the fertile granites occasionally enclose highly fractionated pegmatite pods enriched in rare elements, including tantalum (e.g. Haapala 1966; Goad and Černý 1981). However, they are almost invariably of insignificant dimensions; large-volume mineralization within parental granites is extremely scarce but some of the known occurrences are of prime economic interest.

The two best-known examples are the pegmatitic cupolas of leucogranites at Montebras and Echassières in the Massif Central of France (Aubert 1969; Burnol

1974; Kosakevitch 1976; Cuney et al. 1986). The first occurrence matches the characteristics of the amblygonite subtype of complex rare-element pegmatites, the second locality closely corresponds to a P-enriched lepidolite subtype with significant Li, Be, Sn and Ta reserves (cf. Table 1). The textural, paragenetic and geochemical parameters of these highly fractionated cupolas correspond to an extreme case of the pegmatite phase totally contained within its parent granite, rather than to a mineralized granite in the usual application of this term.

Another example of highly fractionated pegmatitic granite was recently described from the Archean of northeastern Manitoba (Chackowsky 1987; Chackowsky and Černý 1987). An extensive, 6-km-long linear intrusion of leucogranite following an axial fault in a greenstone belt contains in its central parts a subcontact concentration of petalite, lithian mica and columbite-tantalite (Fig. 13). The paragenesis corresponds to the petalite subtype of complex rare-element pegmatites (ii). Geochemical features of the linear intrusion indicate extreme fractionation in its mineralized part; selected indicators of this trend are shown graphically in Fig. 14. The economic potential of this deposit has yet to be determined.

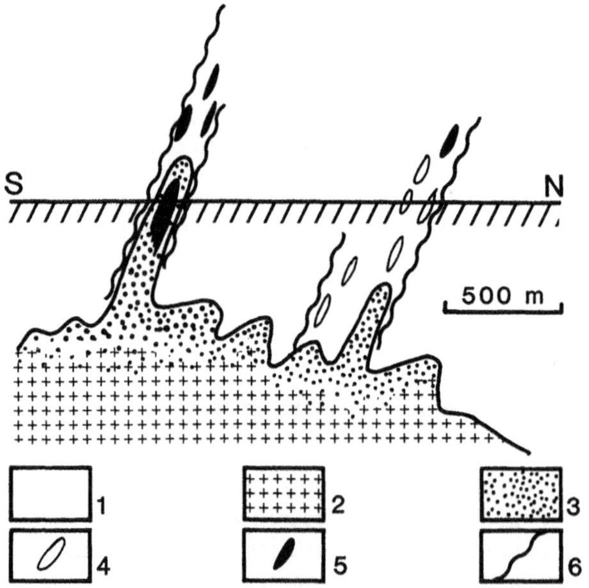

Fig. 13. Schematic section across two east-west-trending fertile leucogranite plus pegmatite systems at Red Sucker Lake, northeastern Manitoba (modified from Chackowsky 1987): *1* greenstone belt (undivided); *2* hypothetical granite parental to both systems; *3* leucogranite, largely pegmatitic, capping the granite; *4* barren (to beryl-columbite subtype) pegmatites; *5* highly fractionated petalite-subtype pegmatites, and a mineralized segment within the southern pegmatitic leucogranite, analogous in composition to the external complex pegmatites outcropping on the northern side of the leucogranite west of the section; *6* fault and shear system

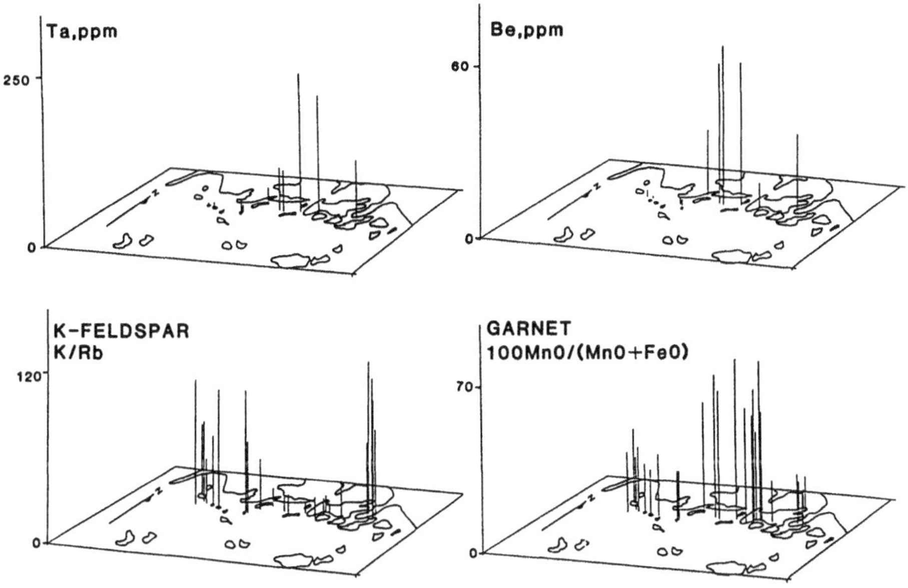

Fig. 14. Geochemical features of the southern linear intrusion of pegmatitic leucogranite at Red Sucker Lake, northeastern Manitoba (modified from Chackowsky 1987): whole-rock tantalum and beryllium contents, K/Rb of blocky K-feldspar from potassic pegmatite pods, and (Mn × 100)/(Mn + Fe) of accessory garnet. The highly fractionated region corresponds to the petalite-, columbite-tantalite- and lithian-muscovite-bearing central part of the intrusion (cf. Fig. 13)

5.2 Pegmatite Deposits

Qualitative identification of pegmatites potentially containing significant tantalum concentrations results from the study of regional zoning patterns. However, the textural, paragenetic and geochemical parameters described in the preceding sections are indicative of only the *qualitative potential* at different fractionation levels that give no guarantee of a sufficiently high *quantitative concentration* of tantalum within a given pegmatite body. A different type of geochemical indicator is required that would reflect the absolute tantalum content within the pegmatite bulk composition.

Odikadze (1958), Heinrich (1962), Beus (1966) and Gordiyenko (1970) collected data on the Nb and Ta contents of muscovite and showed a good correlation between the Nb-Ta content of the mica and the presence of Nb-Ta mineralization. Beus (1966) concluded that the probability of encountering relatively high (≥20 ppm) concentrations of Ta in muscovite is best in the rare-element pegmatites containing columbite-tantalite. High Nb-concentrations (≥200 ppm) are diagnostic of muscovite from columbite-bearing pegmatites. Also the mean Nb/Ta ratio is lower for muscovite from tantalite-bearing pegmatites (4 ± 0.7) than that from the relatively Ta-poor pegmatites with tantalian columbite (9.5 ± 1.5). Gordiyenko's (1970, Table 71) data and interpretation compare well with those of Beus (1966). Gordiyenko also concluded that prospective Ta-rich pegmatites have an average Ta_2O_5 content higher than 80–90 ppm in their muscovite.

Gaupp et al. (1984) recently improved the muscovite-based technique and provided several indicators based on extensive data collected on samples from pegmatites with well-established levels of tantalum mineralization. The reader is referred to Morteani and Gaupp (this Vol.) for an updated review of the original data. However, it must be pointed out that the above studies are focussed at the recognition of economic tantalum concentrations in the most fractionated complex pegmatites (ii) and (iii), and that the conclusions do not necessarily apply to pegmatites of the beryl (i), albite-spodumene (iv) and albite (v) types.

This muscovite "tantalometer" is the last of the different methods that provide a relatively inexpensive and time-saving reduction of target areas in the exploration of pegmatite populations for tantalum deposits. Beyond this treatment, conventional trenching, drilling and assaying must be utilized to establish the grades and reserves, and their distribution within a given pegmatite body. General guidance for this last aspect is provided in Černý (this Vol.).

6 Search For Hidden Deposits

Within an identified regional zone of pegmatites with proven tantalum potential, a search for blind deposits that do not outcrop on the surface may be advisable. Locating dispersion aureoles of mobile, pegmatite-related elements in country rocks is the first step in such a search. Geophysical outlining of a pegmatite intrusion underlying a surface-detected aureole may precede, follow or be conducted simultaneously with test drilling.

6.1 Geochemical Methods: Primary Dispersion Aureoles

Highly mobile alkali elements and volatile components enriched in pegmatite fluids cause extensive contamination of the host rocks by forming exomorphic aureoles. The geochemical detection of such dispersion haloes forms a principal approach in exploration for hidden or blind pegmatites.

Such aureoles are characterized by particular elements or their combinations which relate to the degree of fractionation of the particular type of pegmatite (Table 3). Depending on the particular mobility of a certain element, such an aureole may

Table 3. Pegmatite types with corresponding diagnostic elements enriched in primary dispersion haloes (after Beus et al. 1968)

Pegmatite type	Diagnostic elements
Aplites and barren pegmatites	Boron
Beryl-columbite pegmatites	Boron, lithium, tin and beryllium
Tantalum-enriched pegmatites	Boron, lithium, tin, cesium, beryllium and rubidium
Cassiterite-quartz veins	Boron, lithium, tin, beryllium, cesium, tungsten, antimony and fluorine

be detectable for different distances from its source. This relative mobility becomes useful in multi-element haloes (for example in the case of highly fractionated bodies), and the ratios of such elements combined with a knowledge of their relative mobilization distances can provide a first estimate of depths of residence.

In general, in using geochemical approaches, background values should be determined for any given pegmatite-bearing terrane. Subsequently, definition of an anomaly can be based on features that are, for example, in excess of two standard deviations. Figure 15 shows an example of geochemical anomalies overlying the Tanco tantalum deposit, demonstrating the difference in mobilities of Li and Rb.

Besides the simple concentrations of individual elements, different enhancement techniques have been developed. Multiplicative formulas, occasionally utilizing elements forming negative dispersive haloes (e.g. Ba, Sr) are available that may differentiate between footwall and hanging-wall portions of dispersion aureoles.

Further details on the methods discussed in this section are available in Hornung (1962), Beus et al. (1968), Makagon et al. (1974), Ovchinnikov et al. (1976), Zagorskyi et al. (1979), Trueman and Černý (1982) and London (1986).

6.2 Geophysical Methods: Gravity Anomalies

Excepting radiative properties of some types, pegmatites might best be described as geophysical non-responders. If, however, the anticipated density contrast between a blind pegmatite and its host rocks is adequate, then gravity methods become useful.

Figure 16 shows an example of a gravity survey over the gabbro-hosted Tanco pegmatite. Of interest is the profile which displays a gravity low over wall zones of

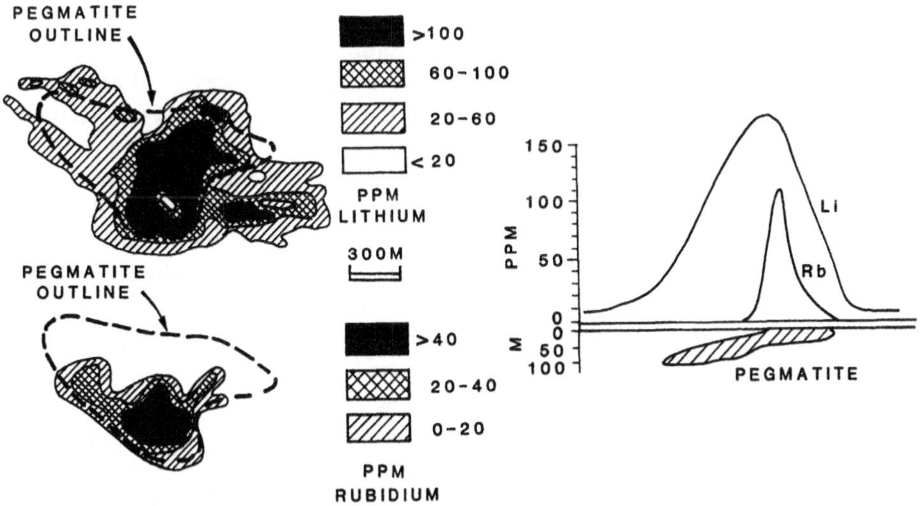

Fig. 15. Lithogeochemical anomalies of Li and Rb in hanging-wall amphibolite over the Tanco pegmatite, southeastern Manitoba, in plan (*left*) and in a north-south cross-section (Trueman and Černý 1982)

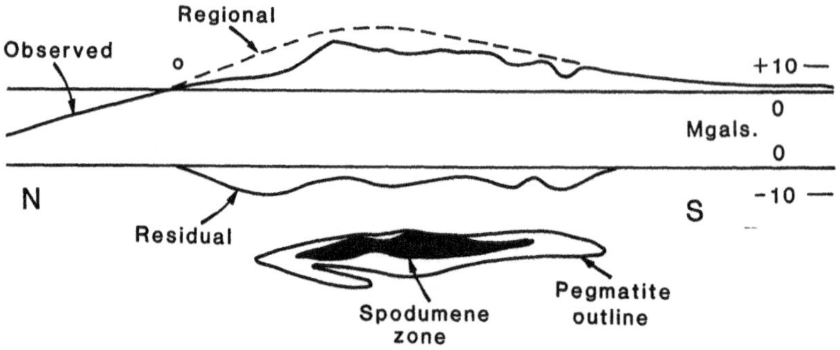

Fig. 16. A Bouguer gravity profile over the Tanco pegmatite, southeastern Manitoba (after Trueman and Černý 1982). Note the effect of spodumene (3.15 g/cm³) enrichment in the central upper parts of the pegmatite (2.68 g/cm³), reducing the gravity contrast to the enclosing amphibolite (3.00 g/cm³)

the pegmatite, but a gravity positive over the lithium zone. A simple quartz + feldspar + mica pegmatite could yield a gravity low, but a spodumene-bearing dyke may manifest a gravity high relative to the same host rock, due to the high density of spodumene (3.15 g/cm³). Discretion in gravity interpretation is necessary therefore and should be accompanied by primary, for example, geochemical data.

7 Exploration in Deeply Weathered Terranes: Secondary Dispersion Aureoles

Lack of bedrock outcrop in regions of deep weathering restricts the exploration to methods that are much less direct than those discussed in Sections 4 to 6. This is particularly true for lateritized terranes that commonly feature quartz cores as the only pegmatite relics that did not succumb to a ubiquitous decay of silicate rocks.

Secondary dispersion haloes of typical pegmatitic rare elements in soils facilitate locating pegmatites, although the contrast between their decomposition products and those of the country rocks is much lower than in unaltered fresh rocks. Nevertheless, Matheis (1981) showed that Li, Be, Rb, Cs, Mg/Li and K/Rb in the B-horizon of lateritic soils effectively outlines pegmatite bodies. Marshall and Herman (1986) examined secondary dispersion aureoles surrounding weathered pegmatites in gneiss-derived sapropelites of Virginia. They used a sequential extraction procedure to measure the distribution of Be, La, Nb, Sn, U, Li and Ni among different fractions of the B-horizon soils, and developed a sequential discriminant function analysis to differentiate between soils formed over primitive pegmatites, complex pegmatites and country rocks. They found the data on Sn, La, U and Li particularly useful.

Panning for heavy mineral concentrates in the argillic weathering products in situ, or in transported and redeposited lithologies is a possible approach, providing information on the weathering-resistant accessory mineral assemblage (which includes all tantalum-bearing oxide minerals). Panning of stream sediments is particularly informative for initial prospecting in virgin areas.

Further information on exploration for mineralized pegmatites in weathered regions is available in Beus (1960), Debnam and Webb (1960), Beus et al. (1968), Reedman (1973), Ovchinnikov (1976), Matheis and Küster (1988) and in the references quoted above. However, only Beus et al. (1968) deal specifically with exploration for tantalum-bearing pegmatites. Recognition of regional zoning and identification of different pegmatite types is not an easy task in residual overburden.

8 Exploration in Glaciated Terranes

In exploration for tantalum-enriched pegmatites, regions affected by continental-type glacial erosion are a study in contrasts. Areas of dominant bedrock exposures, in which differential erosion leaves pegmatitic granites and pegmatites distinctly above the level of metamorphic country rocks, are ideal but scarce. Drift-covered regions are possibly the most difficult ones to deal with, because the pegmatite-derived components can be extremely diluted in the clastic material of the dominant country rocks during extended glacial and/or fluvial transport. However, the classic methods of boulder-chasing and panning for heavy mineral concentrates do produce results, particularly in "average" situations of incomplete drift cover and in the presence of eskers. The resistance of tantalum-bearing minerals to weathering and abrasion preserves them through extensive glacial and fluvial transport.

9 Concluding Remarks

The present review has not exhausted all possible techniques applicable to the exploration for tantalum-enriched pegmatites. The reader is referred to the sources quoted throughout the text for specific details and additional information, such as the biogeochemical and hydrogeochemical methods (Beus et al. 1968). Nevertheless, this review should provide a useful guide through all scales of the exploration task, and may serve as an expanded version of district analysis (Pollard this Vol.), by encompassing approaches to the problem at all levels from metallogenic provinces to individual deposits.

Future development of exploration techniques should be based on better petrogenetic understanding of the fertile granites, and of the relationships among regional metamorphism, tectonic processes, anatexis and fractionation generating these rocks and their pegmatite aureoles. This may considerably restrict the terranes potentially hosting fertile granites, beyond the general structural and metamorphic constraints discussed in Sections 3.1 to 3.3. Further generalization and refinement of the indicators of quantitative Ta accumulation in individual pegmatite deposits is also desirable; it should include moderately fractionated pegmatites and those of the structurally "homogeneous" types (iv) and (v).

Acknowledgements. The author is indebted to D.L. Trueman of Highwood Resources, Vancouver for many years of sharing his experience in pegmatite exploration, and for a critical review of the manuscript. Much of the material presented here is based on our joint 1982 publication. Thanks are also due to R. Alviola, A.A. Beus, R. Gaupp, E. Wm. Heinrich, J.J. Norton and S.-A. Smeds for discussions on diverse aspects of tantalum-bearing pegmatites, and to my students A.J. Anderson, L.E. Chackowsky, B.E. Goad, R.E. Meintzer, E.K. Ucakuwun and M.A. Wise whose work contributed extensively to the development of some exploration criteria. Financial support by the National Science and Engineering Research Council of Canada, Canada Department of Energy, Mines and Resources, Department of Indian and Northern Affairs Yellowknife, and Tantalum Mining Corporation of Canada, Ltd. is gratefully acknowledged.

References

Anderson AJ (1984) Geochemistry, mineralogy and petrology of the Cross Lake pegmatite field, Manitoba. MSc Thesis, Univ Manitoba, Winnipeg

Anhaeusser CR (1976) Archean metallogeny in southern Africa. Econ Geol 71:16–43

Aubert G (1969) Les coupoles granitiques de Montebras et d'Echassières (Massif Central Français) et la genèse de leurs minéralisations en étain, tungstène, lithium et beryllium. Mem BRGM 46, 345 pp

Ayres LD (1978) Metamorphism in the Superior Province of northwestern Ontario and its relationship to crustal development. In: Fraser JA, Heywood WW (eds) Metamorphism in the Canadian Shield. Geol Surv Can Pap 78-10:25–36

Ayres LD, Černý P (1982) Metallogeny of granitic rocks in the Canadian Shield. Can Min 20:439–536

Ayres LD, Thurston PC (1985) Archean supracrustal sequences in the Canadian Shield: an overview. Geol Assoc Can Spec Pap 29:343–380

Baldwin JR, Knorring O von (1983) Compositional range of Mn-garnet in zoned granitic pegmatites. Can Min 21:683–688

Baragar WRA, Scoates RFJ (1981) The Circum-Superior belt: a Proterozoic plate margin? In: Kröner A (ed) Precambrian plate tectonics. Elsevier, Amsterdam, pp 297–330

Beus AA (1960) Geochemistry of beryllium and genetic types of beryllium deposits. Acad Sci USSR Moscow Engl Transl Freeman, London 1966, 401 pp

Beus AA (1966) Distribution of tantalum and niobium in muscovites of granitic pegmatites. Geokhimya 10:1216–1220 (in Russian)

Beus AA, Sitnin AA (1968) Geochemical features of the distribution of tantalum in granitoids. Geokhimya 579–585 (in Russian)

Beus AA et al. (1968) Geochemical exploration for endogenic deposits of rare elements on the example of tantalum. Nedra, Moscow, Engl Transl GSC Libr, Ottawa

Blockley JG (1980) The tin deposits of western Australia, with special reference to the associated granites. Geol Surv W Aust Min Res Bull 12:184 pp

Burnol L (1974) Géochimie du béryllium et types de concentration dans les leucogranites du Massif Central Français. Bur Rech Geol Min Mem 85

Černý P (1975) Alkali variations in pegmatitic beryl and their petrogenetic implications. Jahrb Min Abh 123:198–212

Černý P (1982a) Anatomy and classification of granitic pegmatites. In: Černý P (ed) Granitic pegmatites in science and industry. Min Assoc Can Short Course Handb 8:1–39

Černý P (1982b) Petrogenesis of granitic pegmatites. In: Černý P (ed) Granitic pegmatites in science and industry. Min Assoc Can Short Course Handb 8:405–461

Černý P (1989) Rare-element granitic pegmatites. In: Brown AC (ed) Monograph series on geology of ore deposits (in press)

Černý P, Brisbin WC (1982) The Osis Lake pegmatitic granite, Winnipeg River district, southeastern Manitoba. In: Černý P (ed) Granitic pegmatites in science and industry. Min Assoc Can Short Course Handb 8:405–461

Černý P, Meintzer RE (1988) Fertile granites in the Archean and Proterozoic fields of rare-element pegmatites: crustal environment, geochemistry and petrogenetic relationships. In: Taylor RP, Strong DF (eds) Recent advances in the geology of granite-related mineral deposits, CIM Spec Vol 39:170–207

Černý P, Trueman DL, Ziehlke DV, Goad BE, Paul BJ (1981) The Cat Lake-Winnipeg River and the Wekusko Lake pegmatite fields, Manitoba. Man Dep Energ Min Mineral Res Div Econ Geol Rep ER 80-1

Chackowsky LE (1987) Mineralogy, geochemistry and petrology of pegmatitic granites and pegmatites at Red Sucker Lake and Gods Lake, northeastern Manitoba. M Sc Thesis, Univ Manitoba, Winnipeg

Chackowsky LE, Černý P (1987) A unique lithophile mineralization of pegmatitic granites at Red Sucker Lake, northeastern Manitoba. GAC-MAC Annu Meet Progr Abstr 12, Saskatoon 1987, p 30

Clark GS (1982) Rubidium-strontium isotope systematics of complex granitic pegmatites. In: Černý P (ed) Granitic pegmatites in science and industry. Min Assoc Can Short Course Handb 8:347–372

Clark GS, Cheung S-P (1980) Rubidium-strontium ages from the Oxford Lake-Knee Lake greenstone belt, northern Manitoba: Can J Earth Sci 17:560–568

Condie KC (1981) Archean greenstone belts. Elsevier, Amsterdam

Cuney M, Autran A, Burnol L, Brouand M, Dudoignon P, Feybesse J-L, Gagny C, Jacquot T, Kosakevitch A, Martin P, Meunier A, Monier G, Tegyey M (1986) Résultats préliminaires apportés par le sondage GPF sur la coupole de granite albitique à topaze-lépidolite de Beauvoir (Massif Central, France). C R Acad Sci Paris 303 Ser II, 7:569–574

Debnam AH, Webb JS (1960) Some geochemical anomalies in soil and stream sediment related to beryl pegmatites in Rhodesia and Uganda. Trans Inst Min Metall 69:329–344

Dimroth E (1981) Labrador geosyncline: type example of early Proterozoic cratonic reactivation. In: Kröner A (ed) Precambrian plate tectonics. Elsevier, Amsterdam, pp 331–352

Gaupp R, Möller P, Morteani G (1984) Tantal-Pegmatite: Geologische, petrologische und geochemische Untersuchungen. Monogr Ser Min Dep 23. Bornträger, Berlin Stuttgart, 124 pp

Gibb RA, Walcott RI (1971) A Precambrian suture in the Canadian Shield. Earth Planet Sci Lett 10:417–422

Ginsburg Al, Timofeyev IN, Feldman LG (1979) Principles of geology of the granitic pegmatites. Nedra Moscow, 269 pp (in Russian)

Goad BE, Černý P (1981) Peraluminous pegmatitic granites and their pegmatite aureoles in the Winnipeg River district, southeastern Manitoba. Can Min 19:177–194

Gordiyenko VV (1970) Mineralogy, geochemistry and genesis of the spodumene pegmatites. Nedra, Leningrad, 237 pp (in Russian)

Gordiyenko VV (1971) Concentration of Li, Rb, and Cs in potash feldspar and muscovite as criteria for assessing the rare metal mineralization in granite pegmatites. Int Geol Rev 13:134–142

Gordiyenko VV (1976) Diagrams for prognostic evaluation of rare-element mineralization in granitic pegmatites utilizing compositional variations in potassium feldspars. Dokl Acad Sci USSR 228:442–444 (in Russian)

Haapala I (1966) On the granitic pegmatites in the Peräseinäjöki-Alavus area, South Pohjanmaa, Finland. Bull Comm Geol Finl 224, 98 pp

Hatcher MI, Elliott A (1986) Greenbushes – a new world source of lithium. 7th Int Congr Ind Mineral Proc, pp 217–232

Heinrich EWm (1953) Zoning in pegmatite districts. Amer Min 38:68–87

Heinrich EWm (1962) Geochemical prospecting of the beryl and columbite. Econ Geol 57:616–619

Heinrich EWm (1975) Economic geology and mineralogy of petalite and spodumene pegmatites. Ind J Earth Sci 2:18–29

Henderson JB (1981) Archaean basin evolution in the Slave Province, Canada. In: Kröner A (ed) Precambrian plate tectonics. Elsevier, Amsterdam, pp 213–236

Hornung G (1962) Wall rock composition as a guide to pegmatite mineralization. Econ Geol 57:1127–1130

Hornung G, Knorring O von (1962) Pegmatites of the North Mtoko region, southern Rhodesia. Trans Geol Soc S Afr 65, 2:153–180

Jahns RH, Griffitts WR, Heinrich EWm (1952) Mica deposits of the southeastern Piedmont. US Geol Surv Prof Pap 248-A:102 pp

Kesler TL (1961) Exploration of the Kings Mountain pegmatites. Min Eng 13:1062–1068

Kesler TL (1976) Occurrence, development and long-range outlook of lithium-pegmatite ore in the Carolinas. US Geol Surv Prof Pap 1005:45–50

Kistler RW, Ghent ED, O'Neill JR (1981) Petrogenesis of garnet two-mica granites in the Ruby Mountains, Nevada. J Geophys Res 86, B 11:10591–10606

Kosakevitch A (1976) Evolution de la minéralisation en Li, Ta, et Nb dans la coupole granitique de Beauvoir (Massif d'Échassières, Allier). Serv Geol Nat Bur Rech Geol Min Rep 76 SGN, 316 MGA

Kuzmenko MV (ed) (1976) Rare-element granitic pegmatite fields (geochemical specialization and distribution). Nauka, Moscow, 332 pp (in Russian)
London D (1986) Holmquistite as a guide to pegmatitic rare metal deposits. Econ Geol 81:704–712
Longstaffe FJ (1982) Stable isotopes in the study of granitic pegmatites and related rocks. In: Černý P (ed) Granitic pegmatites in science and industry. Min Assoc Can Short Course Handb 8:373–404
Longstaffe FJ, Černý P, Muehlenbachs K (1981) Oxygen isotope geochemistry of the granitoids in the Winnipeg River pegmatite district, southeastern Manitoba. Can Min 19:195–204
Makagon VM, Nesmelov GG, Poletayev IA, Shmakin BM (1974) Contact alterations and geochemical anomalies around pegmatite veins with tantalum-cesium-lithium mineralization. Dokl Earth Sci Sect AGI Trans 219:221–223
Marshall BT, Herman JS (1986) Trace element distribution in the soils above deeply weathered pegmatites, Virginia, U.S.A.: implications for exploration. Appl Geochem 1:681–690
Matheis G (1981) Trace-element patterns in lateritic soils applied to geochemical exploration. J Geochem Explor 15:471–480
Meintzer RE (1987) The mineralogy and geochemistry of the granitoid rocks and related pegmatites of the Yellowknife pegmatite field, Northwest Territories. PhD Thesis, Univ Manitoba, Winnipeg
Nedumov IB (1964) The process of differentiation of pegmatitic melt and the role of tectonics in the formation of rare-element pegmatites. Int Geol Congr New Delhi 22 Sess Rep Pt VI:116–139
Nurmi PA, Haapala I (1986) The Proterozoic granitoids of Finland: granite types, metallogeny and relation to crustal evolution. Bull Geol Soc Finl 58:203–233
Odikadze GL (1958) On the presence of niobium and tantalum in muscovites from pegmatites of the Dzirulsk crystalline massif. Geokhimya 4:479–485 (in Russian)
Ovchinnikov LN (ed) (1976) Lithogeochemical methods for prospecting rare-metal pegmatites. Acad Sci USSR, Moscow, 79 pp (in Russian)
Reedman AJ (1973) Prospection and evaluation of beryl pegmatites in south-west Uganda. Overseas Geol Min Res 41:86–100
Saggerson EP, Turner LM (1976) A review of the distribution of metamorphism in the ancient Rhodesian craton. Precamb Res 3:1–53
Shearer CK, Papike JJ, Laul JC (1987) Mineralogical and chemical evolution of a rare-element granite-pegmatite system: Harney Peak granite, Black Hills, South Dakota. Geochim Cosmochim Acta 51:473–486
Shmakin BM (1979) Composition and structural state of K-feldspars from some U.S. pegmatites. Am Mineral 64:49–56
Solodov NA (1962) Internal structure and geochemistry of rare-element granitic pegmatites. Acad Sci USSR, Moscow, 243 pp (in Russian)
Solodov NA (1971) Scientific principles of prospective evaluation of rare-element pegmatites. Nauka, Moscow, 292 pp (in Russian)
Trueman DL, Černý P (1982) Exploration for rare-element granitic pegmatites. In: Černý P (ed) Granitic pegmatites in science and industry. Min Assoc Can Short Course Handb 8:463–494
Utchakin YM, Shimanskyi AA, Pauller TI (1962) Rare-alkali content of feldspar of the Sayan pegmatites. Geokhimiya 1962:673–680 (in Russian)
Varlamoff N (1972) Matériaux pur l'établissement des types et de la zonéographie des pegmatites granitiques à métaux rares de Madagascar. Acad R Sci Outre Mer, Sci Nat Med NS 18-6:72 pp
Walker RJ (1984) The origin of the Tin Mountain pegmatite, Black Hills, South Dakota. PhD Thesis, Inst Stud Min Deposits, S Dak School Min Tech, Rapid City, 384 pp
Winkler HGF (1967) Petrogenesis of metamorphic rocks, 2nd edn. Springer, Berlin Heidelberg New York, 334 pp
Wise MA (1987) Geochemistry and crystal chemistry of Nb, Ta and Sn minerals from the Yellowknife pegmatite field, N.W.T. PhD Thesis, Univ Manitoba, Winnipeg
Zagorskyi VYe, Makagon VM, Makagon LD, Chuvashova LA, Arbatskaya SN (1979) Tantalum, niobium, zirconium and hafnium in rare-metal pegmatite exo-contacts. Geokhimiya 1979 4:528–534 (in Russian)

Geochemical Evaluation of the Tantalum Potential of Pegmatites

G. MORTEANI[1] and R. GAUPP[2]

1 Introduction

Besides alkali granites, granitic pegmatites with local Ta_2O_5 concentrations of up to 0.1 wt.% are the most important economic sources for tantalum. Pegmatites very often occur in large swarms scattered over a wide area (e.g. Tantalite Valley, Namibia; Manitoba, Canada). Since not all pegmatites in such swarms are tantalum-bearing, many attempts have been made to distinguish pegmatites of economic interest from barren ones. This chapter presents and discusses geochemical methods based on mineral chemistry which are suitable for such distinction. It is not the intention of this chapter to discuss general prospection methods for concealed pegmatite deposits such as soil or stream sediment geochemistry. The focus here is a method useful for evaluating the tantalum potential of exposed pegmatites in a given pegmatite field.

2 Geology

The confusing spectrum of pegmatite classifications has been recently reviewed by Černý (1982). Ta-bearing pegmatites are usually grouped together with Li-, Cs-, Be-, Sn-, Nb-bearing ones as rare-metal or rare-element pegmatites. The rare-metal pegmatites are predominantly found in high temperature/low pressure metamorphic series particularly in the andalusite-cordierite-muscovite subfacies of Winkler's (1967) classification of metamorphic facies (Ginsburg et al. 1979; Černý 1982). The most important Ta_2O_5-bearing pegmatites are found within ortho-amphibolites in granite-greenstone terranes and schist belts of the Archean shields (Bikita/Zimbabwe, Tanco/Canada, Harding/USA, Wodgina/Australia, Morrua/Mozambique etc.). The pegmatites occur in the marginal zones of the greenstone belts and their adjacent granitoid and gneissic cupolas, or they are aligned within these belts along deep-reaching axial faults or along the contact zones of old intrusions (Moloshag 1974; Anhaeusser 1976; Černý 1982a; Trueman and Černý 1982).

[1]Lehrstuhl für Angewandte Mineralogie, Technische Universität München, Lichtenbergstrasse 4, D-8046 Garching
[2]Salbeiweg 3, D-3000 Hannover 51

Lanthanides, Tantalum and Niobium
P. Möller, P. Černý and F. Saupé (Eds.)

Many of the rare-metal pegmatites show a distinct spatial and geochemical relation to extremely fractionated leucogranitic to quartz-monzonitic intrusions of small to medium sizes (Brögger et al. 1922; Jahns 1955; Varlamoff 1972; Černý 1975, 1982; Černý et al. 1981).

Among the family of rare-metal pegmatites, Ta-pegmatites show the most complex internal structures and a vast variety of texturally and paragenetically different units. The primary zonal structure is often completely obliterated by lepidolitization or albitization, the latter leads to an extensive formation of saccharoidal albite and cleavelandite. Albitization is observed in marginal as well as in central parts of the pegmatite body. The bodies formed by metasomatic replacement are the main hosts of economically important Ta_2O_5 mineralizations; minerals of the columbite-tantalite group, microlite, stibiotantalite and wodginite being the most important Ta-ore minerals (Parker and Fleischer 1968; Hatcher and Bolitho 1982; Trueman and Černý 1982).

An extensive review of the chemistry, structural properties and paragenetic assemblages of tantalum-bearing minerals in different types of pegmatites is given by Černý and Ercit (1985).

3 Geochemistry

Feldspars and white micas record the trace element contents in the primary pegmatite-forming phase or later in the metasomatizing fluids. For this reason the geochemistry of these minerals is a geochemical indicator for Ta-mineralization in pegmatites.

3.1 Albite

Most of the albite in the pegmatites is found in the albite-rich aplitic zones and in the replacement bodies but also in small amounts in all zones of rare-metal pegmatites. Unfortunately, an unequivocal distinction between primary and secondary metasomatically formed albite is only rarely possible. Gaupp et al. (1984) distinguished secondary metasomatic albites, "albite I", from primary albites, "albite II" on textural grounds.

Variation of the most important minor and trace elements in the diverse albite types in a single pegmatite body have been investigated by analyzing samples from the Tanco pegmatite (Gaupp et al. 1984). The Cs, Sc, Rb and Ta contents are found to be the highest in secondary albite, but they show large scatter and overlap with Ta and Cs values characteristic for primary albite.

The albite of the strongly Ta-mineralized pegmatite of Tanco (Fig. 1) shows much higher Ta and Cs + Rb contents than that of poorly mineralized or barren pegmatites, such as those of Noumas (South Africa), Silverleaf, Coe Claim, Odd West, Buck Claim, Central Claim (Manitoba/Canada) and Hagendorf (Germany).

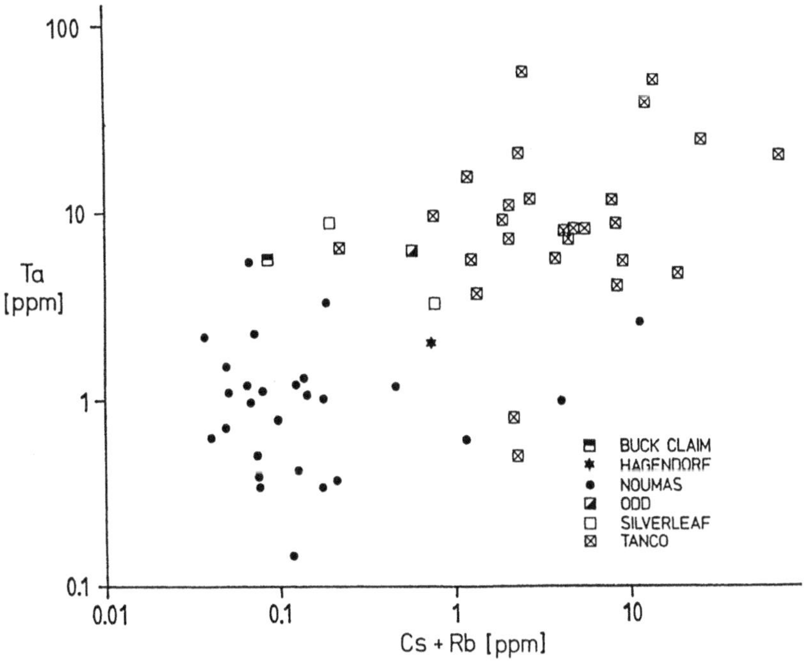

Fig. 1. Ta vs Cs + Rb plot of albite. Saccharoidal albite of type I always shows higher sums of Cs and Rb and mostly higher Ta contents than the associated cleavelandite

3.2 White Mica

White mica formation may occur in nearly all stages of pegmatite formation. As with albite, primary and secondary generations of white mica are not easily distinguishable by morphological and textural criteria.

During evaluation of pegmatites it is difficult to distinguish between the various types of mica. This is particularly true in weakly zoned and/or badly exposed pegmatites. For this reason, no distinction is made between different mica types in the following text. The only exception is that micas with a grain size below 2 mm are generally disregarded. In a K/Rb vs Cs diagram the white micas of Tanco show low K/Rb ratios and high Cs values, whereas micas from poorly mineralized pegmatites show lower Cs and higher K/Rb values, indicating lower degrees of element fractionation (Fig. 2).

Looking at the Ta vs K/Cs ratio the exceptional Ta content of the white micas from Tanco is evident. It correlates with low K/Cs ratios (Fig. 3). The analyses plot with most of the analyses of Noumas, White City and Cross Lake above the "a" and "b" boundaries which mark the Ta contents in white micas from pegmatites with interesting Ta mineralizations as proposed by Beus (1966) and Gordiyenko (1971) respectively. The Ta/W ratio increases with increasing Cs contents of white micas (Fig. 4). As the increasing Cs content is a fractionation indicator in the pegmatite fluid, the Ta/W ratio in white micas can be used as such, too.

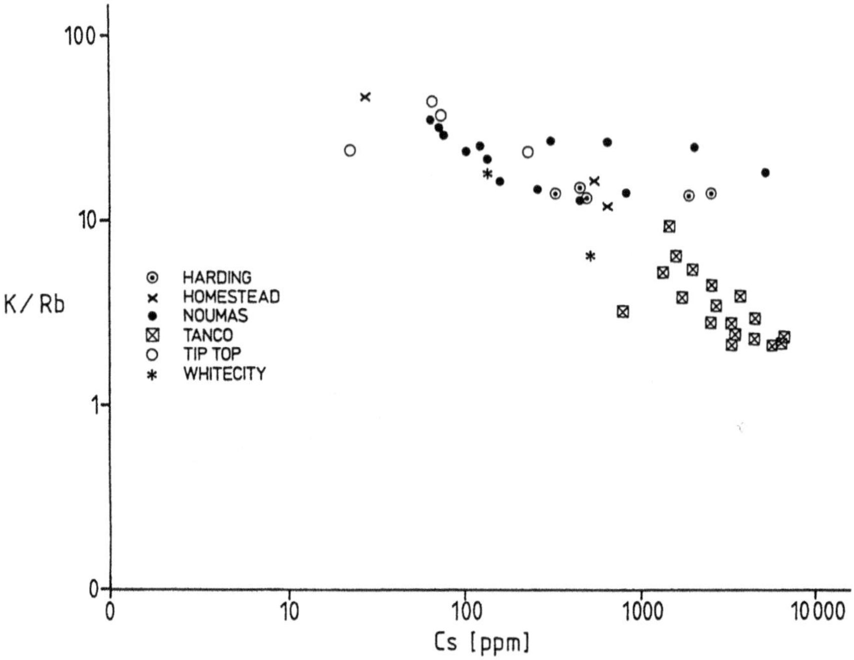

Fig. 2. Plot of K/Rb vs Cs in white mica

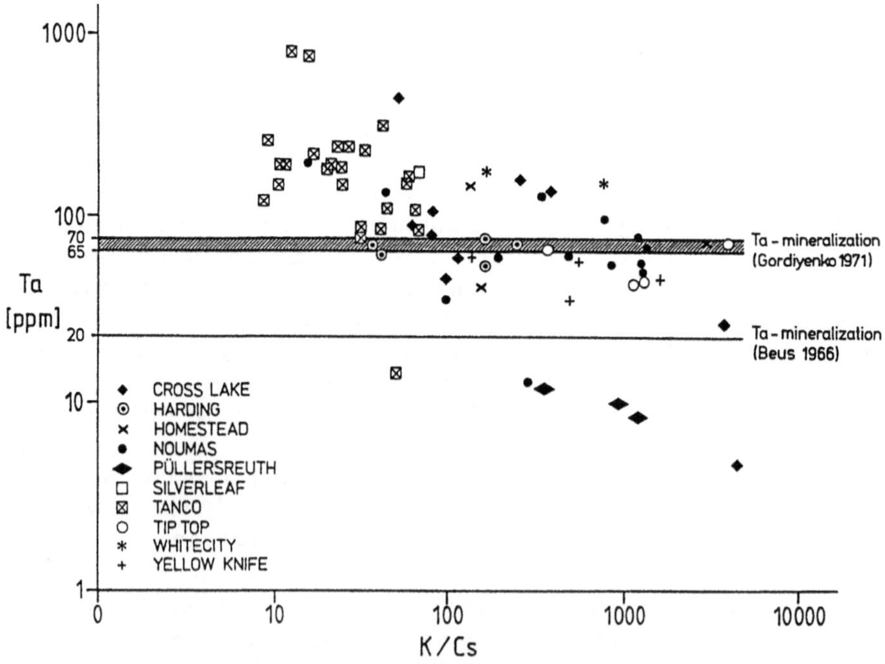

Fig. 3. Plot of Ta vs K/Cs in white mica. The Ta content in white mica increases with decreasing K/Cs ratio. The minimum Ta content for whiteicas in "economically important" Ta-mineralized pegmatites according to Gordiyenko (1971) and Beus (1966), respectively is given

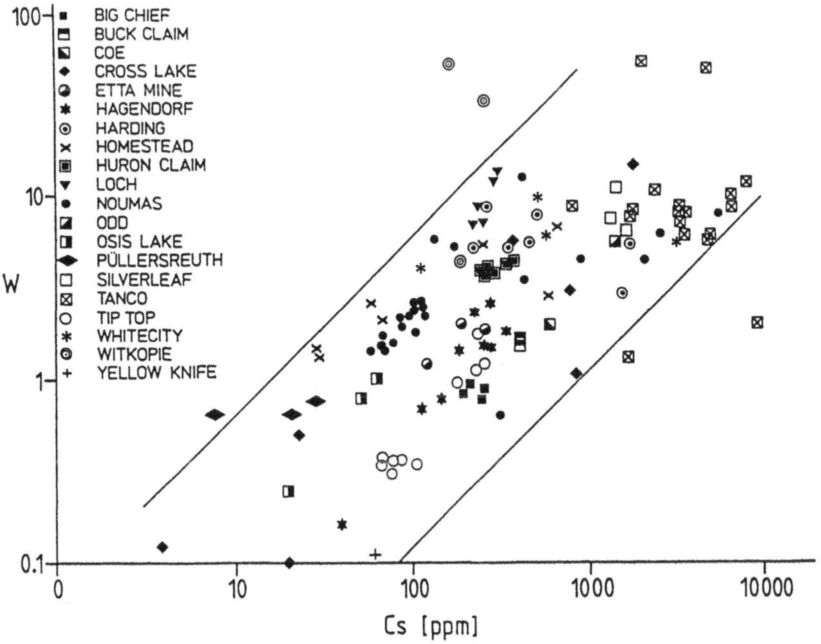

Fig. 4. Plot of the Ta/W ratio vs Cs for white mica. The Ta/W ratio increases with increasing element fractionation

Comparing Ta contents in white mica from barren, poorly and strongly mineralized pegmatites, extreme Ta contents are found in white micas from the most evolved pegmatites, but within individual pegmatite bodies, the Ta contents of white micas show considerable scatter. This extreme scatter of Ta values is predominantly caused by a very variable Ta content in the late (secondary) generations of mica.

3.3 K-Feldspar

K-feldspar belongs to the minerals forming the primary zonal structure of the pegmatites, crystallizing commonly as giant crystals. As the K-feldspar crystallizes in nearly all zones in pegmatites, the rare-metal content of the K-feldspar reflects the changes in the chemical composition of the pegmatite-forming fluids. It is generally observed that especially Rb and Cs show an increasing concentration during crystallization in pegmatites. The concentration of the rare alkali metals increases by a factor of 4 to 5 in K-feldspars, when going from the outer pegmatite unit to the center (Gaupp et al. 1984; Černý and Ercit 1985).

The rare alkali element fractionation trend is shown in diagrams such as K/Rb vs Cs (Fig. 5). The field of the data points of the feldspars from the highly mineralized Tanco pegmatite in this diagram indicates extreme element fractionation in the fluid from which these samples crystallized. The field for the samples from the poorly mineralized Hagendorf Süd pegmatite is found at the other extreme end of the fractionation trend. The central area is covered by samples from pegmatites such as Noumas, Central Claim, Silverleaf and Odd West.

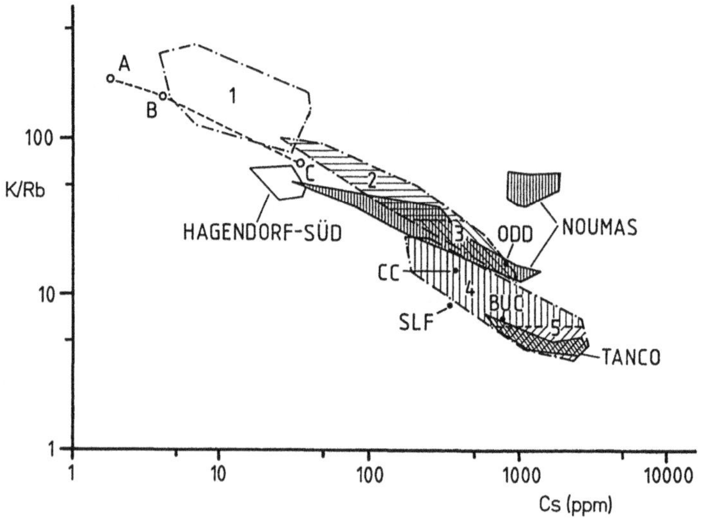

Fig. 5. Plot of K/Rb vs Cs in K-feldspars: Data for: *A* mica pegmatites; *B* ceramic pegmatites; *C* rare metal pegmatites (after Manuylova et al. 1966). *1* Osis Lake pegmatitic granite; *2* Rush Lake pegmatite group (unmineralized); *3* Rush Lake pegmatite group (Li-bearing); *4* Bernic Lake pegmatite group (Li, Be, Rb, Cs); *5* Tanco pegmatite: Li, Cs, Ta, Sn, Be (after Goad and Černý 1981). *CC* Central Claim pegmatite; *BUC* Buck Claim pegmatite; *SLF* Silverleaf pegmatite; *ODD* Odd West pegmatite

4 Exploration Criteria

The degree of alkali fractionation in pegmatite minerals, as shown by diagrams such as K/Rb vs Cs in K-feldspars and muscovite, shows a correlation with the degree of Ta mineralization.

The most interesting mineral for the assessment of the pegmatites' Ta potential seems to be the white mica. Beus (1966) and Gordiyenko (1971) suggested that the Ta contents of white mica alone, at or over of 20 and 65–70 ppm respectively, indicate Ta-oxide mineralization. They, however, did not consider the additional information presented by the simultaneous fractionation of other elements such as Cs, Rb and W. It is the correlation diagram of Ta vs Cs (Fig. 6) in white mica that offers the most reliable possibility to distinguish between low and high Ta contents in white mica at a given degree of element fractionation.

The Ta contents in the primary white mica reflect the over-all concentrations of Ta in the bulk of the fluids from which the pegmatite crystallized and therefore they are best correlated with the potential for Ta mineralization. Confirmation is provided by high F contents, low K/Rb ratios, high Ta/W ratios correlated with high Cs contents and high Ta contents correlated with low K/Cs ratios.

It is the experience of the authors that the scatter in the Ta vs K/Rb diagram is very low for the primary micas of an individual pegmatite. In most cases 10 to 15 samples are therefore sufficient for assessing the tantalum mineralization potential.

It has to be pointed out that high Ta contents in the early formed white mica are much more indicative for a tantalum mineralization if the pegmatite also contains abundant bodies of saccharoidal albite.

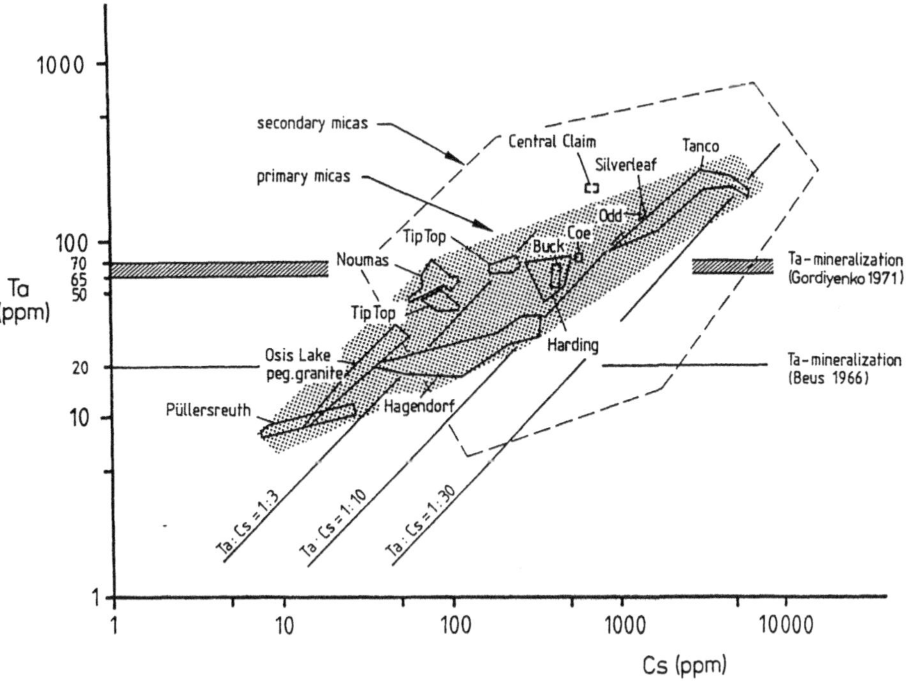

Fig. 6. Ta vs Cs in pegmatite white mica. The *dotted area* represents the field of the primary white mica generations. The field of the late (secondary) white mica is outlined by a *dashed line* and shows the large scatter of the late generations

Acknowledgements. For fruitful discussions we thank P. McArdle, L. van Wambeke, R. Saager, G. Friedrich and P. Möller (Berlin). R. Trumbull (München), who critically reviewed the manuscript and improved the English, is also thanked. The study was supported in part by the Bundesministerium für Forschung und Technologie and by the Deutsche Forschungsgemeinschaft.

References

Anhaeusser CR (1976) Archean metallogeny in southern Africa. Econ Geol 71:16–43

Beus AA (1966) Distribution of tantalum and niobium in muscovites from granitic pegmatites. Geokhimiya 10:1216–1220 (in Russian)

Brögger WC, Vogt T, Schetelig J (1922) Die Mineralien der südnorwegischen Granitpegmatite. Videnskr Selsk Scr Math Naturviss Kl 2:1–151

Černý P (1975) Alkali variations in pegmatitic beryl and their petrogenetic implications. N Jahrb Mineral Abh 123:198–212

Černý P (ed) (1982) Anatomy and classification of granitic pegmatites. Mineral Assoc Can Short Course Handb 8:405–461

Černý P, Ercit TS (1985) Some recent advances in the mineralogy and geochemistry of Nb and Ta in rare-element granitic pegmatites. Bull Mineral 108:499–532

Černý P, Trueman DL, Ziehlke DV, Goad BE, Paul BJ (1981) The Cat Lake – Winnipeg River and the Wekusko Lake pegmatite fields, Manitoba. Manitoba Dep Energ Min, Min Res Div, Econ Geol Rep ER 80:1:216

Gaupp R, Möller P, Morteani G (1984) Tantal-Pegmatite, geologische, petrologische und geochemische Untersuchungen. Monogr Ser Mineral Deposit 23:124

Ginsburg AI, Timofeyev IN, Feldmann LG (1979) Principles of geology of the granitic pegmatites. Nedra, Moscow, pp 1–296

Goad BE, Černý P (1981) Peraluminous pegmatitic granites and their pegmatite aureoles in the Winnipeg River District, southeastern Manitoba. Can Mineral 19:177–194

Gordiyenko VV (1971) Concentration of Li, Rb and Cs in potash feldspar and muscovite as criteria for assessing the rare metal mineralization in granite pegmatites. Int Geol Rev 13:134–142

Hatcher MI, Bolitho BC (1982) The Greenbushes pegmatite, southwest Western Australia. In: Černý P (ed) Mineral Assoc Can Short Course Handb 8:513–525

Jahns RH (1955) The study of pegmatites. Econ Geol 1:1025–1130

Manuylova MM, Petrov LL, Rybykova MM, Sokolov YuM, Shmakin BM (1966) Distribution patterns of alkali elements and beryllium in minerals of pegmatites of the north Baykal pegmatite belt. Geochem Int 1966:309–321

Moloshag VP (1974) On the geochemistry of the rare elements in one of the rare-metal pegmatite fields of Siberia. Vestn Moskow Univ Geol 29:108–110

Parker RL, Fleischer M (1968) Geochemistry of niobium and tantalum. US Geol Surv Prof Pap 612:1–41

Trueman DL, Černý P (1982) Exploration for rare-element granitic pegmatites. In: Černý P (ed) Mineral Assoc Can Short Course Handb 8:463–493

Varlamoff N (1972) Central and West African rare-metal pegmatites, related aplites, quartz veins and mineral deposits. Mineral Deposit 7:202–216

Winkler HGF (1967) Petrogenesis of the metamorphic rocks. Springer, Berlin Heidelberg New York

Prospection for Niobium-Rich Alkaline Rocks

G. MORTEANI

1 Introduction

Nb belongs to the fifth group of the periodic table of the elements. Its atomic number is 41, its atomic weight is 92.9046 and its electronegativity is 1.6. In a six-fold coordination its ionic radius is 0.64, 0.68 and 0.62 Å for the valencies of 5, 4 and 3 respectively. Niobium is virtually identical in size with tantalum as a consequence of lanthanide contraction. Nb and Ta are therefore usually found together in minerals. The chief mineral containing both elements is $FeMn(NbTa)_2O_6$. This mineral is known as columbite or tantalite, depending on which metal predominates. For Nb the most important mineral is pyrochlore $(Na,Ca)_2(Nb,Ta,Ti)_2O_6(OH,F)$. The crustal abundances of Nb and Ta are 20 and 1.7 ppm respectively. These abundances make them comparable with nitrogen (19 ppm), gallium (19 ppm) and lithium (18 ppm), on the one hand, and to arsenic (1.8 ppm) and germanium (1.5 ppm), on the other hand.

Alkaline rock complexes are overwhelmingly important for the production of Nb both today, and in the future; alkaline rocks, e.g. syenites, nepheline syenites and alkaline ultrabasic rocks having the highest Nb-contents of all magmatic rocks (Fig. 1). No other known geological environment shows Nb mineralization equal

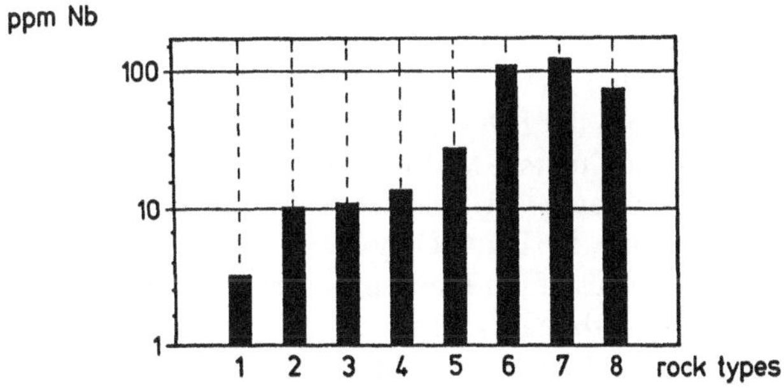

Fig. 1. Average Nb concentrations in various rock types, as given by Kuzmenko (1978). *1* Ultrabasites (dunites, pyroxenites, peridotites); *2* diabases, gabbros; *3* intermediate rocks (diorites; granodiorites); *4* acid leucocratic granites; *5* alkaline granites; *6* syenites; *7* nepheline syenites; *8* alkaline ultrabasites

Lehrstuhl für Angewandte Mineralogie und Geochemie der Technischen Universität München, Lichtenbergstr. 4, D-8046 Garching

Lanthanides, Tantalum and Niobium
P. Möller, P. Černý and F. Saupé (Eds.)

in quality and extent. Alkaline rock complexes and carbonatite complexes have, therefore, the highest priority in the search for new Nb deposits but it must be mentioned that until now there have been no geological, petrological or geochemical criteria known which would allow an estimation of the Nb potential of a whole alkaline rock complex, or the recognition of parts of a complex as highly prospective for Nb on the basis of the analyses of just a few samples.

According to Krauss et al. (1982) only one in ten alkaline complexes may be worth a more detailed prospecting campaign. At the moment there are approximately 280 alkaline rock complexes known in the world but the number of known complexes is quickly increasing.

2 Petrology of Alkaline Rocks and Carbonatites

Carbonatites show features of both intrusive and extrusive rocks. They can be related with ultrabasic rocks, alkaline rocks or kimberlites. Mostly the alkaline complexes are bound to deep fracture zones such as rifting areas of cratons or they are found along the margins of cratonic areas. The very different types of occurrences of the carbonatites and related alkaline rocks has led to very different hypotheses of the formation of these types of rocks. Until today four hypotheses of carbonatite formation have been under discussion:

1. The magmatic formation of the carbonatite was first formulated by Högbom (1895) and later supported by Brøgger (1921). The carbonate substance was supposed to have been derived from a melting of metamorphic marbles by an essexitic magma.

2. Von Eckermann (1948), Strauss and Truter (1951), Holmes (1952) and Pecora (1956) supposed that a carbonatitic magma of more or less undefined origin reacted with crustal rocks forming the wide variety of alkaline rocks associated with carbonatites.

3. A hydrothermal formation of the carbonatites was supported by Smyth and Adams (1923) and Bowen (1924). Until the 1970s the Russian geologists preferred this mode of formation of the carbonatites (see Smirnov 1970).

4. Based on their studies of the South African carbonatites of Spitzkop and Palabora in East Transvaal Shand (1921, 1931) and Daly (1933) suggested that these carbonatites were formed by the anatexis of the Transvaal carbonates. They called this mode of formation a syntexis of carbonate rocks.

The works of Wyllie and co-workers (Wyllie and Tuttle 1960; Wyllie 1966, 1977, 1979, 1981; Wyllie and Huang 1975) have been milestones in the discussion of the genesis of the carbonatites and of the relation of the carbonatites to the alkaline and ultrabasic rocks. In these works it was shown by hydrothermal experiments that in the system CaO-CO_2-H_2O, i.e. in the presence of a water-rich phase, carbonatites form melts down to a temperature of 640 °C. This was vital new information because the dry melting point of calcite of 1340 °C had previously been assumed in discussing the magmatic formation of carbonatites. The experimental work of Wyllie and co-workers has led to the fact that in most recent times only the magmatic hypothesis is discussed seriously.

Given that the magmatic genesis of carbonatites is acceptable, two modes of formation can then be discussed:

1. The carbonatite magma is of a primary origin;
2. The carbonatite magma is of secondary origin being the product of a differentiation of an alkali to ultrabasic magma. King and Sutherland (1960) suppose an alkali peridotitic parental magma; Tomkeieff (1961), a pyroxenitic parental magma; Wimmenauer (1963) and Bailey and Schairer (1964) speak in favour of an olivine nephelinite or an ijolitic magma.

The discussion of the relations of the carbonatites and various ultrabasic to alkaline rocks is given by Tuttle and Gittins (1966), Heinrich (1966), Kukharenko and Dontsova (1962), Tomkeieff (1961) and Verwoerd (1967). In the last few years the discussion on the genesis of carbonatites and related alkaline rocks focussed mainly on the following details:

1. What is the ultimate source of the carbonates and the alkaline rocks?
2. What is the role of the fluid phase, which is mainly composed of H_2O, SiO_2, fluorine, chlorine and sulphur, in the formation of the magmas and what is the SiO_2/H_2O ratio during the history of the magma from first formation to cooling?
3. What is the source of CO_2 and H_2O?
4. What is the limit between the magmatic/metasomatic and hydrothermal processes during the solidification of the carbonatites and of the alkaline rocks?
5. To what extent and how, do the carbonatite and the alkaline melts react with the crustal rocks during their ascent and what elements are lost or gained during these metasomatic reactions?

The problems 1 through 5 are discussed, e.g. in the works of Sørensen (1974), Ernst (1975), Lloyd and Bailey (1975), Samoylov (1975), Le Bas (1977, 1981), Smirnov (1977), Rass et al. (1979), Wyllie (1979, 1981), Bailey (1980), Dawson (1981). From the discussion of the above mentioned problems it can be concluded that the carbonatite magma probably is of primary origin and formed by a partial melting of a peridotite in the mantle at a depth of about 18 km (see Huang and Wyllie 1974; Wyllie and Huang 1975; Koster van Groos 1975). The formation of a partial melt requires small but finite amounts of CO_2 and H_2O in the mantle.

An important role is played by the immiscibility of alkaline melts with carbonate melts as observed experimentally by Koster van Groos and Wyllie (1966, 1968, 1973) and results in at least three coexisting phases:

1. An SiO_2-undersaturated alkali silicate melt poor in carbonate matter and of nephelinitic to ijolitic composition;
2. An alkali carbonate melt containing only small amounts of silicates and;
3. An H_2O-rich fluid phase with alkalis and abundant CO_2 in solution. This later phase gives rise to the fenitization.

That the ultimate source of the carbonatites and related silicate magmas is the mantle is supported strongly by the Sr- and O-isotope geochemistry (see e.g. Taylor et al. 1967; Deines and Gold 1973; Pineau et al. 1973; Lancelot and Allegre 1974).

3 Behaviour of Niobium in Alkaline and Carbonatitic Magmas

Figure 1 shows that the concentration of Nb in rocks of the calc-alkaline type ranges from 5 ppm Nb in the basic rocks to about 22 ppm in more acid rocks. A drastic increase in the Nb contents can be seen only in the alkaline rocks irrespective of whether they are acidic or more basic. In the calc-alkaline rocks the increase in the Nb content is accompanied by an equivalent increase of Ta contents. The ratio is very constant between 10 and 12. This indicates that Nb and Ta in calc-alkaline rocks are not separated during magmatic differentiation, but a separation is observed in alkaline rocks. In these rocks Nb is enriched over Ta, but what produces this separation in alkaline rocks is still under discussion. Samoylov (1974) explains the separation of these two rather coherent elements by their different dissociation constant.

The important role of the alkali elements in the transport of Nb (and Ta) and in the separation of both elements results from the observation of Hansen (1968) who reported variations of the Nb/Ta ratio in the agpaitic nepheline syenites of the Ilimaussaq intrusion in Greenland to be in the range of 17 to 64. This variation is positively correlated with the degree of metasomatic alteration. Very strongly altered rocks in shear zones of the lujavrites reach maximum ratios of about 512.

Often the Nb contents are more or less well correlated with the Ti and Zr contents (Gerasimovski and Karpushina 1966), but such a weak Ti-Nb correlation is missing, e.g. for the syenitic rocks of the alkaline complex of the Motzfeldt complex (Greenland) (Fig. 2).

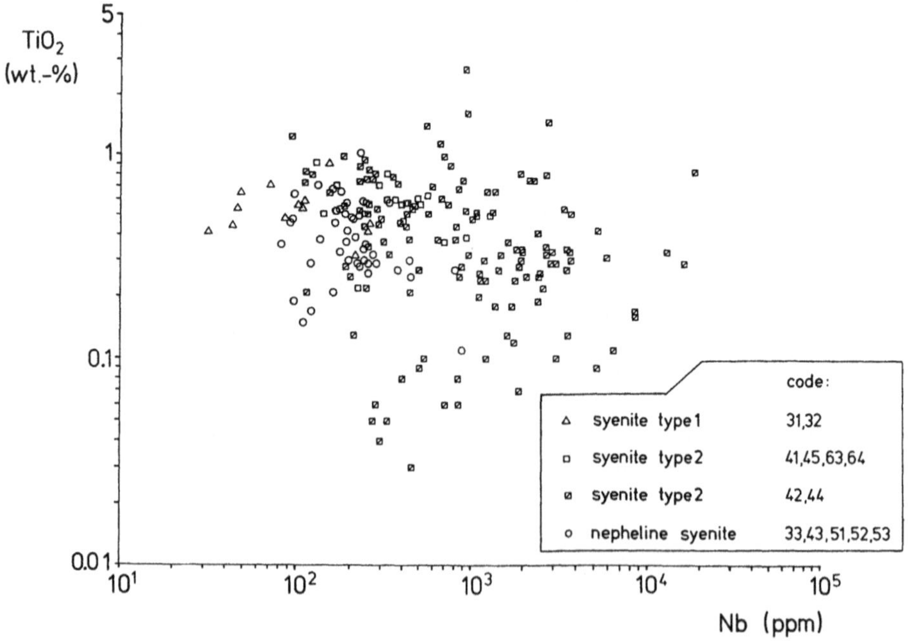

Fig. 2. Correlation diagram of TiO_2 vs Nb for the syenitic rocks of the Alkali rock complex of Motzfeldt Centre (Greenland). The *symbols* represent different syenite types. Data are from Morteani et al. (1986)

In examining the Nb contents of the silicates oxides in alkaline rocks one can see that minerals such as pyroxene, amphibole, biotite, titanite, ilmenite, magnetite and zircon have varied, but nevertheless considerable amounts of Nb (Wedepohl 1969). The highest concentrations of Nb are found in Ti-bearing minerals such as perovskite, titanite and phlogopite, in which possible substitutions include:

Perovskite:	$Ti^{4+} + Ca^{2+}$	$\rightarrow$	$(Nb,Ta)^{5+} + Na^{+}$
Titanite:	$Ti^{4+} + 3Ca^{2+}$	$\rightarrow$	$(Nb,Ta)^{5+} + 2Na^{+} + REE^{3+}$
Biotite:	$2Ti^{4+}$	$\rightarrow$	$(Nb,Ta)^{5+} + (Fe,Al)^{3+}$
Zircon:	$2Zr^{4+}$	$\rightarrow$	$(Nb,Ta)^{5+} + REE^{3+}$
Pyroxenes:	$Al^{3+} + 2Fe^{3+}$	$\rightarrow$	$(Nb,Ta)^{5+} + 2Mg^{2+}$ and
Amphiboles:	$Ca^{2+} + Fe^{3+} + Si^{4+}$	$\rightarrow$	$Na^{+} + (Nb,Ta)^{5+} + Al^{3+}$.

The solubility of Nb under magmatic or hydrothermal conditions is still a mattter of discussion. According to Aleksandrov (1967) and Aleksandrov et al. (1969, 1971, 1975) it can be supposed that:

1. Nb forms complexes in K_2CO_3-$KHCO_3$ solutions at high temperatures and pressures.

2. The solubility of Nb increases with increasing pH value ($8.9 < pH < 9.5$) and decreases when at given pH values the temperature decreases, e.g. from 600° to 300°C.

3. The solubility of Nb increases under hydrothermal conditions if the concentration of HPO_4^{2-} is increased. This is interpreted as an indication of the formation of Nb-PO_4 complexes.

4. The solubility of Nb is higher in potassium-carbonate solution than in a sodium- or calcium-carbonate solution.

Von Maravic (1983) summarized the experimental data of Aleksandrov (1967) and Aleksandrov et al. (1969, 1971, 1975) pointing out that in a hydrothermal solution carbonate, fluoride, phosphate and potassium ions favour the transport of Nb, whereas the presence or admixture of Ca^{2+} and Na^{+} ions lead to the precipitation of Nb as Ca or Na niobates (e.g. lueshite). Projecting this observation into magmatic systems, von Maravic (1983) suggests that the concentration of Nb is maximum in K^{+} rich rocks and is reduced in Na^{+} or Ca^{2+} rich melts.

Comparing the experimental results and the theoretical considerations with field observations, one can say that the highest Nb contents in alkaline complexes are, in fact, mostly not found in the pure carbonatites, but rather in the K-feldspar-rich carbosyenites or in carbonatites rich in silicate inclusions.

A further increase in the Nb contents of the rocks is observed in metasomatic processes, particularly in albitization. Pyrochlore and consequently Nb enrichment is particularly well found in the fenitized contact zones studied by von Maravic (1983) in Lueshe. In case of lateritic weathering the highest Nb contents are found, as in Lueshe, in the laterites overlying the silicate-rich carbonatites, rather than on top of pure carbonatites.

Mantle material, e.g. normal MOR-basalts, is rather poor in Nb. Nb must therefore have been introduced into such alkaline melts by a reaction with the country rocks before or after the liquid-liquid separation of the alkaline silicate melt and the carbonate-rich melt. The most likely place for such a reaction is the lower crust, where the temperatures are high and the chemical disequilibrium between melt and country rocks is large.

On a larger magmatic scale, the process of Nb uptake by reaction with country rocks is described in the study of Morteani et al. (1986) on the Motzfeld alkaline complex. The Motzfeldt complex consists of a sequence of alkaline rocks showing three cycles, each of which start from SiO_2 undersaturation and end with oversaturated syenitic rocks. The changes in melt composition can be very well explained by a repeated cycle of SiO_2-undersaturated syenitic magmas reacting with crustal rocks thus becoming SiO_2-saturated. These magmas show the highest and also the most scattered Nb contents within the whole rock series (Fig. 3). The scatter of the

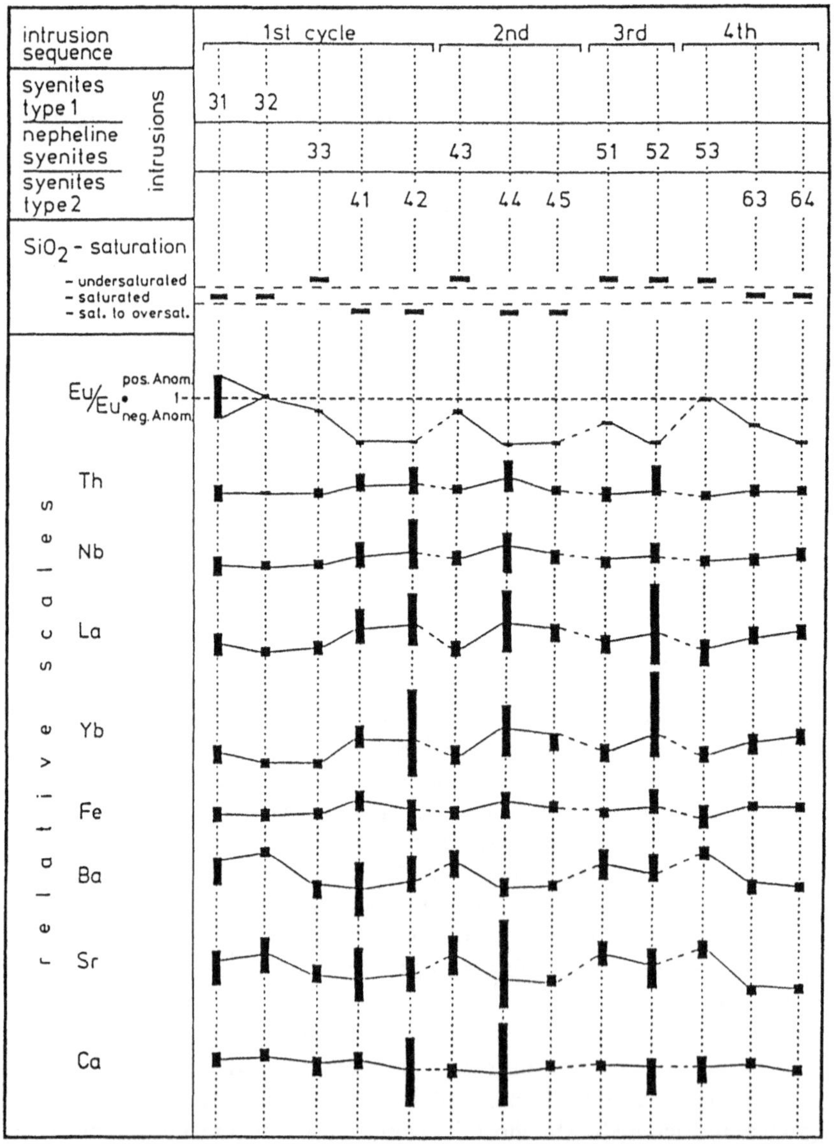

Fig. 3. Chemical evolution of the various alkaline rocks of Motzfeldt Centre (Greenland). Data are from Morteani et al. (1986)

Nb values can be taken as an indicator of an inhomogeneous uptake of Nb during the assimilation of country rocks.

The bulk of Nb in any case goes into the alkali-rich silicate melt and not into the carbonatites. Later introduction of an Na or Ca-rich fluid phase into the alkaline rocks or the silicate melt produces a precipitation of the Nb by lowering the solubility of the Nb complexes in the silicate melt or the solid phases, mainly K-feldspar. This process has been active on a small scale in the fenites formed between the alkali rocks and the country schists as well, as reported for the Lueshe complex by Von Maravic (1983). The stepwise enrichment of alkaline rocks with Nb due to the repeated metasomatic processes at different levels is schematically shown in Fig. 4.

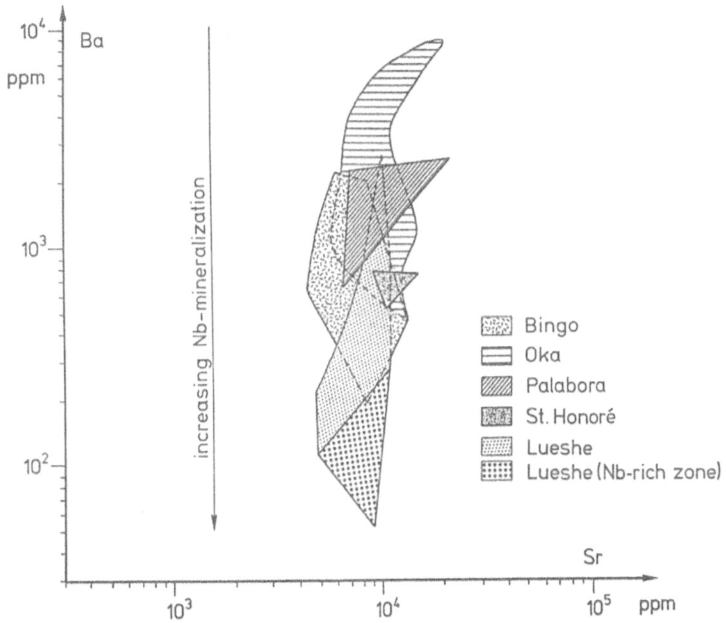

Fig. 4. Plot of Ba vs Sr from calcite of different alkaline complexes. The calcites from Nb-rich calcite-bearing rocks seem to be characterized by low Ba contents

In a plot of Ba vs Sr for the calcites from the alkaline complexes of Bingo and Lueshe (Zaire), Oka and St. Honore (Canada) and of Palabora (Rep. of South Africa) the most Nb-rich complexes, and zones within these complexes, show low Ba contents (Fig. 5). The reason for this may be that Ba isomorphically substitutes for Ca in calcite at high magmatic temperatures and it is expelled from the calcite lattice at low temperatures during metasomatic processes concomitant with the formation of pyrochlore. This late recrystallization also leads to a depletion of the REE contents as demonstrated by their increasingly flat REE fractionation curves.

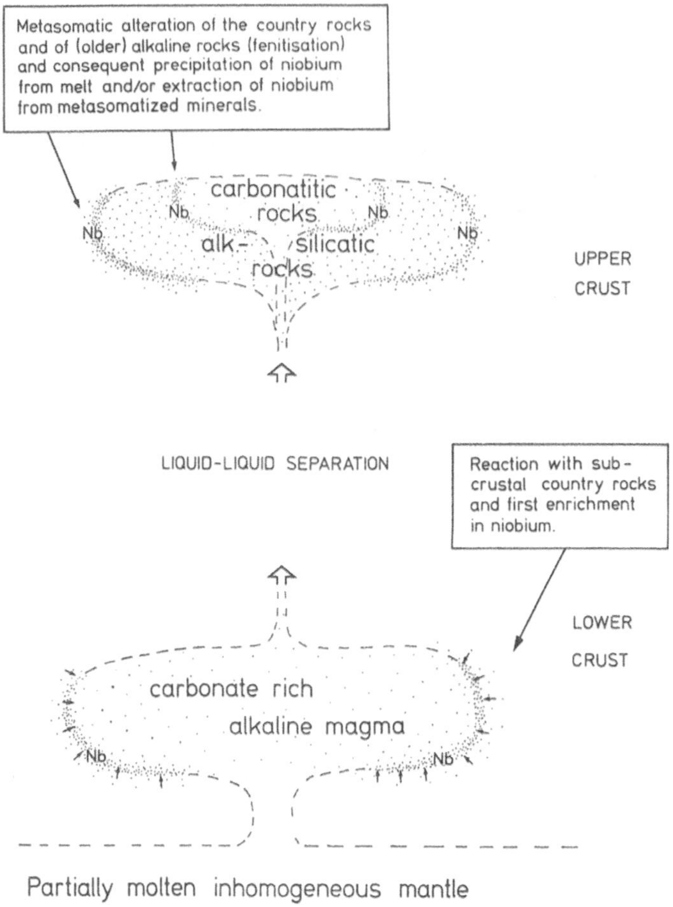

Fig. 5. Schematic diagram showing the evolution of a carbonate-rich alkaline magma and Nb enrichment due to metasomatic processes in the lower and upper crust

4 Exploration Criteria

The results of theoretical and field studies show that in the prospection for Nb in alkaline complexes one has to search for complexes or for areas within complexes characterized by carbonatitic rocks showing high amounts of silicate components. A further important guide is geochemical evidence (major and trace elements) of interaction with crustal rocks. Examples are reduced Ba and REE contents in calcite, or an alternation of undersaturated and saturated rocks.

Furthermore, the most promising are rocks containing K-feldspars which show a late to postmagmatic albitization. Such areas have been found to be very rich in Nb [e.g. the case of the big pyrochlore deposit of Lueshe near the contact of the carbosyenites to the fenitized country rocks (von Maravic 1983)].

Acknowledgements. For their contributions to the discussion at the presentation of this paper in the workshop in Berlin, I thank G. Roethe, G. Friedrich, F. Saupe, L. Van Wambeke. P. Möller (Berlin) and R. Trumbull (München) who critized the manuscript. R. Trumbull improved the English as well. The study was supported in part by the Bundesministerium für Forschung und Technologie.

References

Aleksandrov IV (1967) Niobium in the carbonate solutions and some considerations on the migration of rare elements under hydrothermal conditions. Geochem Int 4, 1:558–566

Aleksandrov IV, Trusikova TA (1973) Niobium and tantalum in fluoride-phosphate solutions. Geochem Int 10, 5:641–648

Aleksandrov IV, Trusikova TA, Tupitsin BP (1969) Niobium and tantalum in carbon dioxide solutions. Geochem Int 6, 1:558–566

Aleksandrov IV, Trusikova TA, Tupitsin BP (1971) Effect of sodium and potassium on migration and precipitation of niobium. Geochem Int 8, 1:229–233

Aleksandrov IV, Trusikova TA, Tupitsin BP (1975) The geochemistry of niobium and tantalum in the carbonatite process. In: Tugarinov AI (ed) Recent contributions to geochemistry and analytical chemistry. Wiley & Sons, New York, pp 335–344

Bailey DK (1980) Volatile flux, geotherms, and the generation of the kimberlite-carbonatite-alkaline magma spectrum. Min Mag 43:695–699

Bailey DK, Schairer IF (1964) The peralkaline residual system: Na_2O-Al_2O_3-Fe_2O_3-SiO_2. Carnegie Inst Washington Annu Rep Geophys Lab 1963, 64:74–79

Bowen WL (1924) The Fen area in Telemark, Norway. Am J Sci 8:1–11

Brøgger WC (1921) Die Eruptivgesteine des Kristianagebietes, IV. Das Fen-Gebiet in Telemark, Norwegen. Norsk Vidensk Selsk Skifter I, Math Naturwiss Kl (1920) 9:1–408

Daly RA (1933) Igneous rocks and the depth of the earth. McGraw Hill, New York

Deines P, Gold DP (1973) The isotopic composition of carbonatite and kimberlite carbonate and their bearing on the isotopic composition of deep-seated carbon. Geochim Cosmochim Acta 37:1709–1733

Dawson JB (1981) The nature of the upper mantle. Min Mag 44:1–18

Eckermann H von (1948) The alkaline district of Alnø Island. Sv Geol Undersök Ser Ca 36:9–176

Ernst T (1975) Erdmantelprobleme – Bericht und Theorie. Fortschr Min 52:106–140

Gerasimovski VI, Karpushina VA (1966) Geochemistry of niobium and tantalum in nepheline syenites. Geochem Int 3, 4:609–621

Hansen J (1968) Niobium mineralization in the Ilimaussaq alkaline complex, south-west Greenland. 22nd Int Geol Congr 7:263–273

Heinrich EW (1966) The geology of carbonatites. Rand McNally, New York, pp 1–555

Högbom AG (1895) Über das Nephelinsyenitgebiet auf der Insel Alnø. Geol Fören Förh 17:100–160, 214–256

Holmes A (1952) The potash ankaratrite – melaleucitite lavas of Nabugando and Mbuga craters, south-west Uganda. Trans Geol Soc Edinburgh 15:187–213

Huang WL, Wyllie PJ (1974) Eutectic between wollastonite II and calcite contrasted with thermal barrier in $MgO - SiO_2 - CO_2$ at 30 kilobars, with applications to kimberlite – carbonatite petrogenesis. Earth Planet Sci Lett 24:305–310

King BC, Sutherland DS (1960) Alkaline rocks of eastern and southern Africa, Part I-III. Sci Progr 48:298–321, 504–524, 709–720

Koster van Groos AF (1975) The effect of high CO_2 pressures on alkalic rocks and its bearing on the formation of alkalic ultrabasic rocks and the associated carbonatites. Am J Sci, 275:163–185

Koster van Groos AF, Wyllie PJ (1966) Liquid immiscibility in the system $Na_2O - Al_2O_3 - SiO_2 - CO_2$ at pressures to 1 kb. Am J Sci, 264:234–255

Koster van Groos AF, Wyllie PJ (1968) Liquid immiscibility in the join $NaAlSi_3O_8 - Na_2CO_3 - H_2O$ and its bearing on the genesis of carbonatites. Am J Sci 266:932–967

Koster van Groos AF, Wyllie PJ (1973) Liquid immiscibility in the join $NaAlSi_3O_8 - CaAl_2Si_2O_8 - Na_2CO_3 - H_2O$, jae Sci 273:465–487

Krauss U, Schmidt H, Kippenberger C, Eggert P, Priem J, Wettig E (1982) Untersuchungen über Angebot und Nachfrage mineralischer Rohstoffe XVI Niob. Bundesanst Geowiss Rohst, Hannover, pp 1 208

Kukharenko AA, Dontsova EI (1962) A contribution to the problem of the genesis of carbonatites. Econ Geol USSR 1, (3-4):31-46 (English Transl 1964)
Kuzmenko MV (1978) Geochemistry of tantalum and the genesis of endogenetic deposits. Moskow (in Russian) In: Krauss et al. (1982)
Lancelot IR, Allegre CJ (1974) Origin of carbonatitic magma in the light of the Pb- U- Th isotope system. Earth Planet Sci Lett 22:233-238
Le Bas MJ (1977) Carbonatite-nephelinite volcanism, an African case history. Wiley & Sons, New York London, pp 1-347
Le Bas MJ (1981) Carbonatite magmas. Min Mag 44:133-140
Lloyd FE, Bailey DK (1975) Light element metasomatism of the continental mantle: the evidence and the consequences, Phys Chem Earth 9:389-416
Maravic H von (1983) Geochemische und petrographische Untersuchungen zur Genese des niobführenden Karbonatit/Cancrinit-Syenitkomplexes von Lueshe, Kivu/NE-Zaire. Thesis, Tech Univ Berlin, pp 1-330
Morteani G, Möller P, Spiegel W (1986) Entwicklung petrologischer und geochemischer Kriterien zur Auffindung von Niob- und phosphorreichen Karbonatiten. EEC Final Rep Contrib MSM-005-D(B)
Pecora WT (1956) Carbonatites: a review. Bull Geol Soc Am 67:1537-1556
Pineau F, Javoy M, Allegre CJ (1973) Etude systématique des isotopes de l'oxygène, du carbone et du strontium dans les carbonatites. Geochim Cosmochim Acta 37:2363-2377
Rass IT, Laputina IP, Bykova VS (1979) Genetic criteria and mineralization of carbonatites. Geochem Int 16:43-50
Samoylov VS (1974) Effect of temperature and pressure on acidity and basicity of carbonatite-forming solutions. Dokl Earth Sci Sec 218:193-195
Samoylov VS (1975) The temperatures of forming of carbonatites (based on geothermometric studies). Geochem Int 12:52-58
Shand SJ (1921) The nepheline rocks of Sekukiniland. Geol Soc S Afr Trans 24:111-149
Shand SJ (1931) The granite-syenite limestone complex of Palabora, eastern Transvaal and the associated apatite deposits. Geol Soc S Afr Trans 34:81-105
Smirnov VI (1970) Karbonatitlagerstätten. In: Geologie der Lagerstätten mineralischer Rohstoffe. VEB Dtsch Verl Grundstoffind, Leipzig, pp 134-150
Smirnov VI (1977) Deposits of tantalum and niobium, carbonatites. In: Ore deposits of the USSR, vol 3. Pitman, London, pp 409-420
Smyth FH, Adams LH (1923) The system calcium oxide-carbon dioxide. J Chem Soc Am 45:1167-1184
Sørensen H (ed) (1974) Origin of the alkaline rocks- a summary and retrospect. In: The alkaline rocks. Wiley & Sons, New York London, pp 535-543
Strauss CA, Truter FC (1951) The alkali complex at Spitskop, Sekuniland, Eastern Transvaal. Trans Geol Soc S Afr 53:81-130
Taylor HP, Frechen J, Degens ET (1967) Oxygen and carbon isotope studies of carbonatites from the Laacher See district, West Germany, and Alnø district, Sweden. Geochim Cosmochim Acta 31:407-430
Tomkeieff SI (1961) Alkalic ultrabasic rocks and carbonatites in the USSR. Int Geol Rev 3:739-758
Tuttle OF, Gittins J (1966) The pyroxenes of the alkaline igneous complex of eastern Uganda. Min Mag 36:5-22
Verwoerd WJ (1967) The carbonatites of South Africa and South West Africa. Geol Surv S Afr Handb 6:1-452
Wedepohl KH (ed) (1969) Niobium. In: Handbook of geochemistry, vol II/4. Element 41, 41-A to 41-0 Springer, Berlin Heidelberg New York
Wimmenauer W (1963) Beiträge zur Petrographie des Kaiserstuhls. Pt: VI Die Karbonatite. Pt VII: Zur Petrogenese des Kaiserstuhls. N Jahrb Mineral Abh 99:231-276
Wyllie PJ (1966) Experimental data bearing on the petrogenetic links between kimberlites and carbonatite Proc 4th Gen meet Int Mineral Assoc 1964, India, pp 67-82
Wyllie PJ (1977) Mantle fluid compositions buffered by carbonates in peridotite – CO_2 – H_2O. J Geol 85:187-208
Wyllie PJ (1979) Magmas and volatile components. Am Mineral 64:469-500
Wyllie PJ (1981) Plate tectonics and magma genesis. Geol Rundsch 70:128-153
Wyllie PJ, Huang WL (1975) Peridotite, kimberlite, and carbonatite explained in the system CaO – MgO – SiO_2 – CO_2. Geology 3:621-624
Wyllie PJ, Tuttle OF (1960) The system CaO – CO_2 – H_2O and the origin of carbonatites. J Petrol 1:1-46

Geochemical Exploration Guides for Rare-Metal Pegmatites – Examples from Nigeria and Sudan

G. MATHEIS and D. KÜSTER

1 Introduction

Primary and secondary dispersion patterns related to late Pan-African (i.e. about 590–520 Ma) rare-metal pegmatites of Nigeria/West Africa have been studied since 1976 in the following sampling types: stream sediment, residual soil, whole rock, heavy mineral concentrate, muscovite and feldspar (Matheis 1979, 1981; Matheis et al. 1982, 1984; Küster 1987). More recently, Holocene wadi sediments and muscovite related to comparable pegmatites from the Bayuda Desert of northern Sudan have been investigated for comparison (Matheis and Küster 1987).

About 95% of the Nigerian tin and columbite output has been produced from the anorogenic Jurassic ring complexes of the Jos Plateau in central Nigeria. However, most of the tantalum (about 53 long tons in the time span 1942–1970 according to available records) was mined from pegmatitic sources of late Pan-African age; Rb/Sr geochronological investigations by Matheis and Caen-Vachette (1983) confirmed emplacement between 590 and 525 Ma for various occurrences in the so-called Older Tinfields of southwestern and central Nigeria. The distribution of these pegmatite fields as well as the Jurassic ring complexes within the Nigerian tin province are displayed in Fig. 1. The rare-metal pegmatites of southwestern Nigeria are confined exclusively to greenstone-dominated metavolcanic-metasedimentary belts and are characterized by distinctly high tantalum concentrations. This feature is typical of the major tantalum-producing pegmatitic source areas worldwide (Canada, southern Africa, western Australia; details in Gaupp et al. 1984). Details on the geological setting of the Nigerian Older Tinfields are provided by Matheis (1987). Pegmatite occurrences of the Bayuda Desert/Sudan were mined until recently for their muscovite books; no rare-metal occurrences have been reported as yet.

Alluvial and eluvial placer concentrates formed the basis of the typical small-scale mining operations in Nigeria, like in many other regions of the tropical belt. Due to the high degree of physical and chemical weathering, the rapid disintegration of the coarse-grained source rocks, combined with short distances of transportation by seasonal streams, resulted in very inconsistent patterns of heavy mineral deposition. This causes serious problems in the recognition of rare-metal target areas at

Technische Universität Berlin, Sonderforschungsbereich 69, Ackerstr. 71, D-1000 Berlin 65

Lanthanides, Tantalum and Niobium
P. Möller, P. Černý and F. Saupé (Eds.)

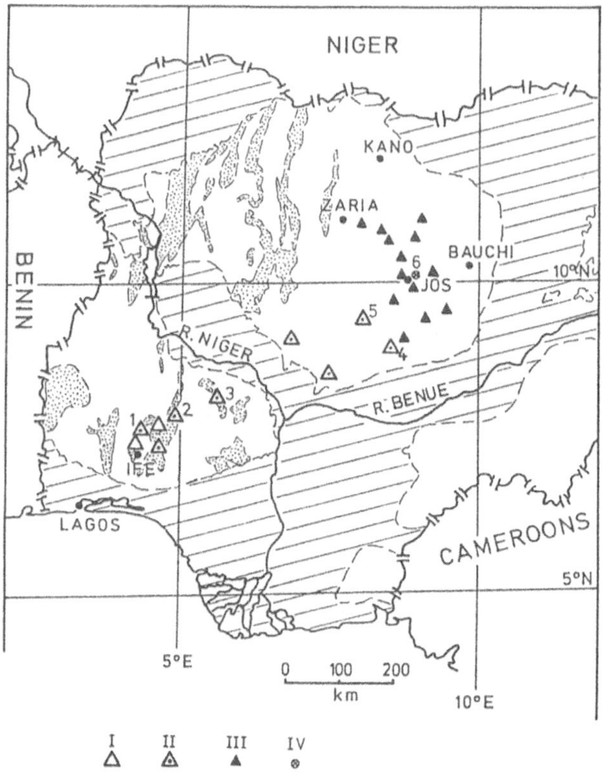

Fig. 1. Geological situation of the Nigerian tin province, including the major units: Precambrian basement complex (*blank*), late Proterozoic metavolcano-sedimentary sequences (*stippled*) and Cretaceous and younger sediments (*hatched*). *I* Barren pegmatite fields; Data of rare-metal-bearing pegmatites (*II*) are reported from locations: *1* Iregun; *2* Ijero; *3* Egbe; *4* Wamba; *5* Jemaa; *III* Jurassic ring complexes of the Jos Plateau; *IV* a rare-metal-bearing basement unit (further details in Matheis, 1987)

the reconnaissance level, which has just been confirmed by Fletcher et al. (1987) on the basis of investigations in the southeast Asian tin province.

Based on the assumption that the rather mobile petrogenetic rare-metal pathfinders like Li, Rb, Cs, F – in contrast to the immobile metallogenic pathfinders Sn, Nb, Ta – are easier to trace at the regional scale, orientation surveys in Nigeria were started first on secondary dispersion media and at a later stage, were accompanied by studies of primary dispersion patterns. Part of the objective was the application of simple analytical tools like the AAS technique. The presentation of results will not follow this working sequence but will start with some basic primary dispersion studies of our own; references to similar investigations elsewhere are quoted in our former publications.

2 **Primary Dispersion** (Whole Rock)

The geochemical investigations were aimed at identifying element indicators for the "degree of specialization" (Tischendorf 1977) of the Nigerian rare-metal pegmatites and to test their usefulness as pathfinders for exploration purposes (Matheis 1979; Matheis et al. 1982). Major and trace element data on fresh pegmatite whole-rock samples from three mineralized and two barren occurrences in southwestern and one mineralized occurrence in central Nigeria (cf. Fig. 1) were analyzed in detail, while specific rare-metal indicators like F and Cs have been analyzed so far only in some of these occurrences (Matheis 1987). However, a detailed study of the central Nigerian rare-metal-bearing pegmatite fields is presently in progress (Küster 1987). In total, the best, and analytically easiest to determine, discriminants are the pegmatite indicator elements Rb, Li, Cs, F and the element ratios Mg/Li and K/Rb.

The various pegmatite fields in Nigeria display distinctly different geochemical signatures which are related to their special geological framework (some selected data are shown in Tables 1 and 2). The Egbe pegmatites, which are emplaced into a tholeiitic metabasaltic sequence, have been studied geochemically in greater detail than the other localities, due to their higher tantalite output. All pegmatite whole-rock samples of the main pegmatite-amphibolite complex display the anomalous enhancement of the easily detectable pathfinders Li, Rb, F and K/Rb which correlate significantly with the whole-rock tin contents. Although reliable Ta data on whole-rock samples are not available, semi-quantitative XRF measurements indicate a correlation with Sn, i.e. an accumulation in the metabasalt-hosted pegmatites relative to Nb; this correlation is also supported by data on the heavy-mineral concentrates (cf. Sect. 4). Despite all the investigated rare-metal-bearing pegmatites, genetic relations to proximal rapakivi-type granite/granodiorite complexes could not be established (Matheis 1987).

Table 1. Frequency distribution of lithium in stream sediment samples of southwestern Nigeria

Concentration ranges		Iregun 46 samples/ 320 km^2	Ijero 121 samples/ 850 km^2	Egbe 108 samples/ 760 km^2
	< 1.0 ppm	87%	66%	56%
	1.0–2.5 ppm	13%	28%	36%
cxLi	2.6–4.0 ppm	- -	3%	5%
	> 4.0 ppm	- -	3%	3%
	Mean value	0.5 ppm	1.4 ppm	1.0 ppm
	< 3.5 ppm	81%	72%	10%
	3.5–10 ppm	17%	24%	48%
Li	10–30 ppm	2%	4%	36%
	> 30 ppm	- -	- -	6%
	Mean value	2.6 ppm	5.4 ppm	12.4 ppm
Reference values in rocks:				
- Barren pegmatites		42 ppm	47 ppm	38 ppm
- Rare-metal pegmatites (after Matheis et al. 1982)		- -	96 ppm	120 ppm

3 **Secondary Dispersion** (Stream Sediment, Soil)

Although chemical weathering progresses rather fast in the tropics, extensive leaching is largely restricted to the major elements Mg, Ca, Na and in part K and Si. Most of the trace elements, however, are retained in their approximate bedrock proportions. As a general result, the primary bedrock concentration ranges of trace elements are not disturbed by weathering processes, but enhancement or depletion takes place according to the original bedrock ratios (Matheis 1981).

To avoid the rather inconsistent distribution patterns of the heavy-mineral fractions at a regional scale, active stream sediment and residual soil reconnaissance surveys were used to trace the rare-metal-bearing pegmatite occurrences. Special attention was given to the petrogenetic indicators of rare-metal specialization such as Li, Rb, F, Mg/Li, K/Rb that give more consistent distribution patterns. At the regional scale of one sample per 7 km^2, the old mining areas in southwestern Nigeria could be singled out, even by cold-extractable Li (cxLi) as indicated in Table 1. As a followup, residual soil samples of the upper B-horizon were used to differentiate bedrock compositions in areas covered by thick lateritic oveburden and dense vegetation: the best contrasts between soils overlying the rare-metal-bearing pegmatites and those overlying the country rocks were obtained for Mg/Li, followed by Li, cxF, Rb, Cs, Be and K/Rb respectively (selected data from two occurrences in Table 2). A similar utilization of secondary geochemical dispersion into pisolitic duricrust in the rare-metal pegmatite district around Greenbushes/Western Australia was recently reported by Smith et al. (1987).

Table 2. Lateritic soil patterns overlying rare-metal-bearing pegmatites, southwestern Nigeria

Localities		Li	Rb	Mg/Li
Ijero area	Areal range	2–71	7–140	250–6
100 soil samples/110 km^2	Pegmatite mean	39	111	15
Rock references:	Barren pegmatites	47	204	23
Rock references:	Rare-metal pegmatites	96	241	12
Egbe area	Aeral range	10–217	45–610	7600–13
120 soil samples/140 km^2	Pegmatite mean	105	580	44
Rock references:	Barren pegmatites	38	415	17
Rock references:	Rare-metal pegmatites	120	1220	10

(all concentration levels in ppm)

With respect to Ta-rich pegmatite complexes emplaced into basic metavolcanic sequences, the most potential exploration targets at reconnaissance scale are outlined by an overlap of anomalous concentration levels of both the pegmatoid (as above) and mafic (e.g. Ni, Cr, Ti) pathfinder elements.

To overcome the erratic nature of depositional events in wadi systems under arid conditions, the remnants of the Holocene paleohydromorphic dispersion, which reflect a more humid savannah-type climatic regime, are considered to be a better sample media at the reconnaissance scale than the traditionally used heavy-mineral particle dispersion (Matheis and Küster 1985). An orientation study, carried out in the late Proterozoic basement complex of the Bayuda Desert of northern Sudan (cf. Fig. 2), confirmed the applicability of such paleo-dispersion patterns for outlining both lithological and metallogenetic source units at the regional scale (Matheis and Küster 1987).

4 Heavy-Mineral Concentrates (Crude, Cassiterite)

In line with the previous statements, investigation of heavy-mineral concentrates is seen as part of the detailed studies within an exploration project; it should follow the secondary dispersion surveys, once a pegmatoid target area with enhanced specialization is outlined.

The crude ore of the Nigerian pegmatite fields is recovered from alluvial/eluvial placers close to the source rocks which ensures that their compositions are representative of the primary mineralization. Within the framework of investigating the mining concentrates of the Nigerian tin province as a whole, i.e. the Jurassic ring

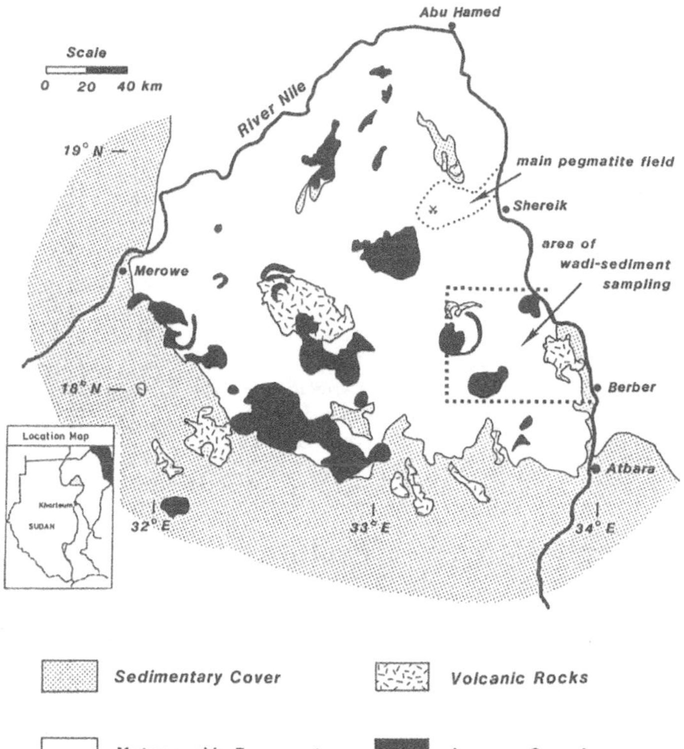

Fig. 2. Major geological units of the Bayuda Desert in northern Sudan; target areas of wadi-sediment sampling and muscovite investigations are outlined (Matheis and Küster 1987)

complex and Pan-African pegmatitic rare-metal mineralization (Matheis et al. 1984), a total of 13 concentrates from 8 pegmatite occurrences have been studied geochemically. Of these, six concentrates from five localities have been analyzed by three methods simultaneously: XRF on milled concentrate, AAS and ICP-ES on HF-leached concentrates, as well as NAS and SSMS on hand-picked cassiterites. The latter attempt was part of a comparative study of worldwide occurrences by Möller and Dulski (1983).

There is a marked difference between the occurrences in southwestern and central Nigeria, with considerable variations even within these regions. Evidently, the element patterns of the hand-picked cassiterites give the most representative picture of the primary mineralization of each mining site. The good correlation of these mineral data with the analytically easier scanning of the crude concentration is reflected in Table 3. As a general trend in minor and trace element distribution, the occurrences in southwestern Nigeria are dominated by Ta, Hf, Ti, Fe and Mn, while the central Nigerian tin fields have higher concentrations of W, Zr and Sc. There is a tendency for a simultaneous increase in Ta and Ti to its maximum in the Egbe pegmatites which are emplaced into a series of tholeiitic amphibolites. These distribution patterns support the conclusion that there are no uniform geochemical characteristics for the pegmatoid rare-metal mineralization and that the host-rock lithology seems to have a considerable imprint on the individual mineralization event.

Table 3. Dispersion patterns in tin concentrates (crude) and cassiterites of the "Older Tinfields" of Nigeria

Locality	Crude %Sn	%Nb	%Ta	Cassiterite Ta	W	Zr	Hf
Central Nigeria							
Jemaa	94.0	4.0	0.3	2400	1800	270	71
Wamba	88.0	3.0	1.0	8950	32	1200	2110
Southwestern Nigeria							
Egbe	80.0	5.7	8.5	7.2%	350	560	220
Ijero	80.0	6.5	4.5	1.8%	80	420	160

-Crude data in % oxide by XRF
-Cassiterite data = element conc. in ppm or % by NAS (Dulski P./HMI-Berlin)

Correlation ($n = 10$):
Ta vs Sn = −0.81
Ta vs Hf = +0.77

5 Rock-Forming Minerals (Muscovite, Feldspar)

Trace-element analysis of pegmatoid muscovites and feldspars have been carried out on samples from the central Nigerian pegmatite fields (Küster 1985, 1987) as well as on muscovites from the Bayuda Desert/Sudan (Matheis and Küster 1987); the general geological situation and target area is outlined in Fig. 2. Due to their relative stability during chemical weathering processes under both humid-tropical (Nigeria) and arid (Sudan) conditions, muscovite is a suitable exploration tool, i.e. for amenable, fast and inexpensive sampling.

Concentrations of Sn, Nb, Ta, Cs, Rb, Li, K/Rb and K/Cs in muscovites, reported from various localities of known mining potential by Gaupp et al. (1984), have been compared to the occurrences in Nigeria and Sudan respectively. Although the Ta values themselves compare well with those of highly economic deposits (cf. Table 4), the distinctly lower mining potential in our study areas is clearly reflected by rather low Rb, Li and Cs values. Comparing the occurrences from Nigeria and the Sudan, the dominance of Rb and Li and the resulting low K/Rb ratio are the most effective indicators of the real mining potential, also in relation to the highly economic guide levels summarized by Gaupp et al. (1984). Obviously, the rare-alkali element patterns in muscovites are more sensitive for exploration purposes than those of the rare-metal elements themselves.

Table 4. Trace element data (mean values) in muscovites from pegmatitic source rocks of Nigeria and the Sudan

Locality	(n)	Sn	Nb	Ta	Cs	Rb	Li	K/Rb	K/Cs
Bayuda, Sudan	(6)	465	225	92	200	2215	155	39	567
Southwest Nigeria	(5)	710	198	69	187	5650	800	15	485
Central Nigeria	(51)	745	238	53	116	3150	505	27	800
Economic Ta-pegmatites (after Gaupp et al. 1984)				> 70	> 250	> 7500	> 500	< 20	< 300

All concentrations in ppm; own analysis by ICP-ES.

6 Conclusions

The search for rare-metal potentials from pegmatoid source rocks calls at the reconnaissance scale for the application of the geochemical pathfinders Li, Rb, Cs, F and the resulting ratios like Mg/Li, K/Rb, and K/Cs. Under humid conditions, the most suitable sample media at this early exploration stage are active stream sediments, white mica flakes and/or soils. Under arid conditions, the use of clay-rich wadi sediments, which are relics of Quaternary climatic phases, may be equally used as white mica. These recommendations combine simple and fast handling with low-cost methods for sampling, leaching and analytical procedures.

Special attention has to be directed to the geological setting and prominent tectonic features: amphibolite-pegmatite complexes of Proterozoic age provide obviously the most effective conditions for the accumulation of tantalum, combined with advanced geochemical specialization.

At advanced exploration stages, detailed geochemistry on heavy-mineral concentrates, whole-rock samples and again rock-forming minerals will determine the final economic potential. It is only at this stage that analysis of the rare-metal elements themselves has its major importance.

References

Fletcher WK, Dousset PE, Yussof bin Ismail (1987) Elimination of hydraulic effects for cassiterite in a Malaysian stream. J Geochem Explor 28:385–408

Gaupp R, Möller P, Morteani G (1984) Tantal-Pegmatite – Geologische, petrologische und geochemische Untersuchungen. Monogr Ser Mineral Deposits, vol 23. Borntraeger, Berlin Stuttgart

Küster D (1985) Trace element distribution in pegmatitic muscovites from central Nigeria. In: Bowden P, Kinnard J (eds) 13th Colloq African geology, 10–13 Sept 1985, St. Andrews. CIFEG, Paris

Küster D (1987) Geology and geochemistry of the Wamba pegmatite field, central Nigeria. In: Matheis G, Schandelmeier H (eds) Current research in African earth sciences. Balkema, Rotterdam Boston, pp 371–374

Matheis G (1979) Geochemical exploration around the pegmatitic Sn.Nb.Ta mineralization of SW Nigeria. Geol Soc Malaysia Bull 11:333–351

Matheis G (1981) Trace-element patterns in lateritic soils applied to geochemical exploration. J Geochem Explor 15:471–480

Matheis G (1987) Nigerian rare-metal pegmatites and their lithological framework. In: Bowden P, Kinnard J (eds) African geology reviews. Geol J 22:271–291, Thematic Issue

Matheis G, Caen-Vachette M (1983) Rb-Sr isotopic study of rare-metal bearing and barren pegmatites in the Pan-African reactivation zone of Nigeria. J Afr Earth Sci 1:35–40

Matheis G, Küster D (1985) Paleo-dispersion in wadi drainage and its potential prospecting applications. In: Jones MJ (ed) Prospecting in areas of desert terrain. Inst Min Metall, London, pp 209–215

Matheis G, Küster D (1987) Geochemical dispersion patterns in the Bayuda Desert/Sudan: evaluations by atomic-spectroscopy (AAS & ICP). Berliner Geowiss Abh 75.3:787–800

Matheis G, Emofurieta WO, Ohiwerei SF (1982) Trace element distribution in tin-bearing pegmatites of southwestern Nigeria. In: Evans AM (ed) Metallization associated with acid magmatism. Wiley & Sons, New York London, pp 205–220

Matheis G, Dulski P, Schreiber W (1984) Geochemische Charakterisierung von Schwermineralseifen und Kassiteriten der nigerianischen Zinnprovinz. Fortschr Mineral 62 Beih 1:149–150

Möller P, Dulski P (1983) Fractionation of Zr and Hf in cassiterites. Chem Geol 40:1–12

Smith RE, Perdix JL, Davis JM (1987) Dispersion into pisolitic laterite from the Greenbushes mineralized Sn-Ta pegmatite system, western Australia. J Geochem Explor 28:251–265

Tischendorf G (1977) Geochemical and petrographic characteristics of silicic magmatic rocks associated with rare-metal mineralization. In: Stemprok M, Burnol L, Tischendorf G (eds) Metallization associated with acid magmatism. Ust Ustav Geol, Prague, pp 41–98

Part V Processing of Ores

Processing of Tantalum and Niobium Ores

G. ROETHE

1 Introduction

Although geochemically closely associated, today's niobium and tantalum mining operations do not have very much in common.

Carbonatites are by far the most important source for niobium. Especially the weathered portions of some carbonatites are mined for their high niobium grade ($\sim$2.5% Nb_2O_5), while ore grades of fresh carbonatites rarely exceed 0.6% Nb_2O_5. Pyrochlore (Table 1) is the only valuable mineral recovered from such deposits. Altogether there are three niobium mines in production in the western world, one in Canada and two in Brazil, which account for some 95% of the western world's niobium production amounting to about 16,000 tons Nb in 1985. The remaining 5% were recovered as a by-product of tantalum and tin mining operations.

The western world's tantalum production is much smaller. It amounted to 1200 tons Ta in 1980 and 850 tons in Ta in 1984, which by magnitude is of the same order as the western world's gold production. Tantalum is recovered from pegmatites and placer deposits often as a by- or co-product to tin. Columbotantalite, wodginite, microlite, and strüverite are the most important tantalum minerals (Table 1). All tantalum minerals contain some niobium values.

In addition, cassiterite may contain up to 4% Ta_2O_5 either in solid solution or as fine to ultrafine exsolutions of tapiolite. A substantial portion of the tantalum production, ranging between 60 and 70% of the total, is therefore recovered from tin smelter slags, which may contain up to 25% Ta_2O_5 depending on the origin of the tin concentrate feed.

Brazil, Australia, and Thailand are major tantalum-producing countries, while smaller quantities are produced in Zaire, Zimbabwe, Mosambique, and Namibia. In the past, Canada was also a substantial tantalum producer (TANCO's Bernic Lake Mine). Ore grades mined are low-ranging from a few ppm to 1500 ppm Ta_2O_5, depending on the type of deposit and the co- and by-product relations.

Since 1980 when tantalum prices reached an all-time high of US$ 126 per lb Ta_2O_5 in concentrates, prices slumped to US$ 50 in 1982 and US$ 30 per lb Ta_2O_5 in 1983 and further declined to the present level ranging from 18 to 25 US$ per lb Ta_2O_5. As a consequence of this dramatic fall of tantalum prices, a number of mining operations were closed.

Bundesministerium für Wirtschaft, Postfach 140260 D-5300 Bonn 1

Lanthanides, Tantalum and Niobium
P. Möller, P. Černý and F. Saupé (Eds.)

Table 1. Tantalum and niobium minerals of economic significance

Formula		% Nb_2O_5	% Ta_2O_5	Sp. gr.
Pyrochlore	$(Ca,Na)_2(Nb,Ta)_2(O,OH,F)_7$	50–70	0.5	3.5–4.6
Microlite	$(Ca,Na)_2(Ta,Nb)_2(O,OH,F)_7$	1–10	68–79	6.4
Columbite	$(Fe,Mn)(Nb,Ta)_2O_6$	30–75	1–40	5.2
Columbotantalite		60–75%		
Tantalite	$(Fe,Mn)(Ta,Nb)_2O_6$	2–30	40–80	7.9
Wodginite	$(Ta,Sn,Mn,Nb,Fe,Ti)_{16}O_{32}$	1–15	45–70	
Bernic Lake		1.8	69	7.0
Strüverite	$(Fe,Mn)(Ta,Nb,Ti)_2O_6$	9–17	5–26	5.4

The niobium market is much more stable, and is largely controlled by the dominant Brazilian producer, the Companhia Brasileira de Metalurgia e Mineracao (CBMM), which alone has a share of 60–70% in the world's niobium market. Producer prices are published for intermediate niobium products like standard grade ferroniobium (US$ 10.30 per kg Nb fob Santos) and niobium oxide. NIOBEC Inc. in Canada is the only producer still selling pyrochlore concentrates to independent converters in the United States and Western Europe, while the Brazilian producers are converting all their pyrochlore concentrates into ferroniobium, niobium oxide, and NiNb-master alloys (CBMM).

2 Processing of Niobium

Pyrochlore is amenable to flotation when using amine collectors and fluorides as activating reagents at weak to strong acid pH conditions. Flotation is the principle method in ore dressing used for both, the weathered carbonatite ore (Araxa and Catalao in Brazil) as well as the fresh carbonanite ore mined at St. Honore in Canada.

Flotation, however, in general, and even more so when amine collectors are used, is seriously impaired by slimes. The effective removal of such slimes used to be a serious problem, especially for the treatment of weathered ores because a substantial portion of the pyrochlore in these ores occurs in the fine to ultrafine particle size range. The problem was overcome by the introduction of small diameter hydrocyclones operated under high pressure. For instance, at Araxa, Brazil, the ore after grinding is deslimed in three stages using hydrocyclones of decreasing diameters. The overflow of the first two stages is passed to the next stage, while the underflow is recycloned after intensive scrubbing to avoid secondary slime generation in the flotation circuit (Fig. 1). In the third stage the minus 5 μm fraction is rejected as final slimes. A special feature of Araxa's flowsheet is the separate flotation circuit for the 5–37-μm fraction of the deslimed ore.

Desliming is also an important pretreatment for the flotation of carbonatite ores to remove carbonate slimes formed in the grinding circuit. The remaining high amounts of carbonates of these ores seriously affect the selectivity of the amine

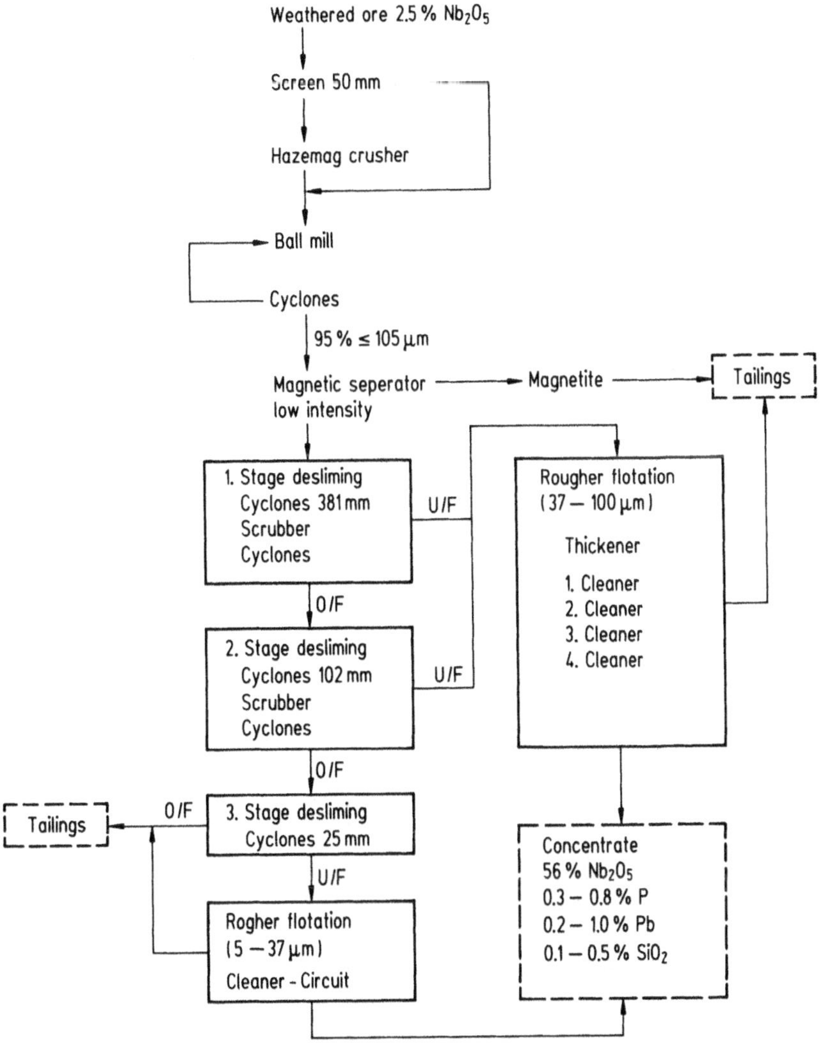

Fig. 1. Flowsheet of the Araxá pyrochlore ore dressing plant. (Anonymous 1982, Stūart et al. 1980)

collectors in the pyrochlore flotation circuit. Therefore the carbonates are removed at NIOBEC's St. Honore mine in Canada by a prefloat using fatty acid collectors. The subsequent pyrochlore flotation yields a concentrate high in sulfide impurities which are floated off in a third flotation circuit. (Fig. 2).

The flowsheets outlined show that the processing of pyrochlore ores is in an advanced state. Niobium recoveries from these ores have been considerably improved during the last 10 years and today range between 60 and 75%. The quality of the pyrochlore flotation concentrates is, however, not acceptable for direct smelting into ferroniobium because of impurities such as phosphorous, sulfur, and lead, the latter being an impurity of the barium pyrochlore at Araxa. Flotation concentrates

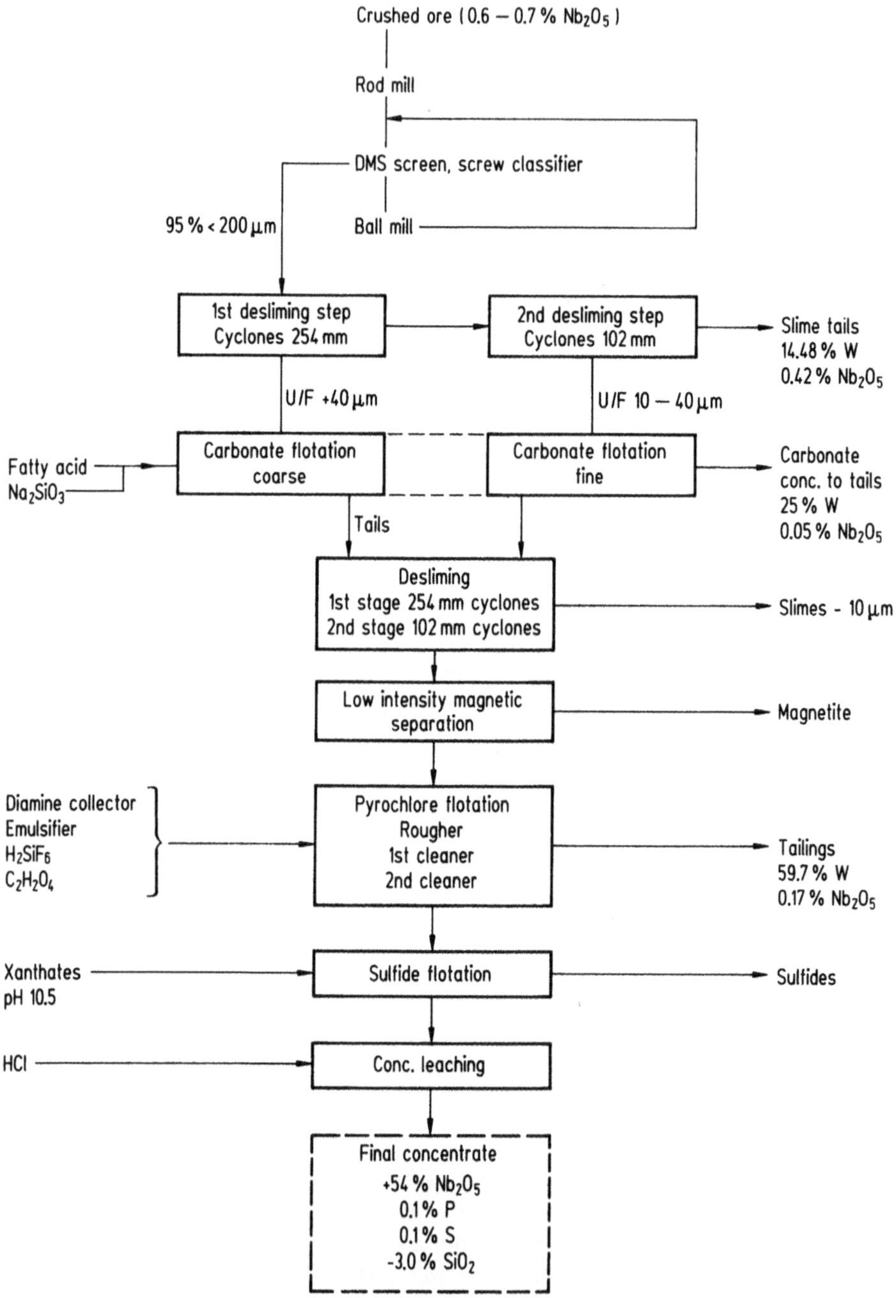

Fig. 2. Flowsheet of NIOBEC's pyrochlore ore concentrator. (After Biss 1982)

therefore are further purified by a roast-leaching process (Araxa) or by leaching with HCl (St. Honore) to meet the specifications set for ferroniobium.

The major part of the pyrochlore concentrates is smelted to standard grade ferroniobium (65% Nb) by the aluminothermic reduction process. At Araxa and Catalao, Brazil, the total concentrate production is further processed at the mine site. In addition to ferroniobium also technical grade niobiumpentoxide and Ni/Nb-masteralloys are produced at Araxa since 1980. Only NIOBEC's St. Honore mine is still selling pyrochlore concentrates to converters in the United States and Western Europe.

3 Processing of Tantalum Ore

Tantalum mining operations are typically small-size operations. Because of the density of the tantalum minerals (Table 1) mainly gravity methods are applied in ore dressing. Equipment used in these operations ranges from jigs and spirals to sand- and slime tables depending on the size distribution of the feed and throughput requirements (Table 2). Gravity concentrates, however, are only in rare cases pure enough to be marketed directly. Usually they represent a mixture of heavy minerals which is further treated by magnetic and high tension electrostatic methods to yield a marketable tantalum concentrate. From a practical point of view, the following types of operations are distinguished.

3.1 Hard Rock Pegmatite Mining

Liberation of tantalum minerals from adherent gangue minerals is a precondition for an effective separation by gravity methods. The grinding circuit has to be carefully controlled to avoid overgrinding because tantalum minerals are extremely friable and tend to disintegrate into fine to ultrafine particles, which below 30 μm are

Table 2. Gravity concentration equipment, size ranges and capacity

Equipment	Size range of feed (microns)	Capacity t p.h.
A Preconcentration		
Jig (coarse roughing)	1500 – 6500	12
Jig (fine roughing)	500 – 1500	6
Spiral	100 – 500	2–4
Reichert cone	50 – 500	70
Bartles-Mozley table	5 – 100	2
B Upgrading		
Jig (coarse cleaning) 12" × 18", 1 hutch	1500 – 6500	0.5
Table (sand)	100 – 500	1
Table (slime)	40 – 100	0.25
Vanner (Bartles crossbelt)	10 – 100	0.45

very difficult to recover by gravity methods. Insufficient liberation and overgrinding are the main causes for tantalum losses in such operations. Therefore successive stages of grinding are a common feature of such hard rock gravity concentrators. The tantalum minerals are separated in a state as coarse as possible. The middlings are reground and retreated. The feed for the various separation sections has to be closed-sized to achieve optimal results. Classification is done mainly by screening even in the fine particle size range because of the high specific gravity of the tantalum minerals. An excellent example of such an operation used to be TANCO's Bernic Lake tantalum ore concentrator in Canada with its spiral circuit, the sand table and the fine ore section, and the slime treatment section (Fig. 3).

Bernic Lake was the first tantalum mine in the western world to introduce flotation in its slime treatment section (Burt 1979). Tantalum recoveries between 72 and 74% have been reported, which is a fairly good result considering the low ore grade (0.13% Ta_2O_5).

3.2 Weathered Pegmatite Operations

Since the tantalum minerals as well as the cassiterite, if present, are already in a liberated state, these ores do not require crushing and grinding. The high clay content of these ores is washed off in drum scrubbers and removed by hydrocyclones. The barren oversize fraction is discarded. The ore is classified and further treated by jigs, spirals, sand- and slime tables. Greenbushes "clay plant" in West Australia may serve as an example to illustrate this type of operation (Fig. 4). Satisfactory recoveries, as well as good concentrate grades, are achieved in the coarse fractions. Substantial tantalum losses occur in the fine and ultrafine particle size range (minus 100 μm) and concentrate grades from these sections are usually low. The resulting gravity concentrates are dried and further treated by high intensity magnetic and high tension electrostatic separators to upgrade and separate columbotantalite from cassiterite and microlite (Fig. 5).

3.3 Placer Mining Operations

Cassiterite placer deposits mined in Malaysia and Thailand often contain strüverite and sometimes columbotantalite as a by-product. The on-shore operations are mainly working with the traditional sluices ("palong"), whereas off-shore dredging operations are nowadays mostly equipped with jigs. The resulting rougher concentrates are further upgraded by tabling and sluicing to marketable tin concentrates. The middlings, mainly containing ilmenite and minor quantities of cassiterite and strüverite, are usually sold to small companies specialized on the retreatment of this material using tables, sluices, magnetic and electrostatic separators.

While columbotantalite concentrates usually contain >25% Ta_2O_5 (or combined oxides >50% $Ta_2O_5 + Nb_2O_5$), tantalite and microlite concentrates are much higher in grade >60% Ta_2O_5 (or combined oxides >80% $Ta_2O_5 + Nb_2O_5$), strüverite concentrates contain only between 8 and 12% Ta_2O_5 and as a tantalum raw material compare with the high Ta-grade tin smelter slags.

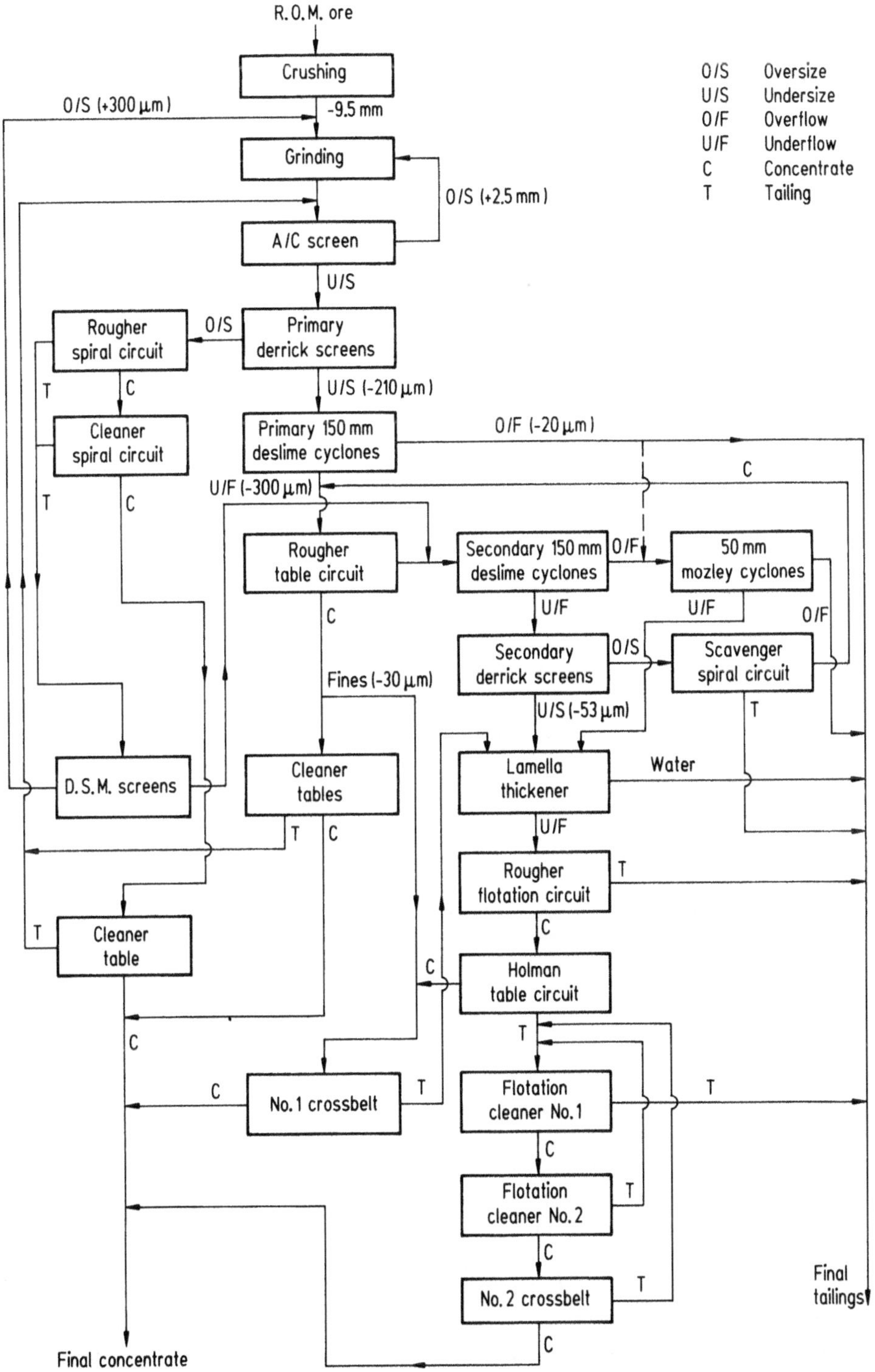

Fig. 3. Flowsheet of the tantalum ore concentrator at Bernic Lake mine, Canada. (Flemming et al. 1982)

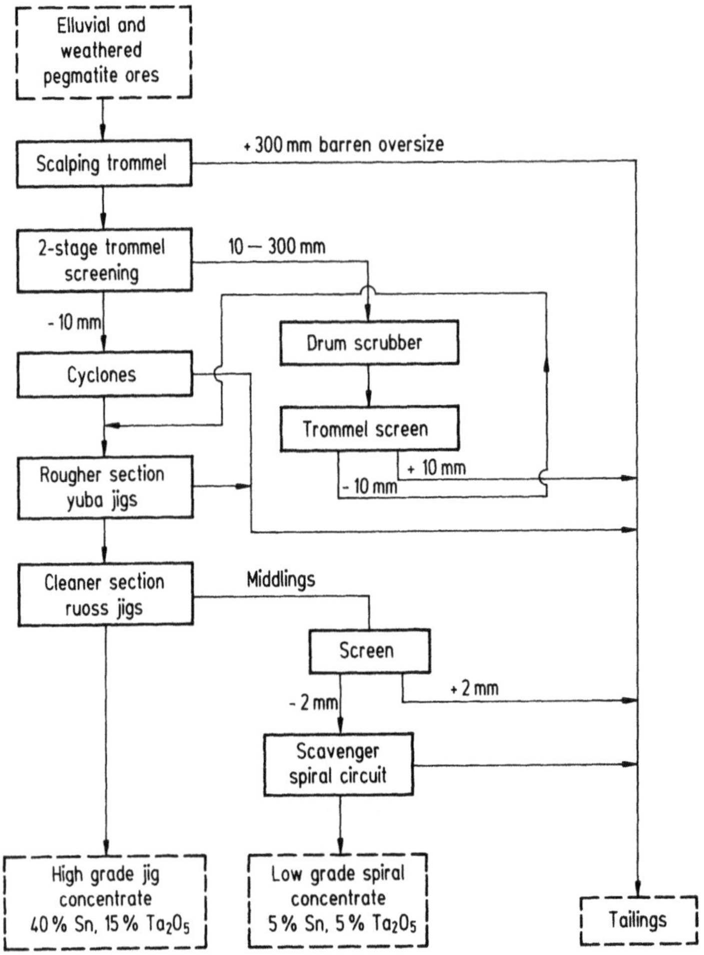

Fig. 4. Flowsheet of Greenbushes "clay plant" tin and tantalum ore dressing operation. (Company information, anonymous 1979)

3.4 Tin Smelter Slags

Cassiterite may contain up to 4% Ta_2O_5 either in solid solution or as fine to ultrafine tapiolite exsolutions. These values are concentrated during smelting in the final slag, which may contain as much as 30% Ta_2O_5, like the Greenbushes tantalum glass concentrate. Usually, however, tantalum grades of such slags are much lower and slags containing 8 or 10% Ta_2O_5 are already considered high grade. The two-stage smelting process for cassiterite concentrates is illustrated by Fig. 6.

For economic reasons low-grade slags are upgraded by an electrothermic reduction process followed by a partial oxidation, which yields the so-called synthetic concentrates. Depending on feed grades, such concentrates contain 50% $Ta_2O_5 + Nb_2O_5$ and more. An intermediate product of this process is the block metal,

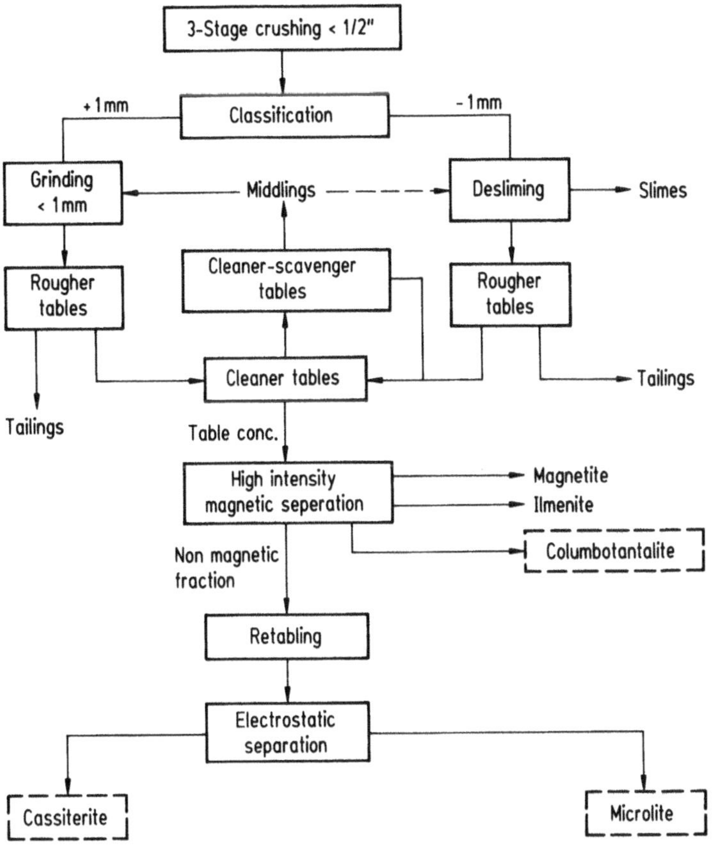

Fig. 5. Flowsheet of the hard rock pegmatite operation at Minas Brasil. (Company information inpubl.)

which is also used as a starting material for the production of high-purity oxides by chlorination.

In times of high tantalum prices, low grade slags containing 2% Ta_2O_5 and even less were processed and substantially contributed in those days to the overall tantalum supply. At current low tantalum prices, the processing of low-grade slags is, however, not economic.

3.5 Processing of Tantalum Concentrates and Tin Smelter Slags

While only a few mines are equipped to process their tantalum concentrates, most of the concentrates and virtually all slags are sold to independent converters. The liquid-liquid extraction process as originally developed by the US Bureau of Mines is nowadays the main process used for the tantalum and by-product niobium extraction from both the natural and synthetic concentrates as well as the high-grade tin smelter slags.

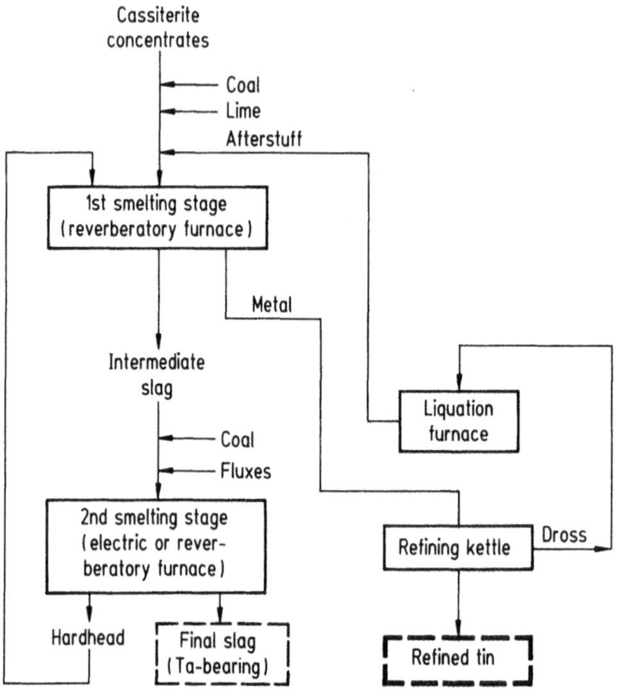

Fig. 6. Flowsheet of smelting cassiterite concentrates.

After digesting the concentrates or slags in hydrofluoric and sulfuric acid, the metal fluoride complexes of the filtrate are extracted into a methyl-isobutyl-keton phase (MIBK) in a series of mixer and settler tanks from which the tantalum and niobium is stripped. From the resulting fluorotantalic acid solution either potassium tantalum fluoride (K_2TaF_6) is precipitated by additions of potassium chloride or fluoride or as tantalum pentoxide by adding ammonia (Albrecht ibid). Potassium tantalum fluoride is the intermediate product for the tantalum metal production, while tantalum oxides are the starting products for tantalum carbides. Niobium oxide is recovered as a by-product and further processed into high purity niobium oxides, niobium metal, and special alloys.

4 Summary

The major source for niobium is carbonatite deposits containing pyrochlore in minable concentrations. Among these deposits the high-grade weathered ore deposits are of major importance. Flotation is the principle ore-dressing method. The introduction of more effective desliming techniques during recent years has greatly improved the efficiency of flotation in the fine and ultrafine particle size range. Pyrochlore concentrates are further processed to ferroniobium, niobium oxide, Ni/Nb-masteralloys, and niobium metal.

Three niobium mines, two in Brazil and one in Canada, supply about 95% of the total western world's niobium demand. The remaining 5% are covered by niobium oxide produced as a by-product to tantalum.

There are a number of tantalum-producing countries. The more important ones are Brazil, Australia, and Thailand. The major part of tantalum production comes from tin smelter slags and concentrates recovered as a by-product of tin mining and smelting operations. Tantalum ore grades are low, often extremely low. Thus the upgrading factor in ore dressing may be as high as 3000. Gravity is the principle method used to recover tantalum minerals like columbotantalite, wodginite, microlite, and strüverite. The concentrates are further upgraded by electrostatic and high intensity magnetic separation methods. Efforts made during times of high tantalum prices in the late 1970's and early 1980's to improve ore dressing efficiency in the low particle size range have been largely discontinued as a consequence of the present low tantalum prices. A number of tantalum mines have been closed down.

Somewhat in contrast to the mining sector, the processing industry is using most advanced technologies for the tantalum extraction and the production of high-purity metal, powders, carbides, and oxides for the numerous areas of industrial applications of this metal.

References

Anonymous (1982) Araxa niobium mine. Mining Mag 132–147

Anonymous (1979) Greenbushes Tin. N.L.: The Greenbushes Mining Operation. Company brochure prepared for the 11th General Assembly of the Tantalum Producers. Tantalum Int Study Centr (TIC) Perth, 19p

Biss R (1982) Pyrochlore ore beneficiation at Les Services TMG Inc. (Niobec) concentrator. Can Min J 103:17–25

Burt RO (1979) Tantalum Mining Corporations gravity concentrator – recent developments. CIM Bull 72:103–108

Flemming J, Mills C, Burt RO (1982) TANCO's rare metal concentrator. Can Min J 103:40–47

Stuart H, de Souza Paraiso O, de Fuccio R (1980) Ferro columbium production at CBMM, Brazil. Iron and Steel Maker 11–18

Part VI Application

Production, Properties and Application of Tantalum, Niobium and Their Compounds

W.W. Albrecht

Abstract

Methods for processing tantalum-niobium ores and synthetic tantalum-niobium concentrates for separating tantalum and niobium are discussed.

For the production of tantalum and niobium compounds such as oxides, chlorides, carbides and metals a variety of different processes have been developed. Among these, liquid-liquid extraction, chlorination, sodium reduction, aluminothermic and carbothermic reduction, electron beam- and arc-melting are the most important ones.

Properties and application of tantalum and niobium and their compounds in electronics, optics, ceramics, superconductivity and alloys are discussed.

1 Introduction

Geology, exploration and mining of tantalum and niobium were the main topics of this workshop. In this presentation I want to explain what happens to the ores after having been mined or to the tin slags after having been provided by tin smelters in Thailand, Malaysia or throughout the world.

As can be seen in Fig. 1, especially Ta is used by about 50% in sophisticated applications like electronics and optics, another 30–40% is used in cemented carbides in the form of mill products or as an alloying additive in superalloys.

Over 90% of the Western world's demand for niobium consists of ferroniobium consumed in the production of steel and alloys. The remaining 10% in the form of niobium oxide is split into about 6–7% which goes into high-purity FeNb, NiNb or CrNb and 3–4% high-purity niobium oxide or niobium metal which is also used as an essential additive to modern electronics, optics and superconductivity (Bureau and Mines Minerals Yearbook 1985; Roskill Informations Services Ltd 1986). These applications will be discussed in this presentation in more detail later, but what can be seen from the beginning: processing and application of tantalum and high-purity niobium is mainly a commodity of countries at the so-called forefront of technology like North America, Japan and Europe.

Hermann C. Starck Berlin, Werk Goslar. D-3380 Goslar 1

Lanthanides, Tantalum and Niobium
P. Möller, P. Černý and F. Saupé (Eds.)

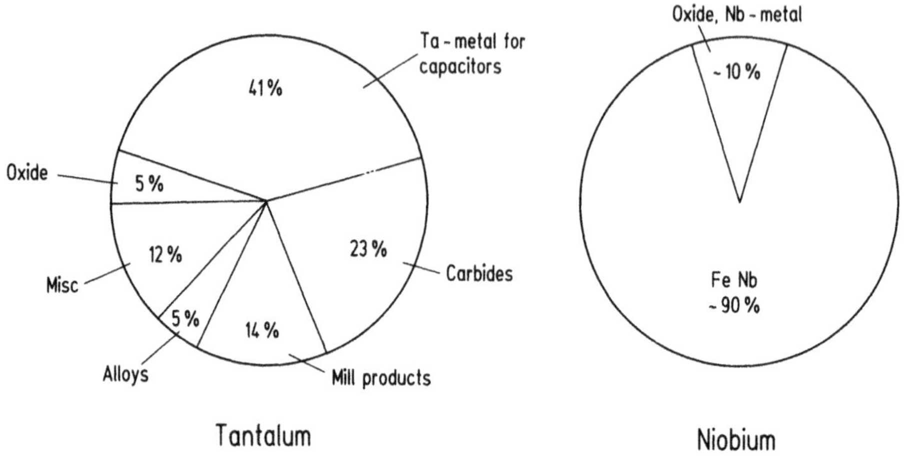

Fig. 1. Application of Ta and Nb

The following topics shall be discussed in more detail:

1. Composition of Ta- and Nb-ores and concentrates mainly used for processing;
2. Main steps of modern processing; quality of Ta and Nb in metals, oxides and carbides produced (intermediate products for further application);
3. Application of tantalum and niobium.

2 Composition of Ta-and Nb-Ores and Concentrates Mainly Used for Processing

The composition of the main types of concentrates used in the chemical process is shown in Table 1.

Numbers 1–4 are the traditional raw materials for the processing industry, satisfying about 50% of the demand of Ta today. The more complex types have occasionally been treated (Borchers and Korinek 1981).

Only a small portion of the tantalites, which are closely associated with the tin mineral cassiterite, can be recovered by ore beneficiation methods. The bulk of this tantalum is collected in the slags of tin smelters in different grades.

Slags with Ta_2O_5 contents of more than 10% have been or are treated directly by chemical means to recover Ta/Nb. The tonnagewise more abundant tin slags below 10% Ta_2O_5 are upgraded by chloro-and pyrometallurgical methods. Figure 2 shows a flow sheet for tin slag processing.

Tin slags are melted in an electric arc furnace together with a flux material; the tantalum-niobium content is collected in the resulting ferroalloy and the tantalum-niobium free slag is poured off. After crushing and dressing, this ferroalloy is converted to a tantalum-niobium concentrate either by means of a second pyrometallurgical process, or by means of a chemical process, such as:

Table 1. Main tantalum minerals

No.	Mineral	Formula	%Ta_2O_5	%Nb_2O_5
1	Tantalite		40–80	2–30
2	Columbite	$(Fe,Mn,)(Nb,Ta)_2O_6$	1–40	40–75
3	Wodginite	$(Ta,Nb,Sn,Mn,Fe)_{16}O_{32}$	45–56	3–15
4	Microlite	$(Ca,Na,)_2Ta_2\ (O,OH,F)_7$	50–70	5–10
5	Strueverite	$(Ti,Ta,Nb,Fe)_2O_4$	7–13	9–14
6	Euxenite	$(Y,Ca,Ce,U,Th)(Nb,Ta,Ti)_2O_6$	2–12	22–30
7	Samarskite	$(Fe,Ca,U,Y,Ce)_2(Nb,Ta)_2O_6$	15–30	40–55

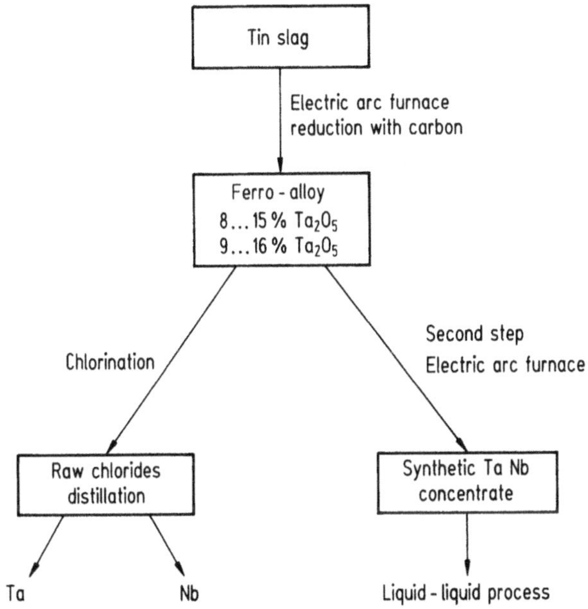

Fig. 2. Flow sheet for tin slag processing

1. Dissolution of the material in mineral acids to produce a tantalum-niobium concentrate.

2. Chlorination of the alloy and separation of the tantalum-niobium chlorides by fractional distillation (Rockenbauer 1981).

Analyses of the alloy and the concentrate produced therefrom by a pyrometallurgical process, based on two different kinds of tin slags of Malayan provenance, are shown in Table 2.

HCST started producing synthetic Ta/Nb-concentrates from tin slags 30 years ago. Today it possesses the world's largest plant of this kind, capable of treating 18 000 tpy tin slags.

Figure 3 shows part of H.C. Starck's plant for the production of synthetic tantalum-niobium concentrates.

Table 2. Enrichment of tantalum in the Starck tin slag process

Composition of slag (Malaysia)	Alloy	Synth. concentrate
4% Ta_2O_5	~15% Ta_2O_5	26.9% Ta_2O_5
4.3% Nb_2O_5	~17% Nb_2O_5	30.9% Nb_2O_5
10.5% TiO_2	(as Metal)	18.6%TiO_2
1.5% SnO_2		0.1% SnO_2
2 % Wo_3		0.1% Wo_3
1.2% P_2O_5		0.1% P_2O_5
15 % Fe O		8% FeO
Ca O, Mg O, Al_2O_3, SiO_2		
2% Ta_2O_5	~ 8% Ta_2O_5	20.8% Ta_2O_5
2.2% Nb_2O_5	~10% Nb	25.6% Nb_2O_5
10.5% TiO_2	(as Metal)	25.0% TiO_2
1.2% SnO_2		0.1% SnO_2
2.6% Wo_3		0.1% Wo_3
1.2% P_2O_5		0.1% P_2O_5
15 % Fe O		8% Fe O
Ca O, Mg O, Al_2O_3 SiO_2		

Fig. 3. A part of HCST's plant for the production of synthetic TaNb concentrates

3 Main Steps of Modern Processing

Decomposition of the ores and synthetic concentrates by means of acids, mostly HF or $HF+H_2SO_4$, and separation of tantalum and niobium by liquid-liquid extraction:

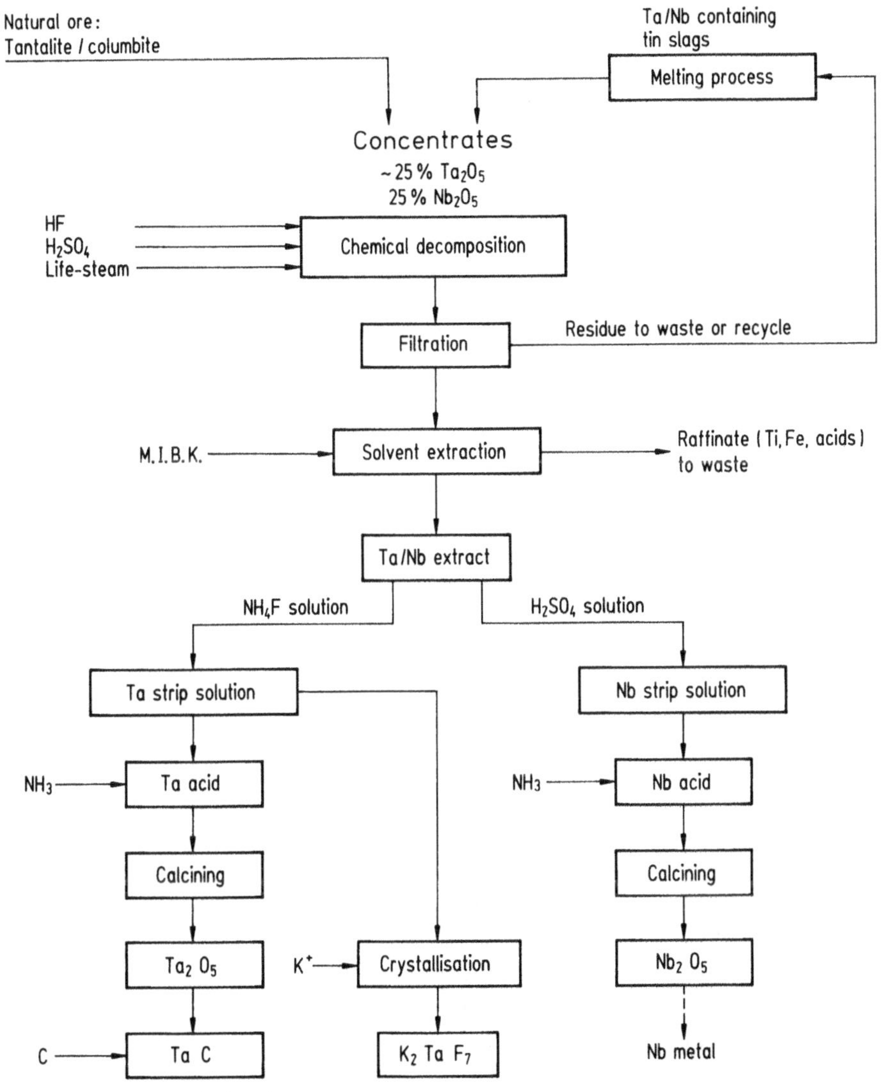

Fig. 4. HCST's Ta/Nb chemistry

Figure 4 shows a simplified flow sheet of a Ta/Nb separation process by liquid-liquid extraction.

The pure products are isolated after liquid-liquid separation. Tantalum and niobium hydroxides are precipitated with ammonia from the corresponding strip solutions and converted to form the oxides by calcining. Potassium tantalum fluoride, the "K_2 salt", is crystallized by adding potassium ions to the organic tantalum phase or the tantalum strip solution.

Chlorination of ores or ferroalloys and the separation of the raw chloride mixture by distillation:

Most of the constituents of a tantalum ore form stable anhydrous chlorides; their boiling points differ over a wide range. Thus chlorination technology offers good possibilities for extraction and purification (Fig. 5).

Ta/Nb ferroalloys were chlorinated in a melt of sodium-iron chloride. The volatilized chlorides were roughly separated in a solid fraction, containing Ta, Nb and other high-boiling-point chlorides, and a liquid fraction, containing the low-boiling-point chlorides of Ti, Si and Sn. Iron chlorides are retained in the salt melt.

The solid chloride mixture is separated by fractional distillation in heated, packed glass columns, using the very small difference in boiling points of $TaCl_5$ 232°C and $NbCl_5$ 245°C. The pure chlorides are converted to oxides by means of a special steam hydrolysis or taken as feed material for the metal reduction.

Because only glass, quartz or Teflon material is in contact with the chlorides and no additional chemicals are used, it is possible to produce extremely pure compounds (Rockenbauer 1981). Final products resulting from these processing steps are as follows:

K_2TaF_7; Ta_2O_5; Nb_2O_5; $TaCl_5$; $NbCl_5$.

In Table 3 the analyses of commercial high-purity products according to the fluoride and chlorination processes are compared.

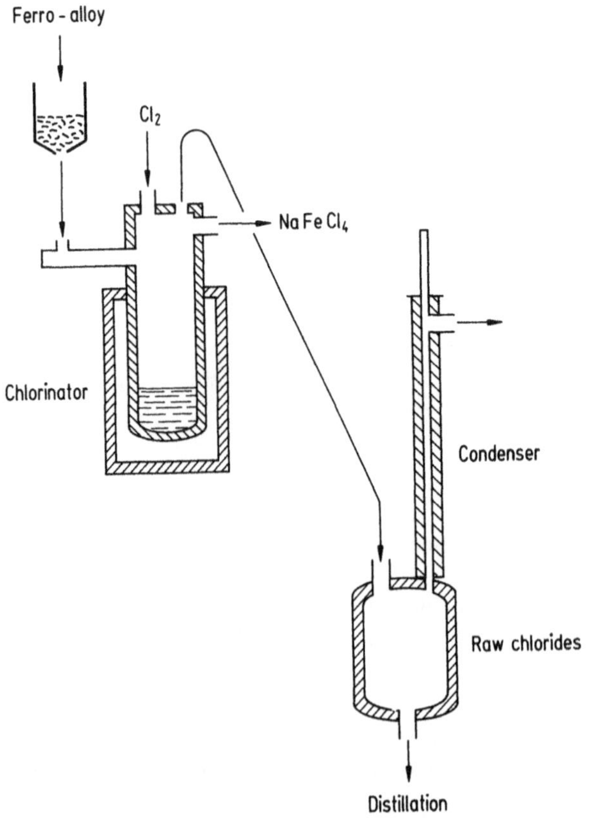

Fig. 5. Ferro-chlorination

Table 3. Typical analysis of Ta/Nb compounds

	Ta_2O_5 NB_2O_5 Crystal grade (ppm)	$TaCl_5$ NbCl Highest purity (ppm)	K_2TaF_7 (ppm)
Al	< 1	< 1	<7
Ca	< 2	<0.5	<7
Cr	<0.5	<0.5	<7
Co	<0.5	<0.5	<7
W	< 1	< 3	<9
Fe	< 1	<0.5	<9
K	< 1	< 1	–
Mg	< 1	< 1	<7
Nb/Ta	< 10	< 10	<9
Ni	<0.5	<0.5	<7
Si	< 10	< 2	<7

3.1 Production of Ta and Nb Metal

Tantalum and niobium metal can be produced from three different compounds: fluorides, oxides and chlorides.

3.2 Tantalum

In case of Ta the sodium reduction of K_2TaF_7 is widely used (Fig. 6).

K_2TaF_7 liquid sodium and inert salts are mixed and charged in the form of a paste into trays which then pass through a continuous furnace. The reaction which takes place is

$$K_2TaF_7 + 5\,Na + \text{Inert Salt}$$
$$\text{Ta-Metal} + 2KF + 5\,NaF + \text{Inert Salt.}$$

The salts are leached off in further steps in order to obtain the Ta-metal powder.

Most of this powder is used in the electronic industry, therefore, the process has to be carried out in Ta-vessels with utmost protection against, e.g. oxygen, in order to obtain the requested level of high purity.

Additional purification of tantalum (and niobium) metal is carried out in tonnage scale via arc and electron beam melting. Today EB-furnaces up to 2 MW are used for purifying and melting Ta-ingots of up to 30 cm diameter and Nb-ingots of up to 40 cm diameter (Albrecht and Ingalls 1983).

3.3 Niobium

The major part of niobium is produced in the form of ferroniobium by alumino-thermic reduction of pyrochlore concentrates. In this batch process (CBMM) (Deutsches Institut für Wirtschaftsforschung Berlin 1982):

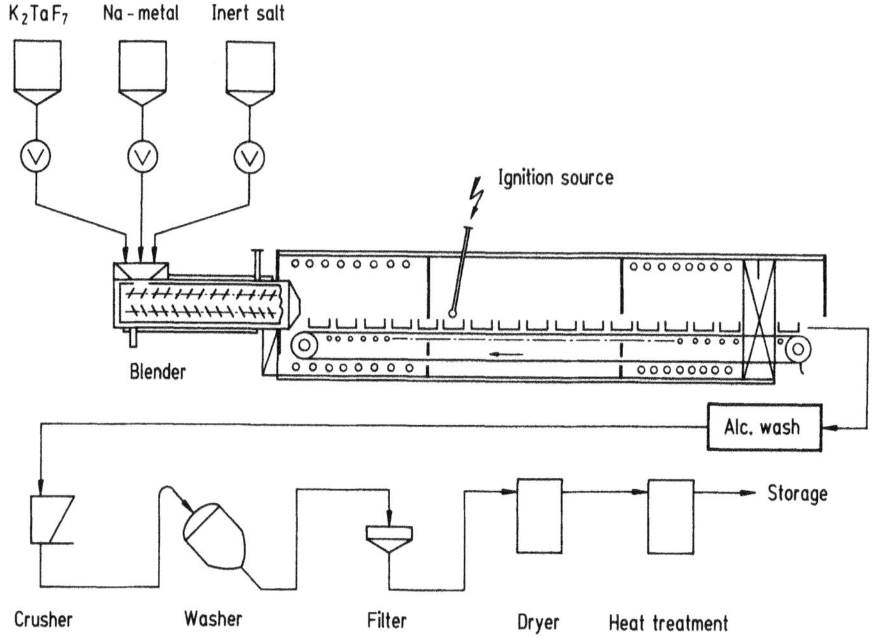

Fig. 6. Na-reduction of K_2 TaF_7

Pyrochlore concentrate	(60% Nb_2O_5)	18 000 kg
Iron oxide	(68% Fe, Hematite)	4 000 kg
Aluminium powder		6 000 kg
Fluorspar		750 kg
Lime		500 kg

are reacted. The batch size increases to 11.5 tons of ferroniobium (66%).

Typical analysis of FeNb:

Nb	66.0%	Al	0.5%
Fe	30.5%	Ti	0.1%
Si	1.5%	P	0.1%

High purity ferroniobium is produced by reducing 99% pure niobium oxide in an alumino-thermic reduction.

Nickel-niobium (54–57% Nb) and chromium-niobium (65–80% Nb) are manufactured by the alumino-thermic process too, using 99% pure niobium oxide. Aluminium-niobium is used in titanium alloys (2–3% Nb).

By far the most common method of producing niobium metal is the alumino-thermic reduction of niobium pentoxide in a process similar to that used for the production of high-purity ferroniobium.

Electron beam melting is the most common process for further purifying this type of niobium, which contains about 4–5% Al.

To some extent niobium metal is produced by carbothermic reduction of niobium oxide. In this process Nb_2O_5 and carbon are mixed, pelletized and converted to form niobium metal by means of a two-step process in a vacuum furnace at 1950°C.

3.4 Production of Ta- and Nb-Carbides

The basic method for producing carbides is quite similar to carbothermic reduction: an oxide-carbon mixture is reacted at temperatures of 1450°–1950°C.

In Table 4 standard purities of Ta-metal (capacitor grade), Nb-metal (EB-melt) and TaC are compared.

4 Application of Tantalum and Niobium

As mentioned before about 41% Ta is used in the form of metal in Ta-capacitors (Fig. 7). They are basic electrical components used to store electrical energy, like a battery.

Basically, one electrode of a capacitor consists of a spongelike, pressed, sintered body of Ta-powder. Formed Ta_2O_5 acts as a dielectric. The counterelectrode is produced by coating the Ta_2O_5 with MnO_2 (Albrecht and Hoppe 1983).

Most attention is paid to producing Ta-powders with the highest surface and highest purity. From the electrical point of view the surface can be described by microcoulomb per gram Ta-powder. The figures available today are in the range of 18000–22000 μC/g.

Worldwide, 4–5 billion Ta-capacitors are annually produced today, consuming about 500 t of tantalum metal and wire.

4.1 Tantalum Carbide

Tantalum carbide powders in the form of TaC, TaNbC, WTiTaC or WTiTaNbC are added to tungsten carbide (WC) – cobalt (Co) cutting tool compositions to improve hardening and to increase crater resistance and high-temperature strength.

Table 4. Typical analysis of Ta/Nb compounds

	Ta Capacitor Grade ex Na-Reduction (ppm)	Nb by electron beam melting (ppm)	TaC (ppm)
Al	< 3	< 3	20
Ca	< 5	< 3	20
Cr	7	< 3	50
W	< 6	20	500
Fe	< 25	10	90
K	< 10	< 3	10
Nb	< 6	–	0.1%
Ni	7	< 3	90
Si	25	6	40
O	1500	100	1300
C	60	15	6.24%
N	100	20	350

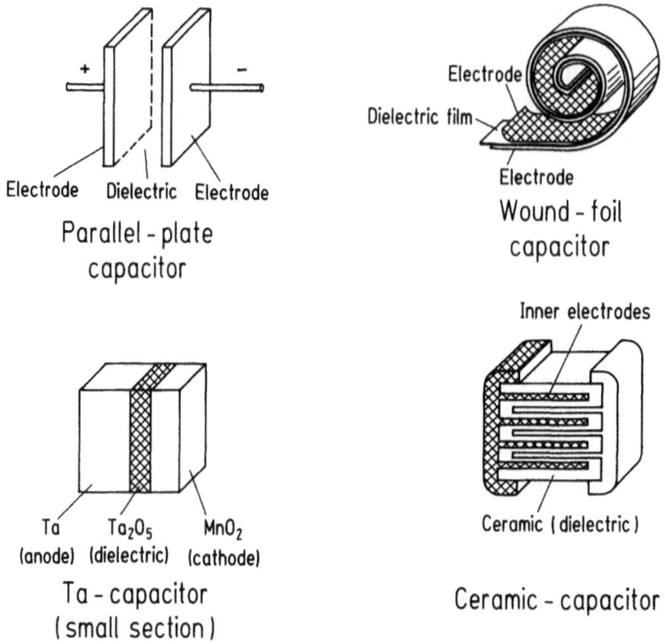

Fig. 7. Construction principles of capacitors

Two specific ranges of WC-TaC-Co compositions have been developed: those with a TaC content of less than 1% by weight and those with a TaC content greater than 2% by weight.

The smaller TaC additions serve as a grain-growth inhibitor. Typical applications are K-grade cutting tools (for nonferrous metals and wood). Compositions with higher TaC content (Iso-P grades), which have good crater resistance and high-temperature strength, are commonly used for machining steels.

The amount of tantalum carbide, ~ 586 000 lbs. Ta_2O_5 obtained in 1984 is decreasing because of increased use of reclaimed carbides and the extended life of cutting tools made possible by coating.

4.2 Ta in Corrosion-Resistant Applications

Fabricated tantalum equipment used in chemical process industries includes lined vessels, heat exchangers, heating elements, evaporators, condensers, coolers, pumps, valves and textile spinnerettes. Tantalum equipment is required for sulphuric acid concentrators, for handling chlorine, in the manufacture of pharmaceutical chemicals, in which contamination must be avoided, and in nuclear applications.

Most chemical equipment is designed and fabricated for specific applications. Likewise tantalum alloys have been and are being developed to provide improved properties in corrosion applications.

4.3 Ta in Superalloys for Aircraft Turbine Engines

Tantalum is a key alloying element in nickel-based superalloys used for the most critical, high-temperature, rotor-blade applications in aircraft turbine engines. It is also an important alloying element for strengthening cobalt-based superalloys used for stationary inlet vanes and in other components for turbine engines.

The use of tantalum is expected to increase in conventional superalloys, particularly in the new, high-performance, single-crystal alloys for blades in military aircraft. The leading single-crystal blade alloy (Pratt and Whitney PWA 1480) contains 12% tantalum. Tantalum-containing aircraft superalloys contain tantalum ranging from 1.5 to 12%.

4.4 Ta_2O_5 and Nb_2O_5 in Ceramic and Optical Applications (Manker 1985)

4.4.1 Optical Glasses

Optical glasses containing up to 30% Nb_2O_5 have very high refractive indices, while maintaining densities similar to ordinary optical glasses. A high refractive index allows lenses to be much thinner and lighter than standard optics. For example, two lanthanum borate glasses may have the same optical constant, but the one containing Ta-oxide and lead oxide will have a specific gravity of 4.63. The other containing niobium pentoxide will have a specific gravity of 4.19 and therefore it will be 10% lighter.

These advantages have led to the widespread use of niobium glasses in lenses for cameras, copying machines and other sophisticated optical applications. Niobium optical glass is also used in eyeglass lenses, where the high refractive index allows lenses which are nearly as light as plastic lenses, but which retain the strength and abrasion resistance of glass lenses.

The niobium oxide used for optical glass must be of very high purity, and only 1–2 ppm of colouring impurities, such as chromium, cobalt and copper, can be tolerated. The main producers of niobium oxide glass are optical companies in Japan and Europe.

Ta_2O_5 is used in smaller amounts, e.g. in phosphate glasses. This type of glass shows a higher transmission index for UV, compared to boron or silica glasses.

One disadvantage of glass containing Nb_2O_5 is that it absorbs ultraviolet radiation near the visible range. If this application is critical, Ta_2O_5 is used instead of Nb_2O_5.

4.4.2 Ceramic Application

Over the last 2 decades multilayer ceramic capacitors (MLC) (Fig. 8) have been well established mainly using Nb_2O_5-doped $BaTiO_3$ dielectrics ($< 1\%$ Nb_2O_5).

The disadvantages of $BaTiO_3$ dielectrics are dielectric saturation and a large grain size of $BaTiO_3$ when produced by solid state reaction (5 to 10 μm). The large grain size makes it difficult to realize a dielectric layer thinner than 20 μm for increasing the capacitance.

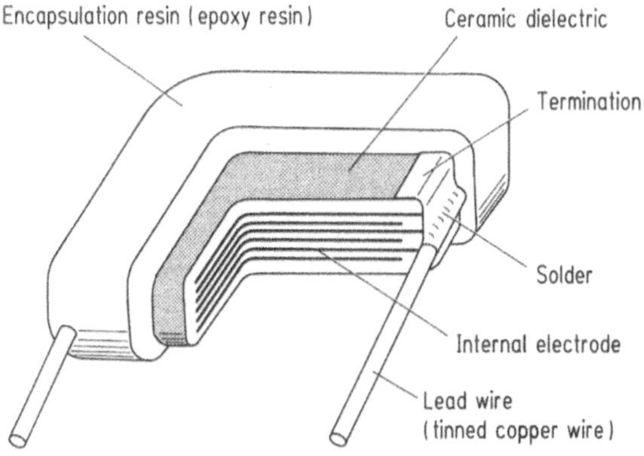

Termination: silver
Internal electrode: silver-palladium alloy (70 Ag - 30 Pd)
Dielectric: sintered ceramic compound based on lead zinc niobate

Fig. 8. Multilayer ceramic capacitor

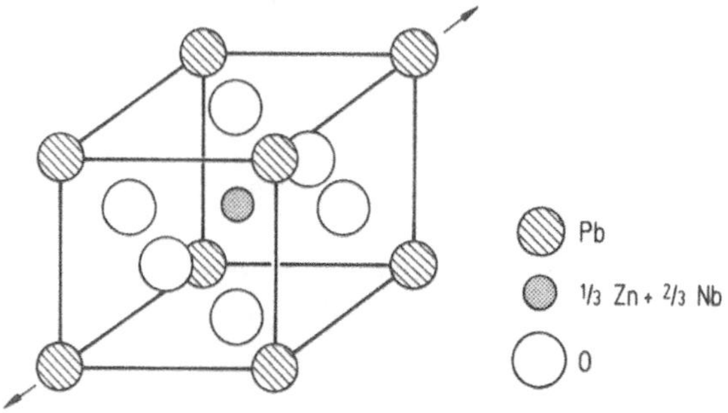

A unit cell of rhombohedral perovskite
Pb ($Zn_{1/3}Nb_{2/3}$) O_3 (a = 4.063 Å, α = 89°55′)

Pb Zn Nb
Pb Fe Nb
Pb Mg Nb

Fig. 9. Lead perovskite

With the introduction of new dielectrics based on a lead perovskite relaxor (Fig. 9) (Yamashita et al. 1985):

$$(Pb_{1-x}\ Sr_x)\ [(Zn_{1/3}\ Nb_{2/3})\ y\ Ti_z]O_3,$$

the disadvantages of $BaTiO_3$ could have been overcome.

Today, Pb Zn Nb
Pb Fe Nb
Pb Mg Nb

relaxor-type perovskites are used with a content of up to 25% Nb_2O_5.

Ceramics with these formulas were basically prepared by blending and calcining chemically pure oxides.

In 1984 26 billion MLC were produced mainly using a $BaTiO_3$-type dielectric. This field of new dielectric materials opens a wide range of additional application of Nb_2O_5.

Lead niobates have some very useful piezoelectric properties. So when lighting a cigarette today or listening to the sound of an alarm clock you most likely make use of lead niobate.

Lithium tantalate and lithium niobate have a larger coupling factor as a piezoelectric transducer. The piezoelectric transducer can produce an acoustic wave from a radiofrequency input, or can convert an acoustic wave to an RF output. A variety of filters, delay lines and other devices can be produced.

For producing these devices $LiTaO_3$ or $LiNbO_3$ single crystals are drawn by inch diameter, sliced into wafers and coated with aluminium electrodes.

These filters are used in e.g., VTRs, colour TV and video discs.

Experimental work has demonstrated the possibility of storing holographic images in potassium tantalate niobate doped with Fe. In theory, billions of holograms could be stored in, and retrieved from, a single crystal. The information storage potential is still beyond current technology but bears a tremendous potential for further application of Nb_2O_5 of highest purity.

4.5 Steel

Ninety percent of the Western world's demand for niobium consists of ferroniobium consumed in the production of steel and alloys. The most important application of niobium is as high-strength, low alloy steel, which currently accounts for about one-third of the ferroniobium consumption. Additions of up to 1% niobium act as a grain-refining and precipitation-strengthening element to increase the hardness and strength of steel (by forming carbides and nitrides). Major uses for niobium strengthened HSLA steels include automobile components and exhaust systems, electrical transmission towers, offshore platforms, pipelines and bridges.

The increasing emphasis on weight reduction, fuel efficiency and pollution control implies that HSLA steels are replacing carbon steels in transport and construction markets. Consumption of HSLA steel in an average car, for example is estimated to be likely to increase from 250 to 375 lbs. between 1984 and 1990.

4.6 Superconductivity

Superconducting magnets fabricated from niobium-titanium alloys are currently used in medical diagnostic equipment. If adopted for particle accelerators in super-colliders, the world consumption of niobium in superconducting magnets could increase from about 100 tpy today to 2000 tpy in 2000.

The alloy Nb-46.5 Ti and the intermetallic compound Nb_3Sn are standard materials for high-field superconductors. They belong, like pure metal, to type II superconductors.

Metallurgy and processing of NbTi and Nb_3Sn are well understood with regard to the extreme requirements concerning critical currents and/or magnetic fields. The unalloyed metal Nb has just recently become a subject of similar interest due to its high frequency, superconductivity for the acceleration of electrons, protons and positrons in linear accelerators and storage rings. Additional energy can be conferred to elementary particles by applying high-frequency fields in cavities.

If superconducting niobium is used as a wall material, a ten-fold reduction in the high-frequency losses is reached. Thermal conductivity in the temperature range of 0.5 to 10 K plays an important role in this application, depending upon the interstitial impurities of niobium. Heraeus succeeded in producing high-purity niobium with interstitial COHN in the range of 65 ppm in total, i.e. a residual resistivity ratio (RRR) of 100 to 140 in 1 ton Nb ingots (Hörmann 1986).

References

Albrecht WW, Hoppe A (1985) Qualitätssicherung hochkapazitiver Tantalpulver. Erzmetall 38, 7/8:369–374

Albrecht WW, Ingalls D (1983) Hochtemperaturreinigung von Tantal und Niob – Raffinationsverfahren in der Metallurgie. GDMB 1983. Chemie, Weinheine pp 229–242

Borchers P, Korinek GJ (1981) Extractive metallurgy of tantalum. 110th AIME Annu Meet, Chicago, pp 97

Bureau of Mines Minerals Yearbook (1985) Columbium and tantalum reprint (1986), pp 4–5

Deutsches Institut für Wirtschaftsforschung Berlin (1982) Niob, Untersuchung über Angebot und Nachfrage mineralischer Rohstoffe, pp 24–25

Hörmann M (1986) Herstellung von hoch-wärmeleitfähigem Niob im technischen Maßstab für die Hochfrequenz-Supraleitung. Z Metallkd 77, 5:338–347

Manker EA (1985) Niobium-exotic application, tantalum Int Stud Center, Brussels

Rockenbauer W (1981) Production of niobium metal and compounds from tantalite-columbite natural ores and synthetic tantalum-niobium concentrates. Proc Int Symp Niobium 81. Metall Soc AIME, pp 133–152

Roskill Information Services Ltd. (1986) 2 Clapham Road, London SW9 Economics of niobium 1986, pp 171–197

Yamashita Y et al. (1985) A new lead perovskite Y5S dielectric for multilayer ceramic capacitor. Jpn J Appl Phys 24:1027–1029

The Industrial Applications of Rare Earth Elements

C. PREINFALK and G. MORTEANI

Abstract

Rare earth elements are used in several high technology industrial products. Their presence in alloys, ceramic and glass compounds and in oxide compounds provide special material properties. In metallurgy, alloing and addition of rare earth elements to steel increase strength and thermal stability. The cracking of petroleum requires catalysts containing various rare earth elements to increase the amount of short hydrocarbon molecules in the product.

In the ceramic and glass industry rare earth element containing compounds are used for high temperature materials, for coatings, polishing, coloring, decoloring and as glass additives. The main other uses of rare earth elements are in permanent magnets, phosphors in color television tubes, x-ray tubes, fluorescent lamps, electronic and computer devices and lighter flints. For all these applications two types of REE products are used:

1. "Mischmetall", a mixture of rare earth oxides and
2. "high purity" compounds containing at least 90% of an individual rare earth element.

1 Introduction

Because of their special properties, the rare earth elements, including yttrium, have a wide range of applications. With their increasing use in high technology, the demand for REE is growing steadily. The annual growth rate of REE consumption also depends strongly on the discovery of new fields of REE application. Essentially there are four domains in which REE are used at present:

- metallurgy
- catalysts
- ceramics and glass
- other uses

The extent of REE consumption in the USA in these fields is shown in Fig. 1.

The publications used to write this paper are given in the reference list. Individual quotes are not cited.

Lehrstuhl für Angewandte Mineralogie und Geochemie, Technische Universität München, Lichtenbergstrasse 4, D-8046 Garching

Lanthanides, Tantalum and Niobium
P. Möller, P. Černý and F. Saupé (Eds.)

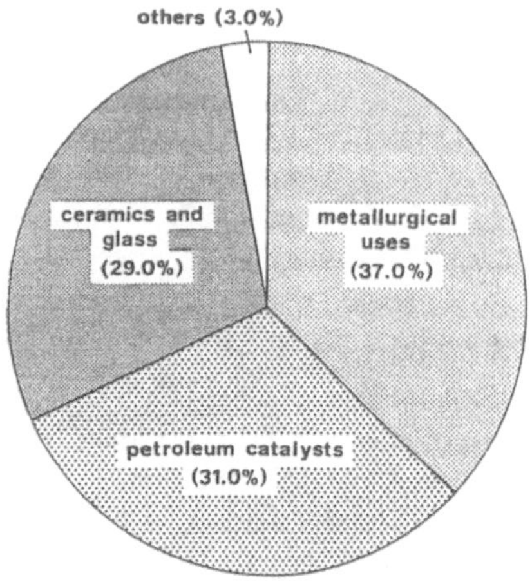

Fig. 1. Uses of rare-earth elements in the USA in 1986

2 Metallurgy

2.1 Steel

The main application of rare earths in the steel industry is the addition of mischmetall or rare earth silicides for deoxidation, and desulfurization. The most important use of REE in metallurgy is sulfide and graphite shape control to modify remaining inclusions and thereby increase the strength of the steel.

HSLA Steel. The addition of mischmetall increases the resistance of HSLA (high strength low alloy) steel to cracking under stress and gives more uniform strength in longitudinal and transverse directions. HSLA steel is a high strength steel used in bumper face bars, reinforcing bars, motor mounting brackets, chassis components, spring supports, bridges, power station foundations, prestressed concrete bridges, road tunnels, and foundations of offshore installations.

Fecralloy Steel. Fecralloy steel is a stainless steel which contains 15–20% chromium, 4–5% aluminum and 0.5–4% yttrium oxides precipitate on the surface of the fecralloy steel parts and give them good resistivity to high temperatures (up to about 1300 °C) and oxidizing gases, for example in combustion engine exhaust systems. Furthermore, it is used in electrical furnace windings, laboratory furnaces and as a substrate for automobile emission control catalysts instead of ceramic substrates, which are heavier and need more volume.

Other Steels. The addition of mischmetall to ferritic stainless steel and cast steel improves the resistivity to oxidation. Low amounts of mischmetall in free machining steels favor the formation of spheroidal graphite in steel with carbon contents of 0.2–0.9%. This avoids the release of poisonous gas in the production process. Rare earth-bearing high carbon steels are further used for agricultural tools, sawblades, and mechanical bearings.

2.2 Ductile Cast Iron

Mostly Ce-bearing mischmetall is used in ductile cast iron as a nodulizing and nucleating agent and to control tramp elements. Further, the addition of rare earth metals raises the weldability, ductility and fatigue resistance of ductile cast iron.

Rare earth doped ductile cast iron is used for transmission housings, motor exhaust pipes, eccentric gears, ingot molds, and brake drums. Markets of increasing importance are pipes for drinking water supply and sewer pipes, as well as quality castings for cars and agricultural tools.

2.3 Pyrophoric Flints

The production of pyrophoric flints was one of the first areas in which RE metals were used, because mischmetall ignites readily when alloyed with iron. The alloys for phyrophoric flints consist of 74% mischmetall, 23% iron, and small amounts of Mg and Cu, which are added to enhance strength.

2.4 Other Alloys

Mischmetall additions strengthen magnesium-based alloys used in aerospace engineering.

In aluminum die casting alloys the use of mischmetall and Ce enhances the fluidity and reduces the material shrinkage. These alloys are used in high temperature-engine blocks. Rare earths added to Al-Mg conducting wire in high voltage transmission lines increase tensile strength, and resistance to heat, vibration, and corrosion.

High performance alloys are used under extreme heat and corrosion conditions. The addition of yttrium and lanthanum to these superalloys increases the resistance to corrosion and oxidation under high temperatures. Gas turbines and vane material as well as industrial furnaces are the main applications of high performance alloys.

Mischmetall (0.25–0.8%) enhances the hot workability and deep drawing properties of leaded brasses. The time stability and creep strength of platinum alloys used in glass production are improved by Y addition.

In vacuum-melted Ti-alloys Y also increases the ductility. Copper alloys with iron and Ce-additions are highly corrosion-resistant at high temperatures, as is necessary for the aerospace industry. Steel is galvanized with a Zn-Al-mischmetall alloy which provides higher weldability, ductility, and corrosion resistance, and is more easily painted. Galvanized steels are used for automobiles, washers, fences, and metal buildings.

3 Catalysts

The main use of rare earth elements is in the field of catalysts. As seen in Fig. 1, petroleum catalysts alone account for about 31% of total rare earth element consumption in the USA. Automobile exhaust catalysts are a rapidly growing market.

3.1 Petroleum Cracking Catalysts

The petroleum cracking catalysts consist of three parts, the silica-alumina refractory binder, an inert filler, and a rare earth element containing zeolite.

Na-zeolites are usually used, which contain a mixture of La, Nd and Pr. The addition of REE enhances the structural stability as well as the cracking capacity and reduces coke production. Further, the acidic properties are retained under high temperatures and the thermal stability in stripping processes increases.

3.2 Exhaust Emission Catalysts

The rare earth elements are used in exhaust emission control as coatings in the most common three-way honeycomb catalysts, where they favor oxidation of ecologically harmful gases.

The rare earth element-bearing anti-knock additives are increasingly replacing the lead-based products, thus avoiding environmental contamination and catalyst poisoning. CeO_2 as a burning product works as a catalyst in the exhaust pipe, markedly reducing the carbon monoxide, oxides of nitrogen, and unburnt fuel emission.

3.3 Other Catalytic Applications

Ce is used in hydrocarbon catalysts, in self-cleaning ovens, as a polymerization catalyst, in ammonia synthesis, and partly in photovoltaic devices and tyre rubber production.

La is applied in hydride catalysts in polyethylene production. Mixtures of rare earth elements may be used for recovering chlorine from hydrochloric acid, the reduction of nitrogen oxides in stack effluents and in fumes from acid and nitrating plants, and the removal of sulfur dioxide from gas streams.

4 Ceramics

4.1 Ceramics in the Electronic Industry

Ceramic Capacitors. REE oxides used in ceramic capacitors cause certain changes in features such as temperature-compensation, dielectricity, and permeability. In particular the addition of LREE such as La, Ce, Pr and Nd yields stable dielectric constants.

Nd-oxide is used mainly in disc-type capacitors. Capacitors with few Ce- or La-oxides have a high dielectric constant, but are less stable under temperature variations. The use of REE oxides also prolongs the life of the capacitors. The main consumption in recent times is in multi-layer capacitors which are replacing electrolytic capacitors steadily.

Other Electronic Ceramics. Small amounts of Y and La are used to dope barium titanate, the main component of positive temperature coefficient thermistors. La-chromite ($LaCr_7$) is used as an electric heating element with high resistivity to temperature and corrosion, and the advantage of high electric conductivity.

4.2 Electro-optic ceramics

La-oxide is a major component (96%) of the PLZT (lanthanum-modified lead zirconate-lead titanate) ceramics. PLZT ceramics are used as a very sensitive nonvolatile and selectively erasable image storage medium. PLZT is used in light memory elements, video displays, storage systems, light modulation elements, light shutter, light valves, and cockpit window panels in military airplanes.

4.3 Ceramic Dyes

Small amounts of particular REE-oxides or mixtures give intense colors in ceramic dyes. The addition of Pr-oxide to a zirconia-silica mixture yields a brillant yellow color, Y-oxide an orange color, Nd-oxide a purple color, Ce-oxide causes a whitish opalizing effect; in double firing processes it gives color to porcelain. The amount of REE-oxides in the mixture is about 100–150 g/m^2.

4.4 Refractories

Y_2O_3 is used in ceramic compounds for special high temperature applications, such as windows for furnaces, microscope lenses for the study of molten material, and crucibles for vacuum melting of reactive materials like Ti.

Ce-sulphides are the raw material for metallic crucibles which are used for all melts of high temperature (> 1800 °C) except Pt-bearing melts, which form intermetallic compounds with Ce.

Because ZrO_2, which is used in partially stabilized zirconia (PSZ) ceramics, forms different polymorphs depending on the temperature, it is necessary to add Y_2O_3 to avoid the ZrO_2 ceramics becoming brittle during cooling, due to polymorphic transition. PSZ is used for cutlery, extrusion dies, nozzles, and machining tools. Further applications are in diesel machines, gas turbines, and fuel cells.

4.5 Sialon

Sialon is a silicon nitride-based ceramic material which is one of the strongest ceramic materials. Y_2O_3 is added as a sintering agent. Sialon is used in cutting tool tips, adiabatic engines, and turboengines, and as a coating on nozzles for high temperature technology.

4.6 Other REE-Oxide-Bearing Ceramics

Gadolinia-doped (6–23%) Ce-oxide ceramics with small amounts of Mg are permeable for oxygen ions. These ceramics show similar permeability as doped zirconia ceramics, but at temperatures 200 to 300°C lower.

Ce-oxide ceramics doped with trivalent oxides like Gd_2O_3 or Y_2O_3 are able to convert hydrogen derived from natural gas and water – or nuclear reactors or coal – into electricity.

5 Glass

The glass industry is one of the major markets for REE, where they are used both as mixed compounds and as separated high-purity materials.

5.1 Polishing of Glass

Ce-oxide is a very effective polishing compound because of chemical dissolution and mechanical abrasion. Glasses with Ce-polish find application in a wide range, for example, ophthalmic glass, precision optics, mirrors, cut crystals, camera lenses and cathode ray tubes, television screens, jewellery, and plate glass. There are other polishing compounds (e.g. Nd-oxide), but Ce-oxide works most quickly, particularly high grade Ce-oxide (90% CeO_2).

5.2 Optical Glass

The main use of rare earth oxides in glass production is in optical glasses. Contents of up to 40% high purity La-oxides (99.5–99.995%) in the glass batch are common. High refractive index, high degree of transparency and light transmission, and good qualities of internal structures are the main features required, e.g., in fiber optics technology. Besides La-oxide, Gd and Y are now added to optical glass with similar effects.

5.3 Glass Decolorizing

La- and Ce-oxide are used to decolorize glass and to increase the index of refraction and dispersion of light. In lead crystal glass Ce-oxide replaces the toxic arsenic. The main application of this glass, also treated with Nd-, Pr-, Er-, or Ho-oxide, is in camera lenses and materials where strong light transmission or low colorizing is needed.

5.4 Glass Colorizing

Only small quantities of REE oxides are used in this traditional domain of REE application. Nd-oxide gives a pink tint with violet reflections to art glasses and special filters which are produced from a melt containing 1–5% of rare earth oxides. Welding glasses contain Nd-oxide to protect the eyes from the yellow flare emitted by sodium vapor.

Er-oxide is used in photochromic glasses and crystal glasses, giving them a pale pink color, and as additive to filter material for X-ray fluoroscopy.

Ce-oxide is only used in combination with different colorizing oxides such as titania to give a yellow tint to ophthalmic glasses and tableware, or titania joined with Mn causes a pink to violet color used in absorptive glasses.

5.5 Radiation Control

The use of rare earth oxides in radiation control is increasingly important. The main effect of the REE is their property of absorbing ultraviolet light from natural and technical sources and the absorption of electrons. The rare earth oxides are mainly used in TV screens, fluorescent lamps, microwave devices, space instruments, nuclear energy technologies, and medical devices. Mostly Ce-oxide is usually added to absorb UV-light and trap electrons. REE-stabilized glasses serve as containers for light-sensitive food, light bulbs, museum panes, instrument windows, welding glasses, and protection shield glasses in nuclear and radiochemical laboratories.

5.6 Fluorescence

Fluorescent glass is used in high-powered lasers when doped with high purity Nd-oxide (99.9%). The fluorescence is caused by the exciting of atoms by the absorption of light.

5.7 Glass Coatings

The coating of glass in scientific lenses and laser cavity mirrors is a rather minor use of high purity REE compounds. Pure Ce-oxide (99.99%), Ce-fluoride, La-oxide, and Nd-fluoride without any Fe content are the compounds necessary.

6 Permanent Magnets

6.1 Samarium-Cobalt Magnets

Since the 1970's, permanent magnets have consumed large amounts of high purity and separated rare earth element, Sm-Co alloys are two times stronger than Pt-Co magnets, and four times stronger than Al-Ni-Co magnets. This provides a reduction in weight and size of the magnets used in watches, microwave tubes, radar missile guidance systems, television and computer communication, and other devices.

6.2 Neodymium-Iron-Boron Magnets

The Nd-Fe-B magnets are the strongest available magnetic material. The high magnetic strength further allows an extreme size reduction, and makes the magnets more resistant to demagnetizing; their production is also cheaper than that of Sm-Co magnets. The main applications are brushless motors, other electric and computer peripheral devices.

In automobiles, the REE-containing magnets allow a considerable weight reduction.

7 Phosphors

Phosphors in cathode ray tubes, X-ray screens, fluorescent lamps and lasers are an important use of rare earth elements. The rare earth elements cause the emission of a precise wavelength light. When excited, Ce gives off ultra-violet light, Dy yellow, Tm blue, Eu red, Sm orange, Tb green, and Nd infrared light.

7.1 Cathode Ray Tubes

In color television screens and color monitors, three different phosphors of red, green and blue primary color are used. Even though rare earth elements are used in the three types of phosphors, the most important application lies in the red phosphors, where Eu-activated Y-oxysulphides or Eu-activated Y-phosphovanadates intensify the brightness.

7.2 X-Ray Screens Phosphors

In X-ray intensifying screens the image is created by inorganic crystalline phosphors. The use of terbium, europium, gadolinium and lanthanum oxides, and oxysulfides reduces the necessary X-ray exposure by about 50%.

7.3 Fluorescent Lamps

Fluorescent lamps work with an emission of light by a fluorescent powder coating the tube walls. The rare earth phosphors half energy consumption and increase lifetime up to five times. Typical rare earth phosphors contain 5% Eu and 10% Y. Recently growing markets are screw-in fluorescent bulbs and fluorescent strips.

Other applications of rare earth phosphors in illumination devices are in high pressure mercury lamps where Eu-activated Y-vanadate yields a more neutral color composition of the light. In oil and gas lamp mantles Ce-oxide addition shifts the emitted light spectrum to a less intensive orange green light.

In carbon arc lamps for searchlights and color motion picture photography and projection rare earth elements increase the bow intensity and produce a light most similar to sunlight color.

7.4 Lasers

Yttrium-aluminum-garnets (YAG's) produce short wavelength laser beams in lasers used for cutting and welding of thin metal materials ($<$ 10 mm), for cutting in textile industry, for medical operations, and drilling.

Nd-doping increases the electrical efficiency and improves the properties of the laser while maintaining the same wavelength.

Newly developed gadolinium-scandium-aluminum-garnet (GSAG) lasers have the ability to emit stronger beams than YAG lasers. Main uses are in fusion reactors and SDI where high energy lasers are needed.

8 Nuclear Uses

8.1 Burnable Poisons

In the field of nuclear technology, high purity rare earth products are required. Gd-oxide of 99.9–99.99% purity serves as burnable poison in uranium nuclear fuels in boiling water reactors. Gd has a high neutron absorption cross-section and reduces the consumption of uranium oxide by improving the efficiency of energy output.

8.2 Neutron Absorbers

Sm, Eu, and Dy are used as neutron absorbers in fast breeder reactors; Eu is the most widely used, because of its high captive cross-section and therefore great capacity to absorb neutrons.

8.3 Neutron Radiography

Gd and Dy are used in convertor thin foils in neutron radiography, which are mainly used in aerospace and nuclear engineering industries. The primary gamma radiation is converted by the rare earth-containing foils to beta-radiation which can be registered on film.

8.4 Target Materials

Small amounts of Tm and Er are used as target materials in some sealed-tube neutron generators.

8.5 Control Armplates

Neutron-absorbing Eu cermet is used to build improved reactor control armplates.

8.6 Paint Additives

High purity Gd-oxide is added to military vehicle paints for defence against neutron bomb radiation.

The consumption of rare earth elements in nuclear technology is in general declining, because of political resistance to the use of nuclear energy.

9 Metal Hydrides

Metal hydrides are the most developed form of a solid state hydrogen storage system. Mainly La-Ni-alloys are used to store high quantities of hydrogen at moderate temperatures. The reaction of hydrogen absorption-desorption mechanism can be expressed in the following form:

$$4LaNi_5 + 13H_2 <---> 2\ La_2Ni_{10}H_{13}.$$

The main applications of hydrogen-stored systems are military purposes, gas chromatography, and heat pumps.

10 Oxygen Sensors

The most common oxygen sensors consist of a galvanic cell and a partially stabilized zirconium (PSZ) electrolyte. The oxygen sensors are catalytic convertors to control the exhaust gases of automobiles, particularly hydrocarbons, nitrous oxides, carbon monoxides, and the content of oxygen. The oxygen sensors are sensitive to variation of oxygen contents of the PSZ which contain about 9% Y_2O_3 and are therefore used to measure the oxygen pressure.

11 Bubble Memory Systems

Gadolinium-gallium-garnets (GGG's) are used in substrates and thin film magnetic bubble memory systems which are used mostly in communication and computer systems. High speed information access and the resistivity to extreme conditions and lack of dependence on continuous power supply favor these memories for military and industrial usage.

12 Other Applications of Rare Earth Elements

The synthetic yttrium-aluminum-garnets (YAG's) are also used to substitute diamonds in jewellery, and are called "diamonaire". The YAG's ($Y_3Al_5O_{12}$) have a Moh's hardness of 8.5 and form cubic clear crystals. The Y-stabilized cubic zirconias, are better substitutes for diamonds with a hardness of 7 to 8.5.

Thulium is used in a new blood irradiator to avoid early rejection of organ transplant. Europium is applied in false teeth to give a better neutral irridescence. Gadolinium is used in nuclear medicine in a scanning device.

Erbium- and yttrium borides are used as solar absorbing coatings in solar energy systems. Yttrium-iron garnets and gadolinium-iron garnets are used as ferrite materials in microwave devices. La-boride is used for nozzle material in hydrogen fluorine chemical lasers. Gd-alloys as components of refrigerants provide very low temperature cooling. Y-oxide is a compound in high temperature crucible mold paints on nonreactive or ceramic surfaces. La-boride crystals of high purity are used in analytical electron microscope guns. Eu-doped oxysulfide phosphors are used for fiber optics temperature measurements. Furthermore La and Ce are used in fiber optics. Fertilizers containing rare earth elements enhance the growth of wheat, rice, peanuts, and the production of latex in rubber trees. Rare earth compounds, particularly Ce, protect textiles from sun bleaching and creasing.

Treatment of glasses with rare earths before silver coating for mirrors inhibits corrosion, particularly for use in sun collectors. In propulsion jets used in aerospace technology yttrium oxide stabilizes Pt-alloys at high temperatures.

Table 1 summarizes the industrial applications of rare earth elements.

Table 1. The industrial applications of rare earth elements

Application	Mixed REE	La	Ce	Pr	Nd	Sm	Eu	Gd	Tb	Dy	Ho	Er	Tm	Yb	Y
Metallurgy															
Steel															
HSLA steel	×	×													
Stainless steel	×	×													×
Other steel	×	×													
Ductile cast iron	×		×												
Pyrophoric flints	×		×												
Nonferrous alloys															
Mg alloys	×	×		×	×										×
Al alloys	×	×			×										
Ti alloys	×	×	×		×			×							
Super alloys	×	×													×
Cu alloys	×	×	×												
Catalysts															
Petroleum cracking catalysts	×	×	×	×	×	×		×		×				×	
Exhaust emission Control	×		×												
Other catalytic applications	×	×	×		×	×		×		×			×		
Ceramics															
Coatings	×	×	×	×	×										×
Enamels, glazes			×	×	×										×
Refractories		×	×					×							×
Glass															
Polishing	×		×	×											
Optical glass		×						×							×
Decolourising		×	×	×	×						×	×			
Colourising			×	×								×			
Radiation control		×	×	×											
Fluorescence and coatings		×	×		×										
Permanent magnets	×				×	×									
Phosphors															
Cathode ray tubes	×	×	×	×	×		×	×							×
x-ray screen p.		×					×	×	×						
Fluorescent lamp			×				×								×
Laser					×			×			×	×	×		×
Nuclear uses															
Burnable poisons & neutron absorbers						×	×	×		×					
Radiography								×		×					
Target material												×	×		
Control armplates							×								
Paint additives								×							
Metal hydrides	×	×													
Oxygen sensors															×
Bubble memories								×							
Other uses															
Jewellery, YAG															×
Medical purposes							×		×				×		
Solar energy syst.												×			×
Microwave								×							×
Chemical lasers		×													
Refrigerants								×							
Microscope guns		×													
Fibre optics		×	×				×								
Fertilizers	×														
Textiles			×												

References

Anonymous (1979) Rare earth industry profile and market review. Ind Min 138:21–61

Anonymous (1986) The economics of rare earths and yttrium. 6th edn. Roskill Information Services Ltd, 270 pp

Griffiths J (1984) Rare earths – attracting increasing attention. Ind Min 199:19–37

Greinacher E (1987) Gewinnung von Sondermetallen und ihre Anwendung in Optik, Elektronik und Kommunikation. Erzmetall 40:65–70

Gschneider KA (1981) Industrial applications of rare earth elements. ACS Symposium series 164:297 pp

Mayer W (1987) Industrielle Anwendungen der Seltenen Erden. Erzmetall 40:327–329

Subject Index

GPSR Compliance

The European Union's (EU) General Product Safety Regulation (GPSR) is a set of rules that requires consumer products to be safe and our obligations to ensure this.

If you have any concerns about our products, you can contact us on ProductSafety@springernature.com

In case Publisher is established outside the EU, the EU authorized representative is:

Springer Nature Customer Service Center GmbH
Europaplatz 3
69115 Heidelberg, Germany

Batch number: 10761697

Printed by Printforce, the Netherlands